CNC Programming

Principles and Applications

Michael W. Mattson
Clackamas Community College
Oregon City, OR

Australia • Brazil • Japan • Korea • Mexico • Singapore • Spain • United Kingdom • United States

CNC Programming: Principles and Applications
Michael W. Mattson

Vice President, Career and Professional Editorial: Dave Garza
Director of Learning Solutions: Sandy Clark
Senior Acquisitions Editor: James Devoe
Managing Editor: Larry Main
Product Manager: Mary Clyne
Editorial Assistant: Cris Savino
Vice President, Career and Professional Marketing: Jennifer McAvey
Marketing Director: Deborah S. Yarnell
Marketing Manager: Jimmy Stephens
Marketing Coordinator: Mark Pierro
Production Director: Wendy Troeger
Production Manager: Mark Bernard
Content Project Manager: Mike Tubbert
Art Director: Bethany Casey

For product information and technology assistance, contact us at
Professional & Carrer Group
Customer Support, 1-800-648-7450

For permission to use material from this text or product, submit all requests online at **cengage.com/permissions.**
Further permissions questions can be e-mailed to
permissionrequest@cengage.com.

Library of Congress Control Number: 2008943676

ISBN-13: 978-1-4180-6099-2
ISBN-10: 1-4180-6099-2

Delmar
5 Maxwell Drive
Clifton Park, NY 12065-2919
USA

Cengage Learning products are represented in Canada by Nelson Education, Ltd.

For your lifelong learning solutions, visit **delmar.cengage.com**
Visit our corporate website at **cengage.com.**

NOTICE TO THE READER

Publisher does not warrant or guarantee any of the products described herein or perform any independent analysis in connection with any of the product information contained herein. Publisher does not assume, and expressly disclaims, any obligation to obtain and include information other than that provided to it by the manufacturer. The reader is expressly warned to consider and adopt all safety precautions that might be indicated by the activities described herein and to avoid all potential hazards. By following the instructions contained herein, the reader willingly assumes all risks in connection with such instructions. The publisher makes no representations or warranties of any kind, including but not limited to, the warranties of fitness for particular purpose or merchantability, nor are any such representations implied with respect to the material set forth herein, and the publisher takes no responsibility with respect to such material. The publisher shall not be liable for any special, consequential, or exemplary damages resulting, in whole or part, from the readers' use of, or reliance upon, this material.

Printed in Canada
3 4 5 XX 12 11

Brief Contents

Table of Contents

CHAPTER 1

CHAPTER 2

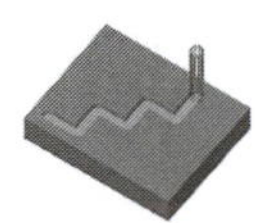

CHAPTER 3

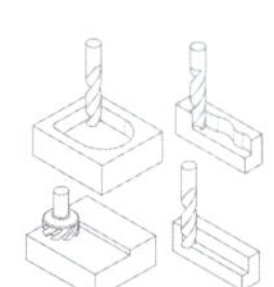

CHAPTER 4

CHAPTER 9

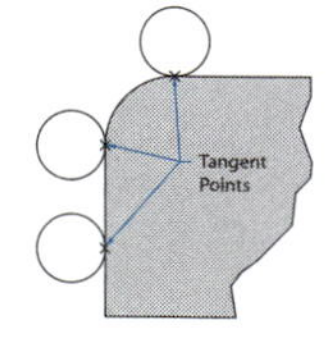

CHAPTER 10

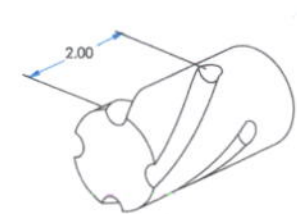

CHAPTER 11

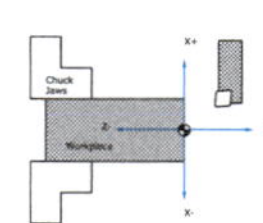

CHAPTER 12

PREFACE

CNC Programming: Principles and Applications is a user-friendly guide for those who are interested in computer-aided machining. This text is suitable for beginning students in high school or technical college programs to learn the fundamentals G & M code programming. CNC beginners will benefit from a comprehensive introduction to machining processes, precision measurement, cutting tools, lean manufacturing, quality control, and shop math.

Seasoned machinists will also find this an accessible resource for making the transition from the concrete world of manual machine tools to the abstract realm of CNC. Complete programming examples lead the learner through the entire manufacturing process from planning through code writing and job setup. Readers at all levels will appreciate the readability of this work.

CNC PROGRAMMING AND PROJECT LEAD THE WAY

The changes you see in this edition resulted from a partnership forged with Project Lead the Way, Inc. in February 2006. As a non-profit foundation that develops curriculum for engineering, Project Lead the Way, Inc. provides students with the rigorous, relevant, reality-based knowledge they need to pursue education in engineering or engineering technology.

The Project Lead The Way® curriculum developers strive to make math and science relevant for students by building hands-on, real-world projects in each course. To support Project Lead The Way's® curriculum goals, and to support all teachers who want to develop project/problem-based programs in engineering and engineering technology, Delmar Cengage Learning is developing a complete series of texts to complement all of Project Lead the Way's® nine courses:

Gateway To Technology

Introduction To Engineering Design

Principles Of Engineering

Digital Electronics

Aerospace Engineering

Biotechnical Engineering

Civil Engineering and Architecture

Computer Integrated Manufacturing

Engineering Design and Development

To learn more about Project Lead The Way's® ongoing initiatives in middle school and high school, please visit **www.pltw.org.**

HOW THIS TEXT WAS DEVELOPED

This series' development began with a focus group that brought together teachers and curriculum developers from a broad range of engineering disciplines. Two important themes emerged from that discussion: (1) that teachers need a single resource that fits the way they teach engineering today, and (2) that teachers want an engaging, interactive resource to support project/problem-based learning.

CNC Programming: Principles and Applications supports project/problem-based learning by:

- Creating an unconventional, show-don't-tell pedagogy that is driven by engineering *concepts*, not traditional textbook content. Concepts are mapped at the beginning of each chapter, and clearly identified as students navigate the chapter.
- Reinforcing major concepts with Applications, Projects, and Problems based on real-world examples and systems.
- Providing a text rich in features designed to bring CIM technology to life in the real world. Case studies, Career Profiles, and Boxed Articles highlighting human achievements show students how engineers develop career pathways and innovate to continuously improve products and processes.
- Reinforcing the text's interactivity with an exciting design that invites students to participate in a journey through the engineering design process.

KEY FEATURES OF THIS EDITION

- Written in an ***easy-to-read, unintimidating style*** to promote faster learning and more complete understanding.
- ***Complete examples and numerous illustrations*** connect well with visual-spatial learners.
- ***Concise content*** provides appropriate depth of knowledge without a lot of extraneous text.
- ***Comprehensive introduction to manufacturing*** makes CNC Programming the only book beginning students will need.

This edition of *CNC Programming: Principles and Applications* provides ***updated and expanded coverage*** of the following topics:

- Types of CNC machine tools
- New chapter on precision measurement and quality
- NC programming process
- NC tooling and machining processes
- Tool and workpiece setup
- Programming concepts and job planning
- Codes for positioning and milling
- New section on blueprint reading

In addition, the new edition's content, pedagogy, and design have been re-developed to complement and support the Project Lead the Way® CIM curriculum and increase this text's effectiveness for all learners. Look for these new features throughout the text:

- **Programing Examples**

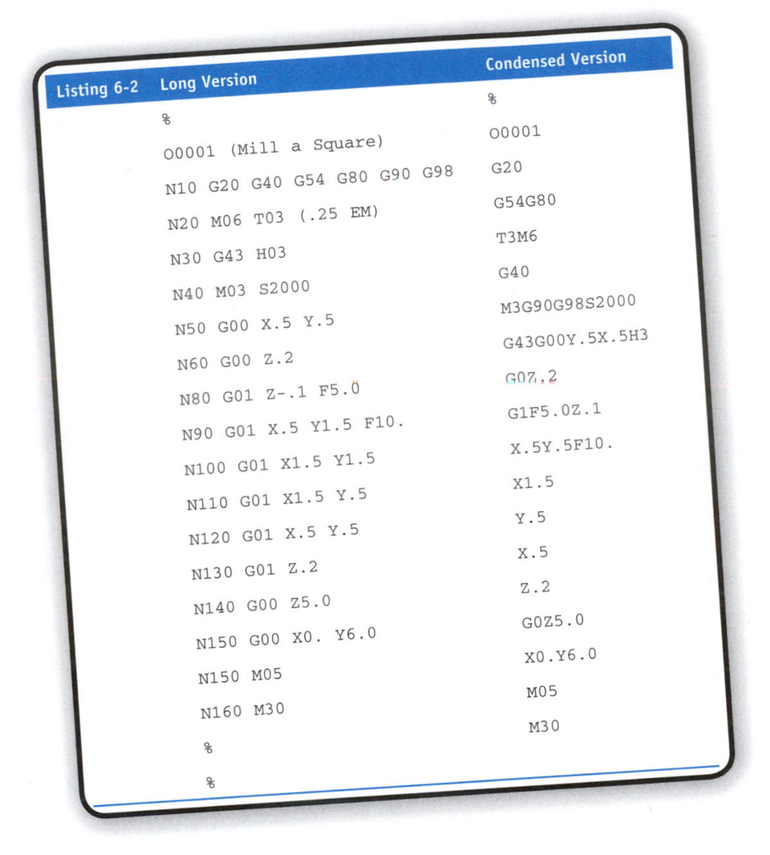

Listing 6-2

Long Version

```
%
O0001 (Mill a Square)
N10 G20 G40 G54 G80 G90 G98
N20 M06 T03 (.25 EM)
N30 G43 H03
N40 M03 S2000
N50 G00 X.5 Y.5
N60 G00 Z.2
N80 G01 Z-.1 F5.0
N90 G01 X.5 Y1.5 F10.
N100 G01 X1.5 Y1.5
N110 G01 X1.5 Y.5
N120 G01 X.5 Y.5
N130 G01 Z.2
N140 G00 Z5.0
N150 G00 X0. Y6.0
N150 M05
N160 M30
%
%
```

Condensed Version

```
%
O0001
G20
G54G80
T3M6
G40
M3G90G98S2000
G43G00Y.5X.5H3
G0Z.2
G1F5.0Z.1
X.5Y.5F10.
X1.5
Y.5
X.5
Z.2
G0Z5.0
X0.Y6.0
M05
M30
```

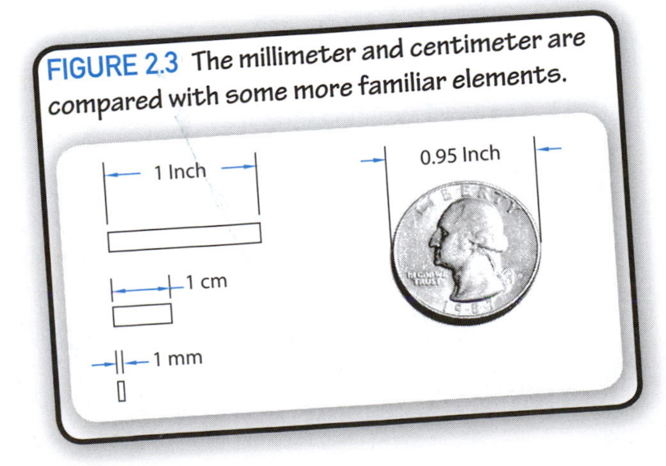

FIGURE 2.3 The millimeter and centimeter are compared with some more familiar elements.

- **More than 120 new or heavily revised graphics** show the latest technologies and help students visualize abstract concepts.

feedback loop:
Electronic signals that are sent back to the control to indicate actual position, velocity, or state of the machine tool. The control will then compare the actual condition to the desired position and make adjustments.

▶ **Key Terms** are defined more systematically to help students master a technical vocabulary with confidence.

Technical Stuff

One horsepower is equal to lifting 550 lbs up to a distance of 1 ft in 1 second. A 200-lb man climbing up a 9,900-ft mountain would use about (200 lb × 9,900 ft = 1,980,000 ft-lbs) of energy. If he could do this in 8 hours (or 28,800 seconds), then he would need to produce about .125 hp to climb up the mountain. Why? Because 1,980,000 ft-lbs/28,800 seconds = 68.75 ft-lbs per second or 1/8th of a horsepower.

So, a man working extremely hard all day is producing about 1/8th of a horsepower. He will use 1,980,000 ft-lbs of energy in the process. Foot-pounds are a pretty small number for measuring energy, so we usually use BTU, which stands for British Thermal Unit. A BTU is equivalent to 778 ft-lbs of energy.

In the metric system, the unit for energy is the joule. 1 BTU is about 1054 joules. For power (power is energy divided by time) the metric unit is watt. One horsepower is about 748 watts. Oddly enough, my German engineering acquaintances tell me that European car engines are still rated in horsepower. I guess it just makes a better visual impression in the mind's eye.

Fossil fuels are very energy dense. They contain huge amounts of energy for their mass. Crude oil contains approximately 139,000 BTU per gallon and coal contains about 12,700 BTU per pound. A pound of coal can do the work of five strong men for a day, at least in theory. The fact is that steam and internal combustion engines are only about 10 to 30 percent efficient. You have to put a lot more energy in than you get out.

▶ **Boxed Articles** highlight fun facts and points of interest on the road to new and better processes.

Motorcycle Design with Mastercam

The *American Chopper* TV show lets viewers look over the shoulders of the designers and machinists at Orange County Choppers (OCC) as they conceive and build one-of-a-kind chopper motorcycles and accessories.

OCC is very high tech. They rely on the latest waterjet system capable of cutting through ceramic, glass, even a 12-inch block of steel with minimal heat and no distortion of the material. They use a mandrel tube bender to create unique snaking exhaust systems without leaving any ripples in the metal. Advanced CNC lathes and mills turn and cut these exotic and beautiful parts.

OCC imports its designs seamlessly from CAD files into Mastercam and translates them into manufacturing programs. Mastercam itself has substantial design capabilities and OCC also creates unique parts within the Mastercam graphic environment.

In 2006, OCC began expanding its manufacturing capacity. The first step was moving its bike manufacturing operations into a new 100,000 sq. ft. facility where it can produce selected bikes on a semiproduction basis. Lead Engineer Jim Quinn said that OCC will have to streamline its manufacturing procedures to achieve output targets.

Jim and his staff will be using Mastercam to refine or redesign tooling and fixtures for increased productivity. For example, an adjustable jig designed in Mastercam and used to fabricate custom bikes will be redesigned for production of limited-edition vehicles. The custom jig makes it possible to adjust the bike's "rise" and "stretch" to make each bike's dimensions proportional to the measurements of the person who will be riding it most often. It's a great timesaver in building custom bikes.

"Once we are ready to make hundreds of bikes with the same 'rake and ride' specifications, we already have that jig designed in Mastercam," said Quinn. "It will be quite easy to go back in and duplicate the dimensions we need, take out a lot of the adjustability, and generate the CNC programs to make our production jig." The same concept also holds for other tooling and fixtures.

The old plant will become the CNC production shop. OCC intends to "cherry-pick" its custom part, accessory, and wheel designs to identify ones that would be most appealing to fans and manufacture these on a production basis. "Nasty Wheel," the engraved Orange County Chopper logo air cleaner cover and the OCC Dagger Shield coil cover will be among the first.

Mastercam will have a central role in optimizing part manufacturing productivity. The OCC Dagger Shield coil cover, the company's signature accessory, is a prime example. It was designed in SolidWorks and imported into Mastercam. Quinn said, "Mastercam allows me to turn, manipulate, and move that part into whatever situation I need to hold it securely and machine it efficiently. Because the work piece is already cut to near net shape with only light cuts required around the perimeter of the shield, the part can be cut at higher speed with less force for a faster production cycle and better surface finish." Mastercam's 3D surfacing toolpath "is wonderful." The polishing they do to get parts ready for chroming is very minimal.

Quinn said that the first couple times this part was made on a prototype basis; the complete process took about six hours. Now after tweaking the various steps, they have the time down to one and a half hours. The part is production-ready.

In just six years, Discovery Channel viewers have watched OCC grow from three employees to 60, and boost its production to 80 commissioned custom choppers in 2005. The new plant will allow OCC to boost overall production to 120 bikes in 2006 and as many as 240 in 2007.

▶ **Case Studies** show students how companies use CIM technology and CNC programming to create real products and improve the speed, quality, and efficiency of their manufacturing processes.

▶ **Career Profiles** provide role models and inspiration for students to explore career pathways in engineering.

Careers in the Designed World

CAREERS IN COMPUTER INTEGRATED MANUFACTURING

Getting with the Program

Julio Ramirez spent most of his career as an old-fashioned machinist, manually making parts for auto manufacturers and cable companies. But he's found a new challenge as a CNC operator for Indian Springs Manufacturing Company, where he's learned the art of programming.

"I have to figure out a lot of things for the program," Ramirez says. "If everything works out right, it makes me smile. I feel like, 'I did it!'"

On the Job

Indian Springs Manufacturing Company makes parts for a variety of customers, from radio components to antennas. The CNC machine does everything automatically—drilling, changing tools, making threads. If Ramirez sets up the program just right, the machine runs at the proper speed and makes every part at the proper dimensions. He adjusts the machine as needed while it runs.

It's not easy to make parts exactly the same size from week to week and month to month. Ramirez has to keep track of everything he's done in the past and do it the same way for each new batch of parts.

Inspirations

As a kid in Cuba, Ramirez wanted to be a firefighter or a policeman. Later, as his friends began finding work in engine factories, he went along. He didn't like the work at first, but soon became intrigued.

"I liked the idea that you could create whatever you want," he says. "With a piece of metal and a blueprint, you can make a motorcycle, or instruments for a hospital. You just need the tools and the machine."

Education

After high school, Ramirez went to Saul Delgado technical school in Havana, learning to work with metals. After that, his education was all on the job.

He came to the United States in 2003 and worked on connectors for cable companies. He went on to operate CNC machines at a company that made brake parts, but he didn't do any programming there.

Ramirez's job at Indian Springs Manufacturing Company now gives him a chance to combine his love of metals with his love of computers.

"Working with a CNC machine is fun once you know how to create the program," he says. "You can play with the machine just the way you play with a computer."

Advice for Students

Ramirez sees CNC machines as the way of the future. "You're just going to need one person to run a machine that can do the job of five people," he says.

He advises students to learn CNC machines methodically.

"Go slow and pay attention to what you're doing in the beginning," he says. "Walk before you run, and you'll learn the steps you need to go faster. By the time you open your eyes, you'll say, man, look at how fast I'm doing this!"

Your Turn

1. Research and create a presentation to the class on the history of NC machining.
2. Dissect the NC equipment in your classroom and be able to explain the key components to the class in terms of operation.

▶ **Your Turn** activities reinforce text concepts with skill-building activities

Capitals or Not?

In the metric system, abbreviations are left in lower case letters. The exception is made when the unit comes from the name of a person. For example, a millimeter is abbreviated mm, but the Newton uses a capitol N because the unit honors a famous scientist.

Examples:

- **25 mm** 1/1000ths of a meter.
- **25 km** Thousands of meters.
- **10 N** Newton is a measure of force.
- **10 kN** A kilo-Newton is a thousand Newtons.
- **30 N•m** A Newton-meter is a measure of torque. The dot is used to show that it is a force (N) multiplied by a distance (m).
- **500 Pa** The Pascal is a unit of pressure honoring Blaise Pascal. The abbreviation is (Pa).
- **2.5 kPa** Thousands of Pascals.

▶ **STEM** connections show examples of how science and math principles are used to solve problems in engineering and technology.

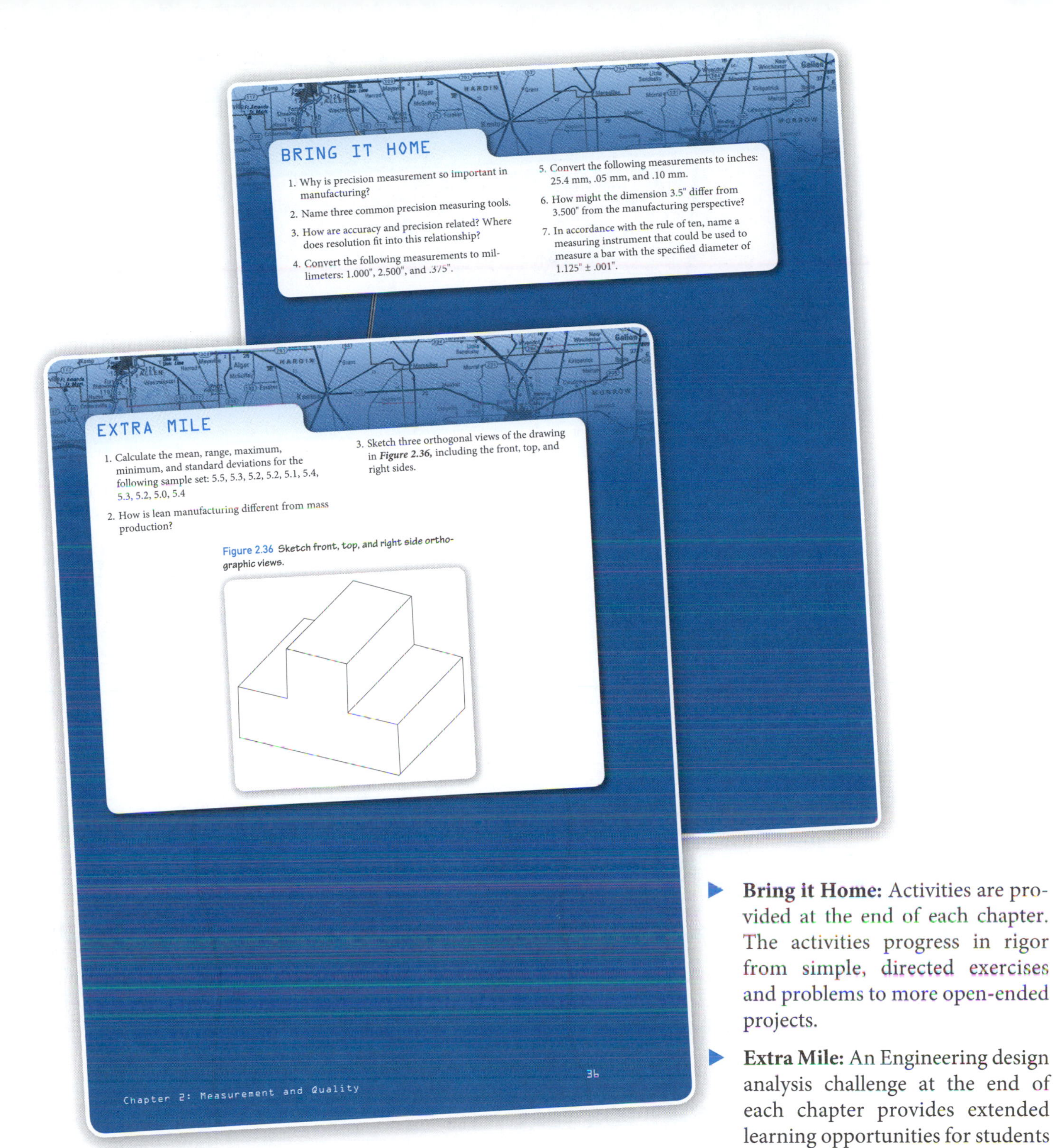

BRING IT HOME
1. Why is precision measurement so important in manufacturing?
2. Name three common precision measuring tools.
3. How are accuracy and precision related? Where does resolution fit into this relationship?
4. Convert the following measurements to millimeters: 1.000", 2.500", and .375".
5. Convert the following measurements to inches: 25.4 mm, .05 mm, and .10 mm.
6. How might the dimension 3.5" differ from 3.500" from the manufacturing perspective?
7. In accordance with the rule of ten, name a measuring instrument that could be used to measure a bar with the specified diameter of 1.125" ± .001".
EXTRA MILE
1. Calculate the mean, range, maximum, minimum, and standard deviations for the following sample set: 5.5, 5.3, 5.2, 5.2, 5.1, 5.4, 5.3, 5.2, 5.0, 5.4
2. How is lean manufacturing different from mass production?
3. Sketch three orthogonal views of the drawing in Figure 2.36, including the front, top, and right sides.
Figure 2.36 Sketch front, top, and right side orthographic views.
Chapter 2: Measurement and Quality
36

- **Bring it Home:** Activities are provided at the end of each chapter. The activities progress in rigor from simple, directed exercises and problems to more open-ended projects.
- **Extra Mile:** An Engineering design analysis challenge at the end of each chapter provides extended learning opportunities for students who want an additional challenge.

- **Revised examples** offer greater clarity and are better supported with graphics.

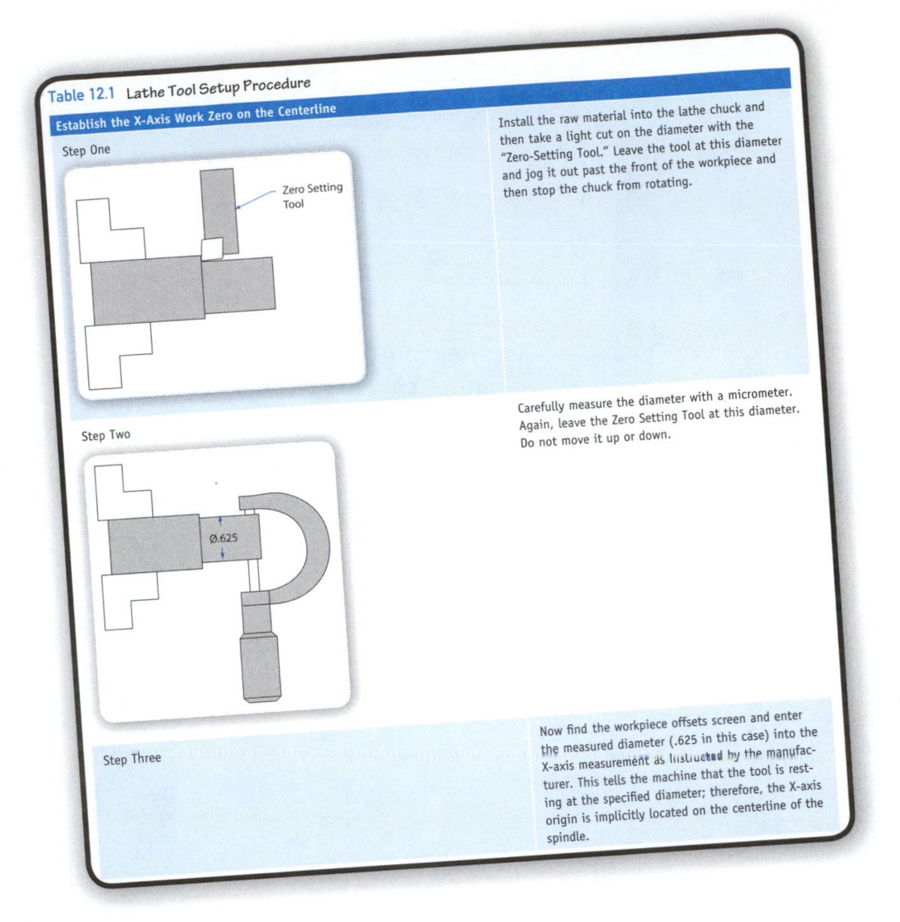

Table 12.1 Lathe Tool Setup Procedure

Establish the X-Axis Work Zero on the Centerline	
Step One (Zero Setting Tool)	Install the raw material into the lathe chuck and then take a light cut on the diameter with the "Zero-Setting Tool." Leave the tool at this diameter and jog it out past the front of the workpiece and then stop the chuck from rotating.
Step Two (Ø.625)	Carefully measure the diameter with a micrometer. Again, leave the Zero Setting Tool at this diameter. Do not move it up or down.
Step Three	Now find the workpiece offsets screen and enter the measured diameter (.625 in this case) into the X-axis measurement as instructed by the manufacturer. This tells the machine that the tool is resting at the specified diameter; therefore, the X-axis origin is implicitly located on the centerline of the spindle.

HOW *CNC PROGRAMMING* SUPPORT STEM EDUCATION

Math and science are the languages we use to communicate ideas about engineering and technology. It would be difficult to find even a single paragraph in this text that does not discuss Science, Technology, Engineering, or Mathematics. This text takes the extra step of showing the links that bind math and science to engineering and technology. A unique STEM icon highlights passages throughout this text that explain how engineers use math and science principles to support successful use of CIM technology. In addition, the Instructors Resource contains a STEM mapping guide to this career cluster.

SUPPORT FOR TEACHERS

A complete Instructors Resource on CD accompanies this edition to help instructors implement 21st-century strategies for teaching CNC programming. The Instructors Resource includes:

- Answers to end of chapter questions
- Instructional Outlines and helpful teaching hints
- A STEM mapping guide
- Computerized test banks
- PowerPoint® lecture outlines
- Image Library

ACKNOWLEDGMENTS

The author and publisher would like to acknowledge the following reviewers and Master Teachers who contributed to the quality and accuracy of this text:

Scott Banister, Pittsford Mendon High School, Pittsford, NY

Chris Hurd, Cazenovia High School, Cazenovia, NY

Terry Nagy, Shenendehowa High School Clifton Park, NY

Wanda T. Staggers, T L Hanna High School, Anderson, SC

Keith Thomas, Abilene High School, Abilene, TX

J. Paul Wahnish, East Lake High School, Tarpon Springs, FL

The publisher extends special appreciation to Project Lead the Way® CIM instructor Terry Nagy for his valuable contributions to this revision, including final revisions and corrections, case studies, illustrations, and activities.

ABOUT THE AUTHOR

Michael W. Mattson is chairman of the Department of Manufacturing Technology at Clackamas Community College, Oregon City, Oregon. He is actively involved in training high school teachers in CNC programming and has completed the summer training intensive with Project Lead the Way®.

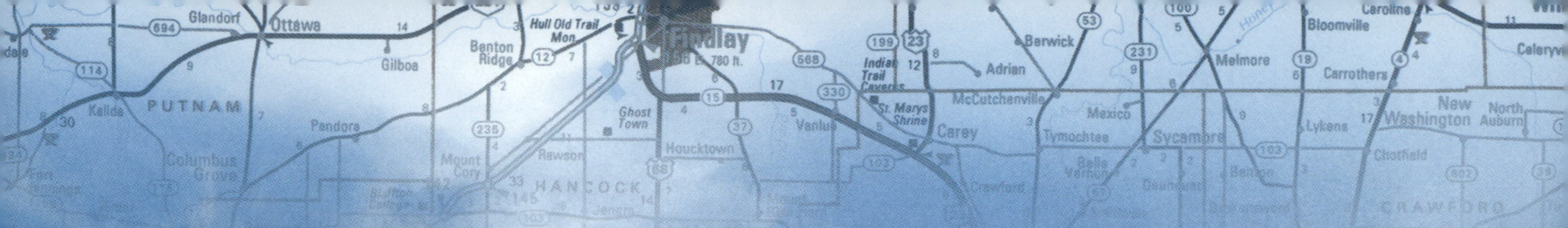

CHAPTER 1

Computer Numerical Control of Machine Tools

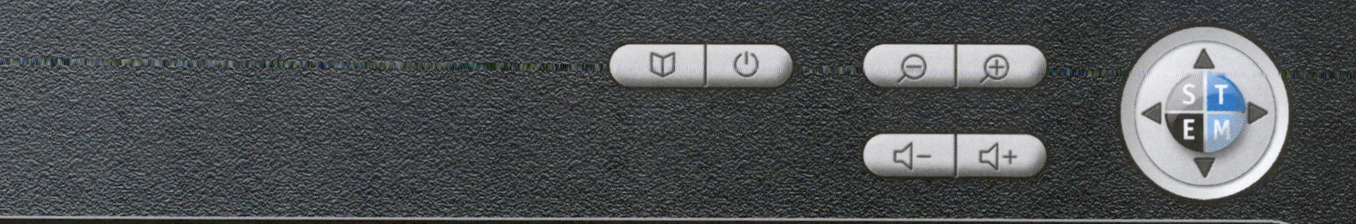

Menu | START LOCATION | DISTANCE | END LOCATION

Before You Begin

Think about these questions as you study the concepts in this chapter:

1. Can you name the common components of a CNC control system?

2. How do inputs and outputs relate to the function of a feedback loop?

3. How have machine tools contributed to our standard of living?

4. How has the modern machine developed from the industrial revolution?

5. What defines a machine tool?

6. What are some of the more common hardware used in controlling of machine tools?

Key Terms

Ball Screw
Computer Numerical Control (CNC)
Control
Encoder
Feedback Loop
Machine Tool
Machine Control Unit (MCU)
Manual Machining
Numerical Control (NC)
Resolution
Servomotors
Stepper motor

1.1 INTRODUCTION TO MACHINE TOOLS

A **machine tool** is a machine that is capable of producing another machine. A machine tool can be used directly or indirectly to produce a car engine or the landing gear for a 747 jumbo jet or even another machine tool. We are not talking about artificial intelligence or the evil plot of a science fiction novel in which the machines take over and enslave mankind. This is much more simple. To illustrate, let's step back a few hundred years.

machine tool:
A machine that can be used to produce another machine. The basic equipment used in precision metalworking.

Before the industrial revolution (1750–1830), products were largely made by hand. Each craftsman produced goods by employing hand tools, visual judgment, some simple patterns, or fixtures and a lot of skill. This is fine if you need a few products. But when you need millions of nails, or thousands of rifles, or hundreds of rail cars, it becomes a problem of productivity.

To scale-up production from a few rifles to a few thousand rifles requires specialization. The craftsman can no longer do the entire job. Specialists are needed to perform a small portion of the production over and over again. They need to produce the same part, the same way, the same size each time. For this we need machine tools.

Figure 1.1 An early engine lathe. The lathe could create round shafts and bore straight holes to a high degree of precision. *Courtesy of McGraw-Hill*

Figure 1.2 An early metal shaper. This machine was capable of creating flat, parallel, and square features. *Courtesy of McGraw-Hill*

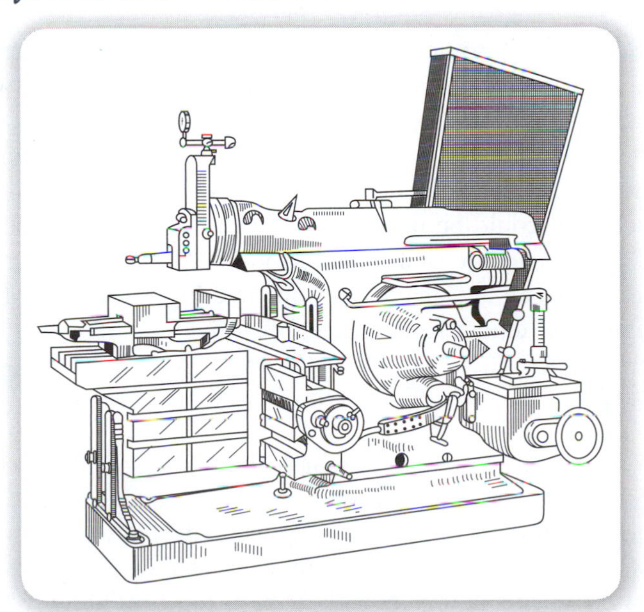

These early machine tools were used to shape metals more accurately than was possible by hand. They allowed manufacturers to consistently produce parts that were nearly identical in size, shape, and form. In fact, the concepts are simple. If you can figure out how to obtain just a few basic geometric forms, then almost any product can be created:

- Round
- Flat
- Perpendicular
- Parallel

Figure 1.3 **The key to creating machine tools was to create accurate geometric forms. Without these basic relationships, products cannot be made consistently.**

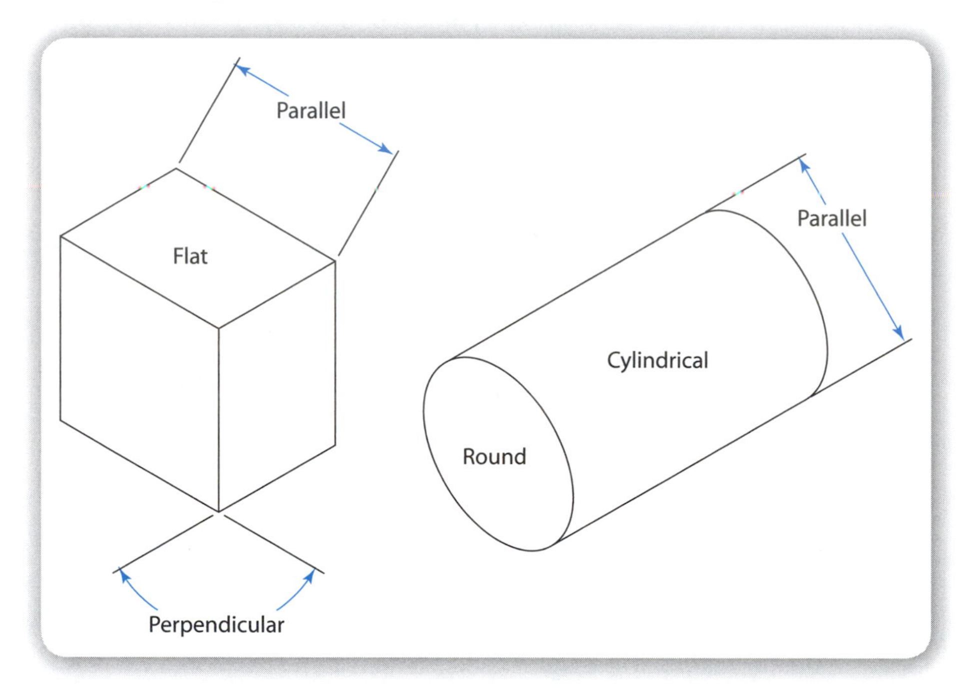

How could one make something round when nothing perfectly round has existed before? How could one make something flat and parallel when nothing perfectly flat and parallel has existed before? These were serious technological challenges to the early industrialists. They were solved mostly through innovations in precision measurement, materials, and techniques. The end result was a quiver of machine tools that could be used to produce parts directly such as rifles and steam engines. Also, machine tools could produce specialized machines to make products, for example, an automatic loom for making textiles or a forming press and die to stretch a car fender from a sheet of steel.

The Rise of CNC

Machine tools stayed roughly the same for 100 years of so. Of course, there were innovations. Steam power was replaced by electric motors, accuracy improved, and machine tools took on various levels of mechanical automation. Advances in "scientific management" changed the way managers thought about work methods. A single discovery in metallurgy quickly tripled the productivity of an average

Figure 1.4 **A punch press is an example of a machine that is used for forming and stamping metal. It is made by a machine tool, but it is not by itself a machine tool.** *Courtesy of McGraw-Hill*

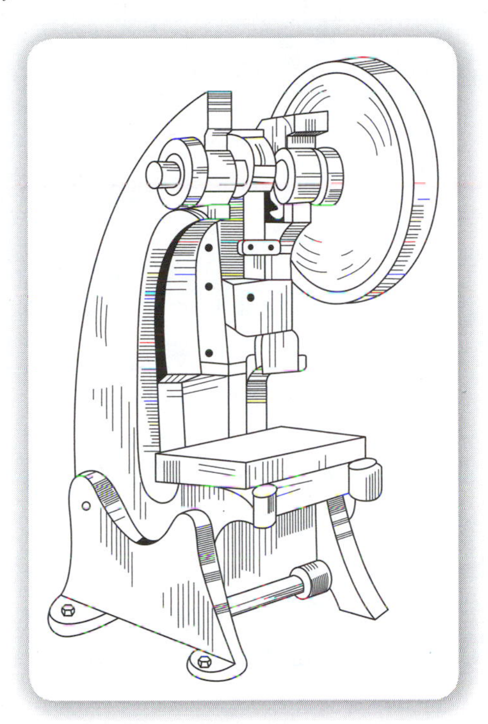

machine tool with high speed tool steel. But CNC was sea change. Few technological milestones could match the level of productivity gained by CNC machining since the industrial revolution.

The history of numerical control and later computer numerical control parallels the history of electronics technology and computer science. The same technologies that made it possible to use a vacuum tube as a logic device made it possible to use that device to instruct a machine tool to move or stop. The development of the transistor and soon after the integrated circuit made it possible to miniaturize devices and make them more powerful and less expensive.

In the early 1950s, the Department of Defense commissioned the Massachusetts Institute of Technology to produce the first workable NC machine. In the early days, there were electronics, but no computers. The machines were crude and difficult to program. They had no ability to perform contouring cuts or interpolate arcs. Every axis movement had to be programmed with a series of holes punched through paper tape. There was no memory as we know it today; the tape was the memory—read then forgotten.

From there, it has been an incremental march toward the technology that we see in manufacturing today. Power, speed, and memory have all made slow strides. CNC technology really only matured in the mid-1980s with the development of relatively cheap microprocessors and electronic memory to replace punched tape. There are not a lot of fundamental differences between a modern CNC machine tool and CNC of 20 years ago. They are better, faster, cheaper, and more accurate, but there has been no paradigm shift.

Today, anyone with $3000, a personal computer, and little ingenuity can build a powerful CNC machine that will fit on a desktop. It will cut curves, lines, and circles with minimal programming. From our vantage point, it all seems so simple, but it took 50 years for the technology to evolve to such a state. In 1950, one couldn't bring up the Internet and buy a motion control card online and have it delivered to the doorstep the next day.

Figure 1.5 NC programs were once stored as holes punched in a paper or plastic tape.

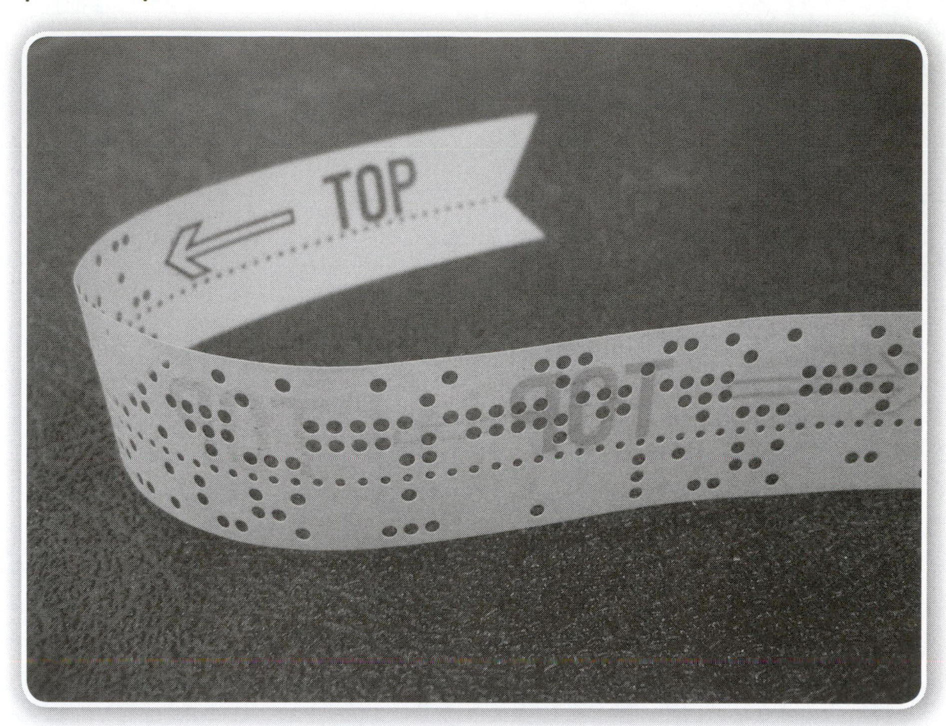

Figure 1.6 Today, inexpensive motion control products are available that work with a personal computer.

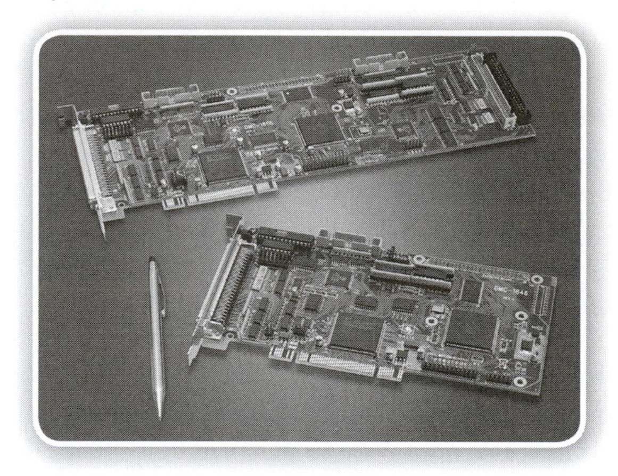

Energy to Power the Industrial Economy

Of course all this great engineering of the Industrial Revolution would have been limited by the bottleneck of energy. Early power sources to run machines and to produce materials were usually a product of animal, wood, or water power, all of which are a result of the Sun's recent energy. The discovery of coal and petroleum reserves provided the energy to fuel the most massive growth of human wealth the world has ever known.

To illustrate the impact of these energy sources, take the case of an exceptional worker without power tools. A very athletic man can perform work at a rate of about one-eighth of a horsepower for an 8-hour day. That amount of energy is contained in just .02 gallons of crude oil or 0.2 pounds of coal. It is clear that without fossil fuels or something similar, we would still be living at a pre-industrial standard of living. We would spend most of our time and energy growing food and very little producing plasma television sets.

You can stand by the railroad tracks in the Midwest and watch trains mile and a half long pass by. The cars are filled to the top with coal to power our electricity generation plants and produce our steel. Its mind boggling to imagine the amount of energy contained in those thousands of tons of coal.

Technical Stuff

One horsepower is equal to lifting 550 lbs up to a distance of 1 ft in 1 second. A 200-lb man climbing up a 9,900-ft mountain would use about (200 lb × 9,900 ft = 1,980,000 ft-lbs) of energy. If he could do this in 8 hours (or 28,800 seconds), then he would need to produce about .125 hp to climb up the mountain. Why? Because 1,980,000 ft-lbs/28,800 seconds = 68.75 ft-lbs per second or 1/8th of a horsepower.

So, a man working extremely hard all day is producing about 1/8th of a horsepower. He will use 1,980,000 ft-lbs of energy in the process. Foot-pounds are a pretty small number for measuring energy, so we usually use BTU, which stands for British Thermal Unit. A BTU is equivalent to 778 ft-lbs of energy.

In the metric system, the unit for energy is the joule. 1 BTU is about 1054 joules. For power (power is energy divided by time) the metric unit is watt. One horsepower is about 748 watts. Oddly enough, my German engineering acquaintances tell me that European car engines are still rated in horsepower. I guess it just makes a better visual impression in the mind's eye.

Fossil fuels are very energy dense. They contain huge amounts of energy for their mass. Crude oil contains approximately 139,000 BTU per gallon and coal contains about 12,700 BTU per pound. A pound of coal can do the work of five strong men for a day, at least in theory. The fact is that steam and internal combustion engines are only about 10 to 30 percent efficient. You have to put a lot more energy in than you get out.

Figure 1.7 An illustration of energy units and the incredible energy content of fossil fuels.

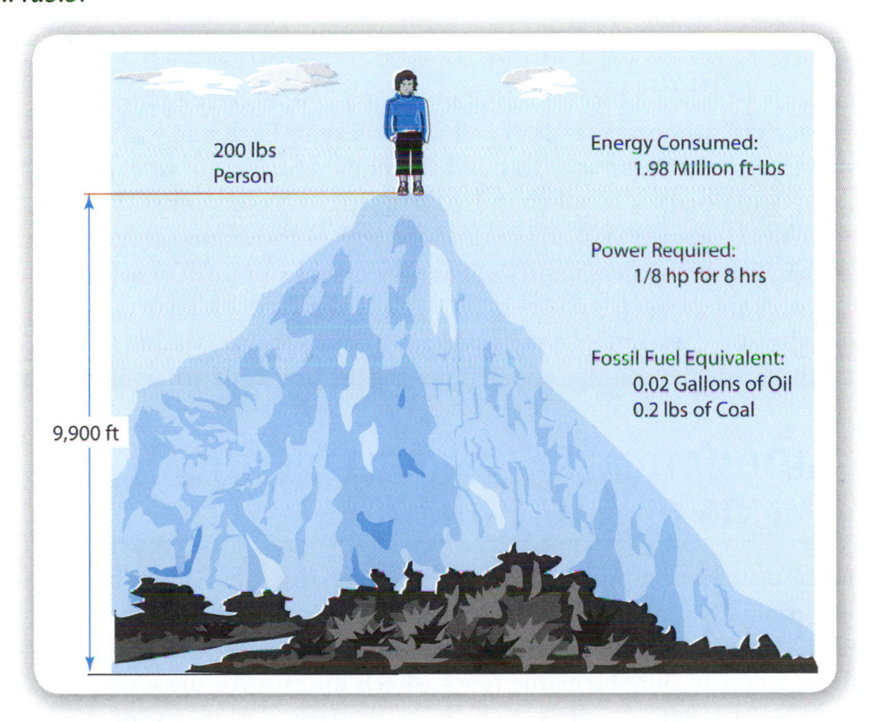

Cheap energy was now available to fuel the industrial revolution. The watt steam engine was the breakthrough technology that harnessed the energy of fire and made it portable. Before the steam engine, the only serious source of industrial power came from running water. Small factories were often located next to a stream or river. They would use the running water to turn a water wheel that in turn could be used to power machinery. This is obviously very limiting. One can't always locate the factory where the raw materials are located.

The steam engine changed all of that. Power was now portable. A steam engine could be put on a locomotive or ship or in a factory on the Plains. The development of the steam engine was critical to the industrial revolution. The engine had to be made precisely, and that spawned one of the first modern machine tools—the Wilkinson boring mill. The boring mill was a machine that could produce a very round and straight hole for the cylinder of the engine.

Figure 1.8 **Timeline of technology.**

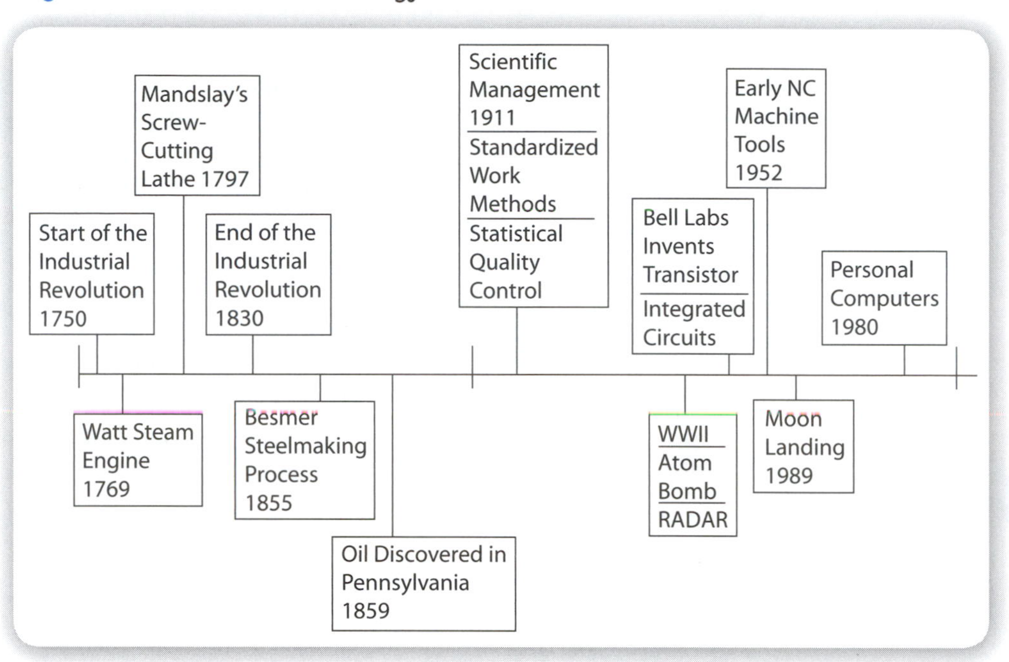

manual machining: Machining that is performed on non-automated machinery. The motion of the tool is guided by the hands of a skilled machinist.

I recently watched a blacksmith pound out a couple dozen nails from red hot iron. It is a very slow process. Each nail was different. Each nail was an individual. A craft, yes, but this is no way to build the houses for a nation. A pre-industrial craftsman could make a couple hundred nails in a day—an entire day's wages for a box of nails. Today, a worker can produce tens of thousands of nails in a single day. We can buy the same quantity of nails for maybe one-fifth of an hour's wages. All the nails are perfect, not an individual in the bunch. That is productivity. Productivity and cheap energy are why we can have nice houses and powerful cars. We can fly across the country on a jetliner for the price of a truckload of rocks.

1.2 AUTOMATICALLY CONTROLLING THE MOVEMENTS OF A MACHINE TOOL

Computer Numerical Control (CNC): A form of electromechanical motion control used on machine tools, whereby a computer and computer program are used to perform machining operations.

Machine tools, as we know them, have been around since the Industrial Revolution. You are probably already familiar with manual machine tools such as milling machines, lathes, and drill presses. These machines are operated by a skilled machinist who carefully turns the hand wheels and moves the levers to position the workpiece or cutting tool in the proper orientation. This type of machining is known as **manual machining**. Manual machining requires a great deal of operator skill and training to produce high-quality parts on a consistent basis. Manual machining is also relatively slow—and relatively expensive.

Today, manual machine tools have been largely replaced by **Computer Numerical Control (CNC)** machine tools. The machines still perform essentially the same functions, but movements of the machine tool are controlled electronically rather than by hand. CNC machine tools can produce the same parts over and over again with very little variation. They can run day and night, week after week, without getting tired. These are obvious advantages over manual machine tools, which need a great deal of human interaction in order to do anything.

This is not to say that manual machine tools are obsolete. They are still used extensively for tool and fixture work, maintenance and repair, and small volume production. However, much of the high- and medium-volume production work is now performed on CNC machine tools. Furthermore, CNC machining has become common in the low-volume *job shops,* where lot sizes of a dozen to several hundred parts are common.

CNC machine tools are highly productive. They are also expensive to purchase, set up, and maintain. However, the productivity advantage can easily offset this cost if their use is properly managed. The decision to produce parts conventionally or by CNC is driven mainly by setup cost and volume. The setup cost to perform a machining operation on a conventional machine tool is rather low; on a CNC machine tool, the setup cost can be rather high. In summation, as the production volume increases, the combination of productivity gains and the ability to spread the setup cost over many parts makes CNC machining the obvious choice.

What is Computer Numerical Control, or CNC? Simply put, CNC is the automated control of machine tools by a computer and computer program. In other words, a computer rather than a person will directly control the machine tool. Before CNC, the motions of the machine tool had to be controlled manually or mechanically (such as with cams or tracer systems).

The axial movements of CNC machine tools are guided by a computer, which reads a program and instructs several motors to move in the appropriate manner. The motors in turn cause the table to move and produce the machined part. Of course, the computer does not know what shape you want to cut until you write a program to describe the part. The conceptual idea of controlling a machine tool electronically is rather simple. A programmer writes a set of instructions describing how he or she would like the machine to move and then feeds the program into the machine's computer (called the **control**). The computer reads the instructions and sends electrical signals to a motor, which then turns a screw to move the machining table. A sensor mounted on the table or on the motor sends positioning information back to the computer. Once the computer determines that the correct location has been reached, the next move will be executed. This cycle repeats itself over and over until the end of the instructions is reached. This is a simplistic but accurate view of a typical CNC machine tool. There are many variations of this model, some of which will be discussed in this chapter.

control:
The computer that operates a CNC machine tool; the machine control unit (MCU).

CNC machine tools are very similar in construction to manual machine tools. The typical CNC vertical milling machine has roughly the same components as a garden-variety conventional milling machine. Each machine has a rigid machine base to hold the moving parts, a spindle in which to mount the cutting tool, and a table that can move left to right and in and out. Attached to the table are accurate lead screws used to position the table to the proper location.

Unlike a manual machine tool, a CNC machine tool also has a motor mounted on each lead screw instead of a hand wheel. A CNC machine tool also has a computer to control the motors, and sensors to keep track of the table position and velocity. Other than that, CNC and manual machine tools can look very similar. In fact, many CNC machine tool manufacturers sell models that are available as a manual machine tool or fitted with a CNsC control system. It is also common for older manual machine tools to be retrofitted with CNC control components that are manufactured by a third-party vendor.

Numerical Control (NC):

The early, programmable electronic control systems that were developed before the availability of inexpensive and compact computers. The grandfather of CNC.

We should also mention that the predecessor to CNC was called **Numerical Control (NC)**. NC machine tools were developed before the availability of inexpensive computing power and therefore had an electronic control but not a genuine computerized control. The programs were loaded by punching code in a paper or plastic tape and then feeding the tape into the machine, where the program could then be executed. If any changes were needed in the program, a new tape had to be punched. In contrast, today's programs are typically stored on floppy or hard disks, and the program can be easily modified directly at the control or at a PC workstation.

The terms NC and CNC are used synonymously today, although actual NC machines are becoming rare in most shops. Regardless of the control type, the code that is used to produce the parts is known as NC code—not CNC code.

1.3 COMMON TYPES OF CNC MACHINE TOOLS

Virtually any machine tool that can be controlled manually can also be manufactured with CNC controls. CNC and manual machine tools share many of the foundation components that make a machine tool a machine tool. Therefore, it is not a great leap in thinking to see that hand wheels can be replaced with motors, micrometer dials can be replaced with positioning sensors, and manual switches can be replaced with electronics. This is true for milling machines, lathes, and many other machine tools and fabrication equipment found in industry. However, the numerous varieties of milling machines and lathes remain the dominant tools in manufacturing.

Milling Machines

The vertical CNC milling machine and the Vertical Machining Center (VMC) are probably the most common CNC machine tools found in shops today. They are really the workhorses of job shops and low-volume production. They are very agile machines and can be easily adapted to a great variety of workpieces.

Vertical CNC milling machines are easy to load, and their operators have good visibility. Vertical CNC machining centers are usually considered good, all-purpose machine tools, but they are not as heavy-duty as their horizontal cousins. Chips can also be difficult to remove, as they tend to fall back into the cutting zone.

So what is the difference between a standard vertical CNC milling machine and a VMC? Probably just a name; if you call a VMC a vertical mill, nobody is going to give you a funny look. It is suspected that the name VMC had more to do with marketing than anything else, but we can observe some common differences between the two. Vertical machining centers are typically made in a bed-type construction, whereas a vertical CNC milling machine uses the knee and column construction. Knee and column machines tend to be fairly small and have a movable quill—a Bridgeport-type milling machine uses knee and column construction. Bed-type construction, like the machine in ***Figure 1.9***, lends itself more readily to larger machines. They most often use a solidly mounted spindle rather than a quill. The solid spindle is inherently more rigid than a movable quill, but it is also heavier and harder to move around.

Vertical machining centers are often manufactured with three linear axes of motion—much like a standard milling machine. However, a fourth or fifth rotational axis is sometimes added. An additional axis can be a simple rotary table or a complex milling head that swivels about as illustrated in ***Figures 1.10*** and ***1.11***.

The Horizontal Machining Center (HMC) is similar in construction to the conventional horizontal milling machine; however, its use is often very different.

Figure 1.9 A cutaway view of a bed-type, CNC vertical machining center (VMC).
Courtesy of Haas Automation

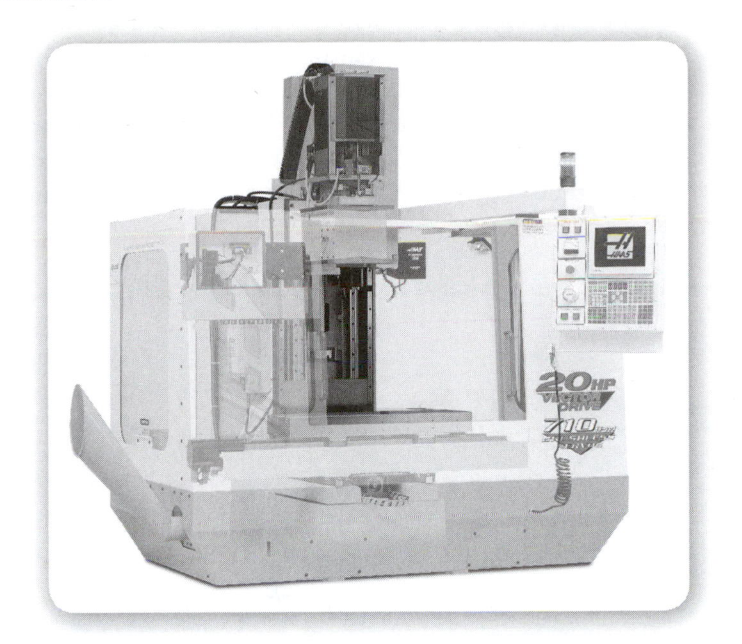

Figure 1.10 A five-axis vertical milling machine is capable of producing complex geometry that is not possible with other methods. *Courtesy of Haas Automation*

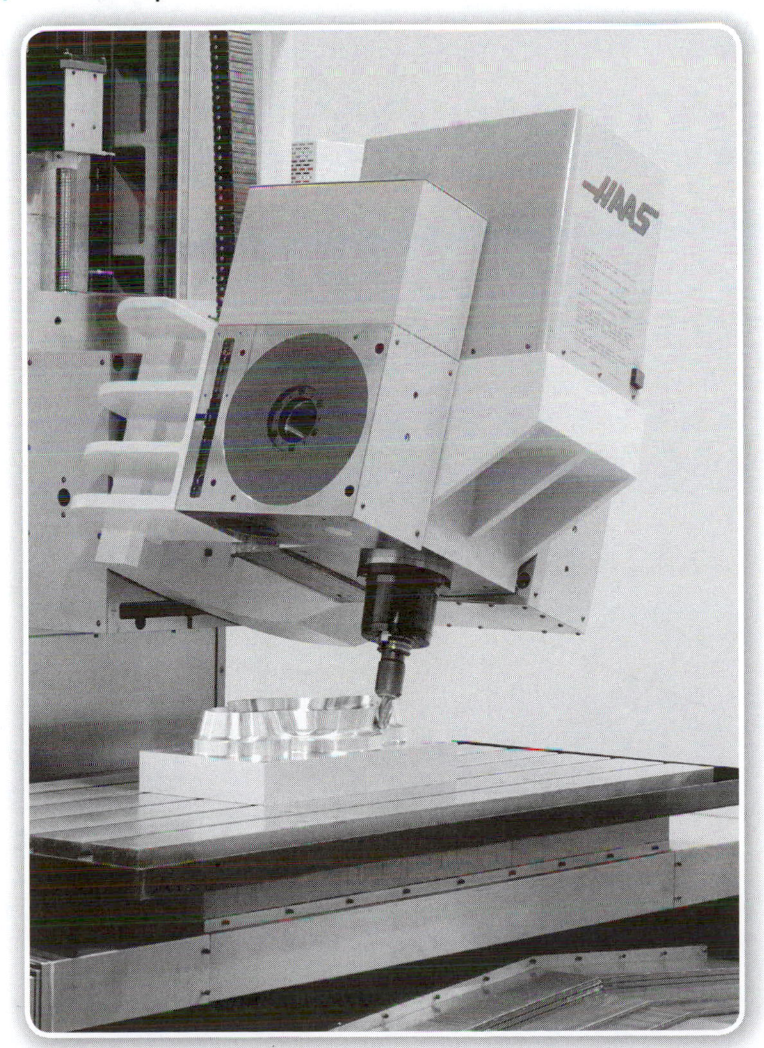

Conventional horizontal mills commonly use cutting tools that are mounted to an arbor. CNC horizontal mills seldom use (or even have the ability to use) arbor-mounted tooling. HMCs usually use spindle-mounted cutting tools in the same way a VMC uses spindle-mounted cutting tools. In fact, an HMC is basically a VMC that has been tipped over 90° as in ***Figure 1.12***.

You may wonder, "Why would you bother?" Well, there are some advantages to the HMC construction and configuration. First, the spindle is mounted to a stationary and extremely rigid base. This rigid construction is ideal for heavy cutting

Figure 1.11 A rotary table can serve as a programmable fourth axis. *Courtesy of Haas Automation*

Figure 1.12 A cutaway view of a horizontal machining center (HMC). *Courtesy of Haas Automation*

and high material-removal rates. This brings us to the second advantage—the tool and the workpiece are mounted horizontally. If we are going to remove a lot of material, then we are going to produce many, many chips. Chips that get stuck in the cutting zone can ruin the finish and destroy the cutting tools. The horizontal configuration allows gravity to help remove the chips.

The disadvantages of HMCs are that the workpiece is often mounted horizontally on a fixture called a "tombstone." It tends to be more difficult for the operator to hold and align heavy workpieces while fighting gravity. Another disadvantage is that it can be difficult for the setup person and the operator to see the tool and cutting zone.

Most CNC milling machines also have an Automatic Tool Changer (ATC), as illustrated in ***Figure 1.13***. ATCs enable the machine to run unattended all the way through the cycle. This frees the operator to perform quality control or deburring, or to run another machine at the same time. The cutting tools are premounted in a tool holder and then stored in a carousel or belt until needed. Tool change time can have a dramatic effect on cycle time (the time it takes to machine the

Figure 1.13 A carousel-type automatic tool changer (ATC). *Courtesy of Haas Automation*

part). Machine tool manufacturers are now making VMCs that can change tools in under a second.

VMCs and HMCs are also available with something called a pallet changer. A pallet is essentially a removable base that fits into a lightweight machining table. Only one pallet can be loaded onto the machining table at a time. The advantage of a pallet system is that while the machine is machining one workpiece, the operator can be setting up the next workpiece on an identical fixture on the other pallet. Another variation is to set up the next job on the second pallet while the first pallet is dedicated to machining the first job. This can reduce the amount of downtime, when the machine is not making chips (or money).

Lathes

The CNC lathe, as shown in ***Figure 1.14***, is another common sight in job shops and production facilities. CNC lathes have some unique advantages over their conventional counterparts. They can cut circular arcs that were all but impossible to cut on a conventional lathe without special tooling. CNC lathes also excel at threading. There are no gears to change or dials to watch while threading. A CNC lathe can cut virtually at any pitch and any number of leads, and it can even cut variable pitch threads without ever selecting or changing the gears. Lastly, CNC makes it easy to maintain a constant surface speed at the cutting edge as the diameter changes. This leads to an increased tool life and consistent surface finishes. The CNC control increases or decreases the spindle RPM to maintain a constant surface speed while the cut is being made. There is no need to stop and change gears as you would on a manual lathe.

Lathe work has historically been a high-volume affair. Because cylindrical parts such as bearings, small fittings, and fasteners are needed in great quantities, many lathe-type machines were automated early on. Examples include turret lathes, which change tools automatically, and screw machines, which are driven by a series of cams. In contrast, milling machines, which tend to produce smaller quantities, prove much more difficult to automate. CNC has largely replaced mechanical automation in turning machines in all but the highest production situations.

Figure 1.14 **A slant-bed CNC turning center.** *Courtesy of Okuma America*

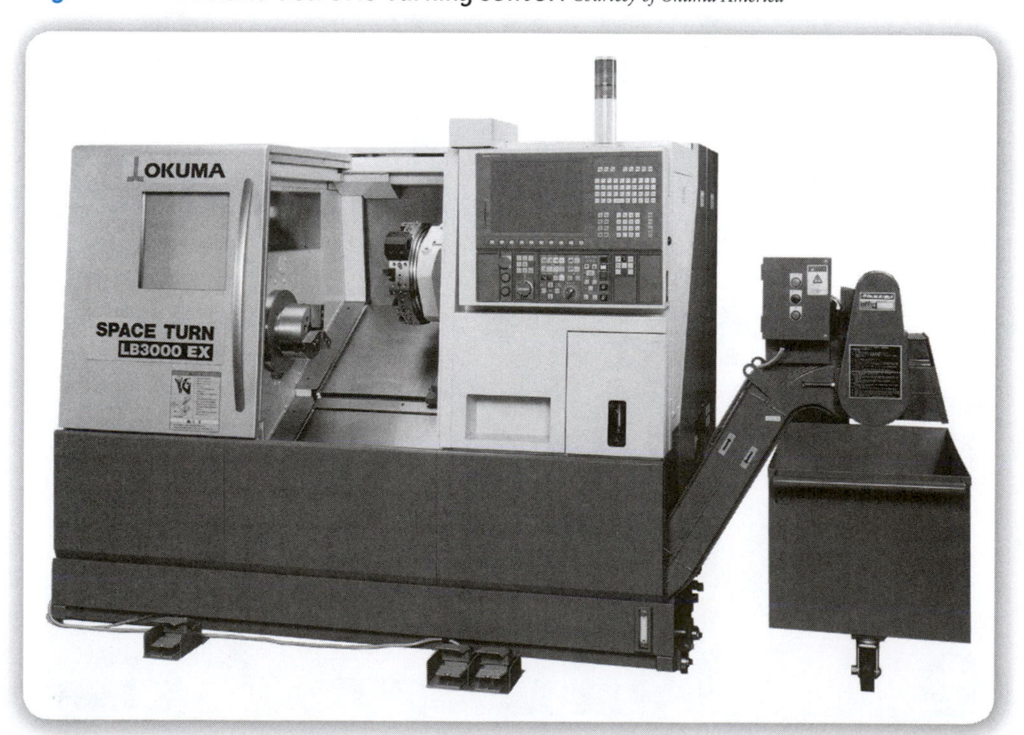

Some of the earlier CNC lathes were nothing more than standard, flatbed lathes fitted with a CNC system. Today this type of construction is rare, with the exception of very inexpensive or very large lathes. Most CNC lathes use a fully enclosed, slant-bed construction, and the manufacturers now call them CNC turning centers (no initials or acronyms please). The slant-bed construction is very rigid and offers excellent loading access and chip removal. CNC turning centers are almost always enclosed with heavy sheet metal and have steel bars across the windows for safety.

The cutting tools on a CNC turning center are usually mounted in a device called a turret. The turret is then mounted on a heavy cross slide and carriage similar to that of a conventional lathe. The turret can be indexed very rapidly to automatically change to the next tool. Another design for automatic tool changing is the tool magazine. A tool magazine carries the tools on a belt and loads each tool onto the tool post when needed. This is slower than a turret, and each tool must be mounted in a special holder, which adds to the cost. The main advantage of a magazine is that it can hold several dozen or more tools (several hundred is not unheard of). A typical turret can only hold 16 tools.

Many CNC turning centers are also equipped with some enhanced features to make them more productive. One example is multiple turrets. CNC turning centers with multiple turrets can perform two operations at the same time. For example, the top turret can cut a groove on the outside diameter while the bottom turret performs a drilling operation on the face of the workpiece. Another productivity enhancement is *live tooling*. This so-called live tooling is in fact a small, light-duty milling spindle mounted in the tool holder. Live tooling can be used to drill cross holes or hole patterns on the face of the part, or in milling operations such as cutting a keyway or milling a wrench flat into the workpiece, as shown in ***Figure 1.15***. Live tooling can dramatically reduce the number of secondary operations that are required. Rather than going from the lathe to the milling machine, the part might be able to go directly to the customer.

There has also been a trend recently toward automated part handling and unattended operation (i.e., no humans needed to load parts and push buttons). Multiple-spindle CNC turning centers help to accomplish this. A multiple-spindle

Figure 1.15 **Live tooling can be used to mill and drill a workpiece while still mounted in a lathe.** *Courtesy of Haas Automation*

turning center is really two lathes that face each other, except that it is all the same machine. There are a spindle and turret on the left and a spindle and turret on the right. The idea behind this is that a stack of raw material (stock) is loaded into a staging area and every few minutes a completely finished workpiece comes out of the other side of the machine. One problem for many turned workpieces is that they require two setups: one for the front side and another for the back. This usually means that all of the front sides will be turned, and then the machine will set up again to turn the back sides. By the very nature of this traditional method, a lot of material will be work in progress (WIP), which to managers means wasted handling, wasted storage space, and wasted money. Besides, many customers want a steady flow of product just in time for assembly—not the whole bunch at once. Multiple-spindle CNC turning centers are normally equipped with a robotic parts loader. The parts loader will first load the part into the first spindle and then move it to the other spindle for the second operation. When the second side is finished, the workpiece is unloaded from the machine and the cycle begins again.

A number of smaller CNC lathes have also come to market in recent years to compete with the turret lathes and automatic screw machines. One such machine is called a gang-style CNC lathe. The gang-style lathes are built with the flatbed or slant-bed construction. Their main attribute is found in the style of their cross slide and tool holding system. The cutting tools on a gang-style lathe are mounted in tool blocks to a rather long cross slide. The cutting tools are many of the same tools used in turret lathe and screw machine work—making it easier to make the transition to CNC. This type of machine is designed for short, small-diameter parts. In fact, the cross slide is not designed to go past the face of the workpiece; therefore the cutting tool holder must protrude several inches from the cross slide. The primary advantage of this configuration is that the tools are always close to the workpiece, thus saving time in long production runs.

Grinders

Surface grinders and cylindrical grinders were probably the last frontier for CNC systems. The tight tolerances required in grinding meant they had to wait for the motion control technology to catch up. Today, the positioning systems are capable of the 50-millionths of an inch tolerances that are common in precision grinding. It is now common to see CNC surface grinders and CNC cylindrical grinders with some impressive features. For example, most CNC precision grinders will compensate for wheel wear and automatically redress the wheel when it becomes too worn. CNC precision grinders that automatically dress special shapes into the wheel and create profiles on the workpiece are also available.

A less common but important category of CNC machine tools falls into the two-axis class of machines. These machines are designed to work on relatively thin, flat stock such as sheets and plates. They do not have the ability to perform cutting operations in three dimensions as do milling machines. Examples include CNC laser cutting, CNC water jet cutting, CNC plasma and oxy-fuel cutting, and CNC turret punches.

Electrical Discharge Machining

EDM or electrical discharge machining is now making great advances into the realm of CNC machine tools. There are two popular types of EDM machine tools: wire and sinker.

Sinker-type machines use a graphite electrode to "burn" a shape into a metal workpiece. The process uses high-frequency sparks to melt microscopic droplets of material away from the workpiece. It is a very slow process but allows for intricate details in hardened metals. The process is used extensively in the plastic molding industry to create mold cavities with otherwise impossible shapes. They are usually light-duty and slow machines, which is acceptable for a process that does not involve any cutting forces from a tool.

Wire EDM (***Figure 1.16***) uses a thin wire (.001–.012" in diameter) to slowly cut through metal up to several inches thick. Sometimes it is described as an "electric band saw." The wire is held taut and then moved along the prescribed path by the CNC control system. Progress is measured in inches per hour. The main advantage of wire EDM is that it can produce intricate shapes for punches, dies, and molds that do not require any expensive grinding. Tooling is often cut in the hardened condition to tolerances that rival grinding processes (±.0001").

The CNC controls systems on such machines are pretty basic as far as axial movements are concerned. However, the sophistication of the electrical feedback system is amazing. The control senses how the cut is progressing. It maintains a small gap between the cutting electrode and the workpiece to produce an efficient cut with the desired surface finish.

Figure 1.16 A modern wire EDM cuts with a very thin electrically charged wire.

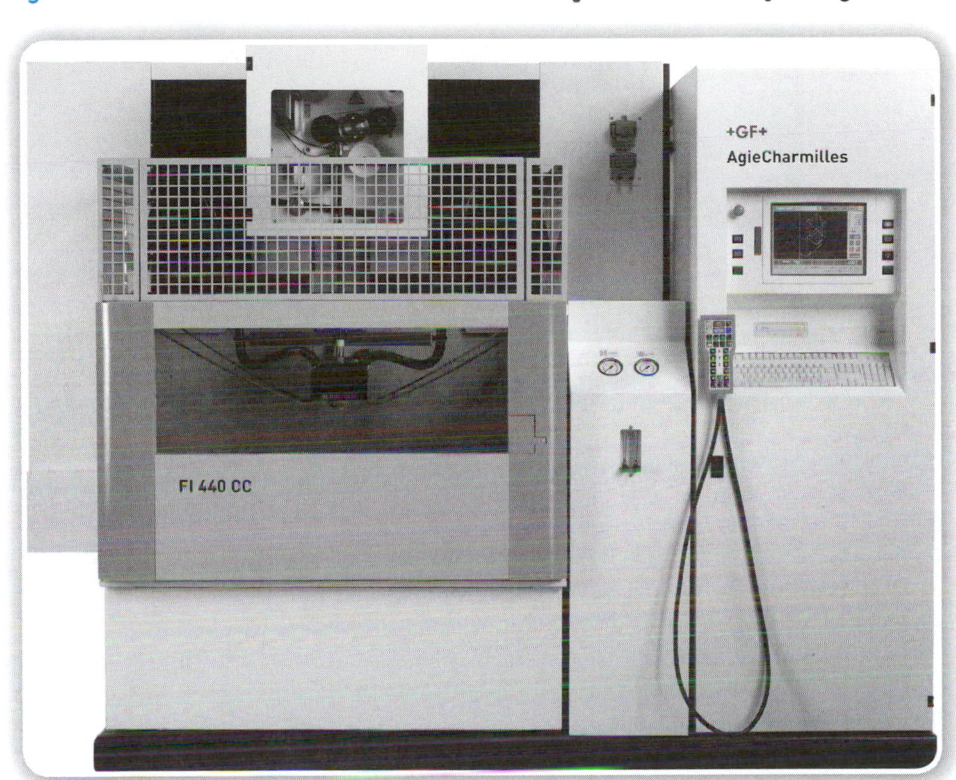

CNC for Precision Metal Fabrication: Lasers, Plasma, etc.

CNC has also found its way into the precision metal fabrication industry. This is mostly in the form of two-dimensional cutting machines powered by a cutting head that uses high-energy lasers or plasma to cut out shapes from flat sheets of material.

Laser cutting is a very fast and precise process that cuts out parts from sheet metal for precision flat patterns, soldering masks for circuit board assembly and virtually any shape with amazing precision. There are even multi-axis lasers that can cut odd-shaped materials such as tubing and metal stampings. Metal-cutting CNC lasers are expensive are expensive with introductory units costing more than $200,000.

Less powerful CNC lasers are capable of cutting plastics, gaskets, and a variety of nonmetallic materials. Additionally, they are used for laser engraving on the surface of products.

A special class of CNC laser is used to directly form a three-dimensional shape out of a solid block of material much the way it might be machined or formed with sinker EDM. In fact, the German manufactures of such a machine stated that EDM sinker process was the target of their technology. The laser is aimed at the surface of the metal, and a nearly invisible blue light evaporates a tiny spot of material. The process is like building a sand castle by removing a grain at a time from big pile of wet sand. The progress is still slow, but the advantage is that you do not have to produce any electrodes. We have still to see three-dimensional (per earlier) laser machining come into the mainstream.

Figure 1.17 A gantry-style CNC plasma cutting machine.

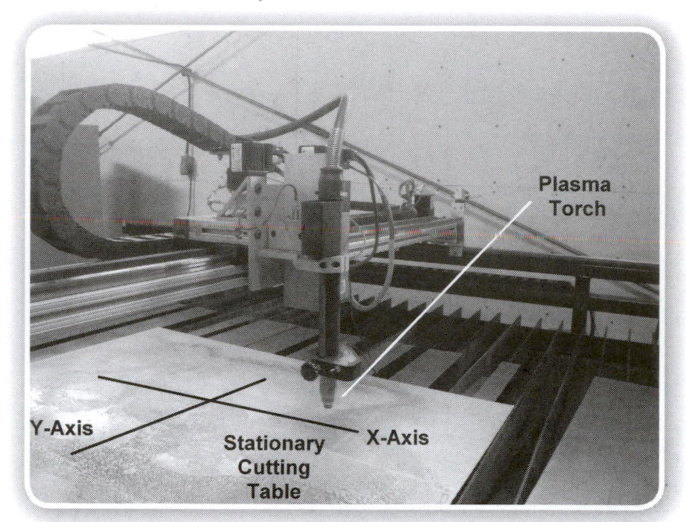

A process similar to a laser is called CNC plasma cutting. The CNC control system for CNC plasma is roughly the same as for any other light-duty platform, and it can position itself to very accurate locations (***Figure 1.17***). The nature of the plasma "flame" (a highly energized stream of ionized gas) is the limiting factor in the precision of the process. The kerf is wide and tends to wander around a bit. So CNC plasma resides in the realm of heavy fabrication. It can rapidly produce elements of a structure that is destined to be welded. This process is great for producing gussets, holes, odd shapes, and any other workpiece that might have formerly been cut out manually with an oxy-fuel torch.

CNC can be applied to virtually any manual process that required manual intervention or operation. We see this in all other areas of precision metal fabrication such as with press brakes, turret punches, and welding machines. These are all important tools for our industrial economy and are fully mainstream tools for fabrication.

CNC Routers

CNC routers are finding their way into industry at an amazing speed. These function essentially the same way as a router that one might use in home woodshop, except that the router is high-speed, quiet, and attached to a CNC-controlled machine frame (***Figure 1.18***). Often the machines are large enough to hold an entire 4 × 8-ft sheet of material.

Routers are exceptionally good at cutting wood, plastic, and even aluminum. They tend to be light-duty machines that are fundamentally very similar to a CNC vertical milling machine, except that they are for cutting soft materials at high speeds. They are used extensively for furniture and cabinet manufacturing, model-making, and display work and increasingly in the aerospace industry for cutting shapes out of thin aluminum and composites.

Two unique features of CNC routers are a vacuum table and multiple heads. Routers are usually cutting soft, flat materials such as plywood of fiberboard. These materials do not generate the high cutting forces that one would see with tradition machining. Therefore, the workpiece is usually held down with on a vacuum table. The cutting bit can then move in a path that is unobstructed by clamps or tooling to cut the entire perimeter of a workpiece.

Figure 1.18 A multi-head CNC production router.

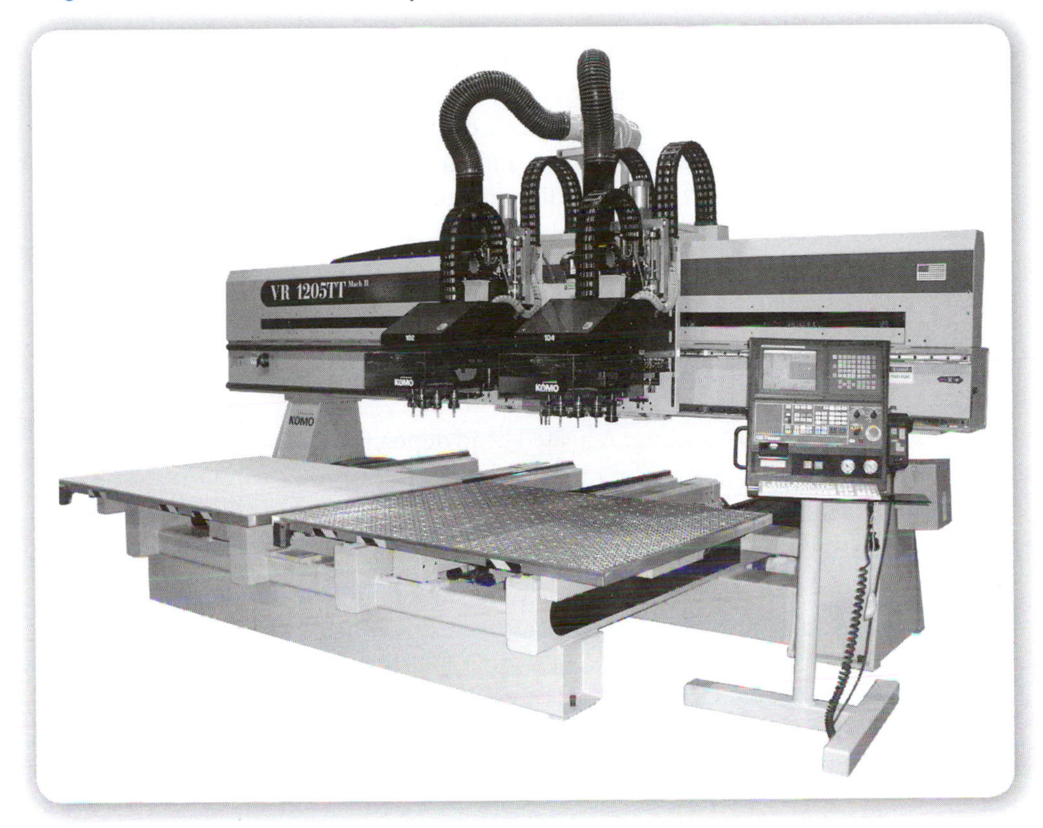

Computer software is used to "nest" parts onto a single piece of material. This leads to much less waste (*Figure 1.19*).

Figure 1.19 Nesting software makes the most efficient use of material on a router, laser, or plasma CNC machine.

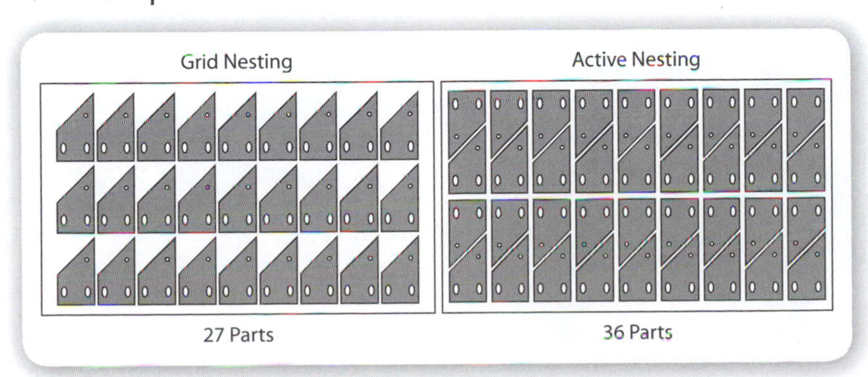

The other unique feature is the multi-head router. The cutting heads are attached in one unit and move in unison. This is great for productivity—a machine that makes three or more of the same parts at the same time!

Rapid Prototyping

A rapid prototyping (RP) machine is a special class of device that uses the same motion control technology as any CNC machine tool. However, the use of the machine is to create a workpiece through additive processes rather than removal processes (such as machining). The machines are used make tangible models from CAD drawings. In other words, the engineer or designer can quickly create a real workpiece that can be held, examined, or tested before it goes into production. The main advantages of having parts made with a rapid prototype machine versus a machined or molded part is cost and time. The engineer can have the part in a few hours rather than days or weeks, and it is much less expensive (***Figure 1.20***).

Most rapid prototyping machines work on the concept of slice and build. A software program takes the CAD model and slices it up into paper-thin sections. The thin sections are effectively two-dimensional cross sections of the workpiece at any point. Then the machine goes to work by taking each "slice" and building it up layer after thin layer. The result is a three-dimensional model that represents the computer-designed workpiece.

Common Rapid Prototyping Technologies

- Fused Deposition Modeling: A heated head deposits minute droplets of molten plastic onto a base layer. The model is built-up drop-by-drop until complete. Think of a hot glue gun. This is really the same concept, but much smaller and more accurate. The finished plastic models are strong and usable.
- 3-D Printing: An inkjet printer head deposits a liquid binder onto a powdered layer of construction material that has been spread flat by a roller. The binder hardens the powder, and then a new layer is printed every .003 inch until the workpiece is finished. The finished models are really just for show because they do not have much strength. However, some can be used for casting processes.

Figure 1.20 Rapid prototyping (RP) can create any shape that can be designed in a CAD system.

- Selective Laser Sintering: A laser strikes and fuses a specially prepared metallic powder. Very similar to three-dimensional printing, except that the binding is done by melting the powder with the heat of a laser. The main advantage of this process is that the prototype is very strong and can be for tooling such as the cavity of a mold.
- Stereo Lithography: The part is built inside a vat of light-curing liquid polymer. Two low-power lasers are aimed from different angles at a small spot in the liquid. Where they intersect, enough energy is present to harden the plastic. This is an expensive process, but the advantage is that relatively large parts can be created.

1.4 SYSTEMS VIEW OF CNC

A CNC machine tool differs from a conventional machine tool only in respect to the specialized components that make up the CNC system. The CNC system can be further divided into three subsystems: control, drive, and feedback. All of these subsystems must work together to form a complete CNC system.

Control System

The centerpiece of the CNC system is the control (*see **Figure 1.21***). This is the computer that stores and reads the program and tells the other components what to do. The control also acts as the user interface so that the operator can set up and operate the machine. Technically the control is called the **Machine Control Unit (MCU)**, but the most common names used in recent years are controller, control unit, or just plain control. Controls come in two basic flavors: proprietary and PC-based (i.e., IBM-compatible personal computer), with the former being by far the most common.

Machine Control Unit (MCU):

The main control computer of a CNC system; control or controller.

Proprietary controls have a closed architecture. The systems are custom-built by the manufacturer and often contain closely guarded circuits, algorithms, and control programs. You cannot run down to the local megastore and buy parts for your proprietary control or load the latest Microsoft operating system onto its hard drive. This type of control is expensive but offers rock-solid reliability.

In recent years, there has been a push toward so-called open architecture controls—that is, controls made from commonly available components and software. To some control builders this has meant PC-based controls. In fact, there are dozens of models of CNC controls that run on the same components that drive your desktop PC. You can go to the local computer vendor and buy parts for these controls. However, the reliability of these systems is somewhat suspect at the moment. In the future, expect great improvements and widespread adoption of the open architecture concept based on common PC hardware.

To other CNC control manufacturers, open architecture means developing and adopting an industry standard that is open to all manufacturers—not necessarily based on a Windows/PC platform. There has been some work on a standard in recent years, but no wide-scale adoption has occurred.

CNC controls are a hybrid of what has come to be known generically as motion control. Motion control systems are found in other applications such as robotics, avionics, and many industrial automation systems. CNC systems are simply a bit more specialized in respect to programming, axis definitions, and preprogrammed functions. Regardless of the end use, all motion-control systems perform roughly the same function. They read a program supplied by the user, generate an electronic positioning profile from the instructions, and then send signals to a motor to produce the desired motion. The more sophisticated systems (practically all modern CNC systems) use a **feedback loop** to compare the desired position to

feedback loop:

Electronic signals that are sent back to the control to indicate actual position, velocity, or state of the machine tool. The control will then compare the actual condition to the desired position and make adjustments.

Figure 1.21 The CNC control interface allows the operator to load programs and control the operation of the machine tool. *Courtesy of Haas Automation*

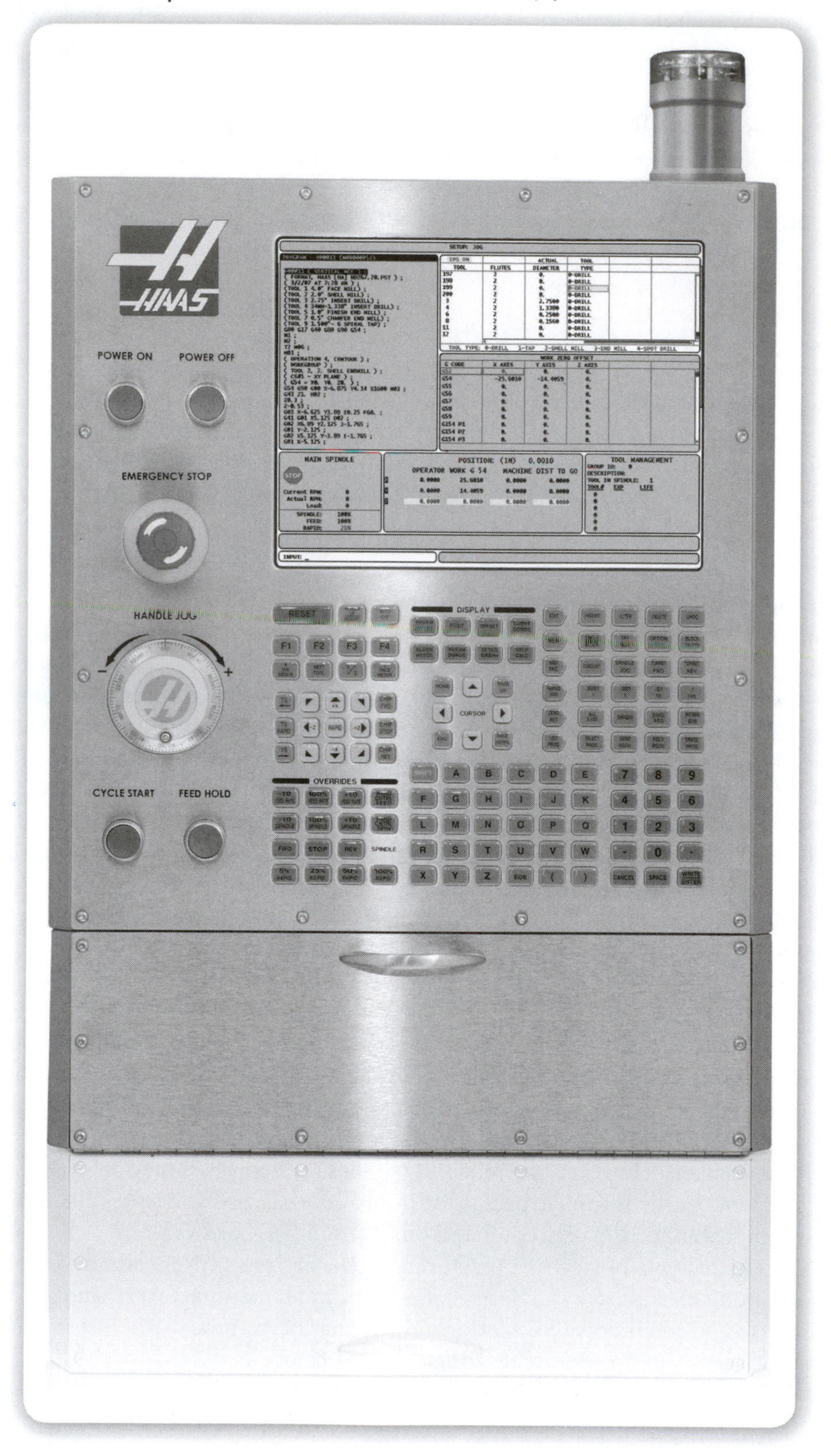

the actual position and make adjustments if any discrepancies are found. We will discuss feedback in more detail later in this section.

A computer program within the control (called the control program) handles the job of interpreting the part program and turning it into a series of electrical signals. The control program will read an instruction within the part program and generate a minute electrical signal to turn on a motor. However, the control is a

Careers in the Designed World

CAREERS IN COMPUTER INTEGRATED MANUFACTURING

Getting with the Program

Julio Ramirez spent most of his career as an old-fashioned machinist, manually making parts for auto manufacturers and cable companies. But he's found a new challenge as a CNC operator for Indian Springs Manufacturing Company, where he's learned the art of programming.

"I have to figure out a lot of things for the program," Ramirez says. "If everything works out right, it makes me smile. I feel like, 'I did it!'"

On the Job

Indian Springs Manufacturing Company makes parts for a variety of customers, from radio components to antennas. The CNC machine does everything automatically—drilling, changing tools, making threads. If Ramirez sets up the program just right, the machine runs at the proper speed and makes every part at the proper dimensions. He adjusts the machine as needed while it runs.

It's not easy to make parts exactly the same size from week to week and month to month. Ramirez has to keep track of everything he's done in the past and do it the same way for each new batch of parts.

Inspirations

As a kid in Cuba, Ramirez wanted to be a firefighter or a policeman. Later, as his friends began finding work in engine factories, he went along. He didn't like the work at first, but soon became intrigued.

"I liked the idea that you could create whatever you want," he says. "With a piece of metal and a blueprint, you can make a motorcycle, or instruments for a hospital. You just need the tools and the machine."

Education

After high school, Ramirez went to Saul Delgado technical school in Havana, learning to work with metals. After that, his education was all on the job.

He came to the United States in 2003 and worked on connectors for cable companies. He went on to operate CNC machines at a company that made brake parts, but he didn't do any programming there.

Ramirez's job at Indian Springs Manufacturing Company now gives him a chance to combine his love of metals with his love of computers.

"Working with a CNC machine is fun once you know how to create the program," he says. "You can play with the machine just the way you play with a computer."

Advice for Students

Ramirez sees CNC machines as the way of the future. "You're just going to need one person to run a machine that can do the job of five people," he says.

He advises students to learn CNC machines methodically.

"Go slow and pay attention to what you're doing in the beginning," he says. "Walk before you run, and you'll learn the steps you need to go faster. By the time you open your eyes, you'll say, man, look at how fast I'm doing this!"

computer, and computers operate on very low voltages and currents. The motor, on the other hand, is a high-powered device that requires a different kind of signal that is much stronger. Therefore, an intermediate device called an amplifier must act as an interpreter between the control and the motors. This is not much different from the way your transistor radio turns a small electrical signal from the antenna into an amplified signal that you can hear on the speaker.

ball screw: A specialized lead screw that uses close-fitting ball bearings to reduce friction and backlash between the screw and nut. Ball screws are used to transmit motion from the servomotors to the machining table.

Drive System

The drive system comprises of the motors and screws that will finally turn the part program into motion. The first component of the typical drive system is a high-precision lead screw called a **ball screw**, which is shown in ***Figure 1.22.*** You

Figure 1.22 **Ball screws are used to transmit motion to the machining table.**

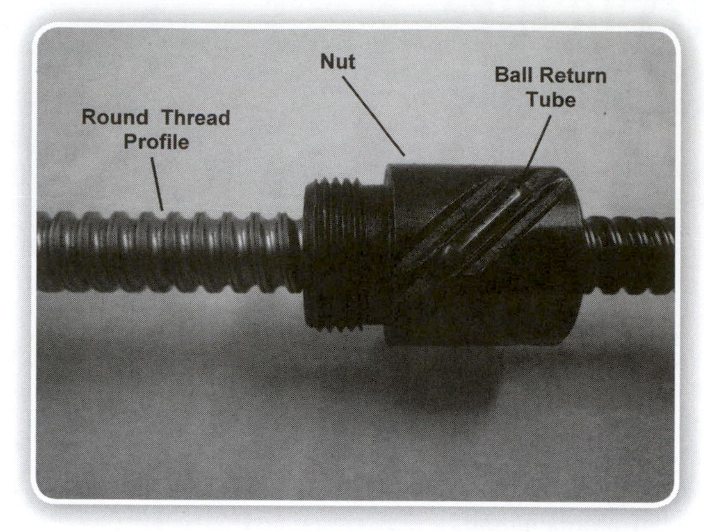

probably know that, on a manual machine tool, turning a hand wheel attached to a lead screw moves the table. A nut is mounted on the underside of the table so that the table moves when the screw is turned. CNC machine tools use a similar setup, but the lead screw has been improved by added ball bearings that ride between the nut and the screw. The bearings recirculate through a tube on the nut. The resulting performance is outstanding with very low friction and zero backlash.

Backlash in a ball screw is eliminated because a slight interference fit is created when the screw, ball bearing, and nut are assembled. In other words, the space between the nut and the screw is a little smaller than the balls that must fit in between them. Eliminating backlash is very important for two reasons. First, high-precision positioning cannot be achieved if the table is free to move slightly when it is supposed to be stationary. The control will constantly try to adjust for any positioning errors, so a loose table will drive the feedback system crazy as it moves around. Second, material can be climb-cut safely if the backlash has been eliminated. Climb cutting is usually the most desirable method for machining on a CNC machine tool. However, climb cutting pulls the workpiece into the cutter. If there is backlash in the screw, the material will jump into the tool and may cause the tool to shatter or possibly damage the workpiece.

Drive motors are the second specialized component in the drive system. The turning of the motor will turn the ball screw to directly cause the machining table to move. Several types of electric motors are used on CNC control systems, and hydraulic motors are also occasionally used. There are many different types of electric motors, but they all work on the same basic principle of magnetic attraction and repulsion. You are probably already familiar with the way two magnets will attract or repel each other. Electric motors work by creating a moving magnetic field with an electromagnet within the motor to attract or repel another set of magnets. This attraction and repulsion causes the shaft of the motor to rotate. Numerous variations of this construction have been developed over the years, but the principle remains the same.

stepper motor: A specialized motor used in low-end motion control systems that rotates a predefined angle with every electrical pulse.

The simplest type of electric motor used in CNC positioning systems is the **stepper motor** (sometimes called a stepping motor). A stepper motor rotates a fixed number of degrees when it receives an electrical pulse and then stops until another pulse is received. Stepper motors are constructed with several different windings to produce the proper combination of attraction and repulsion for a step. A separate electronic circuit called a stepper motor translator is needed to create the proper signal to produce forward or reverse rotation. The stepping characteristic makes stepper motors easy to control. In fact, a simple push-button switch can be connected to the translator to produce a single step. The simplicity and ease of control make stepper motor motion control systems very popular.

A typical stepper motor construction might allow 0.7° per pulse. If this motor were attached to a lead screw with a pitch of 0.250 inches, the table would move 0.7/360ths of the pitch, or about 0.00048 inches. This is called the **resolution**, or simply the smallest movement that is possible within the system. A lead screw with a finer pitch could be selected if a finer resolution were required. For example, using the same stepper motor with 0.7° rotation per pulse and a lead screw with 0.100-inch pitch would result in a resolution of about 0.00019 inch per pulse. However, the motor turns at a constant rate, so the speed of positioning would be slower.

resolution:
The degree to which measurements can be differentiated from one size to another. The scale of the measuring instrument.

It is relatively easy and inexpensive to design CNC systems with stepper motors. Stepper motor systems can operate without expensive and complicated positioning feedback. (Feedback is information sent back to the control by sensors.) In other words, the control can simply tell motor "A" to turn 100 pulses and motor "B" to turn 75 pulses, but there is no guarantee that the table is actually where you expected it to be. All the control knows is that it told the motor to turn, and hopefully it did. If an inexpensive, low-precision system is needed, then stepper motors may be the proper choice. A more sophisticated system can include positioning feedback.

Stepper motors have many drawbacks that make other varieties more desirable in high-precision systems. Stepper motors tend to jerk as they are pulsed. This can lead to poor surface finishes and excess wear resulting from the constant acceleration and deceleration. Furthermore, stepper motors tend to produce a stair-step effect when attempting to produce angled lines and circular arcs. Modern industrial CNC machines seldom use stepper motors, but you will find them on low-priced CNC trainers and on assembly systems.

It is more common to use **servomotors** in CNC systems today (*see Figure 1.23*). Servomotors operate in a smooth, continuous motion—not like the discrete movements of the stepper motors. This smooth motion leads to highly desirable machining characteristics, but they are also difficult to control.

servomotors:
A specialized motor used in motion control systems that can deliver continuous motion at various speeds.

Figure 1.23 **Servomotors provide smooth, continuous motion for CNC machine tools.** *Courtesy of Haas Automation*

Specialized hardware controls and feedback systems are needed to control and drive these motors. The extra hardware makes the system more expensive and complex, but it also provides stop-on-a-dime control and smooth acceleration curves.

Servomotors differ greatly from stepper motors in the way they are constructed and controlled. When an electrical signal is sent to a servomotor, it will turn continuously until the current is stopped. The speed of a servomotor can be controlled by varying the strength of the signal or by varying the frequency of the signal, depending on the particular construction. Servomotors are available with several different constructions and electrical characteristics. Some of the earlier servo-based CNC machine tools used direct current (DC) servomotors. DC is the type of electrical current that flows from a battery. DC servomotors were powerful, but DC current has some inherent inefficiencies and limitations. As control technologies evolved, machine tool builders started to use alternating current (AC) servomotors. AC is the type of current that is available in your wall socket. AC oscillates up and down much like a wave at regular time intervals referred to as frequency. Engineers soon realized that they could manipulate these oscillations to control the speed and motion of a motor. Although the idea of the AC motor goes back to the early part of the century, it was much later before the control systems reached a level of sophistication at which they could economically use AC servomotors.

AC servomotors are currently the standard choice for industrial CNC machine tools. The current trend is also toward a design called the AC brushless servomotor. Brushes are the electrical contacts that make the moving magnetic field possible. The electromagnets within the motor must be switched on and off based on their relative position in order for the motor to rotate. The switching is accomplished by a set of carbon brushes that make physical contact with the rotating shaft of the motor. Brushes tend to wear out and cause problems with the dust they create. The electromagnets in a brushless motor are switched electronically, thus eliminating many problems and increasing reliability.

Feedback System

The function of a feedback system is to provide the control with information about the status of the motion control system, which is described in ***Figure 1.24***. The control can compare the desired condition to the actual condition and make corrections. You use a feedback system every time you drive a car or ride a bike. Your brain tells your hands to move the steering wheel in one direction or another, and then your eyes give feedback to your brain about your trajectory. If you happen to steer off the road, your brain will take corrective action.

The most obvious information to be fed back to the control on a CNC machine tool is the position of the table and the velocity of the motors. Other information may also be fed back that is not directly related to motion control, such as the temperature of the motor and the load on the spindle—this information protects the machine from damage.

There are two main types of feedback systems: open-loop and closed-loop, as illustrated in ***Figure 1.25***. An open-loop feedback system is really a no-feedback system. It does not have any device to determine if the instructions were carried out. For example, in an open-loop feedback system, the control could give instructions to turn the motor 10 revolutions. However, no information can come back to the control to tell it if it actually turned. (Someone could have unplugged it, and the control would not know.) All the control knows is that it delivered the instructions. Open-loop feedback is not used for critical systems, but it is a good choice for inexpensive motion control systems in which accuracy and reliability are not critical. There is one famous incident of open-loop feedback on a water

Figure 1.24 A typical motion control system found on a CNC machine tool.

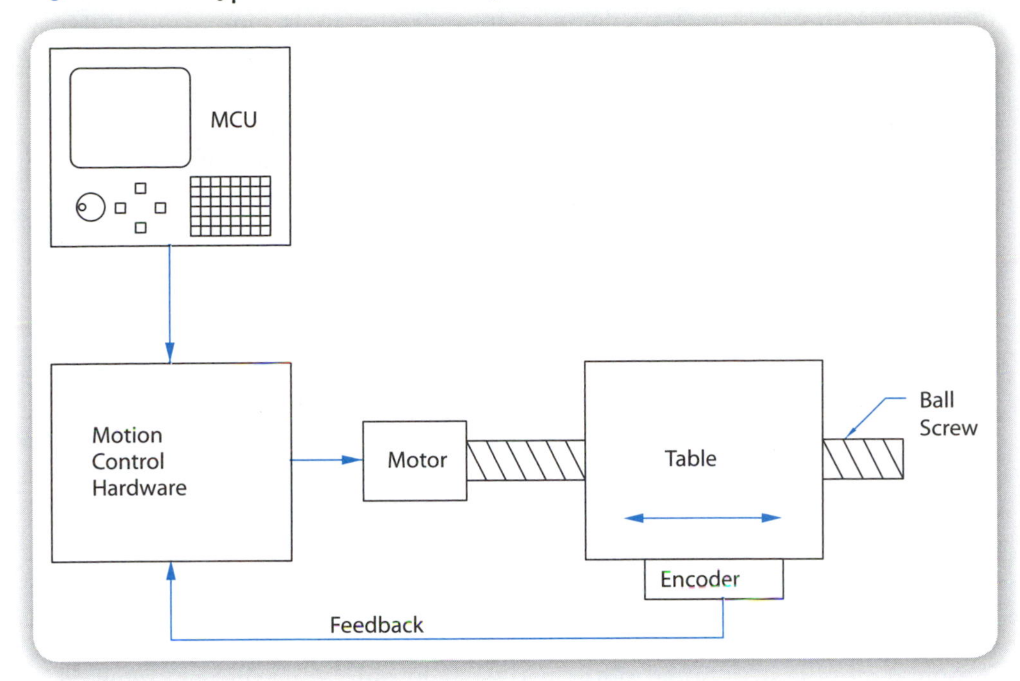

Figure 1.25 Two models for feedback loops. A closed-loop feedback system can correct for errors because the desired input is compared to the actual output.

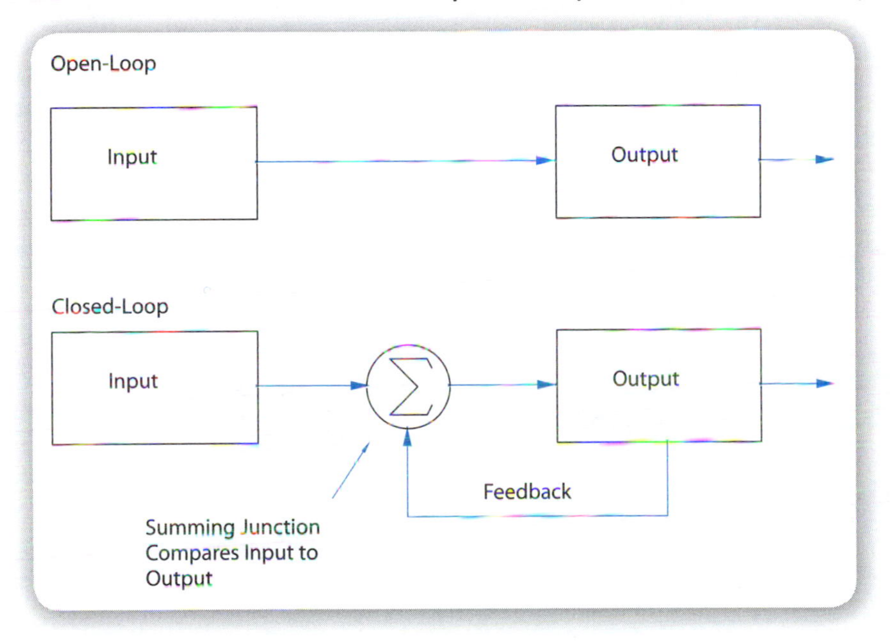

valve at a nuclear power plant. The control registered only that the button had been pushed to operate the valve but no sensors measured the actual status, and it caused an apparent core meltdown.

Closed-loop feedback is a little more meaningful. Closed-loop feedback uses external sensors to verify that certain conditions have been met. ("Yes, the landing gear *is* down.") Of course, positioning and velocity feedback is of primary importance to an accurate CNC system. Feedback is the only way to ensure that the machine is behaving the way the control intended it to behave.

Servomotor CNC systems generally require a closed-loop feedback system in order to track position, velocity, and acceleration. It is possible to build an open-loop servo-system. For example, if you know how fast the motor will turn, the control could simply instruct the motor to turn on for a certain amount of time and then turn off. Nevertheless, the likelihood of positioning the table to machining tolerances with such a system is doubtful. In contrast, stepper motor CNC systems can operate on an open-loop feedback system. The control simply keeps track of position and velocity by counting the number of pulses and the frequency at which they were sent. This method works well if there are no mistakes; the control only knows where the table *should* be, not necessarily where it actually is.

A variety of sensors are used with close-loop feedback systems to report information about the position of the machine and the velocity of the motors or spindle. A sensor called an **encoder** is used to measure linear displacements of the machining table. Sometimes a less expensive tachometer will be used in conjunction with encoders to measure the angular velocity of the motors (the RPM). We might also find a variety of less sophisticated sensors, such as limit switches, that determine certain conditions are safely met before operation.

encoder:

Any device that is used to give positioning feedback to the control. Digital linear and rotary encoders are common on CNC machine tools.

Currently there are two main categories of encoders in use with CNC systems: rotary encoders and linear encoders, as illustrated in ***Figure 1.26***. A rotary encoder is a type of sensor that when mounted to the shaft of the motor can detect its angular position. You will usually find these conveniently mounted to the backside of the motor. Furthermore, many motors come equipped with a rotary encoder as an integrated part of the design. A rotary encoder is made from a light source, an optical sensor, and a disk with small slits cut into it like spokes. The slotted disk is mounted to the shaft of the motor, and the light source and the light sensor are mounted on opposite sides of the disk. As the disk turns, the light shines through the slits onto an optical sensor. The optical sensor acts like a switch that turns on and off as the light strikes it. (These are called encoder ticks.) In this way, the optical sensor sends an electrical signal back to the control.

A rotary encoder with only one set of slots (or only one sensing element) has a fatal defect—it cannot tell which direction it is turning. It only knows that the switch keeps turning on and off. Another set of slots on the same disc and another

Figure 1.26 Rotary (left) and linear (right) encoders use opto-electronics to sense a change in position. Both constructions are available in either relative or absolute measurements. *(Photo by Dustin Piersall)*

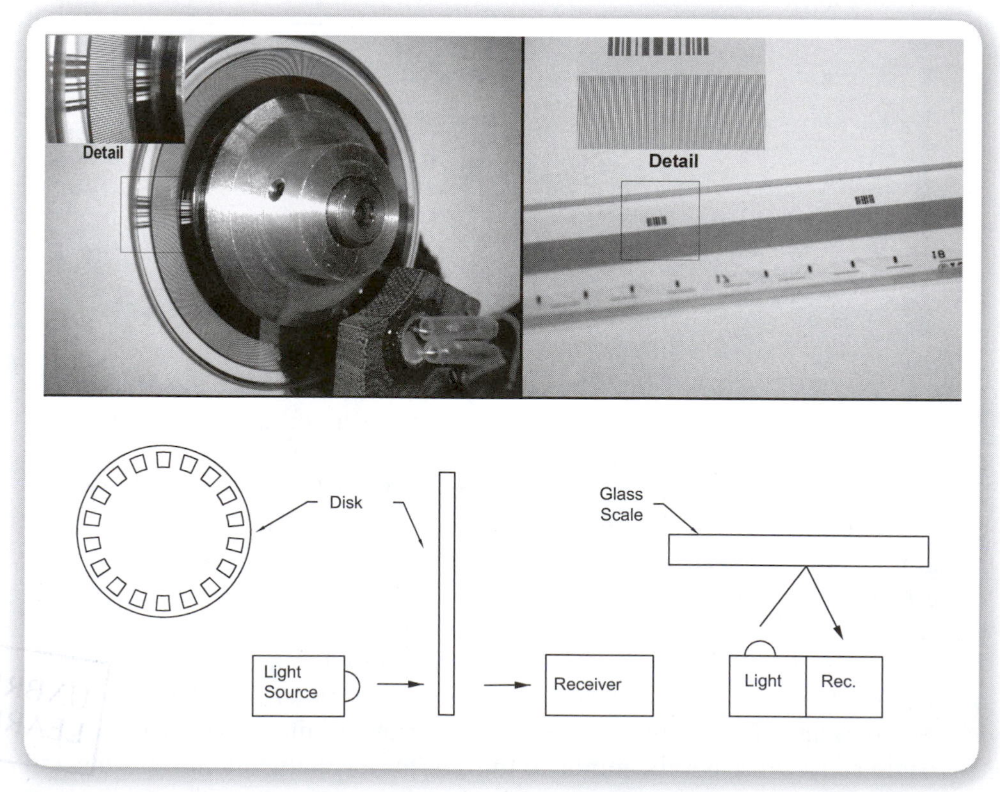

optical sensor is needed to determine the direction of rotation. This is accomplished by setting the second set of slots out of phase by 90°. ***Figure 1.27*** shows the disk and timing graph. The graph shows that while the disk is turning clockwise, sensor A leads sensor B. In other words, when sensor A turns on, sensor B will turn on shortly afterward. If we reverse direction, sensor A will trail sensor B. If this seems confusing, don't worry—it is easy for the control to keep track.

Figure 1.27 A directional encoder disk and its timing graph. The two sets of slots are out of phase, and this will allow the control to determine the direction of rotation. An alternative construction is to use one set of slots with two sensing elements that are placed out of phase.

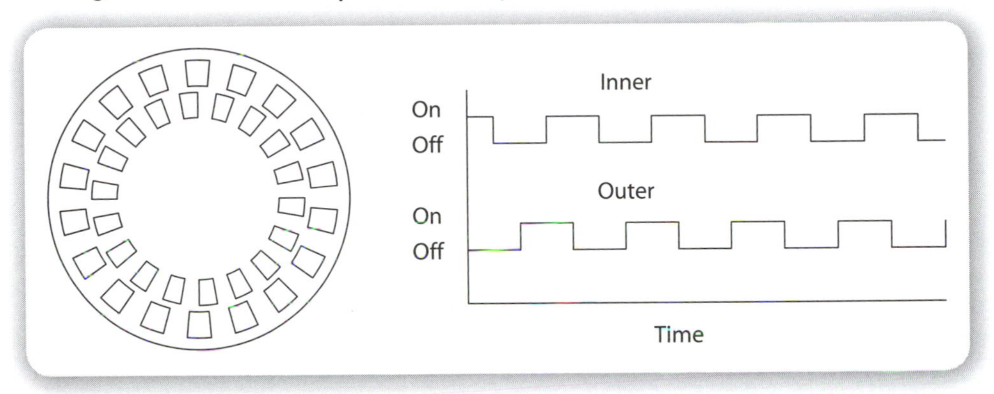

The rotary encoder described above can give only a relative position based on the number of times the light turns on and off and in what direction the motor is turning. It cannot give the actual, angular position of the motor—the control must keep count. Because of this design, the machine must be moved to a *hard* reference position every once in a while so that the counter in the control can reset itself in case any errors have accumulated. There is another rotary encoder design, called an absolute encoder, that can actually determine its real angular position. This is accomplished with multiple sensors on a specially slotted disk. These encoders have many more wires to connect, which makes the system more complex. Furthermore, the control must still count the number of revolutions.

Rotary encoders are subject to a small degree of uncertainty. Because they are mounted on the motor, they do not directly measure the table position—they give a calculated position based on the pitch of the ball screw. The position will be incorrect if there are any errors in the screw.

Another type of positioning sensor called a linear encoder was developed to solve this problem (*see* ***Figure 1.26***). Linear encoders are much more expensive but have become the standard for many CNC machine tools.

Linear encoders are constructed by etching fine graduations into a glass beam at specific intervals. The glass beam is mounted directly to the machining table and essentially becomes a very accurate ruler. Resolutions of 0.00005 inch (1/2 of 1/10,000 of an inch) are commonplace, and better resolutions are available for an exponentially higher price. A stationary encoder head that contains a light source and an optical sensor reads the ruler. A beam of light is shown onto the scale, and an optical sensor catches the reflection (or lack of reflection). Again the signal is sent back to the control. A couple of variations of this basic idea are available, including an incremental variety that simply counts the ticks, and an absolute positioning variety that can determine the actual position along the scale.

CNC systems also need to obtain information about the angular velocity of the motors and the linear velocity of the table. This can be accomplished in two ways. First, the control can calculate the velocity of the table or motor by simply dividing the number of encoder ticks by the elapsed time. For example, if each tick translates to 0.001 inch, and the control is counting 200 every second, then the linear velocity is 0.200 inch per second (or 12" per minute). The control computer can also make some simple calculations to measure acceleration and deceleration with these quantities. The velocity and acceleration are important to the accuracy and quality of the CNC system. An accurate velocity will ensure that the proper surface finish is maintained and that the correct load is on the cutting tool. Acceleration and deceleration are important for positioning. The control must start to decelerate as it approaches the programmed coordinates, otherwise it will overshoot the coordinates and possibly ruin the workpiece. This is analogous to approaching a stop sign. If you don't start slowing down (decelerating) soon enough, you will end up in the intersection.

Tachometers are used to measure angular velocity (RPM). A tachometer is a simple sensor that can, directly or indirectly, measure the number of revolutions the shaft will make every minute (RPM). The spindle RPM is a critical quantity that every machinist and NC programmer is acutely aware of—the spindle RPM will greatly affect the surface finish and tool life during machining operations. Some motion control systems will use tachometers rather than the positioning sensors to calculate the linear velocity of the components. This is a design decision that is based on the computing power available and the time response that is required. A slow computer may not be able to make the calculations quickly enough to matter when the system must operate in real time.

The CNC System

By now you should be able to identify the three subsystems needed to complete a CNC system: control, drive, and feedback. It is important to note at this point that the selected components must all work together as one whole CNC system. Each subsystem must be properly matched to work within the CNC system. For example, if the control system is programmed and designed for servomotors, then you cannot use stepper motors. If the control system is designed for incremental encoders, then you cannot use absolute encoders. This idea is of little consequence when we talk about a factory-made CNC machine tool; the engineers at the manufacturer have already thought it through. However, at some point you may become involved with a repair or retrofit of an obsolete machine tool. You will then have to put this knowledge to work.

Let's now take a look at how the entire CNC system works together. The CNC system we will use as an example uses servomotors and incremental rotary encoders, but these ideas can be extrapolated to other varieties. We will keep the system simple by having only two axes to move the table; each axis will have an encoder, and the spindle will have a tachometer.

A system with feedback can be described in a block diagram. The block diagram is quite generic and does not describe any of the physical components. It describes only the input, output, and feedback of a system. Nonetheless, the system block diagram can help us understand the CNC system. Let's start with just one axis of our system. The control sends a signal into the motor (input) and the motor turns (output). Next, the encoder provides feedback to the control. The feedback is compared to the input at the summing junction (Σ); the difference between the two is called error. The input is then adjusted, which in turn changes the output. Hopefully, the error is smaller this time.

For example, the input is to turn the motor 1,000 clicks. The motor turns for a while, and the encoder sends feedback indicating that it only turned 990 clicks. The two values are compared and the error is determined to be 10 clicks. So a new input is sent to turn the motor 10 clicks. The motor turns for a while, and this time the encoder indicates that it has turned 12 clicks. The cycle then repeats itself over and over again at a very rapid rate until it is *close enough*. The system will never be perfect and there will always be error. In fact, the industry standard for a servomotor system is actually about four encoder clicks of error. However, this can often be translated into a system resolution of less than 50 μ/in.

We could add the other axis and the spindle to our system, but the block diagram for the entire CNC system would remain the same. However, it may be easier to conceptualize the system if we create a hybrid block diagram that describes the components of each system. You may want to review the explanation of the Cartesian coordinate system in Chapter 3 before digging into this next section.

So how is the input of the system generated in the first place? The input starts with the part program that the NC programmer wrote. The part program gives instructions about where to move and how fast to turn the spindle. The control will then read the instructions and generate a motion profile. The motion profile can be represented by a graph of where the table should be located at specific time intervals. For example, imagine that a cut in the shape of a quarter circle must be performed. We can break the arc into equal time intervals. (We want the milling cutter to be moving at a constant rate around the arc.) We can then calculate where the X-axis should be located at every time interval.

The graph of the X position will form a curve called a sine curve. This will then become the input for the drive motor on the X-axis. Of course, the feedback will be used to determine if output (the table location) is actually correct. You already know that there will be errors, so the actual profile might be a little different. The corrections tend to oscillate back and forth just as you would oscillate slightly back and forth as you drive down the road.

We could also take the Y-axis into account. The graph of the Y-axis will also be a sine curve, just a different part of it. Moving the two axes simultaneously results in a circular arc.

1.5 A STANDARDIZED PROGRAMMING LANGUAGE

CNC machine tools produce only the movements that are described in a part program (a program that describes the machining operations and dimensions). Therefore, a programmer must write a part program in a language that the control can understand. The most popular language for NC programming is G & M code programming, based on the Electronics Industries Alliance and International Standards Organization standard (EIA/ISO). This is a language standardized by American and European industry/standards organizations in an attempt to make NC programming uniform.

The standards that define NC programming code are ANSI/EIA 274D-1988, and from the international community, ISO 6983. This may eventually be replaced by a new ISO standard called STEP-NC that would allow a direct exchange of information between the CAD/CAM model and the CNC machine tool.

The control manufacturers are not required to follow this standard, but many have adopted the most popular codes. One of the early adopters/creators of the EIA/ISO standard was Fanuc. Their controls became quite popular, and now the terms *G & M codes* and *Fanuc-style* are virtually synonymous. Today the standardization of G & M codes makes the transition from one machine to another a relatively easy process. The machinist does not need to learn a whole new language—just a slightly different vocabulary. In this book, we will concentrate on G & M code programming.

Many other NC programming schemes and languages are available, but they tend to be either obscure or proprietary. In fact, almost every control manufacturer has a so-called *conversational* control (***see Figure 1.28***) or language that attempts to make programming easier. Conversational languages can be very productive, especially for new operators. A conversational language uses English-like statements and prompts to create a program, often directly at the control. The operator

Figure 1.28 A conversational control does not require knowledge of G & M codes. Instead, the user follows a series of menus to create the program. There are some similarities to a CAD/CAM program.

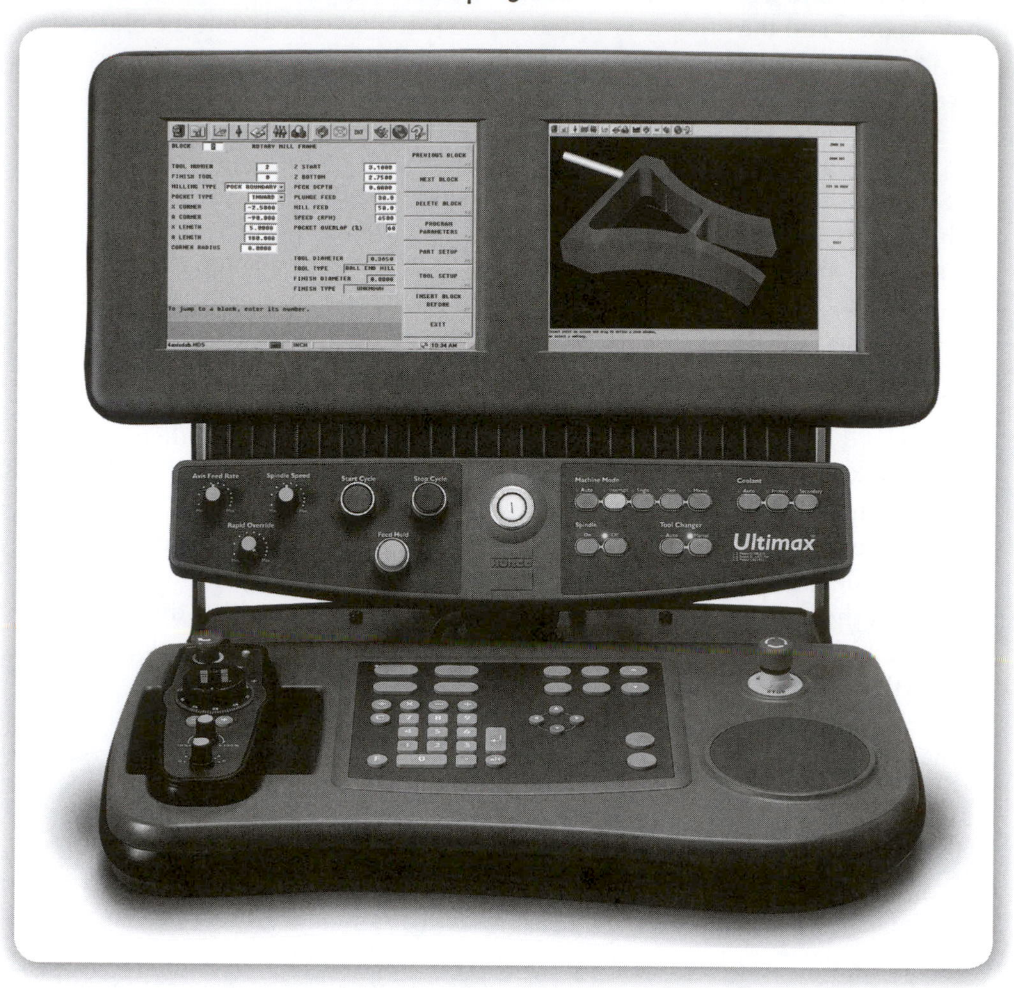

creates a program by selecting the proper function from the screen menu. He or she is then guided through a series of prompts to fill in, such as the start and stop point or the feed rate. This system allows an inexperienced operator to create NC programs very quickly and with very little effort.

1.6 CAREER PATHS IN MANUFACTURING

The beautiful truth about the job of machining is that your job always seems to lead to something else. You can start working in a machine shop from "right off the street," as some human resources people put it. You can start off sweeping the floors, running the saw, or as a machine operator with little or no experience. As you gain experience and begin to prove yourself, then the opportunities start to open. Your success is based on your attitude, skill, knowledge, curiosity, and attention to detail. Eventually, with education and training many machinists become operations managers, engineering technicians, and even full-fledged manufacturing engineers. Or many prefer the day-to-day challenges of being a machinist and spend an entire career at the craft. It is a noble and satisfying profession. The only sure thing is that anyone working in manufacturing will need to constantly learn and grow just to keep pace.

I constantly hear manufacturing managers remark that they just need "someone who can show up on time and follow instructions." There is time to learn the craft of machining over a number of years. In fact, time is a requirement. Learning takes time, and there are no shortcuts in a complex field such as CNC machining and programming.

First you show up on time and follow directions. Then you learn to be a machinist. Not a machine operator, but a machinist. A machinist understands geometry and the steps that go in to making a machined part. A machinist understands metal and how it interacts with the tools that are used to cut it. A machinist understands measurement and the sometimes puzzling relationship between cause and effect.

If you are a real machinist, then the career opportunities will abound over the next decades. A recent industry study[a] reports "machinist" as one of the top five highest-demanded jobs by 2012. The demographics are changing as the Baby Boomers retire, except that it will be a completely new landscape. It will not take armies of machinists, but a fraction of the former workforce that has been highly trained and has access to incredibly productive CNC machine tools and CAD/CAM software. It is an exciting time.

Your Turn

1. Research and create a presentation to the class on the history of NC machining.
2. Dissect the NC equipment in your classroom and be able to explain the key components to the class in terms of operation.

[a]Data from Deloite report, 2005. Published as an article in Career Builder 2006.

CHAPTER SUMMARY

- Machine tools are machines that can be used to make other machines.
- The industrial revolution was fueled by advances in machine tool technology and cheap energy. It is largely responsible for the high standard of live that we enjoy today.
- The development of NC and CNC machine tools started in the 1950s and matured in the mid-1980s. It closely followed the development of electronics and computer technology. As technology became available, it was soon adapted to machine tools.
- A CNC control system can be applied to virtually any type of machine tool. Mills, lathes, EDMs, lasers, routers, and forming equipment are just a few examples of mainstream CNC technology.
- Computer numerical control (CNC) replaces manual controls with electronic systems to perform machine movements. Virtually any machine tool can be controlled with a CNC system.
- The movements of the CNC machine tool are accomplished with a set of instructions called an NC part program that is written by an NC programmer. This program is commonly written with G & M codes, but other languages and methods are available.
- The CNC system consists of three subsystems: control, drive, and feedback.
- The control system is a computer that interprets the part program and generates an electronic signal to the drive system.
- The drive system uses motors and actuators to turn the electronic signals into tool movements and other machine functions.
- The feedback system uses sensors such as rotary and linear encoders to send positioning and velocity information back to the control to maintain the proper positioning.
- Careers in manufacturing are filled with many pathways. A person can start at the bottom and move up through many jobs of higher responsibility. This vertical movement results after education and experience.

CHAPTER QUESTIONS

1. What is a machine tool and how is it different from any other types of machines such as tractors or cars?
2. How is a CNC machine tool different from a conventional machine tool? How is it similar?
3. How are ball screws different from standard lead screws?
4. What are the advantages of stepper motors?
5. What are the advantages of servomotors?
6. What is the function of an encoder?
7. Explain the difference between a stepper motor and a servomotor.

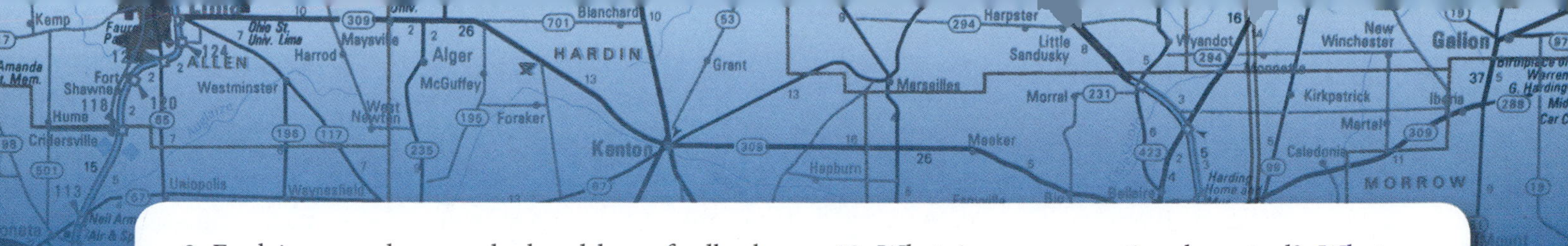

8. Explain open-loop and closed-loop feedback systems. Which feedback system will we most commonly find on a modern CNC machine tool?
9. Name two types of encoders found on CNC machine tools and explain the differences between them.
10. What is a conversational control? What are some advantages over a standard control? Disadvantages?
11. What does it mean when a control is described as proprietary?
12. What does it mean when a control is described as having open architecture?

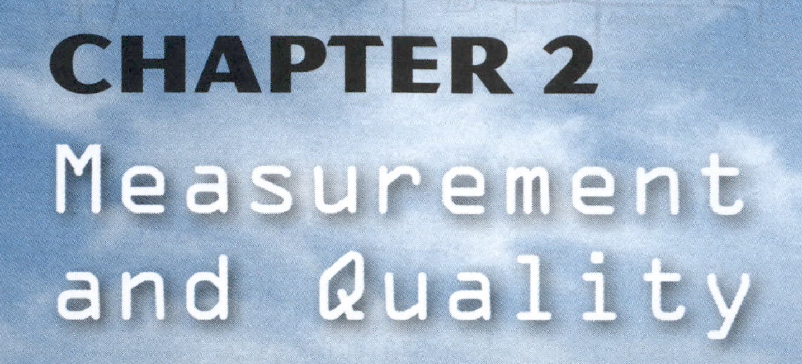

CHAPTER 2
Measurement and Quality

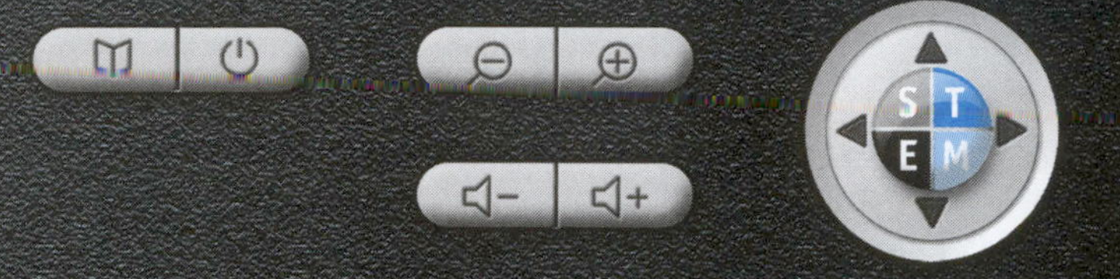

Menu

START LOCATION | DISTANCE | END LOCATION

Before You Begin

Think about these questions as you study the concepts in this chapter:

Why do we need to use both U.S. units and metric units?

How are statistics used to describe and predict the results in manufacturing operations?

What are the philosophies and techniques of lean manufacturing?

Why is it important to be able to read and interpret an engineering drawing?

How is the micrometer used as a precision measurement tool?

Key Terms

Accuracy
Continuous Improvement
Dimension
Distribution
Engineering Drawing
Histogram
Isometric
Lean Manufacturing
Linear
Mean
Orthographic
Population
Precision
Process Capability
Repeatability
Resolution
Rule of Ten
Sample
Section View
Standard Deviation
Statistical Process Control (SPC)
Statistics
Tolerance
Variability

2.1 MEASUREMENT TAKEN FOR GRANTED

Today, we take for granted that the things we buy will be about as standardized as one could expect from modern manufacturing. When we go to the local mega-depot to buy two-by-fours, we are pretty much assured that every board in the stack will be the same width and thickness as every other board. Every ¼-inch bolt will look like all the others, and every ¼ inch nut will fit the same way on each ¼-inch bolt. The idea is simple, boring, and brilliant.

Until very recent history, we could not count on products having any meaningful degree of standardization (i.e., sameness). Every stone block, every nail, every sword, and every chariot wheel was a unique item. In the ancient world, every product was a custom job.

continuous improvement:
A belief and method by which systems can be improved incrementally through rational analysis and observation.

Interchangeable Parts

The manufacture of weapons for war fueled the need for *interchangeable parts,* or, the idea that the parts of one rifle should fit on another. This was not possible before in a system without precision measurement.

Shortly after the American Revolution, Eli Whitney developed a manufacturing method utilizing jigs and patterns by which each rifle part could be formed with a mechanical filing machine until it was the same size as all the others. This process still lacked standards in precision measurement but worked nonetheless.

Around this same time, there were efforts started in France to establish a standard meter. Modern scientific methods have now standardized measurement by referencing the wavelengths of light to linear units.

To the modern person, it seems like common sense that we should have a system of standard measurements to define our world. This was not a new idea even to the ancients, but it took some time for the technology to catch up with the thinking. The ancients could wish all they wanted to be able to make every timber the same thickness, but unless a society has the systems of measurement and highly reliable measuring tools, it is only wishful thinking.

Measuring systems of the past were often based upon anatomical attributes. Measurement included the width of a finger (an inch), the distance across a hand (a palm), the size of a shoe (a foot), and the length of an arm span (a fathom). In fact, you can do pretty well by using anatomical measurement even in modern craft and construction (*see Figure 2.1*).

distribution:
A concept in statistics that describes how a group of measurements will have variation from the average. A "normal distribution" is a common pattern in nature as a result of chance.

engineering drawing:
A technical drawing that formally describes the shape, size, and specifications of a product. Sometimes called a blueprint.

histogram:
A graphical representation of statistical distribution. Individual measurement are grouped into "bins" of a specified size range. The quantity in each "bin" is then displayed on the vertical axis.

Figure 2.1 **In the time before precision measurement, the human body was the instrument of choice—a finger, a palm, a foot, or a forearm. It is a workable method as long as the items don't need to be exactly alike.**

linear:
Along a straight line.

The measurement we have today came from ancient people's need to have standardized units. They did the best they could with their technology. Of course, the problem with this system is that people come in all different sizes. If I'm buying 10 fathoms of rope, it won't necessarily be the same length from seller to seller. Of course, for a length of rope the measurement is not that critical. What if it is more important, such as in the case of cannon balls? A cannon ball has to fit the barrel of the cannon just right. A system of precision measurement is needed to ensure that a foot is a foot and an inch is an inch.

2.2 UNITS OF LINEAR MEASUREMENT

Linear measurement is the quantity that is related to length. In other words, a linear measurement is the distance between two points. The shortest distance between two points is simply a line. This line is where we get the term **linear** from. There are other quantities of geometric measurement, such as angles (***Figure 2.2***) and volume, but linear measurement is fundamental.

section view:
A drawing view that simulates the cutting and removal of a portion of the object in order to visualize details within the interior.

orthographic:
A method of representation for three-dimensional objects by which the object is displayed at a right angle to the viewer. The viewer must construct a mental picture of the object by looking straight at the object from several different orientations.

isometric:
A common form of pictorial representation in technical drawings that represents an object as viewed by rotating about each axis equally (effectively a corner view).

Figure 2.2 **The most common measurements in manufacturing include linear and angular. Linear units measure the length between two points, and angular units measure the degree of rotation between two lines or planes. Area and volume are both derived from linear units.**

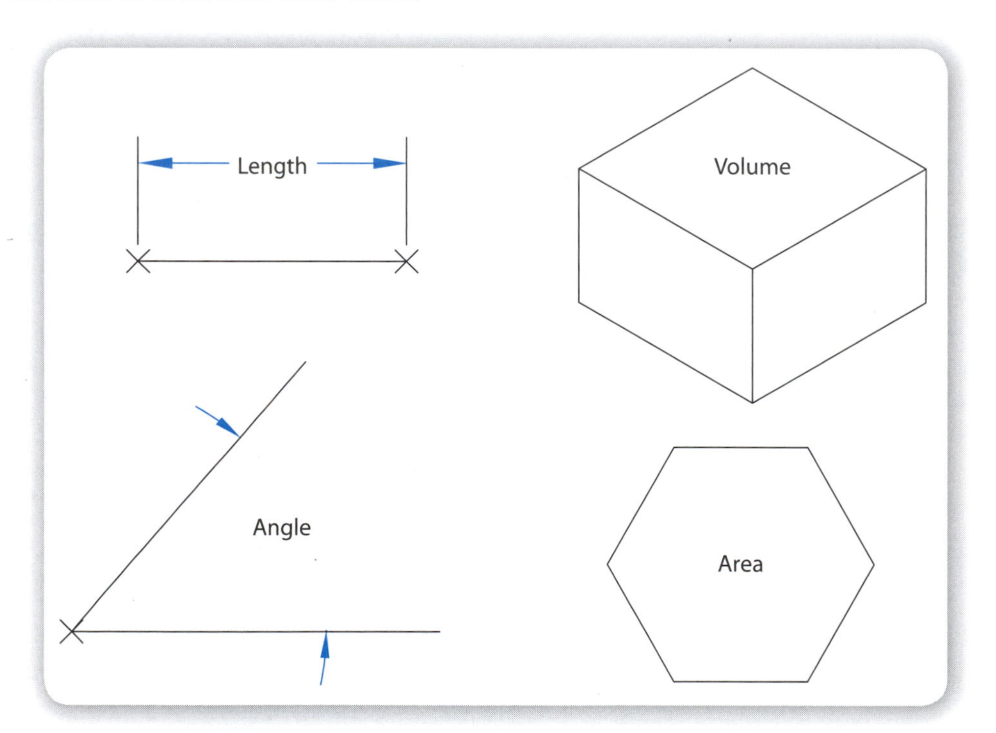

2.3 COMMON LINEAR UNITS

The inch is the most commonly used unit in the U.S. Customary system of measurement in the typical manufacturing environment. The inch is just about the right size for describing many of the regular-sized things we might make, without using a lot of extra numbers. Inches are great for describing the diameter of a piston or the thickness of a brick. Inches are too small a unit of measurement for some things, such as land development and navigation. An inch is not a good unit to describe the layout of a subdivision. "We bought a 6.27-million-square-inch lot (one-acre) with a 2504.5-inch frontage" does not exactly flow off the tongue.

American Engineers Adopt the Millimeter?

Did you know that in 1994 the American Society of Mechanical Engineers (ASME) used the millimeter exclusively to illustrate a very important standards document that sets the rules for engineering drawing in the United States*? This must truly be the death knell of the inch. I wonder why I still can't find a metric tape measure at my hardware store.

*ASME Y14.5M 1994 & Y14.100 2000

The industry typically uses the unit of millimeter (mm) when the metric system (or *Système International (SI)*, as it is known) is employed. The millimeter is a small unit when compared to the inch, so the centimeter is sometimes easier to understand. There are about two and a half centimeters in an inch. These measurements might be easier to understand by comparing their sizes next to each other and as is illustrated in ***Figure 2.3.*** Some common units for manufacturing are also shown in Table 2.1.

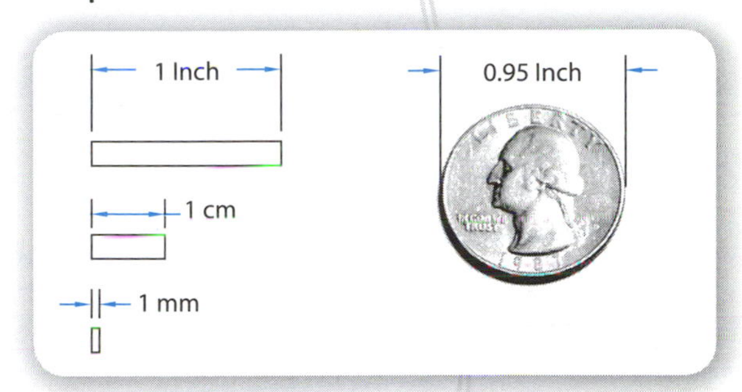

Figure 2.3 The millimeter and centimeter are compared with some more familiar elements.

Table 2.1 Comparative Chart Showing the Most Common U.S. Customary and Metric (SI) Units Found in Precision Manufacturing

U.S. Customary	SI Unit with a Similar Scale	Conversion	Common Use
1 inch	1 cm	1 inch = 2.54 cm .394 inches = 1 cm To convert inches to centimeters: Multiply inches by 2.54	Gross Measurements (Width of a machine, length of a pipe, height of a wall)
1/16th inch	1 mm	1/16th inch = 1.59 mm	Rough Measurements (Cutting stock, Clearance fits, woodworking)
0.001 inch	.01 mm	.001 inches = .0254 mm .00039 inches = .01 mm Common Shop Language: .01 mm is a little less than "Half a thousandth"	General Precision Work (Machine parts, fasteners, close fits)
.0001 inch	.002 mm	.0001 inches = .000254 mm .000079 inches = .0002 mm Common Shop Language: .002 mm is a little less than one tenth (1/10,000")	Very High Precision (Bearing fits, tool and die)
Microinch (μin.)	Micrometer (μm) Also called a micron	1 μin. = .0254 μm 39.4 μin. = 1 μm Common Shop Application: A typical, medium surface finish of 63 μin. is equal to about 1.6 μm.	Surface Finish Specifications and microscopic manufacturing (Semiconductor circuits, MEMS, optics)

Capitals or Not?

In the metric system, abbreviations are left in lower case letters. The exception is made when the unit comes from the name of a person. For example, a millimeter is abbreviated mm, but the Newton uses a capitol N because the unit honors a famous scientist.

Examples:

- **25 mm** 1/1000ths of a meter.
- **25 km** Thousands of meters.
- **10 N** Newton is a measure of force.
- **10 kN** A kilo-Newton is a thousand Newtons.
- **30 N•m** A Newton-meter is a measure of torque. The dot is used to show that it is a force (N) multiplied by a distance (m).
- **500 Pa** The Pascal is a unit of pressure honoring Blaise Pascal. The abbreviation is (Pa).
- **2.5 kPa** Thousands of Pascals.

2.4 PRECISION MEASURING TOOLS

If you have ever worked around the house or in a shop, then you have probably used a tape measure. This is a good tool for measuring the length of a board or the width of a driveway, but we need to be more precise in manufacturing.

A tape measure can be used accurately by a skilled craftsman to measure within 1/16th inch. Although this is really good if you are building a house, 1/16th inch is considered a crude measurement by manufacturing standards. Machinists are expected to work within a tolerance that is much smaller—usually a part must be accurate within ±.003 inch or better. That is about the thickness of a human hair (*Figure 2.4*). Even that small number is considered crude in some industries. Some skilled workers who make molds, stamping dies, and other high-precision equipment are expected to work within ±.0001. That is 1/30th the thickness of a hair. This is so small that a few degrees change in temperature can cause the metal to shrink or grow by that amount.

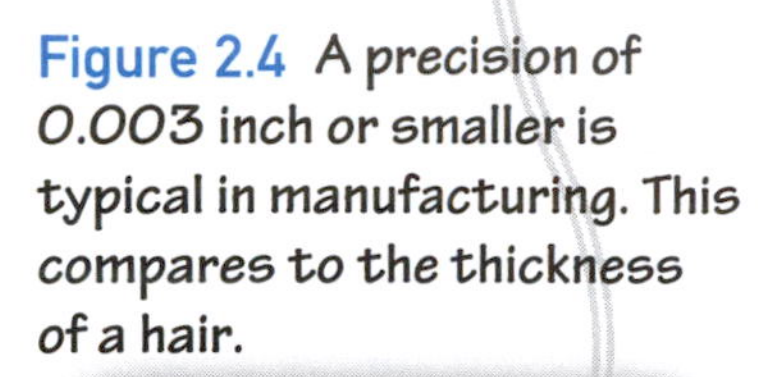

Figure 2.4 A precision of 0.003 inch or smaller is typical in manufacturing. This compares to the thickness of a hair.

process capability: The measure of a process' ability to produce products within the specified tolerances.

2.5 READING MEASURING TOOLS

Tape Measures and Rules

Tape measures and steel rules are common measuring tools that most manufacturing personnel carry with them on a regular basis (*Figure 2.5*). These tools have an accuracy that is good for a variety of work in manufacturing. They are often used to get a rough estimate of size very quickly and without the trouble of using more-precise measuring tools. You might use a rule to check a piece of stock to see if it is 1½ inches and not 1¾ inches. You would not use a rule to check a feature that it is 1.500 inch. When a dimension is given a decimal, it usually means that there is some greater precision involved.

Rules and tape measures are great for verifying that your measurements are within a reasonable range. Tape measures are commonly used for measuring stock to cut on a band saw. Many machinists will use a rule to double check assumptions that were made with high-precision measuring tools. It is very easy to misread a micrometer by a full turn (.025") or for a caliper to be significantly out of adjustment. This is especially true with the new digital tools. All it takes is the push of a button to reset the reference point. This can lead a machinist or inspector to make a series of errors with great speed and precision.

Figure 2.5 A tape measure is the standard low-precision instrument used for rough measurements. Most machinists will use more accurate tools any time a precision of greater than 1/8" is required.

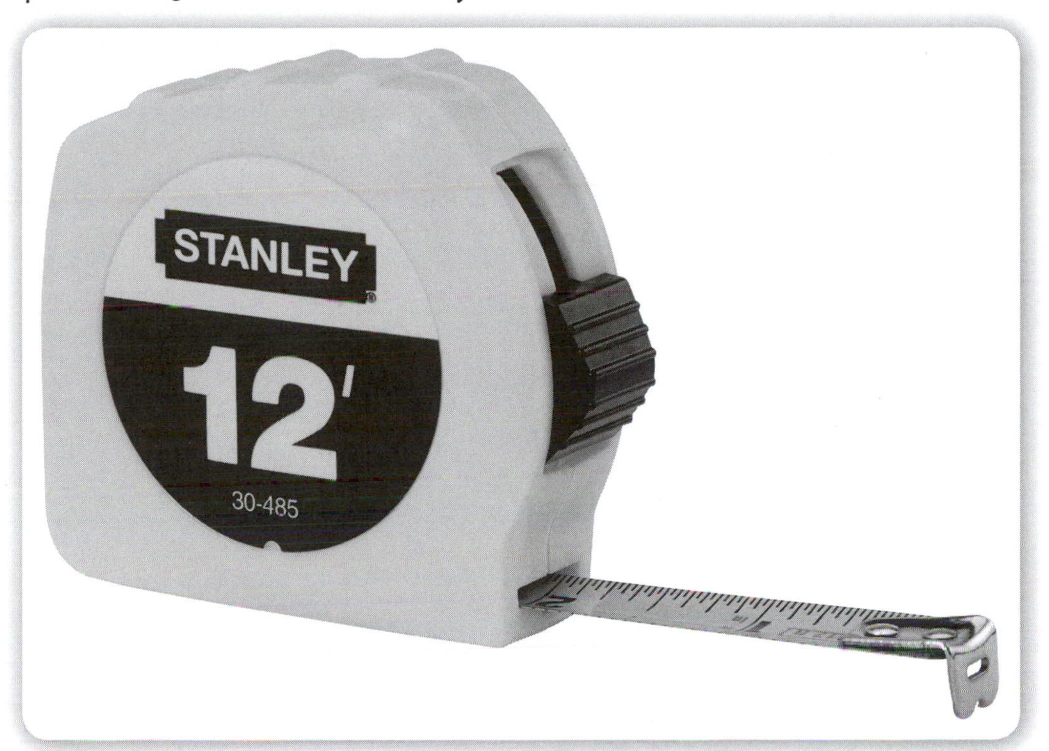

Rules are available in fractions, decimal inches, and metric. The rules shown in *Figure 2.6* are both in inch units. The bottom rule has divisions of 32nds and 64ths, while the top rule has the lesser known decimal divisions of 50ths (.020") and 100ths (.010") of an inch. These divisions are probably unfamiliar for anyone outside of the machining business. The larger divisions on the decimal scale indicate 1/10th of an inch.

repeatability:
The ability of a machine to return accurately to the same position time after time.

Figure 2.6 Steel rules with decimal (top) and fractional (bottom) graduation.

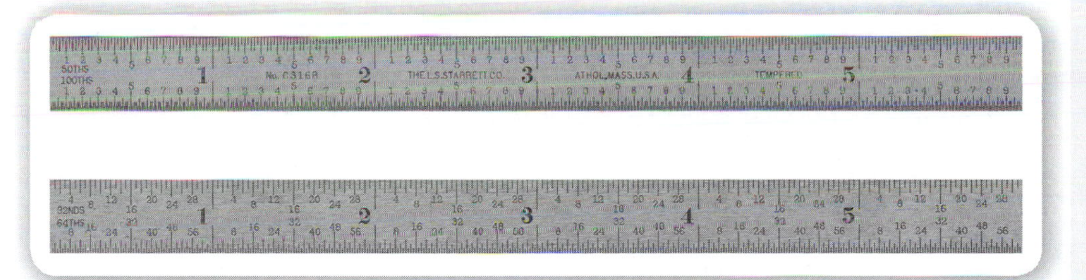

sample:
In statistics, a small group of individuals that are selected for measurement from a much larger population in order to make generalizations about the dimensions of that population.

To read a rule, we simply add all the marks together. For example, "A" in *Figure 2.7* is three full 1/10th marks and two full 1/50th marks from the left-hand side. The decimal equivalents of these measurements are .300 and .040 inch, respectively. Point "C" is on the fractional scale with one full inch and then five full 1/32nd marks. These are calculated below.

standard deviation:
A specific statistical measurement that shows the variability of a sample (or population) from the average.

Measurement A		**Measurement C**	
3 × 1/10th	.300"	Whole inches	1"
2 × 1/50th	.040"	5 × 1/32nd	5/32"
Total	**.340"**	**Total**	**1 5/32"**

Figure 2.7 Read the measurements at points A to D.

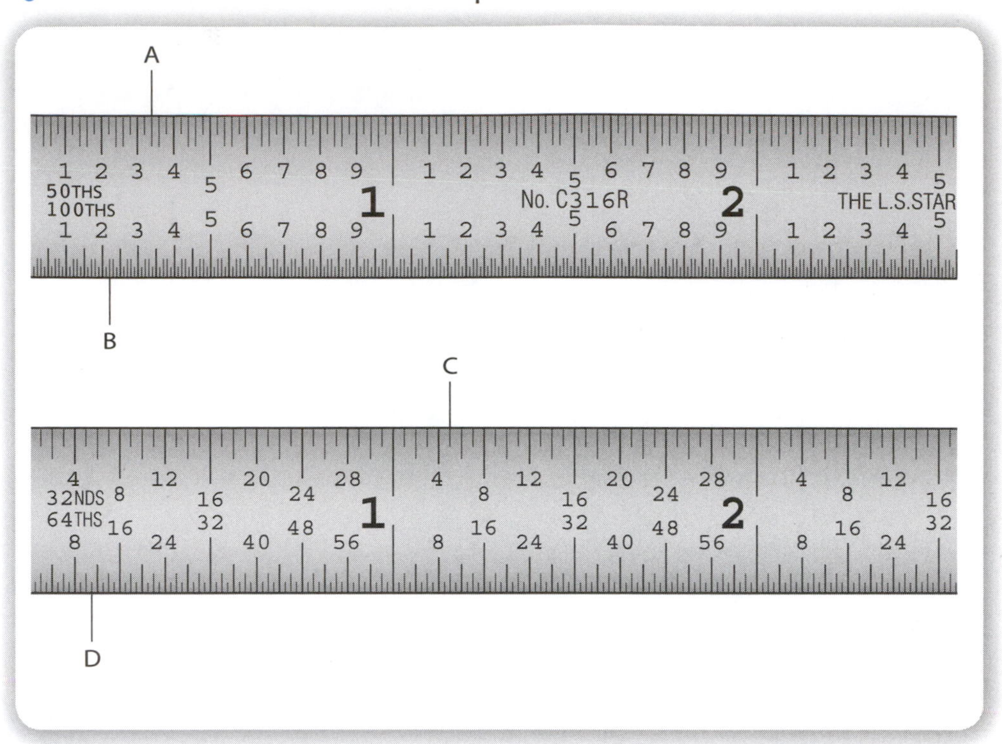

Caliper

A caliper is a middle-of-the-road tool. In the hands of a skilled machinist, it can be used to determine if dimensions are within close specification (±.003). It is also a very versatile tool that can be used quickly with very little training. A caliper is designed to measure outside, inside, steps, and depths (*Figure 2.8*).

Common Fraction-to-Decimal Conversions

1/8"	=	.125"
1/16"	=	.0625"
1/32"	=	.03125
1/64"	=	.0156"
1/50"	=	.020"
1/100"	=	.010"

Statistical Process Control (SPC): A system of dimensional data gathering and statistical analysis used to ensure the quality of a manufacturing process.

Figure 2.8 A dial caliper is capable of measuring many different features ranging from inside and outside diameters to steps and shoulders. The caliper should not be used for high-precision measurements. Use a micrometer when a reliable measurement of greater than ±.003 is required.

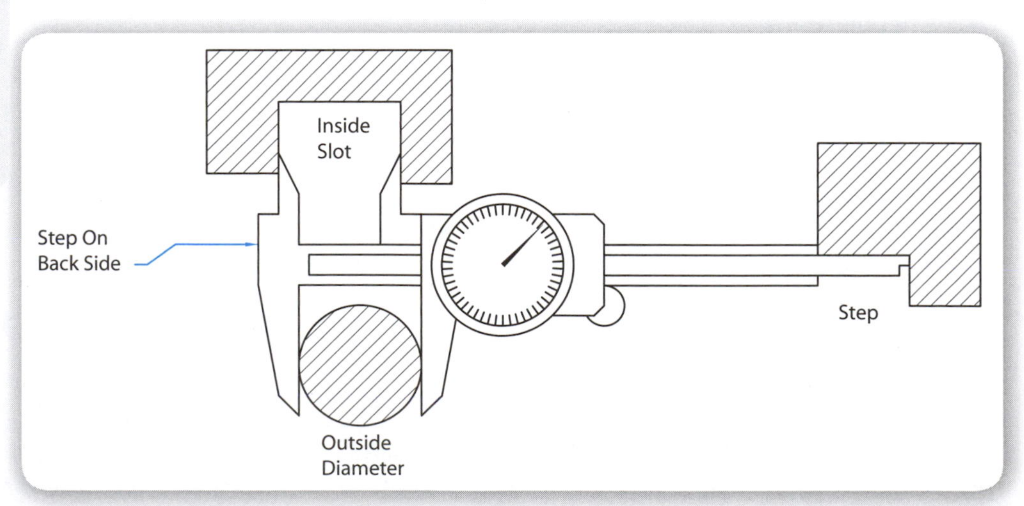

Figure 2.9 What is the measurement on the caliper? In this case the beam shows one full inch plus 1/10th". The dial is resting at .050; therefore, the total measurement is 1.150".

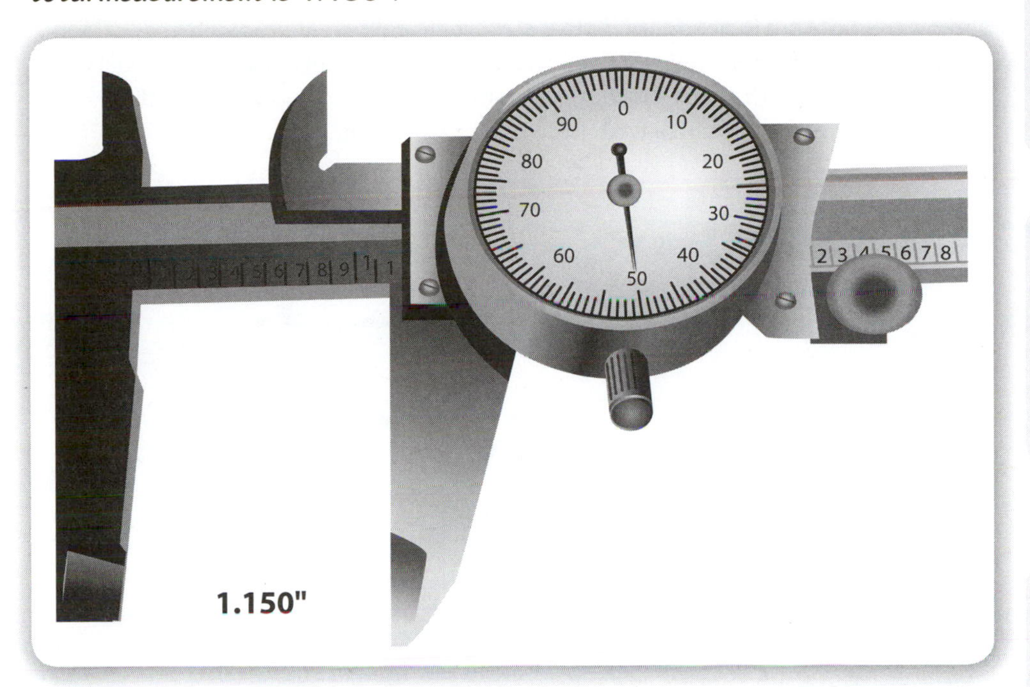

A dial caliper is read by reading the whole inch and 1/10th inch from the middle beam. Then we read the 1/1000ths on the dial itself. We add the three numbers together to get the total reading. We can, for example, see that in ***Figure 2.9*** the following measurements are available:

1.000 One Whole Inch from Beam

0.100 One 1/10th Inch from Beam

0.050 The Dial is Pointed to 50/1000ths

1.150 inch Total

The basic construction of a caliper has remained largely the same for many decades. What has changed is the display mode of the tool (i.e., how the user reads the measured value). Calipers are classified as either mechanical or electronic or an old style called a vernier caliper (***Figure 2.10***). It took people a long time to trust the electronic version of this versatile tool. Machinists are distrusting creatures by nature. After all, if a mistake is made, nobody is going to blame the measuring tool—they will blame the machinist for not checking the tool!

Micrometer

The micrometer is the workhorse in the machine shop for high-precision measurement. Micrometers are designed and manufactured to be very precise and accurate. They use a precision-ground screw to measure the dimensions of a workpiece very accurately. The basic micrometer screw has been adapted to many different shapes including C-frames for outside measurement, bases for depth measurement, and rods for internal diameters as is illustrated in ***Figure 2.11***.

How to Read a Standard Micrometer

A standard micrometer has two scales that are used for reading the measurement to a precision of 1/1000th of an inch. The first scale is located on the sleeve and divides the inch into 40 parts, or .025 inch. This is a result of the construction of

resolution:
The degree to which measurements can be differentiated from one size to another. The scale of the measuring instrument.

rule of ten:
The concept that a measuring instrument should have ten times the resolution of the tolerance of the measured feature.

statistics:
Measurements that describe the attributes of a group.

tolerance:
The deviation in size that is allowed from a specified dimension as measured from maximum to minimum.

variability:
The degree to which a group of measurements might be spread out from the average. The "spread of the curve."

Figure 2.10 The basic construction of the caliper has remained constant for years. It has, however, evolved to be much easier to read. The oldest type is on the top and uses a vernier scale. In the middle is the workhorse dial caliper and on the bottom is a modern digital electronic caliper. High-quality, durable electronic calipers are now more reliable and less expensive than their mechanical counterparts. They also have the advantage of switching between U.S. Customary and metric units with the push of a button.

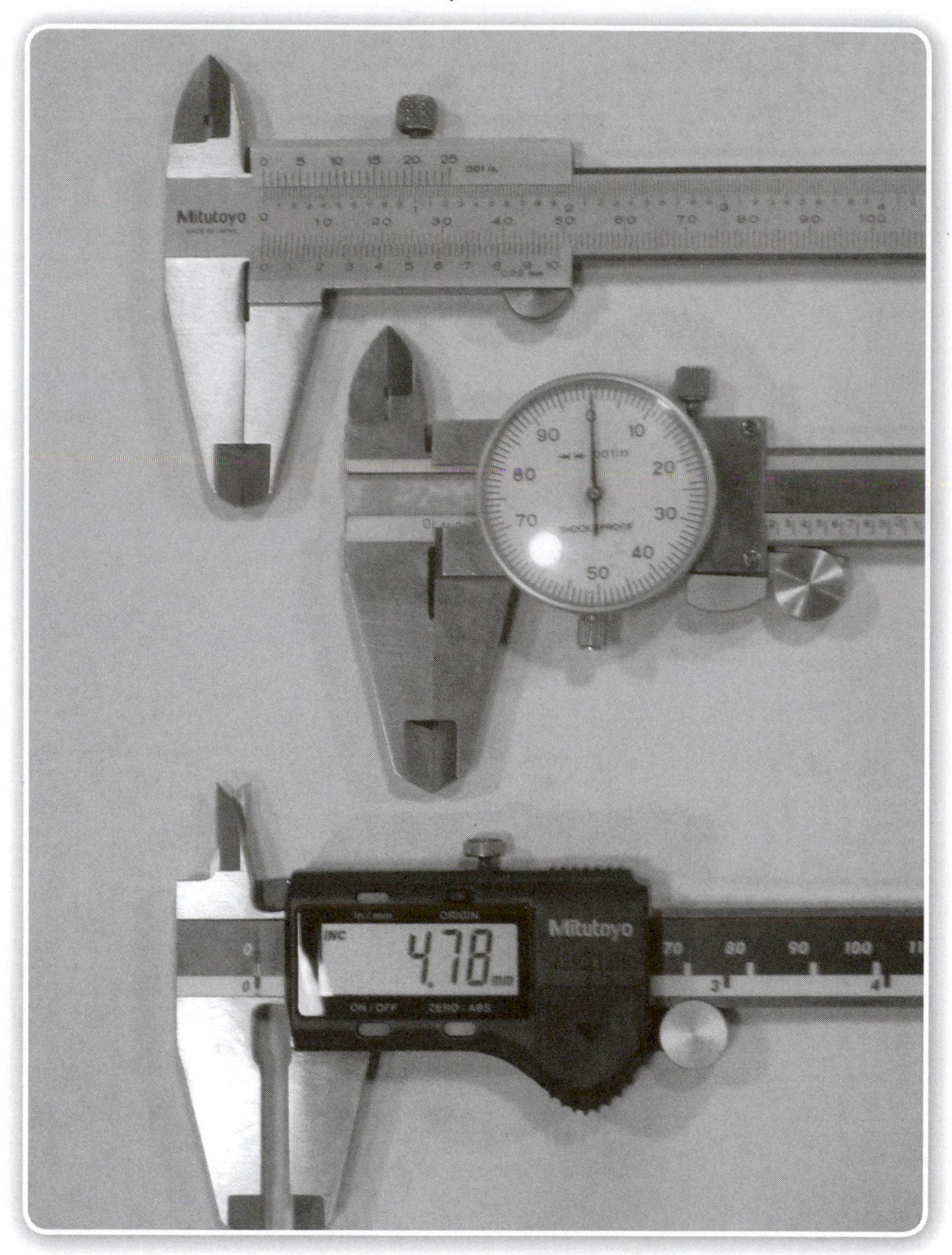

the micrometer screw, which is manufactured with 40 threads per inch. In other words, it takes 40 turns of the screw to open it up to one inch. This first scale is usually marked by .100s and then .025s to make it easier to keep track of the measurement. The second scale is engraved on the thimble as in *Figure 2.12*. The scale divides one rotation in 25 parts. In sum, one full rotation equals .025 inch, and each rotation is divided 25 times, so we can now read the micrometer to 1/1000th of an inch.

Figure 2.11 Micrometer construction is based upon an accurately ground screw thread. The basic micrometer screw can be adapted to a variety of tool styles. On the left half are two outside micrometers, one for inches and the other for millimeters. On the right, a depth micrometer is on top and an inside micrometer is on the bottom.

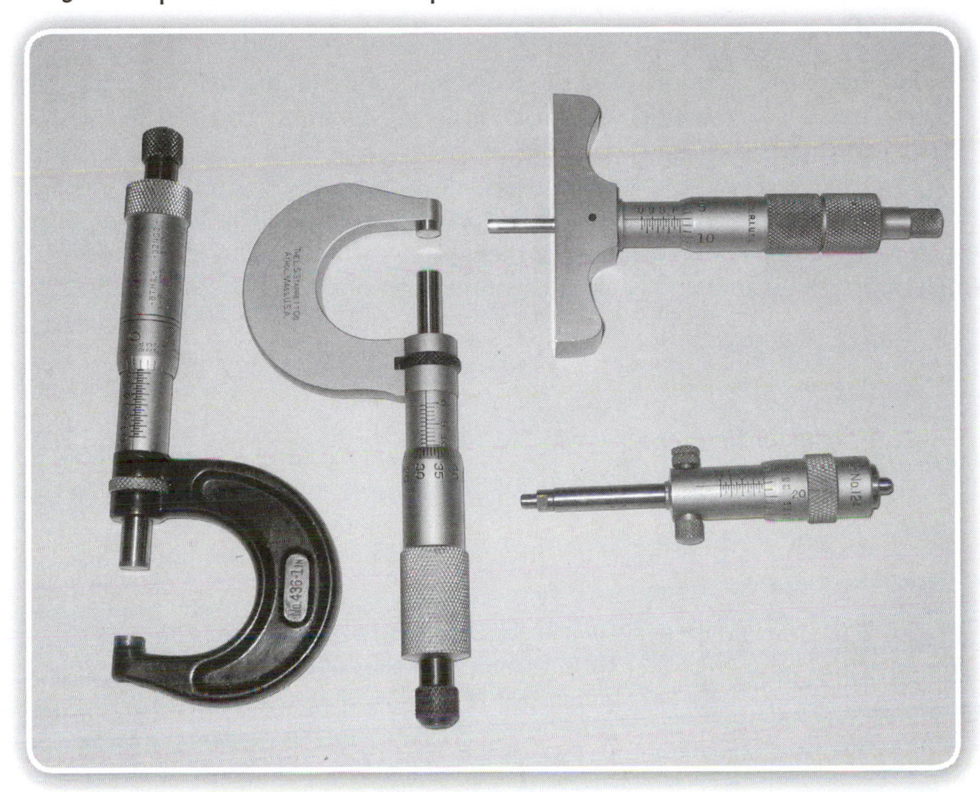

Figure 2.12 A micrometer is read by adding the values from the sleeve and thimble. In this case the measurement is 0.142".

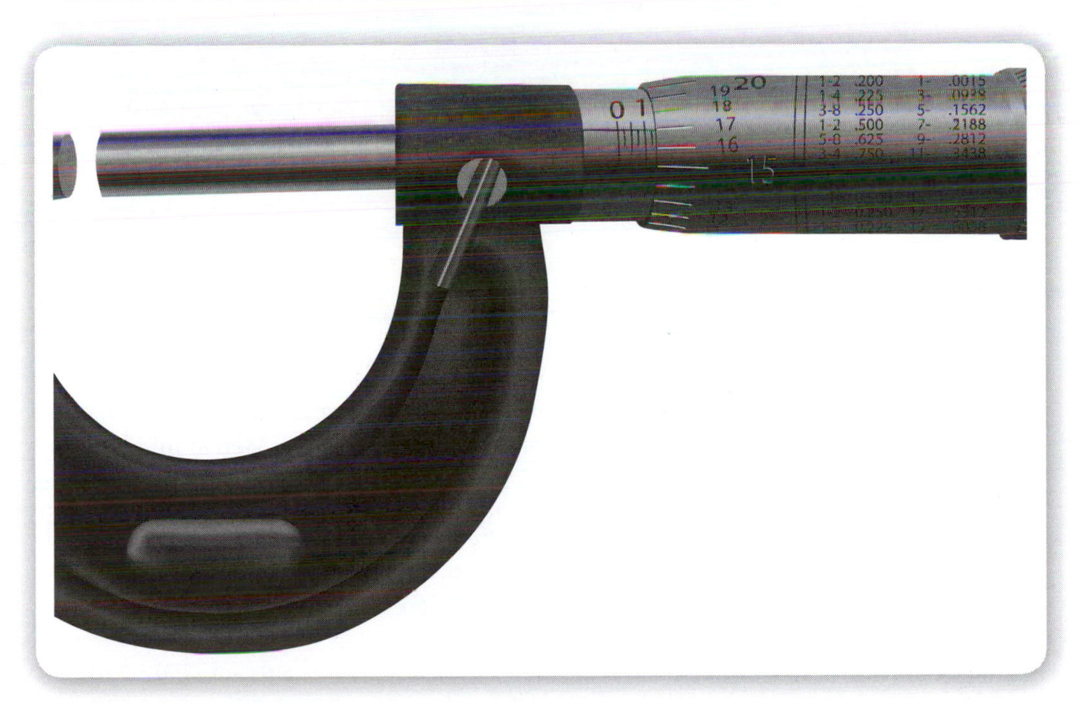

Example from ***Figure 2.12*:**

Reading from Sleeve (5 full-turn lines) *.125*

Reading from Thimble (the "17" line) *.017*

Total ***.142 Inch***

For more precise measurements the "thousandth" is broken down ten more times. This allows a resolution of 1/10,000th inch or commonly referred to a "tenths." A vernier scale for .0001 inch is located on the back side of the sleeve. One horizontal line on the sleeve will be closest to alignment with a number on the thimble. This will be the "tenths" reading. Take ***Figure 2.13***, for example. The "4" on the vernier scale matches one line on the thimble; therefore, we will add .0004 inch to the overall measurement (assuming that we are continuing from the previous example):

Reading from Sleeve *.125*

Reading from Thimble *.017*

Reading from Vernier Scale ***.0004***

Total *.1424 Inch*

Figure 2.13 **The vernier scale can be used to read a micrometer to 0.0001". The "tenths" value on this scale is aligned with the "4" mark.**

Metric Micrometers

Metric micrometers are designed to read to 1/100th millimeter. Remember, a millimeter is much smaller than an inch. So 1/100th millimeter is only .00039 inch (see Table 2.1 for a comparison).This is less than half of the standard "thousandth of an inch" that is used so frequently in American manufacturing. Metric micrometers will often include a vernier scale to read to 2/1000ths of a millimeter. A standard metric micrometer with a range of 0 mm to 25 mm is shown in ***Figure 2.14***. This micrometer does not have a vernier scale.

Dial Test Indicator

A dial test indicator is a precision measuring tool that measures a change in distance with a lever-type tip. An indicator is used for measuring differences between one point and another. For example, in ***Figure 2.15***, the indicator is set at zero on a

Figure 2.14 A metric micrometer is constructed to move ½ mm per revolution. The sleeve has a scale for every whole and half millimeter, and the thimble has 50 divisions on the diameter. Therefore, a basic metric micrometer can measure to 1/100 mm. The measurement in this photograph is 10.30 mm.

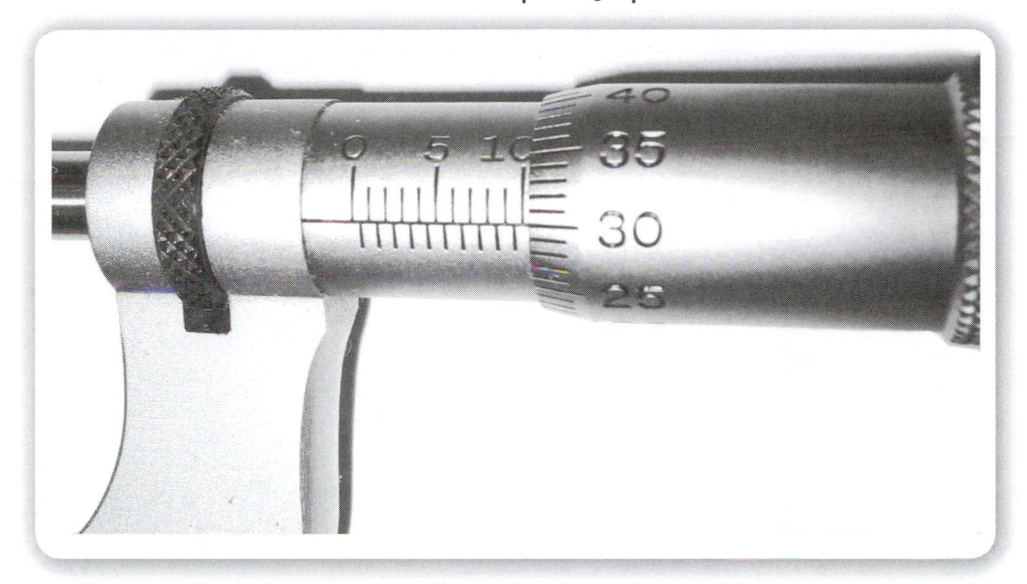

Figure 2.15 Dial test indicators are good for measuring the relative difference between two points. This indicator has a resolution of 0.0001". On the left, the indicator is referenced from a very accurate gage block that measures 1.0000". To the right the indicator shows that the work piece is 0.0003" smaller than the gage block. Therefore, we know that this part is 0.9997" tall.

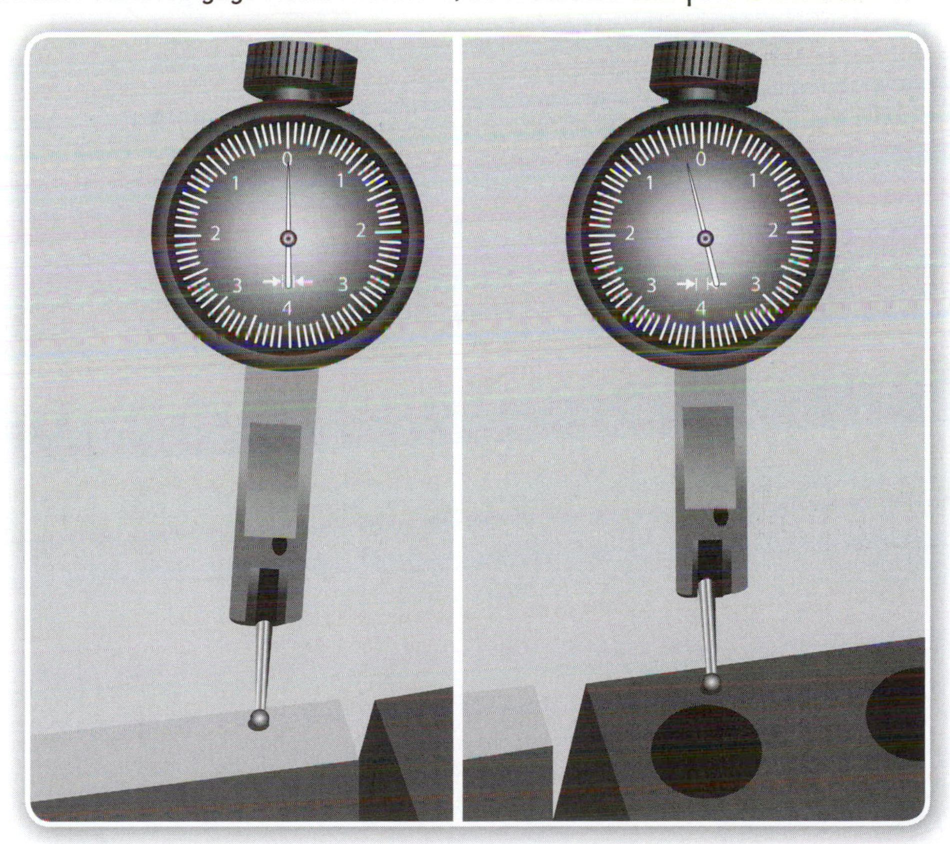

standard 1.0000 inch gage block. The indicator is then moved to the adjacent workpiece to be inspected. We can see that there is a slight difference between the two, as is measured by the indicator. This method is commonly used to measure features accurately to a tolerance of ±.0001 inch or less.

Figure 2.16 The dial indicator is a versatile tool for locating the centerline of an existing hole. In this case, the indicator is held in the spindle of the machine tool. It is swept around the hole while the table is adjusted by the machinist. When the spindle is directly over the centerline, the indicator will show zero difference from side to side and front to back.

A dial test indicator can also be used to align tooling on a machine tool or to center a spindle over a hole in order to machine a new feature on the exact centerline (*Figure 2.16*). Because the indicator measures only the relative change from one point to another it is useful for aligning a vise, which involves setting the indicator on one side of a jaw and then moving the milling table from side to side to check the difference. A similar concept allows a machinist to precisely align the machine spindle with the center of a hole. When the hole is centered, there will be no difference between the indicator reading in any quadrant of the hole as the spindle is turned by hand.

2.6 ACCURACY AND PRECISION

Do you know the difference between accuracy and precision? These terms are often used interchangeably in the nontechnical world. Precision is related to the certainty of the measurements. There is always uncertainty in measurement, and that is an expression of precision. **Accuracy** is the difference between the measured value and the actual value. ***Figure 2.17*** illustrates this concept from the standpoint of target shooting.

accuracy:
The difference between a specified dimension to the actual dimension that is produced.

Which is better, a digital watch or an analog watch with moving hands? The digital watch is certainly more precise. We might read 20 seconds on either watch, but in fact it could be 20.6 seconds, 19.8 seconds, or some other value. On average, we will get less uncertainty in our measurements with the digital watch.

It is easy to read the precise time to a single second with a digital watch (i.e., it is either 20 seconds past, or it is not—there is no guessing). This isn't so easy on an analog model, which might not even have a second hand. The second hand might also be in between two numbers (***Figure 2.18***).

Figure 2.17 The quality target illustrates the concept of precision and accuracy. A precise rifle will be able to shoot a "tight group" even if the sight is misaligned. With some adjustment, the rifle can be both precise and accurate.

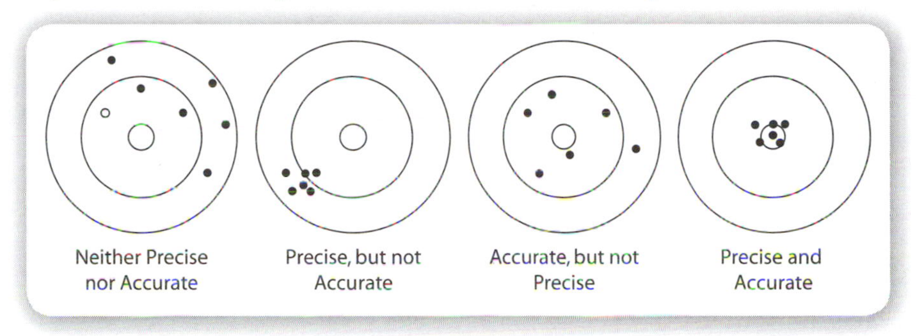

Figure 2.18 The fine resolution of a digital instrument is not an indication of accuracy.

On the other hand, the digital watch might be less accurate than the analog version. Let's say that the digital watch loses 45 seconds each day, and the analog gains two seconds. Is the digital watch still better? Which is more accurate? The digital watch can certainly give you a precise time, but it will be inaccurate.

We can also think of accuracy and precision in a statistical sense. You see, in the real world we are never really sure of a measured value. We have to assume that there is some error due to the limitations of the measuring instrument or

precision:
The degree of certainty in a measurement.

mean:
The average of a set of measurements.

population:
An entire group of objects to be measured.

methods. Therefore, engineers and scientists often think in terms of the probability of having the correct measurement. *Figure 2.19* illustrates the concept of probability. The vertical line is the theoretical target value. Each "bell curve" represents a natural spread (variability) of measurements that might be taken with different measuring instruments, the center being the average. This spread can be thought of as the **precision**. The distance from the curve to the target value is the accuracy of the measurement.

It is common to be lured into thinking that because an instrument is digital or electronic it is always more accurate. It can be true, but there isn't necessarily a connection between the two. For example, the digital clock on my computer is always

Figure 2.19 **Accuracy and precision can be represented with a probability curve. Curves 1, 2, and 4 have the same precision, but 2 is, on average, more accurate. Curves 2 and 6 share the same accuracy, but 6 has much more variation. How would you compare curves 1 and 5 and 3 and 6?**

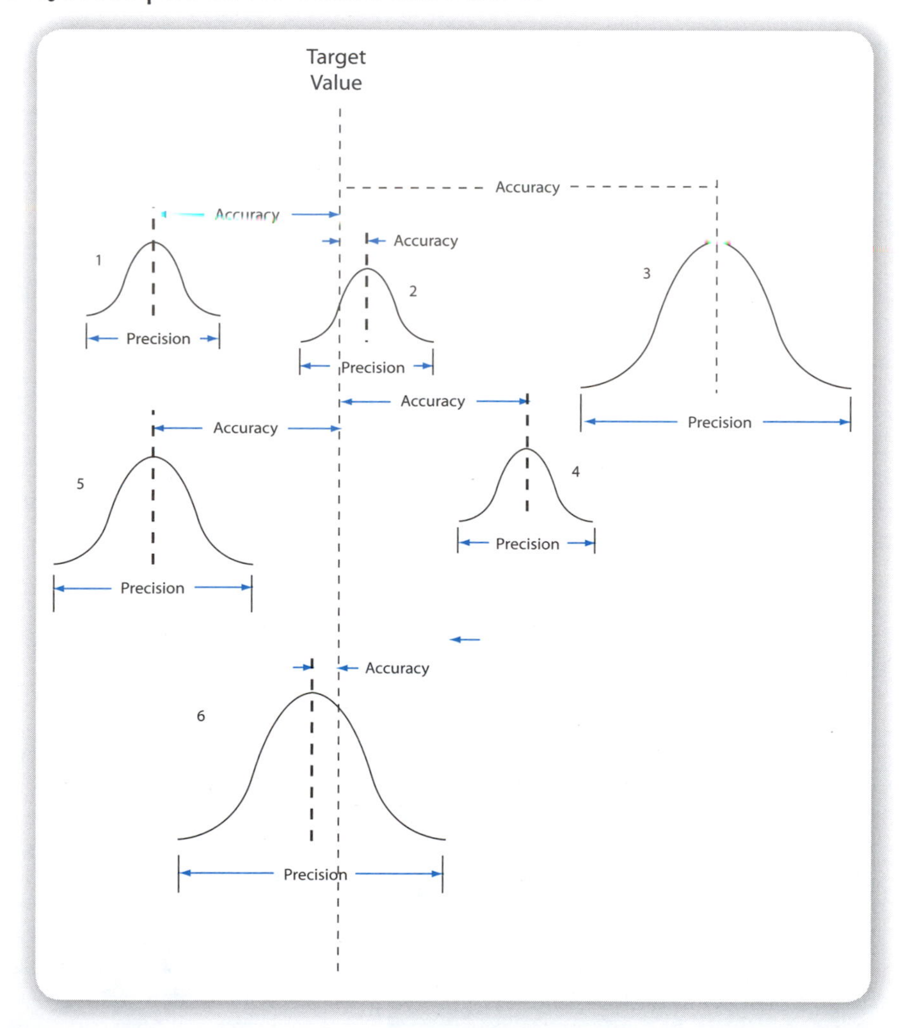

What Time Is It?

Did you know that the U.S. government keeps the nation's time with a very accurate atomic clock?

The U.S. Naval Observatory in Washington, D.C. operates a cesium atomic clock that is accurate to 10 billionths of a second. It turns out that radioactive cesium atoms vibrate at a very stable frequency. This vibration can be measured electronically with microwaves and converted into time.

This precise and accurate timing is used for satellite navigation, astronomy, and a variety of military technology.

You can also hear a constant update on the time if you have a short-wave receiver. Just tune into 5,000 KHz and listen to the tone tapping out the seconds all day and night around the world.

running slow. I can always read the incorrect time to the precise second. "How can this be? It is a computer, after all." Well, it turns out that the timing mechanism on a PC simply isn't up to par with even the cheapest of quartz watches.

Resolution

Resolution is a term that is used to describe the units of a measuring instrument. The resolution is simply the smallest unit to which a measuring tool can be read (it can *resolve* the difference between one measurement and another). For example, the top scale in ***Figure 2.20*** has a resolution of 1/32 inch. We can easily distinguish between a measurement that is 4/32 or 5/32 inch. However, there is no way to tell without guessing if the measurement is 9/64ths or 10/64ths. What is the resolution of the bottom scale?

A micrometer with a vernier scale will have a resolution of .0001 inch. A typical digital caliper will have a resolution of .0005 inch.

Resolution should not be confused with accuracy. A digital tool might have a resolution of .000050 inch, but that does not imply that the measurement is accurate or precise for that matter. It is just that the inaccuracy will be doled out in chunks of .000050 inch.

"Rule of Ten"

There is a principle of metrology that states that any measuring instrument should be at least 10 times as accurate as the required measurement. For example, a shaft is specified to have a diameter of 1.000 ± .003 inch. Therefore, the selected measuring tool should be capable of accurately measuring to 1/10th of that tolerance or ±.0003 inch.

Figure 2.20 The top scale of this rule has a resolution of 1/32 inch.

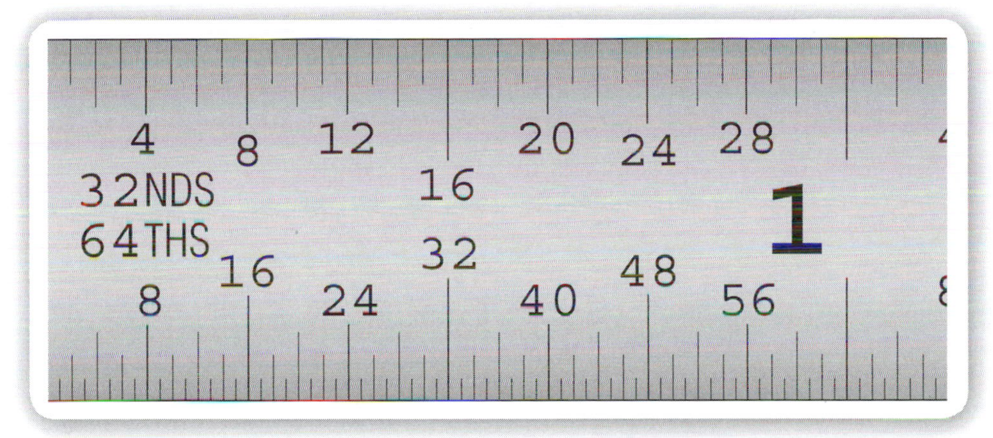

2.7 DIMENSIONS AND PRINTS READING

Engineering drawings are the language of technical communications. These drawings are also commonly referred to as *prints, blueprints,* or *drawings.* Regardless of the name, they are the same document. In this text we will simply refer to engineering drawings as *blueprints.*

Blueprint Basics

Fundamentally, an engineering drawing (blueprint) describes the shape and size of an object to be manufactured. A blueprint provides the necessary documentation to make the product within certain size guidelines, called tolerances. They also communicate information about the material and specific manufacturing processes.

The first order of business for the blueprint is to describe the shape of the object. This is done by providing a series of views from the front, top, and side. The views are taken from the viewpoint of a person standing squarely aligned with the object and looking directly at it. If you have ever read a road map, then you have experience with this concept. A road map is a top-down view that assumes the reader is in the sky and looking straight down on the earth.

A front view is selected to show the most prominent shape of an object. Take *Figure 2.21*, for example. The front view shows that the object has an "L" shape. The top and side views, in this case, show little more about the shape. It may be that other views also show important features of an object, so the drafter must use his or her best judgment in selecting the front view.

Figure 2.21 **A product is represented on a blueprint in views using orthographic projection. An isometric projection makes it easy to visualize the shape and features.**

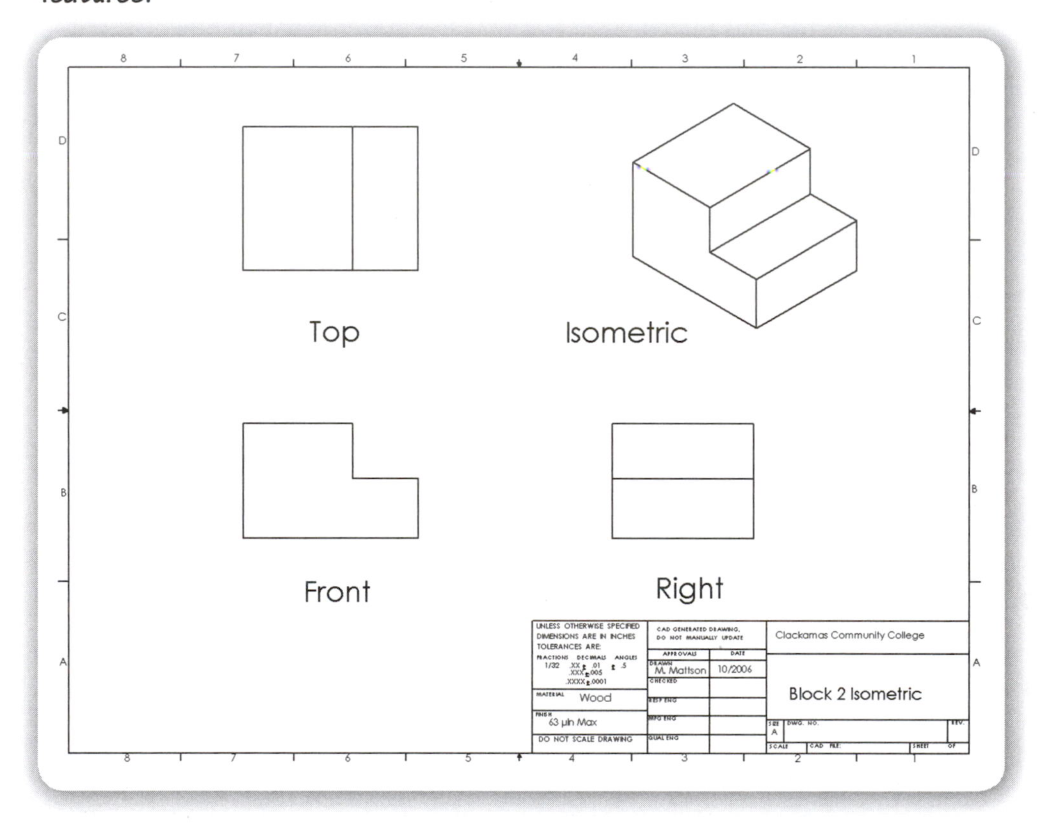

Orthographic and Isometric Projection

When we look at orthographic projections, we are viewing the face of the object from a right angle rather than from a corner. The problem with orthographic projection is that the brain has to take the two-dimensional views and morph them into the three-dimensional object that the designer intended. This takes brain power and practice. Isometric projection helps us to visualize a workpiece quickly by presenting the drawing as viewed from a corner rather than straight on as is illustrated in the "isometric" view in *Figure 2.21*. This is a powerful representation that allows our brains to quickly translate the two-dimensional lines on paper to three-dimensional representations in our heads. Unfortunately, drafters are often stingy with the isometric views, so anyone who frequently needs to use blueprints will need to develop the visualization skills to be able to look at a series of orthographic views and determine the shape of an object.

Figure 2.22 **Hidden lines are used to illustrate features that are obscured by solid material.**

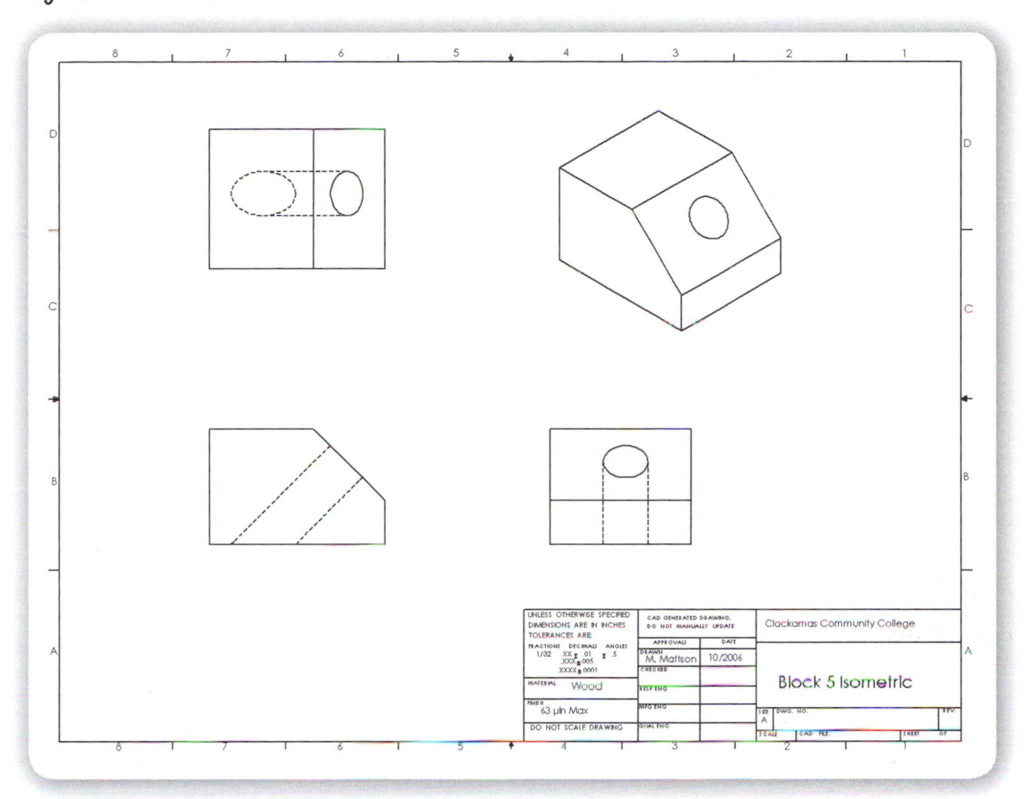

Hidden Lines and Sections

Many times features that exist on an object might be invisible to the viewer in a standard orthographic projection. For example, in *Figure 2.22*, the diagonal hole extends completely through the part. If we were to view this part from the front, then the hole would be completely hidden from view. To solve this problem, the obscured feature (the hole) is drawn with hidden lines. These lines are constructed from short dashes and indicate that there is a hidden feature that cannot be seen.

Hidden lines are drawn only when they are necessary to describe the object. The simple blueprints that we have seen in the chapter are easy to read; however, blueprints can become very complicated. On complex blueprints, unnecessary hidden lines can make for a jumbled mess that is difficult to read.

Some features are difficult to represent with hidden lines. In those cases, the drafter might use a *section view* (*Figure 2.23*). A section view can be thought of as if we took the part and sawed through it along a designated line (section line). The sectioning would then expose previously hidden features to our direct view. There are many different types of section views, but the easiest to understand is the full section. The full section cuts along a straight line. We then view the section by looking squarely in the direction of the arrows that are shown on the section line itself.

Title Block and Notes

The title block provides some guidelines for interpreting a blueprint as well as general information about the actual drawing and the part to be produced, as is illustrated in *Figure 2.24*. Some of the features include:

- Units and Tolerance Information
    - U.S. or Metric?
    - Tolerance Based upon Dimension Precision

Figure 2.23 Section B-B is a full section. Imagine that the work piece is cut in half on a saw. We then view the section by pointing our eyes in the direction of the arrows on section line B-B. The diagonal hatch lines represent the solid material through which the part has been cut.

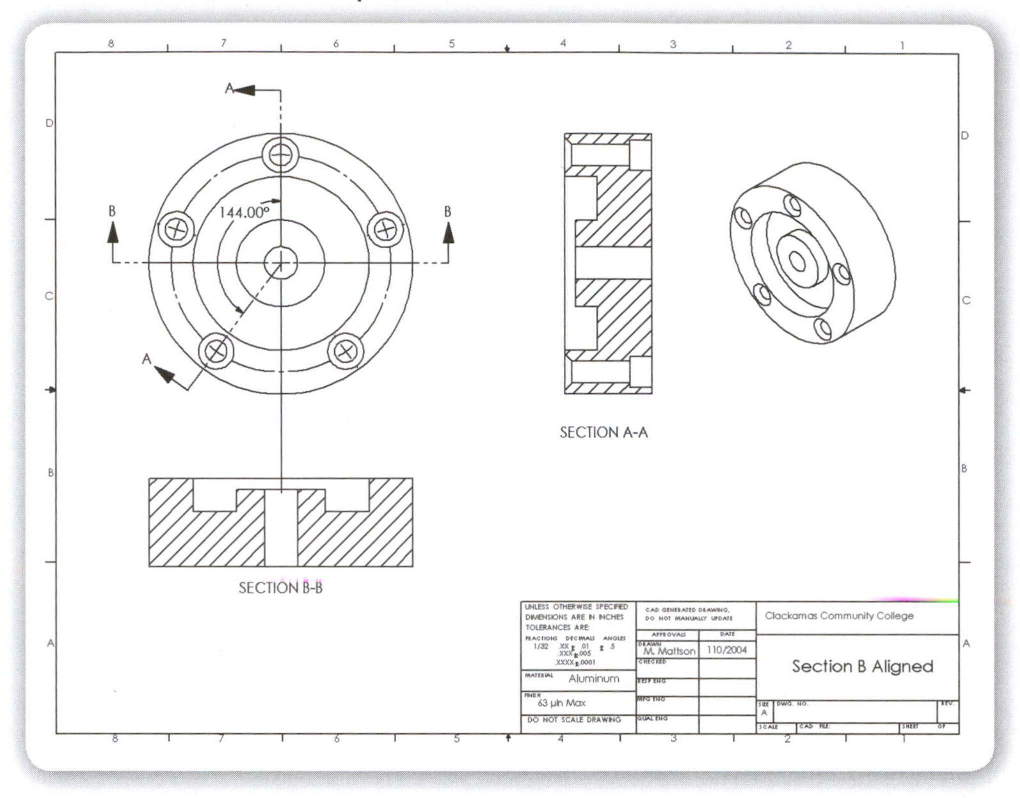

Figure 2.24 A title block communicates a variety of information, including tolerances, material, product name, drawing number, and revisions. It is critical that you carefully read the title block before planning any manufacturing operations.

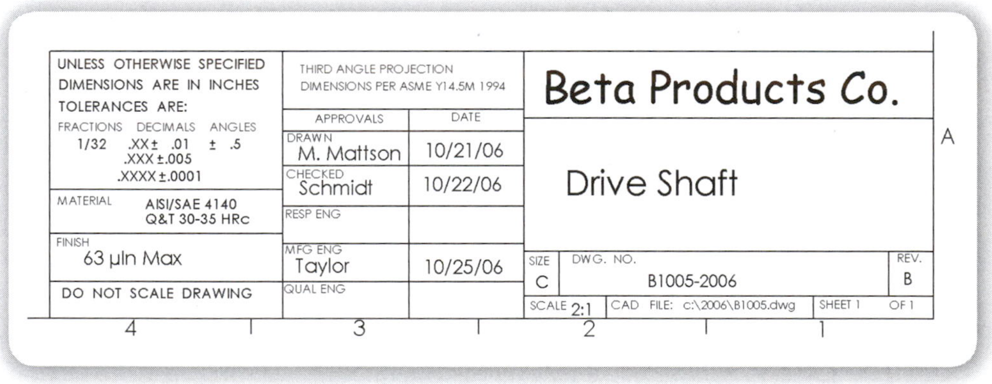

UNLESS OTHERWISE SPECIFIED DIMENSIONS ARE IN INCHES TOLERANCES ARE: FRACTIONS 1/32 DECIMALS .XX± .01 .XXX ±.005 .XXXX ±.0001 ANGLES ± .5	THIRD ANGLE PROJECTION DIMENSIONS PER ASME Y14.5M 1994		Beta Products Co.
	APPROVALS	DATE	
	DRAWN M. Mattson	10/21/06	Drive Shaft
MATERIAL AISI/SAE 4140 Q&T 30-35 HRc	CHECKED Schmidt	10/22/06	
	RESP ENG		
FINISH 63 µIn Max	MFG ENG Taylor	10/25/06	SIZE C DWG. NO. B1005-2006 REV. B
DO NOT SCALE DRAWING	QUAL ENG		SCALE 2:1 CAD FILE: c:\2006\B1005.dwg SHEET 1 OF 1

- Drawing Information
  - Company and Drawing Name
  - Approvals and Dates
  - Drawing and Revision Number
  - Scale of the Drawing
- Material
  - Type and Condition

Figure 2.25 Dimensions define the size of the part on a blueprint. In this case we can see that the part has the overall size of 4" wide, 2 5/8" tall, and 3 1/4" thick.

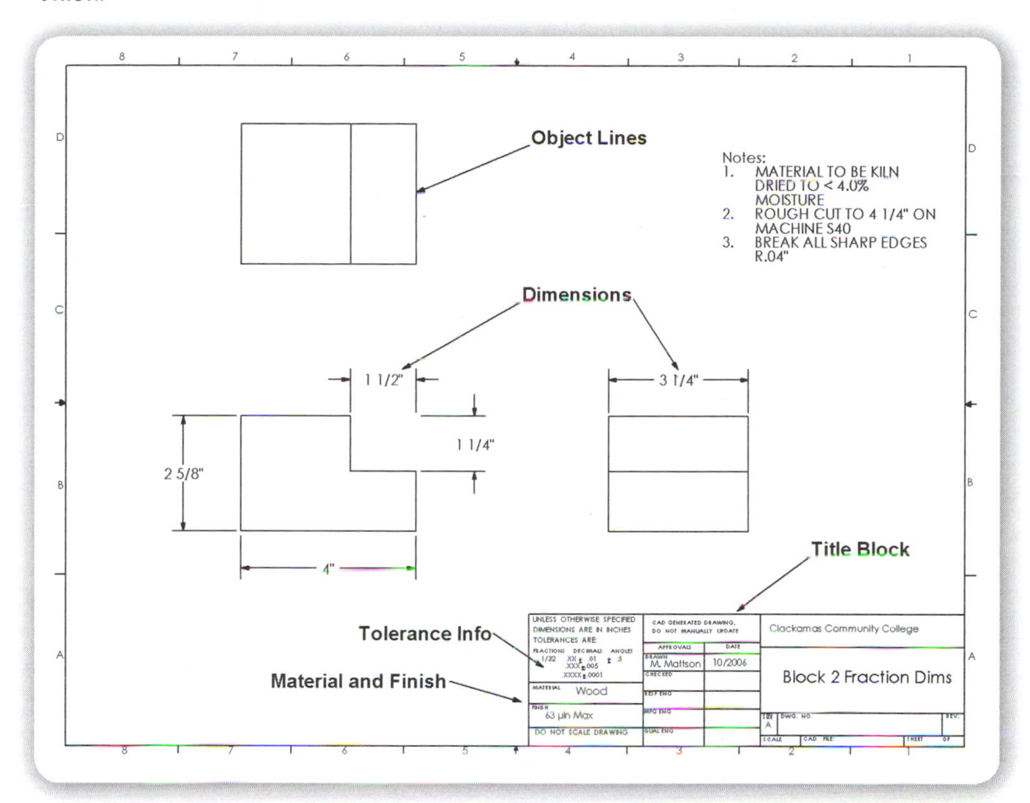

A lot of other information can be included in the title block to communicate to manufacturers. It is well worth the time to take a close look at it to be sure nothing important is missed.

Machinists should also pay close attention to any notes that are given on a blueprint. Notes are used to elaborate on specifications and to give details on particular manufacturing processes. The notes in *Figure 2.25* give insight into the material condition and manufacturing practices. The slight details can often cause expensive cost overruns if missed. For example, if the note calls for all edges to be broken (smoothed) and this process is not completed, then the customer is likely to reject the whole lot. This could cost the company cash, time, or even a customer.

Dimensions and Tolerances

The second question to answer after we have determined the overall shape of an object on the blueprint is, "What size is the part?" This is a story that is told by **dimensions**. Dimensions are simple designations for the size of a feature in the appropriate units. Linear dimensions are the most common of all. If we look at *Figure 2.25* again, we see the same L-shaped part that appeared in Figure 2.21, but this time each feature has a dimension attached to it. These are linear dimensions that determine the length of a feature.

dimension:

The measure of geometrical attributes. In manufacturing, dimensions are used to specify the linear size of features and angular relationships.

The dimensions are the ideal size that the engineer has designated for the particular feature. In reality, there will always be some small error or variation in the manufacturing process. To allow for this error, the designers always account for a **tolerance.** The tolerance is the total amount of variation that is allowed on a feature. For example, if a round bar is specified to have a diameter of 1.000 inch, then diameter might be given a tolerance of plus or minus .005 inch. That is, the bar could be as big as 1.005 inch or as small as .995 inch. If the bar is measured to be 1.006 inch, then it is out of tolerance and will need to be reworked or scrapped.

Technically, the tolerance is the total amount of variation, which in our example was .010 inch. It is, however, common in industry to split the tolerance into even portions and use a plus/minus designation (this is called a bilateral tolerance). On a blueprint, the tolerance for a dimension can be specified in several ways. A general tolerance is usually specified in the title block. A general tolerance is called out by the precision of the specified dimension. For example, in ***Figure 2.26,*** the top two dimensions are mathematically equal at 4 inches. Yet dimension "A" is specified with two decimal places as 4.00, and "B" is written to three decimal places as 4.000. Is there a difference? Yes, and it can be very expensive to have a tighter tolerance than is necessary.

On a typical title block there is a general tolerance specification. Take the title block in ***Figure 2.24*** as an example and look at the tolerances for the different precision's of dimensions:

Fractions ± 1/32"	*Example: 4"*
2-Place Decimals ± .010	*Example: 4.00*
3-Place Decimals ± .005	*Example: 4.000*
4-Place Decimals ± .0001	*Example: 4.0000*

The difference between 4 inches and 4.0000 could potentially be 10 times the manufacturing cost or more. Four inches is really "sloppy," as they might say in a machine shop, and you might expect that this part was made by any old machine with very little attention. It might even be a saw cut, depending on the geometry. Conversely, a dimension of 4.0000 requires very close accuracy and careful operations. This part will likely need to undergo expensive surface grinding.

Sometimes a dimension requires a tolerance that is not the same as the general tolerance. In that case, the tolerance is given along with the dimension. This can be done by simply adding a tolerance to the dimension, as in "C" of ***Figure 2.26***, or by stating the dimension in terms of a maximum or minimum measurement, as in "D."

The Art and Skill of Drafting

Any person studying in a technical field should be encouraged to take at least one formal class in drafting. Practice and study in drafting, whether by hand or computer, will make you a better user of graphical information.

Figure 2.26 **A seemingly equal dimension of 4 inches can mean something very different to a machinist depending on how it is written. The title block often contains information regarding default tolerances.**

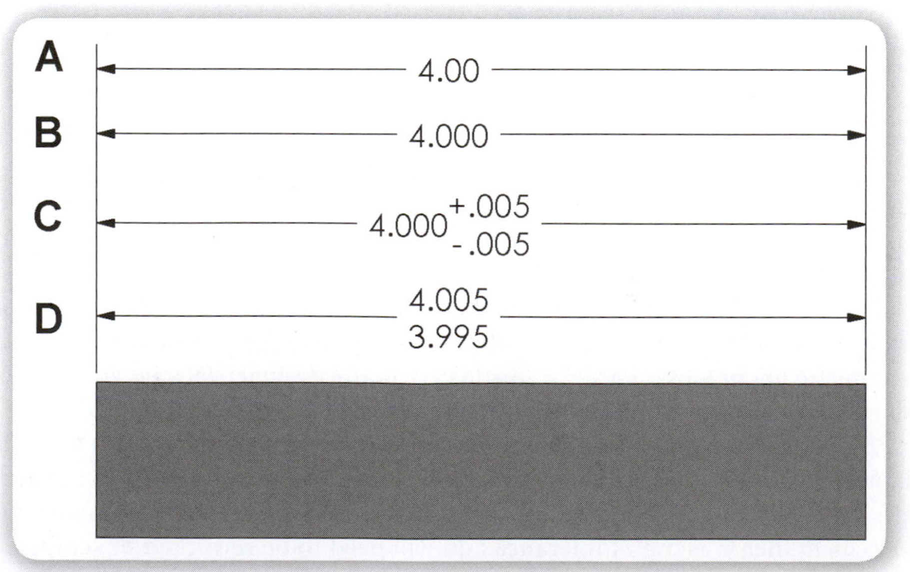

Case Study

What does a husband do when his wife's birthday is less than a week away, and he hasn't shopped for a gift yet? One procrastinator turned to Michael's Designs, 5 days before his wife's birthday, and asked them to design and fabricate a special diamond necklace.

The client's request for a platinum necklace set with black and white diamonds was one of the first orders Michael's received after purchasing ArtCAM. The design was challenging because the necklace had to lie at a certain angle on the neck, and the diamonds were not all the same sizes.

"Without ArtCAM we would never have been able to complete this job because we would never have been able to hand-make all the different sample links to accommodate the different stone sizes," said business owner Michael Buckley. Buckley also had to create each link with a different angle to ensure the necklace was the right size and lay at the right angle.

The process of making compensations for shrinkage and making sure the stones fit after the platinum was cast would have taken more than a week on its own. ArtCAM allowed Michael's to make changes to the model without having to spend the day redoing the links by hand. An example of this was the small matter of changing the stone size.

Buckley said, "In ArtCAM we would spend 2 minutes making the change and an hour later we would have the piece in our hands. We were able to produce this piece set and polished to deliver to our client in 4 days. A new record in our shop!"

"We have been handmaking jewelry for 35 years," said Buckley. "I am a second-generation jeweler, my father made jewelry for the queen of England. With all my experience, I've never seen a necklace made of this high quality in such a short time."

2.8 STATISTICS AND QUALITY CONTROL

Statistics are used in manufacturing to track the quality of products, to make predictions about reliability, and to determine which machinery is capable of producing the required accuracy for a product. A special application of statistics is called statistical process control (SPC). It is one of the fundamental ideas of the post-war quality movement that actually has roots in the earlier emergence of industrial engineering and the work of Fredrick Winslow-Taylor, who sought to make a science of measurement. But it was Walter Shewhart at the Western Electric Hawthorne plant in Chicago who first applied statistics to quality. The control charts used in SPC are often referred to as *Shewhart Charts*. SPC and the quality movement were propelled to fame in the latter part of the 20th century by the prolific Edwards Deming. SPC is the most likely application in which you will be face to face with statistics in the workplace.

Descriptive Statistics

Statistics are numerical data that describe a group of people or things. For example, we might cite the statistic that the average height of men worldwide is about 69 inches. We might also say that the middle 50 percent are between 66 and 71 inches tall—this is also a statistic.

Statisticians often report descriptive statistics to give the user a quick idea about the measures of a group. For example, mean, median, mode, range, and standard deviation are common descriptive statistics that tell a lot about the magnitude and spread of a group of measurements.

Let's say that we were given the following descriptive statistics of a manufacturing process for the number of valves produced per day for the previous month:

Mean = 540 Units

Maximum = 580 Units

Minimum = 520 Units

Range = 60 Units

The mean (also called the average) is really the core piece of information about a group of measurements. The mean tells us that the magnitude is around 540 units—not 20 or 80,000. We can see right away the scale of the operation.

The range tells us that there is some variation in the daily production quantity. We can see that the difference between the highest day of production and the lowest is 60 units. Graph A in ***Figure 2.27*** shows how we might display only the mean and range.

The minimum and maximum show us that the distribution is an uneven distance from the mean. We are more likely to have extremes toward the high side of production.

We might also find it useful to graph a distribution curve of the production, as in ***Figure 2.27***, Graph B. A distribution curve tells us the number of days that the production was at a particular level. We can see in Graph B that on the majority of days, the production was concentrated around 540 units. There were a couple of days when production was below 530 units or above 550 units. This graph also shows the mode or most frequent level of production. The mode and the mean are often the same, but there are exceptions as in C where the mode is shifted to the left in what we call a skewed distribution.

Another important statistic is the standard deviation (STD). The standard deviation gives the user a good picture of how the measurements are spread out from the mean. We will see how to use the STD shortly with statistical process control.

We mentioned earlier in this chapter that the normal distribution is a common pattern that occurs in nature and manmade processes. The normal distribution shows us that on average, most results are likely to occur close to the center and that the extreme results are few and far between. For example, many men are between 5'6" and 5'11" tall. However, very few men are over 7'0" tall or under 4'1" tall. The distribution of heights tends to follow a normal curve. ***Figure 2.28*** shows the bell-shaped normal curve.

You might have seen a probability generator machine, as shown in ***Figure 2.29***. When the ball is dropped from the top center it has an equal chance of going left or right. On the next level, the ball hits the pin and also has an equal chance of going left or right. This continues all the way to the bottom. The accumulation of balls in the bins is an illustration of a probability distribution that approximates a normal curve.

Figure 2.27 ***Graphical displays of descriptive statistics can rapidly tell us about a group of measurements.***

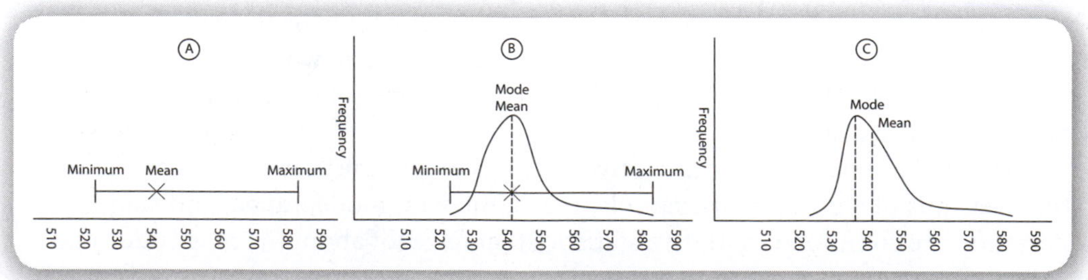

Figure 2.28 A normal curve shows the probability of a result falling within a specific distance of the mean. 68% of results will fall within ±1 STD, 95% within ±2 STD, and 99.7% within ±3 STD.

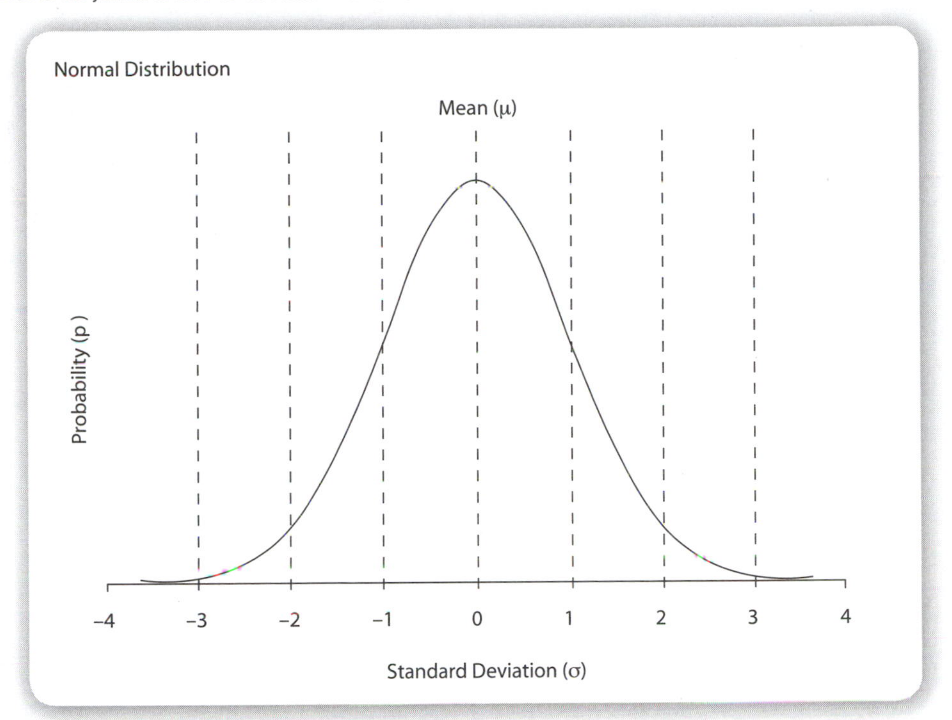

Figure 2.29 A probability machine illustrates the idea of an accumulated probability. Notice that this closely resembles the normal curve.

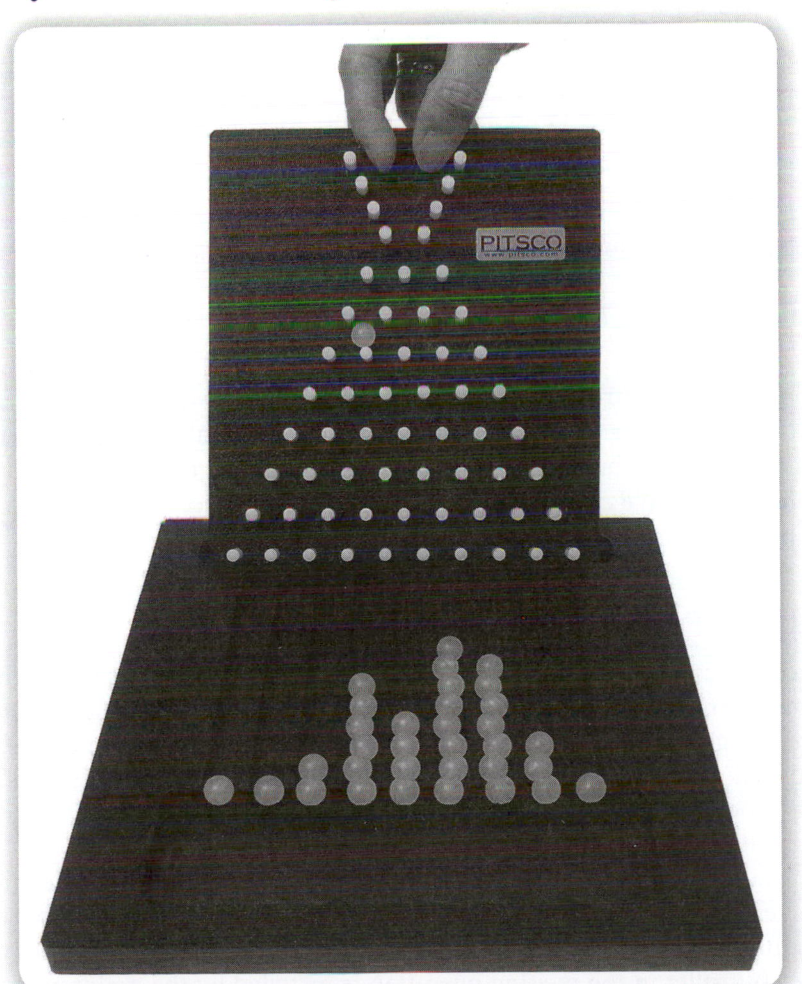

Terms in Statistics

Mean—The sum of the measurements (Σ) divided by the total number measured (n). This is also called the "Average."

Variability—The amount by which a group of measurements deviates from the average.

Normal Distribution—A continuous mathematical function that is roughly based upon the equal chance of a random event falling one way or another. Many natural phenomena are distributed along a *normal* curve. As we might expect, the greatest number of measurements is grouped toward the center.

Standard Deviation (STD)—A measurement of variability in a *normally* distributed population. A predictable proportion of the measurements will fall within an STD above or below the mean. Fewer will fall outside of that, and even few will fall farther out.

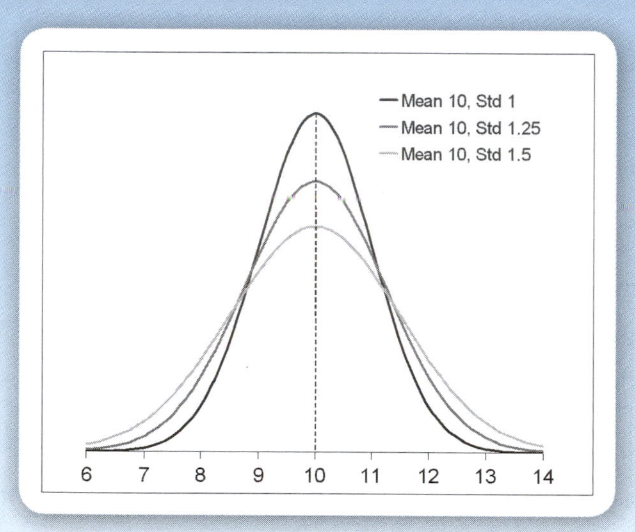

Figure 2.30 **Normal curves with the same mean, but different standard deviations.**

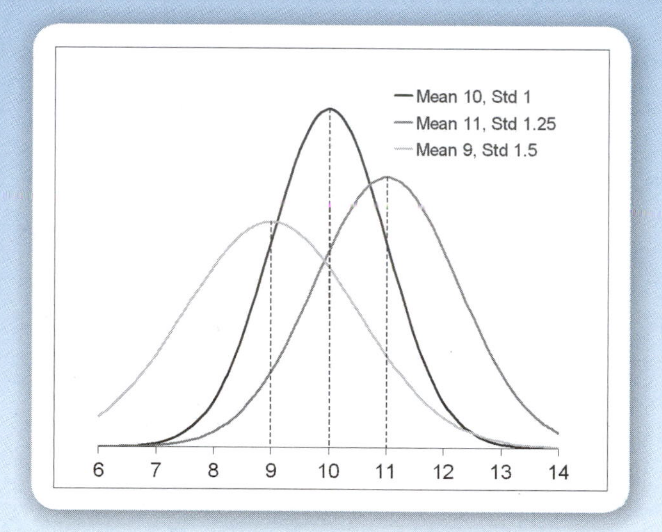

Figure 2.31 **Normal curves with unequal means and standard deviations.**

Population—Every single individual in a group to be studied. In practice, we cannot measure every individual. We usually use a sample or small portion of the population.

Sample—A fractional portion of a larger population in which we wish to make statistical measurements and predictions.

Range—The distance between the biggest and smallest measurement in a sample.

This idea of measurement or results being distributed in a predictable, normal pattern is the basis for SPC. We might know that a machine will make parts to a certain specification, but that process will have some slight variation that is normally distributed. We can use this information to make decisions about manufacturing processes and quality control.

Statistical Process Control (SPC)

Statistical process control is a very specific application of statistics to the endeavor of manufacturing. SPC attempts to use statistics to predict how a manufacturing operation is going to perform. SPC gives an engineer or technician an idea about the "normal" variation in a process so that he or she can make judgments about when a process is running within the regular system variability or whether it is "out of control."

For example, let's say that a marble factory is used to making marbles all day to a diameter of .750 inch. There is *always* some variation in the process, but 99.99% of the time the marbles are between .747 and .753 inch in diameter. One day, a new furnace controller in the melting operation is installed by the maintenance crew.

Suddenly, the marbles have a great deal of variability. The engineers cannot hold the size of the marbles to the historical range. Instead, only 60% of the marbles are in the .747 to .753 inch range. This process is *out of control* and must be stopped until a solution is found.

Using a Small Sample to Judge a Group

In our daily lives, we often use samples to make judgments about larger populations. For example, I've been bitten by only a small fraction of the dogs on the planet, yet I've been able to accurately predict that German Shepherds are, on average, much more likely to bite than Golden Retrievers. It turns out that if we have enough experiences (data points), then it is statistically sound to make such statements with a decent degree of accuracy. This concept is used in industry for all sorts of purposes related to quality.

At the marble factory we make 10,000 marbles per day—this is the entire population of marbles. It isn't practical to measure an entire population, so we usually select a sample set to measure. The sample lets us make some predictions about the population without actually measuring all 10,000 marbles. For example, once per hour we can randomly grab 15 marbles off the production line and measure their diameters (*see **Figure 2.32***). There is always some uncertainty about the conclusions we draw from the sample. The bottom line is that the larger our sample becomes the greater is the chance we have of accurately estimating the attributes of the entire population of marbles. Let's see how this might work.

The first step is to measure the diameters of all 15 marbles. Here are the results:

Marble #	Diameter (inch)
1	.7503
2	.7495
3	.7501
4	.7500
5	.7515
6	.7497
7	.7501
8	.7491
9	.7494
10	.7499
11	.7522
12	.7512
13	.7502
14	.7505
15	.7489

Next, we will calculate some descriptive statistics to tell us about the marbles we have collected. Let's start with the mean. But before we go there we need to understand that the mean for a sample is different from the mean of the population. We designate the *sample* mean as $\bar{X}$ (pronounced X-bar) and designate the *population* mean with the Greek letter μ (mu). In practice, we will always be using the sample mean ($\bar{X}$) in our calculations.

Calculate the sample mean (X-bar):

- Add up all the measurements of the diameters:
  - Σ (the sum) = 11.2526

Figure 2.32 A sample group is drawn from a large production run of marbles. There will be variations in the size—that is a guarantee. This variation is measured through statistical process control (SPC). The statistics will help to predict when the production process is creating a normal run of marbles or when it has gotten into an out-of-control state.

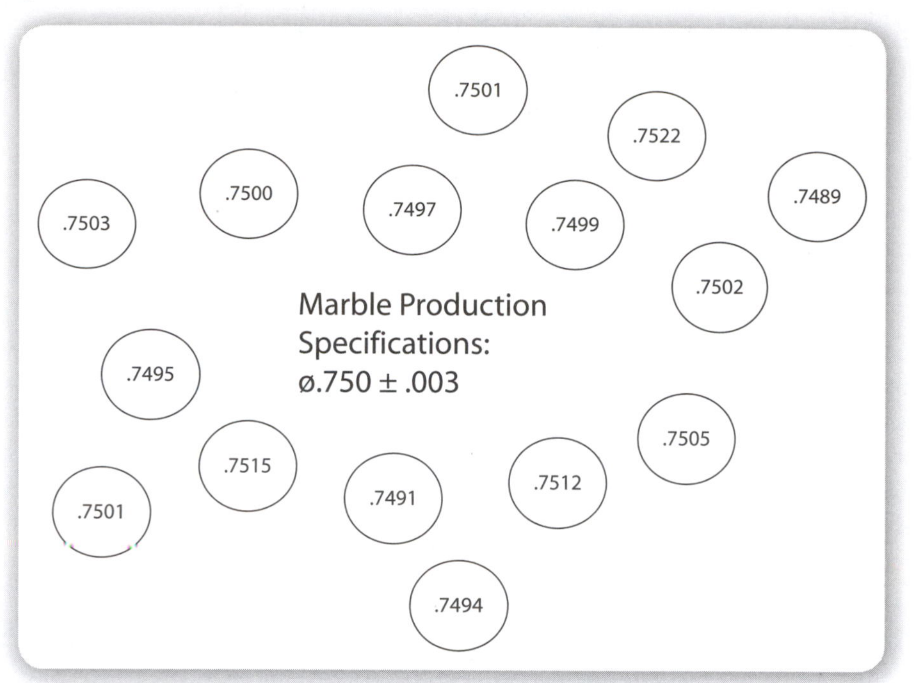

Political Poll

Have you ever read a political opinion poll in a newspaper or magazine?

If so, you might have noticed that every poll comes with a disclaimer in small print that the poll results might be in error by a certain amount.

"40% of voters believe the senator is a big fat liar."*

Sometimes the margin of error is large enough to make it difficult to predict the results with any certainty.

"Exit polls indicate that the Green Party is trailing the Social Democrats 48% to 52%."*

If we polled the next group of voters at the same election, we might find the Greens leading 53% to 47%. The margin of error is larger than the difference.

The margin of error is an acknowledgement that the sample results are an educated guess and are not exactly the same as for the population—this is a fact of life. We can reduce the uncertainty by polling more people. Larger polls take more time and money, which is why newspaper polls often have large margins of error. Very large sample sizes result in the smallest margins of error.

*margin of error: plus/minus 10%.
*margin of error: plus/minus 6%.

- Divide the sum by the number if marbles to find the mean: number (n) = 15
  - 11.252600 / 15 = .7502
  - $\overline{X}$ = .7502

Find the maximum, minimum, and range (R):

- The largest marble is .7522 and the smallest is .7489.
- Subtract the smallest from the largest to get the range.
  - .7522 − .7489 = .0033
  - R = .0033

Calculate the standard deviation. Note that there is also a difference between a *sample* standard deviation (STD) and a *population* standard deviation σ (the Greek lowercase Sigma). In practice, we will be working with the sample standard deviation (STD) in our calculations.

- This is a labor-intensive process to do by hand with more than a few numbers. You will probably want to use a spreadsheet or calculator. See Appendix B for a longhand example of the STD calculation as well as instructions on how to use the statistics functions on your calculator.
- STD = .00089

  Summarize the Results

$\bar{X}$	.7502
Max	.7522
Min	.7489
R	.0033
STD	.00089

Next we will display this information graphically in the form of a histogram, as shown in ***Figure 2.33***. A histogram takes the sample measurements and sorts them into groups (bins) based upon a size range. The information is displayed as a bar graph with frequency (number of marbles in this case) on the vertical axis. The horizontal axis is arranged by bin size. In other words, we sorted the marbles by their diameters. The maximum size span for each group was about .00089 inch, which, for convenience, was the same as the STD.

We have gathered a lot of information about the state of this manufacturing operation. How will we use these statistics? Well, we can look at the data and make several judgments about the process. In general, we are looking to answer the following questions:

Figure 2.33 **The histogram displays the number of marbles whose diameters fell within a maximum or minimum dimension. In this case, the bins were arranged by standard deviation. The greatest number of marbles was between 0.7493" and 0.7502", or 1 STD to the left of the mean.**

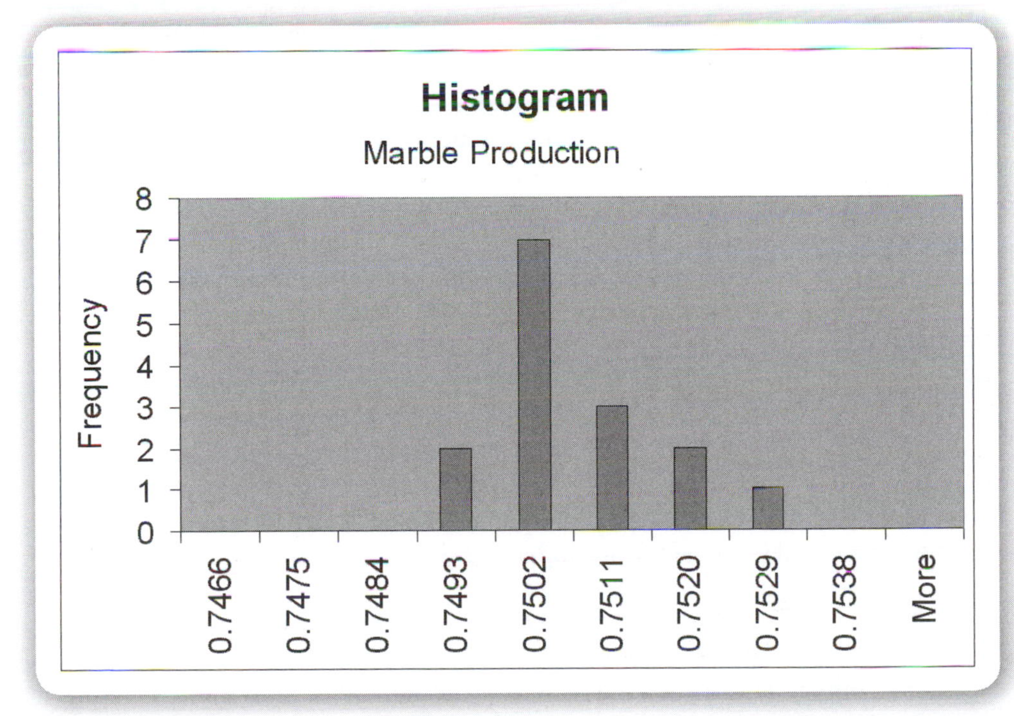

- How would we describe the normal process?
  - Over a long period of time we can use the data to determine the historical mean, range, and standard deviation of the process. This is what the process should look like, so we will know what it shouldn't look like.
- Is this process capable of meeting the quality goals?
  - Analyze the normal variation of the process to see if it is within the specified tolerance of the product. For example, the marbles were specified as Ø.750 ± .003 inch. Our process had an STD of .00089 inch; therefore, it was capable of making 99.7% of the marbles to Ø.750 ± .00267 inch or better. For marbles, this level of quality is probably acceptable.
  - The above explanation was simplified and assumed a stable, centered process. In reality, process capability is more sophisticated than described. There are other measures of capability, such as Cpk, that will give a quality engineer better insight in to a process. I suggest the book *Quality Control,* 7th edition, by Dale Besterfield, if you would like to read more about SPC.
- When is the process under control or out of control?
  - Once we know how the process should look, then we can monitor it via SPC methods to ensure that the quality is maintained. An operator on the production floor would follow a sampling plan and record measurement at a specified period. The typical setup is shown in ***Figure 2.34***. The data might be recorded graphically with control charts to track the mean and range and with a histogram to make it easy to read and to see trends. If the process reaches an out-of-control condition, then actions are taken to bring it back to acceptable levels.

2.9 THE EVOLUTION OF LEAN MANUFACTURING

The fundamental problem in manufacturing is how to make a high-quality product more efficiently than your competition. The manufacturing arena is fierce, and companies are fighting to stay afloat in the face of low-wage, global competition.

When a society takes raw material and processes it into something useful, then that society has created wealth. When we take stones and build a foundation, then we have created wealth. When we turn trees into walls and a roof, then we have created wealth. When we can build the second house with half the effort as the first, then we become richer by becoming more efficient. Ultimately, we must become more efficient every year or the competition will pass us by.

The manufacturing competition for America in recent history has come from Asia—specifically Japan. Starting in the late 1970s, the United States has been most notably engaged in competition within automotive manufacturing. American auto manufacturers were caught up in the old mindset of mass production. They were nearly buried when the Japanese producers started exporting a great variety of high-quality, fuel-efficient cars to the United States. The lessons learned on this front were painful but ultimately propelled us to be globally competitive.

Today the American producers have nearly caught up with the production methods and quality of their Asian competition. America seems to have accepted that dominance is no longer a forgone conclusion and that we will have to work very hard to stay in business. Still, the public had been left with a sense of shock that the rug could be pulled from under our feet at any moment. We still didn't seem to understand what had happened.

In the early 1990s, a group at Massachusetts Institute of Technology (MIT) set out to examine the global automotive industry for the first time and to present an objective comparison of the American, European, and Japanese manufacturers.

They did a bang-up job at dissecting the state of automotive manufacturing worldwide. As it turns out, it wasn't cheap steel, illegal trade practices, or government subsidies that led to the Japanese success; it was their methods, their way of thinking, and their innovation that led to success.

The MIT group described the American system as still being based in mass production, while the Japanese system was so completely different that they coined a new term to describe it: lean production. Since then, American manufacturers have scrambled to learn this system. Much of what the group described has since become ingrained in our manufacturing culture and vocabulary. An alphabet soup of acronyms and abbreviations has emerged to categorize every activity remotely related to the Japanese methods.

Recently, **lean manufacturing** has been the preferred expression to describe what began in Japan as the Toyota Production System. Implementation of this system has fundamentally changed manufacturing worldwide in the past decade.

This new idea of lean manufacturing in America has roots that are almost 60 years old. The thinking and practice that led to this new system came out of the company that later became Toyota. Kiichiro Toyoda and Taichii Ohno were largely responsible for the innovations. Toyota, out of post-war necessity, concentrated on ways of doing more with less. They found that by keeping low inventories, they could more easily identify weaknesses in their manufacturing system. This forced them to think about how things were done. They found that it didn't make sense to fix a defective car at the end of the manufacturing process. Instead, they concentrated on designing parts that were easy to assemble and that would not cause errors in the first place. An improperly assembled car is simply not allowed to continue down the assembly line. If there is a problem, then the line is stopped, and a team comes to fix it as quickly as possible.

Figure 2.34 A typical shop-floor SPC setup. The electronics caliper is equipped with a data port that connects to a processing device. The processor can store data and calculate descriptive statistics.

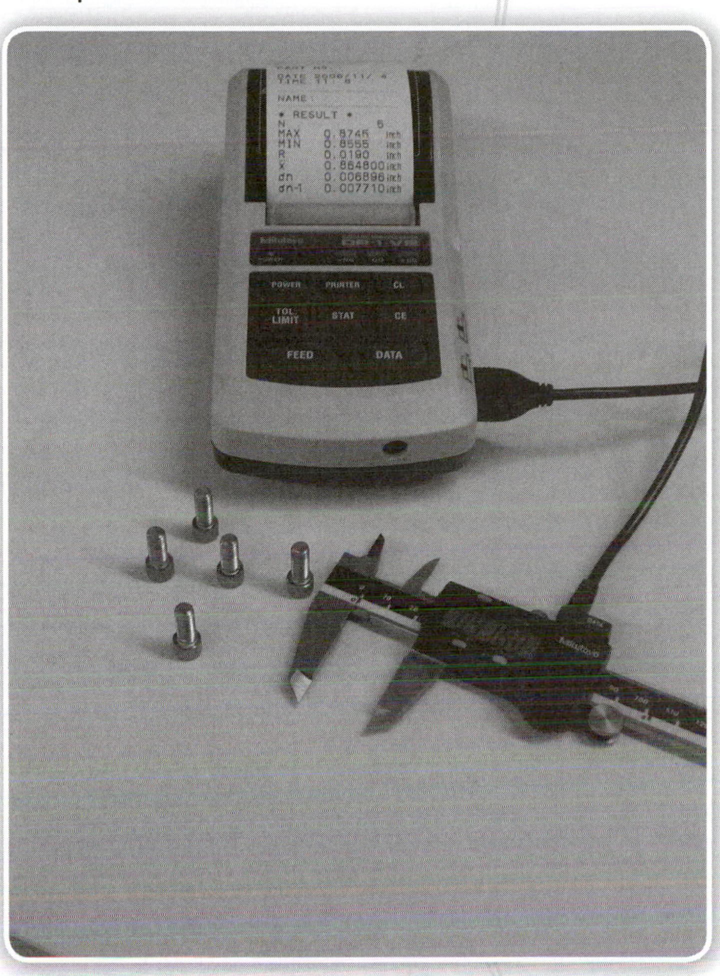

lean manufacturing: The practice, tools and philosophy of manufacturing efficiency first developed in Japan by Toyota Motors.

Ohno and his company also made great strides in flexibility. Their work on quick-change tooling made it possible to stamp out a small quantity of parts as effectively as Ford could make hundreds of thousands. This advance made it economically feasible for small producers to compete.

Low inventories, flexible systems, root cause problem analysis, team environment, and quick changeover are some of the lasting legacies of the Toyota Production System.

What Is Lean Manufacturing?

Lean manufacturing has a lot of small pieces that can make it difficult to identify the core elements. The following is a short description of what three different production systems might look like to an outside observer that should help you see what a lean environment is aiming to be.

CRAFT PRODUCTION Highly skilled craftsmen with flexible tools produce one-of-a-kind products. Workers are fulfilled by the nature and style of the work, and the products are expensive. Example: A custom furniture maker.

MASS PRODUCTION Unskilled workers produce massive quantities of identical products with dedicated tools. The system is robustly tuned to keep the production line moving regardless of problems. Inventories are kept high to provide the assembly line with materials and to provide a buffer for any schedule interruptions. Quality problems are handled through post-production re-work. Workers typically perform the same process repetitively several times per minute and experience low job satisfaction. Workers are easily replaced. Machinery is custom-built for the particular process. Product changeovers can take months. Example: Henry Ford's assembly line.

LEAN PRODUCTION Cross-trained workers produce large or small quantities to the highest quality upon demand on flexible but productive tools. Emphasis is placed on continuous improvement—mistakes are not made twice. Visual queues provide instant information, inventory is kept extremely low, and products are not built until they are ordered by the customer. Workers are expected to be fully engaged team members and may find that the lean environment is challenging. Rapid tooling changeovers make it possible to efficiently produce a large product mix, resulting in more innovation and consumer choice. Example: Toyota vehicle manufacturing.

So the main differences between the old system of mass production and lean productions are those of production line flexibility, low inventories, the culture of continuous improvement, and worker engagement.

When you walk into a modern manufacturing plant that has fully implemented lean manufacturing, you will see a striking contrast to the old. The lean environment is surprisingly empty. The inventories are extremely low. You don't see pallets of materials or unfinished or completed parts just sitting around taking up space. The machines tend to be flexible and programmable. They are arranged into cells where one worker can keep track of several processes. Automation is used sensibly—robots perform only the truly rare repetitive processes in which productivity can offset the cost. The floors and benches are free and clean of clutter. There is yellow tape on the floor outlining where the garbage can or cart is to be parked. Posters document the before and after states of every improvement project as if to say "here is what we learned" because they take pride in continuous improvement (***Figure 2.35***). There are boards to hang only the tools that are needed and no hooks for anything else. Parts for assembly are all visible in neat little plastic bins. All the shelves are slanted to shuttle boxes in on one side and out the other. Colorful laminated cards seem to be everywhere. The cards cue the workers on the "what," "when," and "how many" of the production schedule without any input from a supervisor. Companies take pride in displaying.

It seems that we put a lot of faith in complex automation to become competitive in the 1990s. In these factories, there always seems to be some nostalgic vestige of attempts to modernize before lean manufacturing was truly understood. On the back wall a giant automatic storage and retrieval system—designed to replace order pickers--towers 30 feet high and sits empty and idle. In the corner, an articulated robot waits to assemble 30,000 identical printers when only 300 have been ordered. Tracks in the concrete hide the guide wires that would lead an army of unmanned vehicles to carry inventory from place to place in the factory without human intervention. The lean philosophy compels us to believe that we don't even need to own the inventory much less move it around on expensive automatic guided vehicles.

Lean manufacturing also uses automation, but its core is about organization and methods. It is low-tech and visual and employs a lot of simple, common sense elements. For example, the Japanese idea of *poka-yoke,* or fool-proofing, is an important element of lean manufacturing. With poka-yoke, you design processes so that an operator cannot improperly create a part; the part only fits in the fixture one way!

Lean manufacturing is a subject that is perhaps learned through experience. The lean culture is here to stay, so you might as well embrace the philosophy, master its concepts, and enjoy a long and lucrative career.

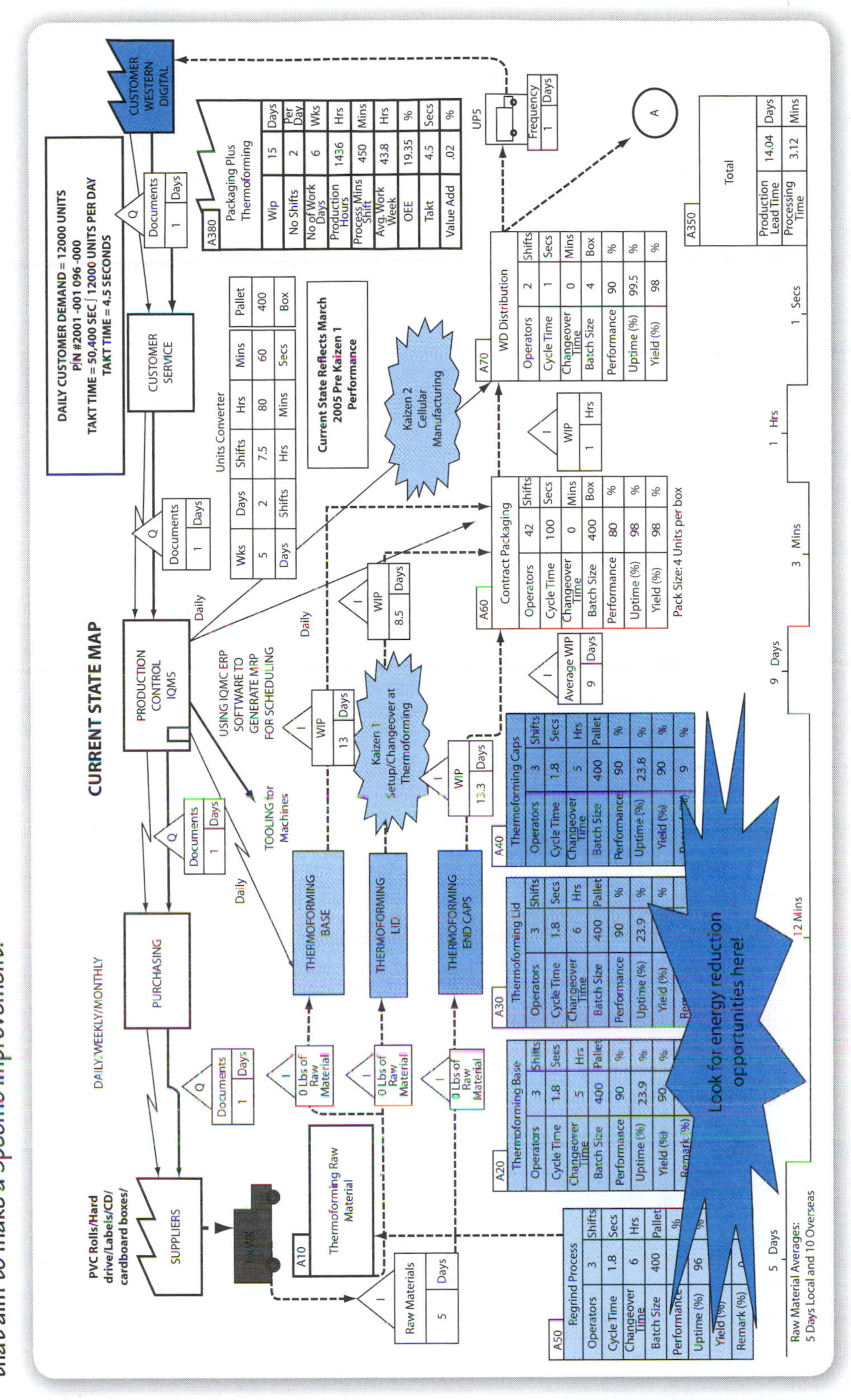

Figure 2.35 Kaizen events are an important activity of "lean" companies. The events are usually focused, short-term projects that aim to make a specific improvement.

Your Turn

1. Use precision measurement tools to reverse engineer a common classroom object and then create a set of engineering drawings using your classroom software.
2. In a group use descriptive statistics to compare and contrast similar classroom objects. Report the findings to the class.

CHAPTER SUMMARY

- Precision measurement is an important part of manufacturing. You must be comfortable using measuring tools and understanding systems of measurement to be successful in manufacturing.
- The calipers, micrometers, and dial indicators are perhaps the most common precision measuring instruments in the shop. It is important to select the most appropriate measuring tool for the job. The rule of ten tells us that our measuring instruments need to be of an order of magnitude that is greater than what is required for the feature.
- Engineering drawings are the primary tool by which technical information is shared in industry. Drawings should be studied carefully to completely understand the customer requirements. The drawings often contain details about the required tolerances, materials, and manufacturing processes.
- Engineering drawings use orthographic projections to define the form of three-dimensional elements. Dimensions, both linear and angular, describe the size of the part, and tolerances define the allowable variation from the dimension.
- Statistics are numbers that describe some attributed of a person, event, or product. In manufacturing, there is always variation, and statistics are used to describe it. Mean, range, and standard deviation are important statistics for understanding the dimensions of a group of measurements.
- Statistical process control is used in manufacturing to predict and measure the outcome of manufacturing operations. SPC uses a small portion of a production run to make judgments about the entire process.
- Lean manufacturing is a name given to describe the manufacturing systems developed by the Japanese in the past several decades. Lean manufacturing differs from mass production by being flexible, keeping low inventories, embracing improvement, and putting much more responsibility for quality on the average worker.

BRING IT HOME

1. Why is precision measurement so important in manufacturing?
2. Name three common precision measuring tools.
3. How are accuracy and precision related? Where does resolution fit into this relationship?
4. Convert the following measurements to millimeters: 1.000", 2.500", and .375".
5. Convert the following measurements to inches: 25.4 mm, .05 mm, and .10 mm.
6. How might the dimension 3.5" differ from 3.500" from the manufacturing perspective?
7. In accordance with the rule of ten, name a measuring instrument that could be used to measure a bar with the specified diameter of 1.125" ± .001".

8. What are the measurements at B and D in *Figure 2.7*?
9. Which measurement in *Figure 2.19* is the best? Worst? Explain why.
10. What are five pieces of information that you can typically find in the title block of an engineering drawing?
11. Where would you look for information concerning manufacturing processes of special instruction on a drawing?
12. Name three statistical measurements that would give you useful information about a group of manufactured parts.
13. What process is used on the manufacturing floor to gather and report manufacturing statistics?
14. What is it called when we measure only a portion of the products and use that information to predict the quality of the entire group?

EXTRA MILE

1. Calculate the mean, range, maximum, minimum, and standard deviations for the following sample set: 5.5, 5.3, 5.2, 5.2, 5.1, 5.4, 5.3, 5.2, 5.0, 5.4
2. How is lean manufacturing different from mass production?
3. Sketch three orthogonal views of the drawing in *Figure 2.36*, including the front, top, and right sides.

Figure 2.36 **Sketch front, top, and right side orthographic views.**

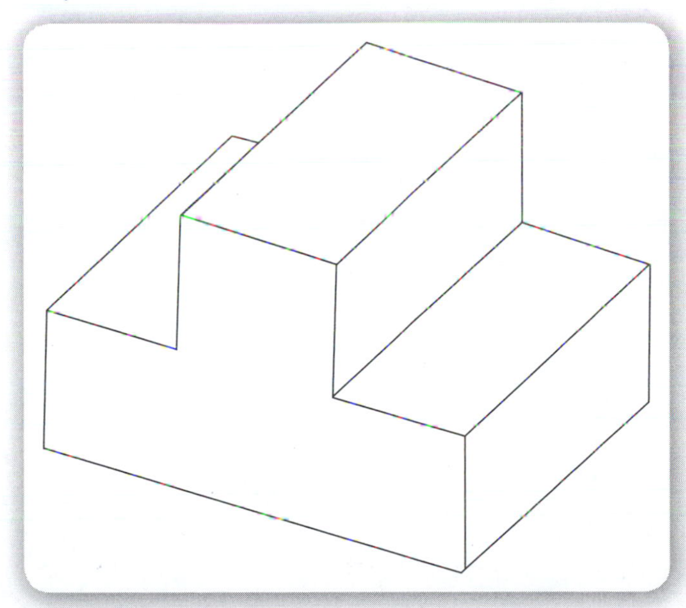

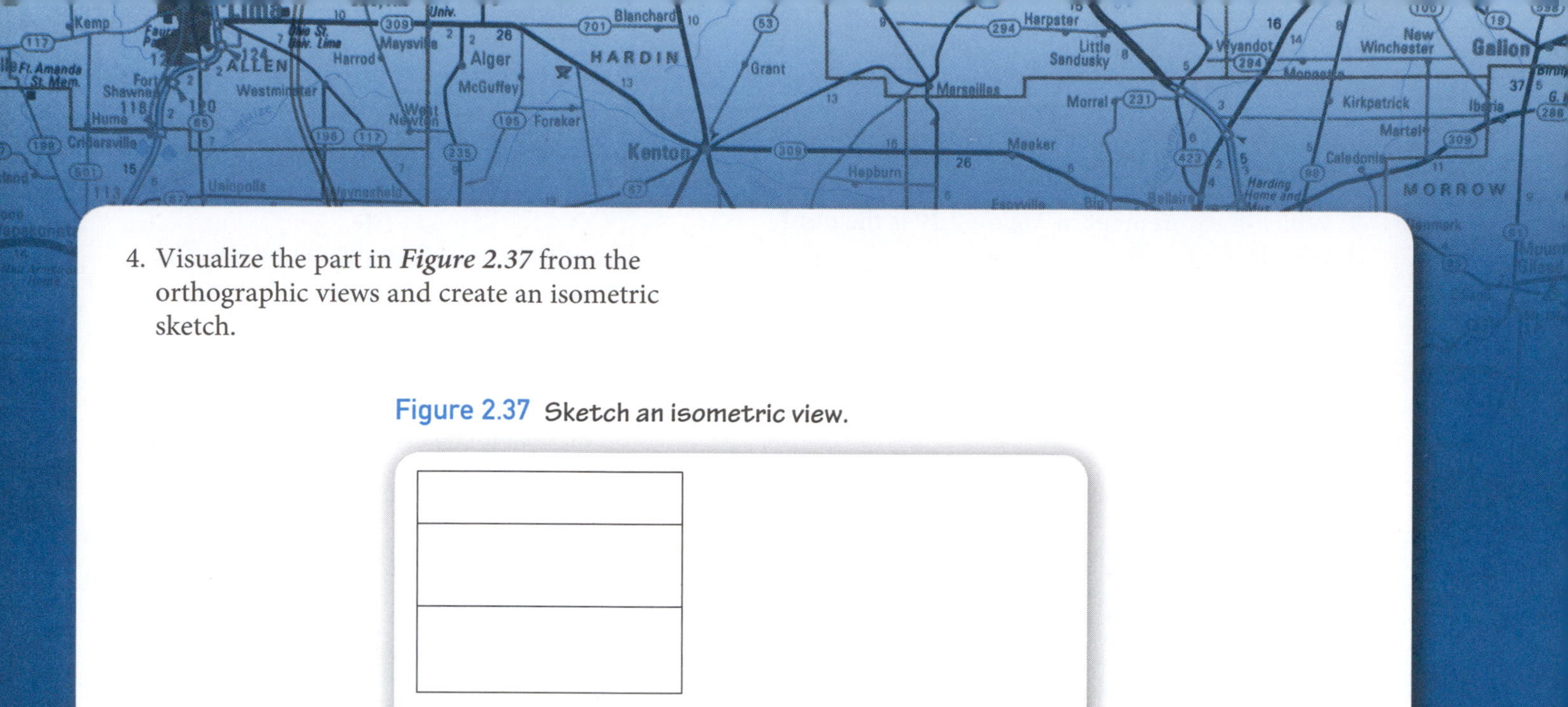

4. Visualize the part in ***Figure 2.37*** from the orthographic views and create an isometric sketch.

Figure 2.37 Sketch an isometric view.

Kemp
Ohio St. Univ. Lima
Maysville
Harrod
Alger
McGuffey
HARDIN
Grant
Harpster
Little Sandusky
Wyandot
New Winchester
Galion
ALLEN
Westminster
Fort Shawnee
Hume
Cridersville
Foraker
Kenton
Marseilles
Morral
Kirkpatrick
Iberia
Martel
Caledonia
Hepburn
Meeker
Uniopolis
Waynesfield
Espyville
Big Island
Harding Home & Mus.
MORROW
Denmark
Neil Armstrong Air & Space Mus.
AUGLAIZE
St. Johns
La Rue
New Bloomington

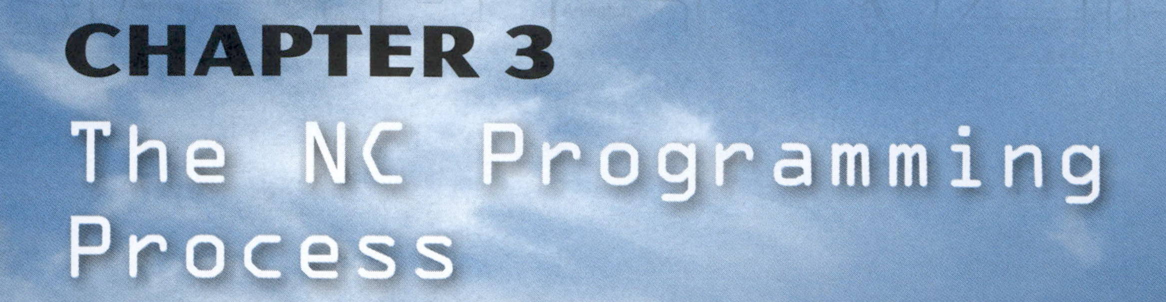

CHAPTER 3 The NC Programming Process

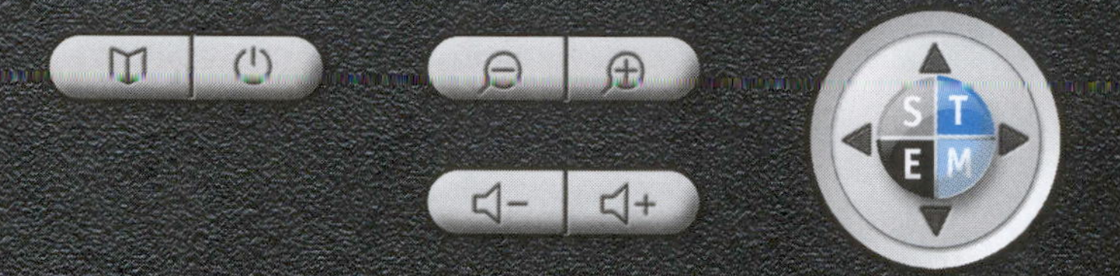

Menu

START LOCATION | DISTANCE | END LOCATION

Before You Begin

Think about these questions as you study the concepts in this chapter:

1. How does the computer turn the ideas from a blueprint to an NC program?

2. Why is offline programming preferred over shop floor programming?

3. What are the major methods for transferring files from the computer to a CNC machine tool?

4. What are some different file formats and encoding schemes for data?

5. How do you safely test part programs offline and on the machine tool?

Key Terms

- ASCII Encoding
- Backplot
- Communication Software
- Distributed Numerical Control (DNC)
- EIA/ISO Encoding
- End of Block (EOB)
- File
- Local Area Network (LAN)
- Offline Programming
- Online Programming
- Serial Port
- Single Block
- Solids Verification
- Text Editor
- Text File

3.1 WRITING AN NC PART PROGRAM

In this chapter, we will jump right into NC programming by writing a short NC part program. Then we will see how to test the program, how to transfer it to a CNC machine tool for proofing, and finally how to run the program in automatic mode to create the workpiece.

The first step in creating an NC program is to plan all of the different points that the tool will have to pass through to create the desired shape. This is called a toolpath. We will start by examining the workpiece in ***Figure 3.1*** and then writing the NC code to produce the toolpath for the stair-step slot.

The NC code for an entire part program can be somewhat complicated and may confuse a beginner. To remedy this, we are going to look at only the few blocks (lines of code) that are actually responsible for moving the tool to produce the toolpath. Specifically, we will study and modify blocks N50 through N130 in Listing 3-1.

Figure 3.1 **A simple toolpath to move the cutting tool to each point to produce a series of steps (left). We start the programming process by planning the toolpath and finding the coordinates of each point (right).**

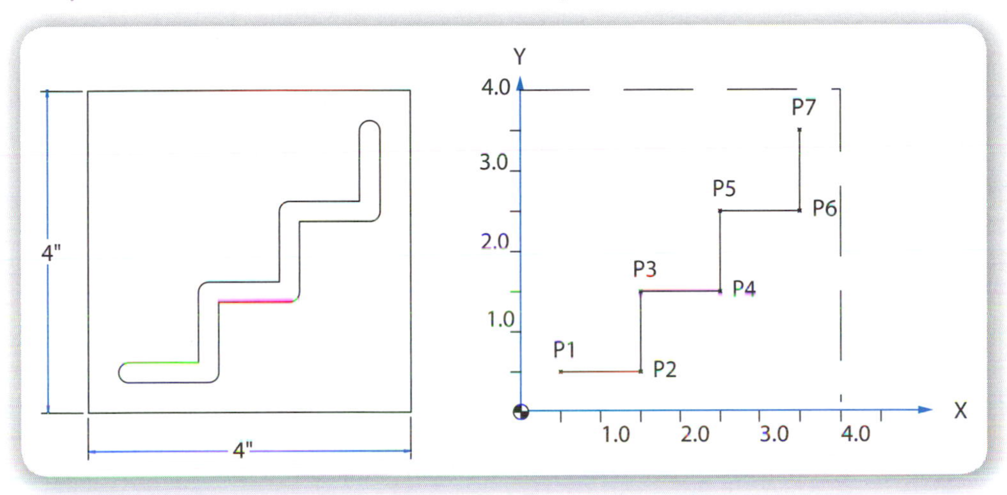

Listing 3-1 Code	Explanation
%	
O0201 (STEPS)	
N10 G20 G40 G49 G54 G80 G90 G98	
N20 M06 T05 (.50 END MILL)	
N30 G43 H05	
N40 M03 S1200	
N50 G00 X_._ Y_._	Position to point 1
N60 G00 Z.2	
N70 G01 Z-.25 F5.0(PLUNGE CUT)	
N80 G01 X_._ Y_._	Move to point 2
N90 G01 X_._ Y_._	Move to point 3
N100 G01 X_._ Y_._	Move to point 4
N110 G01 X_._ Y_._	Move to point 5

ASCII encoding: American Standard Code for Information Interchange. A standard encoding scheme for simple text files with nearly universal portability.

communication software: Application programs that are designed to control the flow of information though, primarily, a serial port.

Distributed Numerical Control (DNC): A distribution architecture that allows numerous machine tools to communicate with a centralized computer. Not to be confused with Direct Numerical Control, which attempts to run many computers from a central controller.

EIA/ISO encoding: An encoding scheme that is similar to ASCII and is most commonly found on CNC controls.

End of Block (EOB): The termination of one complete line of code in a G & M code NC part program.

file: The smallest unit of storage for related information on a computer system. Each individual NC part program is usually stored as an individual file.

Local Area Network (LAN):
A method of connecting local computers so that they may share information with each other. The network is closed and private.

Listing 3-1 *continued*

```
N120 G01 X_._ Y_._              Move to point 6
N130 G01 X_._ Y_._              Move to point 7
N140 G01 Z.2
N150 G91 G28 X0.0 Y0.0 Z2.0
N160 M05
N170 M30
%
```

To machine the slot, the first step is to position the end mill to point P1 and plunge the tool to the proper depth. Next, we need to move the end mill to each of the next six points to produce the slot. Finally, we pull the tool out of the slot at P7 and move the tool to an out-of-the-way position.

The Origin

When we work with a two-dimensional coordinate system, there must always be a point where the horizontal axis crosses the vertical axis. We call it the *origin*. The coordinates of the origin are X0.000 and Y0.000.

We use a symbol that resembles crosshairs to indicate the origin (*see Figure 3.2*)

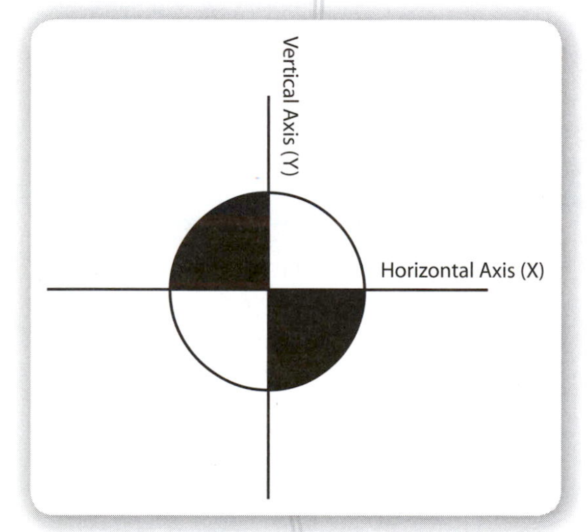

Figure 3.2 *Crosshairs are used to indicate an origin. The coordinates of the origin are X0.000 and Y0.000.*

We usually begin this process by finding the location of each point on an X-Y grid and by making a table of these values, as in Table 3.1. If you are unfamiliar with the Cartesian coordinate system, read the explanation in Chapter 5. For now, we will have to have faith that the coordinates below are correct.

NC part programs are constructed from codes that instruct the machine how to behave and where to move. In line N50, we tell the machine to move quickly to the specified X and Y positions.

```
N50 G00 X_._ Y_._
```

We can then fill in the values for X and Y to give the completed code. The next two blocks tell the machine to move down to a specific height and then to plunge down into the workpiece.

```
N50 G00 X0.5 Y0.5
N60 G00 Z.2
N70 G01 Z-.25 F5.0(PLUNGE CUT)
```

Table 3.1 **Cartesian Coordinates of Points 1 to 7**

Point Number	X Position	Y Position
P1	.5	.5
P2	1.5	.5
P3	1.5	1.5
P4	2.5	1.5
P5	2.5	2.5
P6	3.5	2.5
P7	3.5	3.5

Now that the tool has been plunged into the workpiece to the proper depth, we are ready to move to the remaining points to machine the slot. The code below shows the block of code without the X- and Y- coordinates in the left column and the completed code on the right.

N80 G01 X_._ Y_._	N80 G01 X1.5 Y.5
N90 G01 X_._ Y_._	N90 G01 X1.5 Y1.5
N100 G01 X_._ Y_._	N100 G01 X2.5 Y1.5
N110 G01 X_._ Y_._	N110 G01 X2.5 Y2.5
N120 G01 X_._ Y_._	N120 G01 X3.5 Y2.5
N130 G01 X_._ Y_._	N130 G01 X3.5 Y3.5

Finally, we can pull the tool out of the slot and move it out of the way to a safe location. This is shown in the completed blocks below.

```
N140 G01 Z.2
N150 G91 G28 X0.0 Y0.0 Z2.0
```

The code can then be assembled into a complete and working NC part program as shown in Listing 3-2. We should note that only the text in the left column is part of the part program. The text under the column titled "Explanation" would not actually be placed in the code. It is there only to illustrate the function of the program.

Listing 3-2 Code	Explanation
%	
O0201 (STEPS)	
N10 G20 G40 G49 G54 G80 G90 G98	
N20 M06 T05 (.50 END MILL)	
N30 G43 H05	
N40 M03 S1200	
N50 G00 X0.5 Y0.5	Position to point 1
N60 G00 Z.2	
N70 G01 Z-.25 F5.0(PLUNGE CUT)	
N80 G01 X1.5 Y.5	Move to point 2
N90 G01 X1.5 Y1.5	Move to point 3
N100 G01 X2.5 Y1.5	Move to point 4
N110 G01 X2.5 Y2.5	Move to point 5
N120 G01 X3.5 Y2.5	Move to point 6
N130 G01 X3.5 Y3.5	Move to point 7
N140 G01 Z.2	
N150 G91 G28 X0.0 Y0.0 Z2.0	
N160 M05	
N170 M30	
%	

serial port:
The standard RS-232 communications port that is found on most desktop computers and CNC machine tools. The RS-232 can send and receive data over a distance of 100′, which is much farther than a parallel port. The port may be accessed through a 9-pin or 25-pin sub-D connector.

single block:
A mode of operation that causes the CNC control to stop at the end of each block of code.

solids verification:
Computer-generated three-dimensional representations of CNC toolpaths and the resulting workpiece. Used to aid in visualization of the machining process.

text editor:
Application programs that are capable of saving simple text files.

text file:
A computer file that contains only raw text and line spacing that is usually encoded in the ASCII format. NC part programs are often created as text files prior to being moved to a CNC control.

Offline Programming on a Desktop PC

We can see how the NC part program is constructed, so how do we actually produce the program? NC part programs must be typed into a computer at some point before we can use them. There are two common methods used to produce NC part programs:

- The code is typed in at the control of the machine tool. This method is referred to as online or shop floor **programming**.
- The code is typed on a PC, saved as a text file, and then moved to the machine tool. We call this style **offline programming**.

online programming: The act of entering code on the console of the CNC control.

offline programming: NC programming that is completed on a computer system, which is separate from the CNC control.

Generally speaking, it is not very efficient to stand on the shop floor and key a program into the machine tool. The keyboards and control are difficult to use compared to a modern, desktop personal computer (PC). Furthermore, a typical control lacks the advanced editing features of a text editor or word processor. A more efficient method of creating NC part programs is to type them offline and then download the code to the machine tool.

NC part programs can be typed on any standard text editor or word processor that can save the file in the proper format. For example, Notepad (***see Figure 3.3***) is a text editor that ships with the Windows operating system. Notepad is a simple program that is used to create files that are encoded in the standard text format, which can then be downloaded to almost any CNC machine tool. Notepad can be accessed using the "Start Menu" under "Accessories."

File Formats and Encoding

We don't have to have a sophisticated word processing program to create NC code. Most word processors save their files in a special format that contains information about margins, font size, typeface, and other formatting specifications. NC code does not contain any such formatting. NC code needs only to be the raw text characters that can be produced according to the American Standard Code for Information Interchange (ASCII) text file standards. ASCII is a simple, open standard (or "language") that any word processor or editor can read and write.

If we decide to write a program on a word processor, we can still save the code as an ASCII text file. However, we must be certain to save the file in the correct format by selecting "Text" or ".TXT" when we save the file. If we fail to save the file to the proper format, then the file will be rendered useless if it is transferred to a machine tool.

Machine tools operate in a format that is defined by EIA/ISO standards for G & M codes. The codes are raw text that is similar to ASCII; however, there are a few minor differences in the encoding. For example, ASCII code uses the line feed character followed by the carriage return character to indicate the end of one line of code. The same operation would be encoded in EIA/ISO code with the use of the end-of-block character (;) to indicate the end of the line of code. Fortunately, the differences are usually reconciled automatically during the file transfer.

CNC Programming and Simulation Software

You can download AutoEditNC software for free from the Internet. The software is your quickest path to writing G & M code. The free package includes a text editor with code creation tools, programming templates, a toolpath simulation module, trigonometry and feed calculators, and much more.

To get the software just go to the CNC Programming: Principles and Applications website and follow the links to AutoEditNC. Follow the download and installation instructions and you will be ready to start programming in minutes.

CNC Programming: Principles and Applications Web site: *www.delmar.com/mattson*

Figure 3.3 NC part programs can be written offline on a desktop computer. Notepad (shown) is a simple text editor that ships with Windows and is adequate for editing NC code.

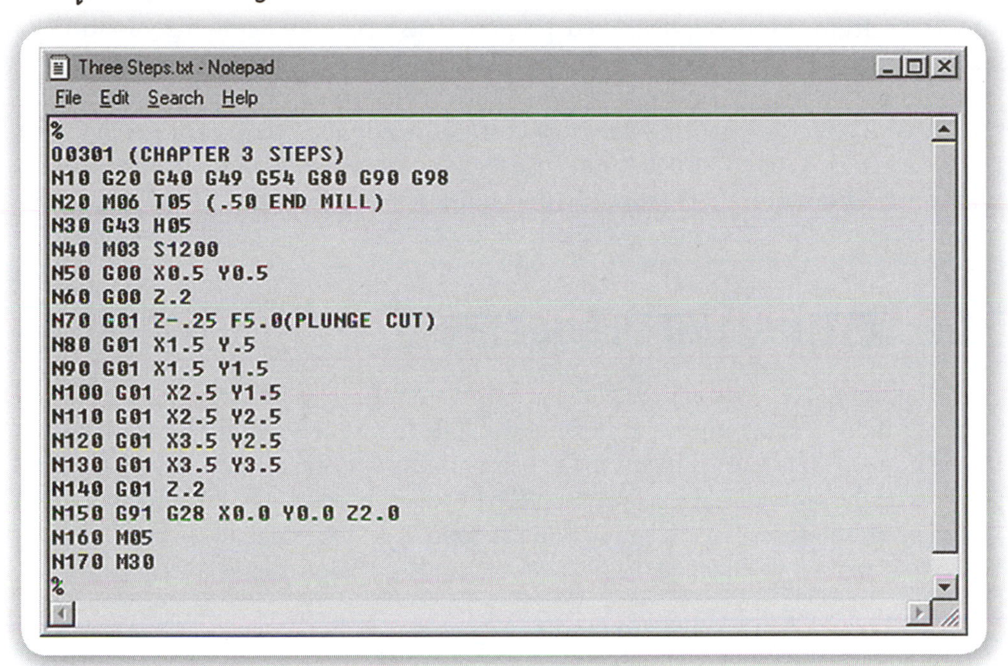

Figure 3.4 There are also a number of specialized editors designed specifically for NC programming. These editors usually contain tools that make it easier to create and edit NC programs.

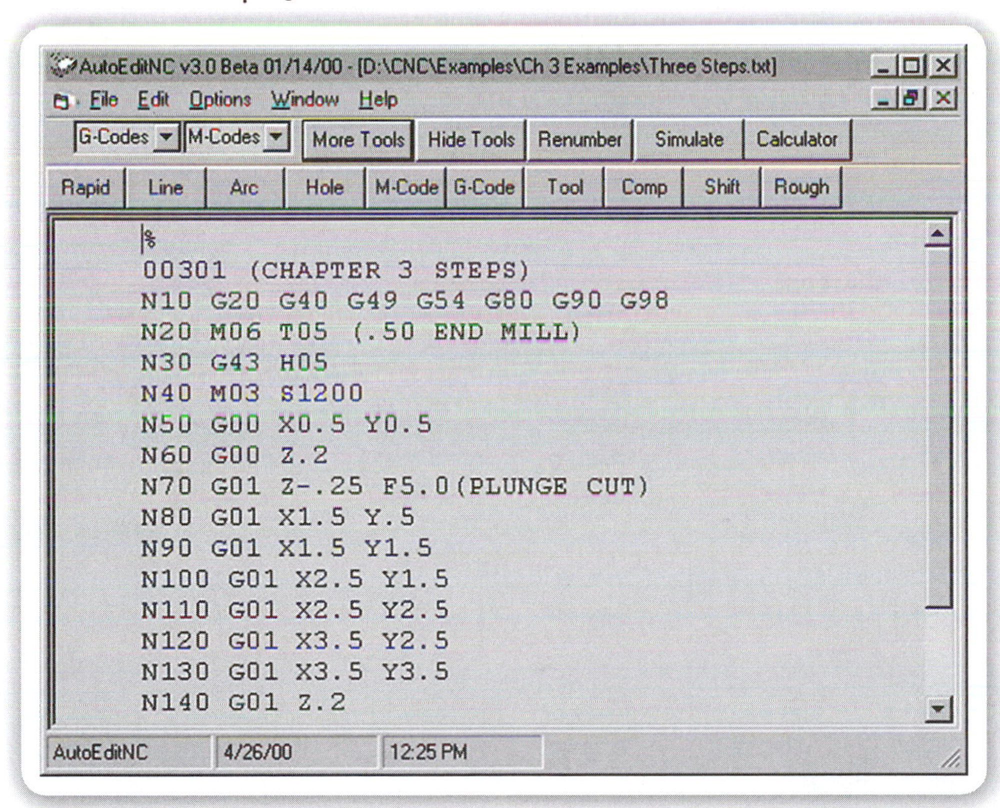

Specialized text editors are also available that are specifically designed for NC programming (*see Figure 3.4*). These programs often have features that are related to NC code, such as the ability to renumber the program or to add or remove spaces between each block of code. These features would not have much relevance to a standard text editor or word processor but are quite useful in NC programming.

Some specialized code editors also have built-in features that insert generic code automatically upon the click of a button. This can make it easier to learn to write code and reduce the number of mistakes we make by giving us an example to work from. More sophisticated programs also have integrated tools to help solve shop math problems or tools to test the program on the PC. Many of these programs are readily available on the World Wide Web. Simple editors can be licensed as freeware or for a very low cost. More sophisticated, commercial software can cost $300 to $1,000 for license fees, but they also tend to include powerful verification tools that can justify the additional cost. See the World Wide Web resources for some links to downloadable programs.

3.2 TESTING AND DEBUGGING

In addition to NC code editors, there are also a number of software programs available to the NC programmer for testing the code before sending the code on to the machine tool. It is more difficult and expensive to fix a mistake on the shop floor than it is to fix the same mistake before the NC part program leaves the PC. Therefore, the wise programmer will use a simulation program to test the code first.

Commercial verification software comes in two major classes:

1. Toolpath backplotting
2. Solids verification

backplot:

Simple graphical representations of a toolpath that is made up of lines and arcs and used to verify the accuracy of an NC part program.

Backplotting software will give a graphical display of the centerline toolpath for each block of code. The programmer can watch the **backplot** and decide if the code will be produced in the manner that was intended. For example, let's imagine that our code from the programming example contained an error in line N100. The line was supposed to read as follows:

```
N100 G01 X2.5 Y1.5
```

Figure 3.5 A quick verification shows that a mistake was made in the programmed toolpath. It is easier and less expensive to fix a programming error before the program is sent to the machine tool.

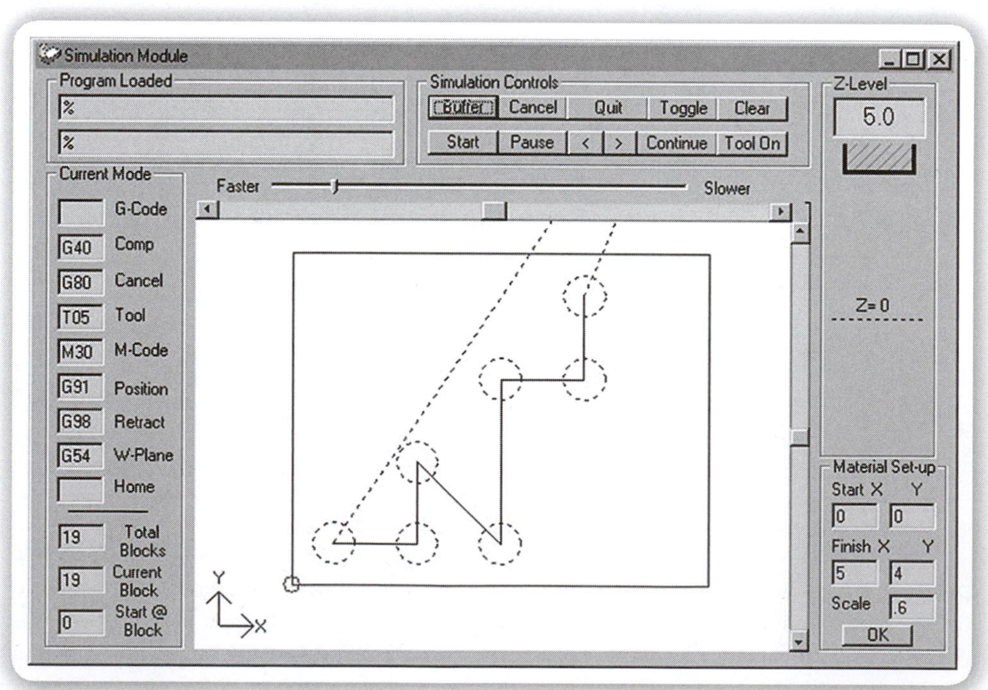

Figure 3.6 The corrected toolpath. The code is now ready to be downloaded to the machine tool.

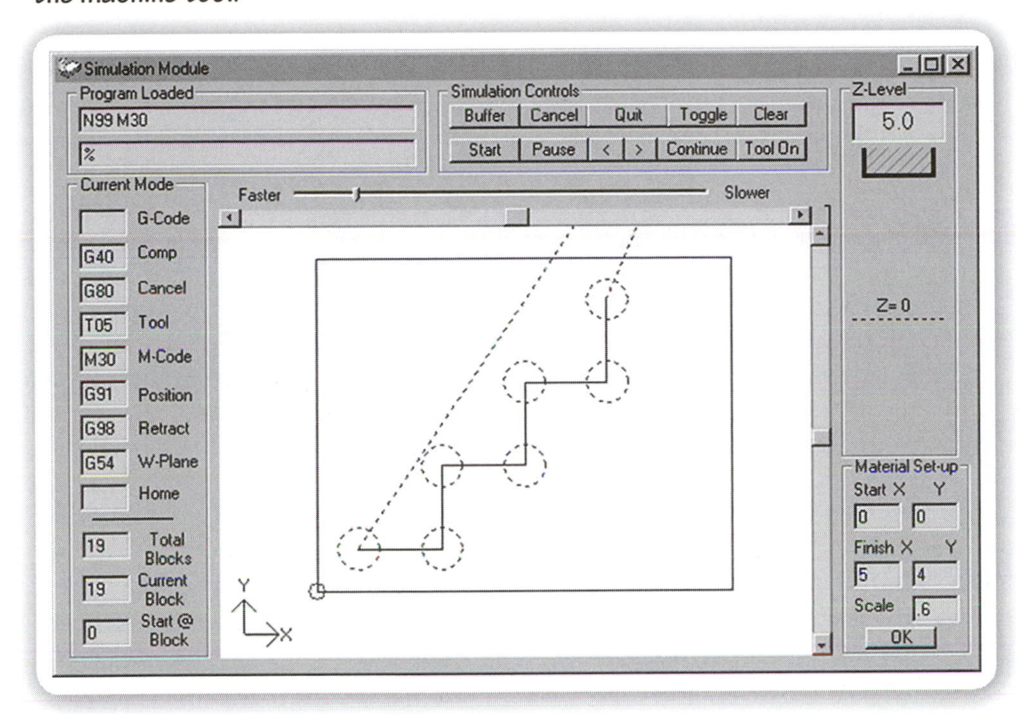

Instead, the programmer made a mistake and typed an incorrect value into the Y-coordinate:

```
N100 G01 X2.5 Y0.5
```

This kind of typographical error can be very difficult to catch when checking the code. Certainly the mistake would be found when the operator attempted to run the program and make the workpiece, but by that time, it would have become an expensive mistake. It would be much better to catch the mistake by cutting pixels on the PC rather than by cutting metal at the machine. A quick backplot, as shown in ***Figure 3.5***, can reveal a potentially expensive error. The error can then be easily changed and tested again to show the proper toolpath as in ***Figure 3.6***.

Figure 3.7 Solid model verification is available with some high-end programming software. Solids are much easier to visualize than back-plotted toolpaths.

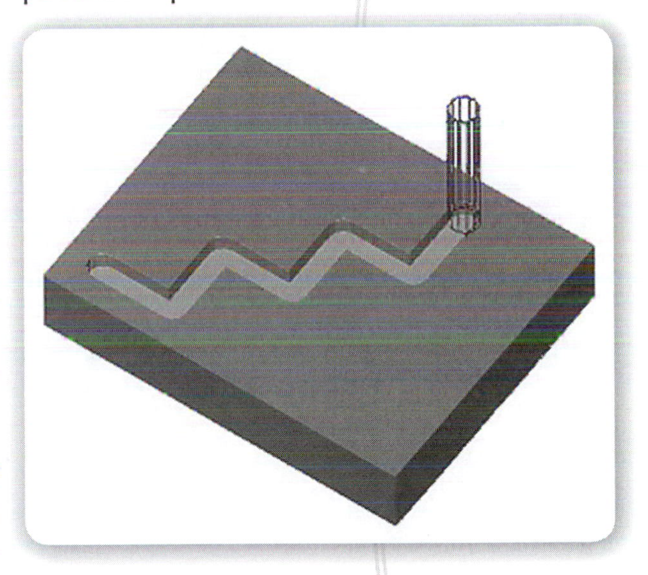

We mentioned earlier that the second type of testing software involves solids verification. This class of software represents the current state-of-the-art technology that presents the programmer with a fully rendered, three-dimensional representation of the programmed toolpath and resulting workpiece. The solid model view makes it much easier for the programmer to visualize and verify that the correct geometry has been machined. Take, for example, the solids verification for our stair-step slot in ***Figure 3.7***.

Solids modeling and verification have become prevalent in the last few years as a result of advances in computer technology and because of one popular *modeling kernel*. This kernel is software that acts like the engine to produce the model. The developers of this software have not kept the kernel under lock and key as one might expect. Instead, they license the kernel to other companies that are free to develop their own user interface. Licensees must pay a fee for every copy they sell, so the price for this class of verification software tends to be expensive (no free samples). However, the power of visualization makes it an obvious choice for many CNC machining enterprises.

3.3 TRANSFERRING THE PROGRAM TO THE MACHINE TOOL

Once we have typed and tested our CNC part program on a PC, we must then move the program to the CNC machine tool before we can machine our workpiece. This is often accomplished through a serial communications port. The serial communications port is built in-to most PCs and is sometimes referred to by its EIA standards number, RS-232. Most CNC machine tools are also equipped with a serial port to facilitate data transfer, as illustrated in ***Figure 3.8***.

The transfer of data over a serial cable is accomplished by running a communications program on the PC and setting up the control to interact with the communications software. Serial communications require that we have the machine and the computer configured identically on each end. Both the communications software and the machine tool control can usually be set up to interact in a variety of protocols. You may want to consult the operator's manual before attempting to change the configuration of the machine tool.

There are numerous serial communications programs available on the World Wide Web and from CNC suppliers. Again, they range from freeware to shareware to fully commercial software with advanced features that are worth the price. If we are just moving programs to and from the machine, then a basic program will work just fine (***see Figure 3.9***).

A more sophisticated software is needed if we want to perform advanced operations such as Distributed Numerical Control (DNC). DNC is an architecture in which all program distribution is controlled by a central computer. Another popular, advanced function is to *drip feed* a program that is too large for the control's memory. Drip feed is a technique used on CNC machines that do not have enough memory for extremely large, computer-generated NC part programs. The communications program *drips* the NC part program into the control one block at a time, and then the control discards each block once it has been executed. Of course, it takes a more sophisticated communications software to perform these operations.

Serial communication to CNC machines is probably going to become extinct as the controls become more advanced. Many controls now come with PC-based controls, large hard drives, floppy disks, and zip drives that lessen the need to send a program over a wire. Furthermore, the serial technology is slow compared to Ethernet communication. Most businesses now maintain a Local Area Network (LAN) of their computer systems. This network is a minimum of 50 times faster than the best serial communications. The natural evolution is to equip the CNC control with an Ethernet card (sometimes wireless) so that the computers can directly communicate with the network. This is, of course, already an option on many CNC machine tools.

Figure 3.8 ***Serial communications between a PC and CNC machine tool.***

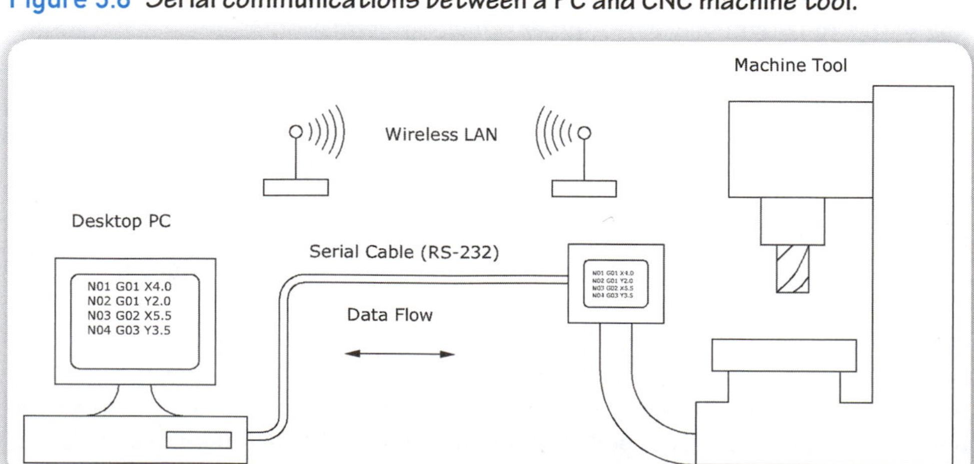

Figure 3.9 A serial communications program allows us to transfer NC part programs between the desktop PC and the machine tool through a serial port. Newer machines are sometimes equipped with a floppy disk drive or an Ethernet adapter card.

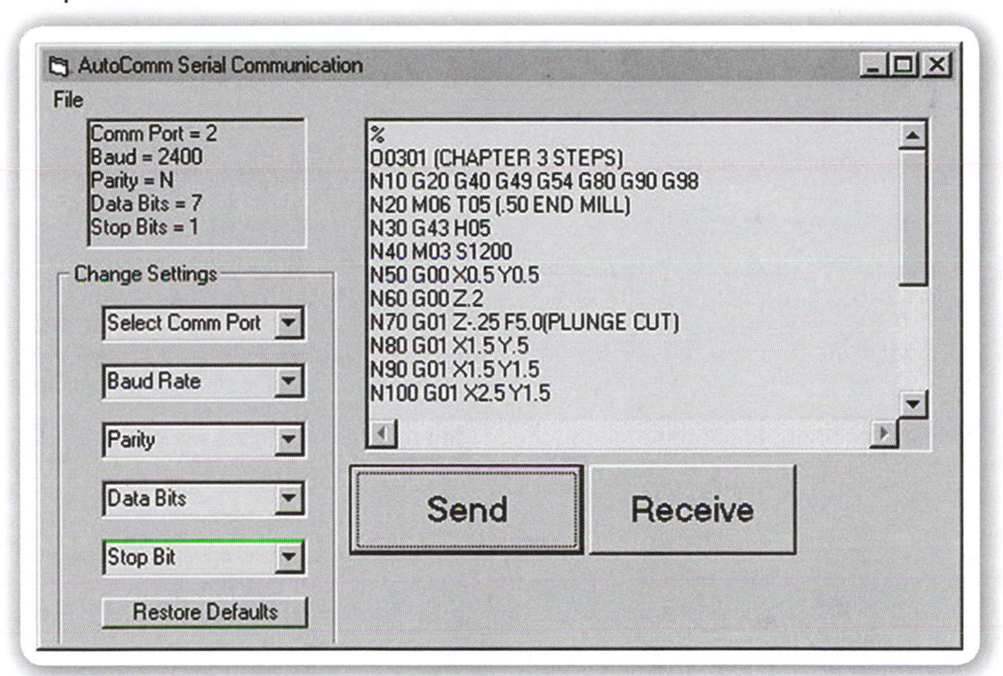

3.4 PROVING THE NC PROGRAM ON THE MACHINE TOOL

Now that we have written an NC part program and transferred it to the machine, we are ready to perform a live test of the program. It is critical that every program be safely tested before we allow it to run at full speed through a block of real material. Do not be lulled into a state of over-confidence in which you believe that your programming skills are so good that there is no chance of having made a mistake. Even if you have simulated the program on a PC, there can still be dangerous errors in your code. These errors might not be caught until the program is run on the actual machine tool.

Table 3.2 details a few examples of mistakes that any programmer or setup person can make. All have the potential to result in a broken cutter, a scrapped workpiece, machine damage, and personal injury.

The following is a listing of the steps that are generally followed when proving a NC part program for the first time. You may want to read these steps to become familiar with the process. We will discuss many of the specific modes and processes afterward.

Steps in Program Proving

1. Verify the code on a PC with backplotting or solid simulation software.
2. Complete a program testing checklist.
3. Load the program onto the machine tool, and run a graphical simulation if it is available.
4. Perform a dry run:
   - Enable the Single Block mode.
   - Turn down the feed and rapid traverse rates.
   - Set the workpiece Z-offset to a safe level above the workpiece.
   - Repeat the process with single block disabled.

Table 3.2 *Common Mistakes and Consequences in NC Part Programs*

Mistake	Result/Consequence
Spindle was not turned on before the cut.	Broken tool and ruined workpiece
Tool was not returned to a safe position above the part before moving to the next pocket.	Scrapped workpiece
Coolant was not turned on before the cut.	Tool was ruined
Fixture or vise interfered with toolpath.	Broken tool and damaged fixture
Toolpaths out of order.	Holes were tapped before they were drilled, resulting in a crash
Incorrect tool installed.	Part made to the wrong dimensions
Wrong tool offset.	The tool will be too high or too low, which can result in a crash or incorrect dimension
Improper speed or feed entered.	Burned or broken tool Workpiece pulled from fixture

5. Return the Z-level to normal and perform a test cut on a setup piece or prototyping material with reduced speeds and feeds.
6. Inspect the workpiece and adjust any offsets to cut the proper dimensions.
7. Run the first *real* part in automatic mode.

Testing Modes

A typical CNC control has a number of features designed to make program proving safe and easy. The first is the *dry run* mode. A dry run will generally run through the program at a reduced rate. Sometimes only the X- and Y-axes will move, while the tool will stay at a preset Z position and the spindle will be turned on. There can be vast differences between the behaviors of machine tools from different manufacturers, so consult the operator's manual for specific information.

A test run of the program might also be made in regular automatic mode. This will give the programmer or setup person the best idea of how the machine will actually behave. The feed and rapid traverse rates are, however, usually turned down.

Our test run will be live, so we want to be sure that the tool will not come into contact with any material. We can handle this by moving the Z-offset of the workpiece up by several inches or by the greatest Z-depth found in the program, as shown in ***Figure 3.10***. Moving the Z-offset will effectively make the machine think that the top of the workpiece is higher than it really is. This will allow us to observe the toolpath without contacting the workpiece. If the height is set too high, then it can be difficult to determine the boundaries of the toolpath.

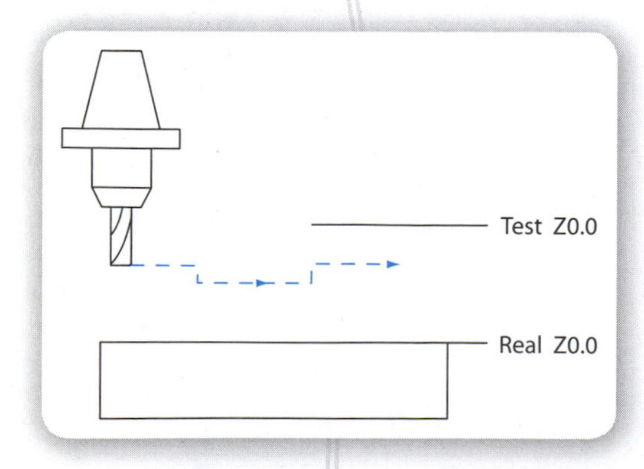

Figure 3.10 It is common practice during testing to set the Z-offset to a level above the top of the workpiece during a dry run. We can then observe the toolpath at a safe distance and determine if there are any major mistakes in the code.

We might also be careful about having any material in the fixture during the test run. The actual workpiece material should be replaced with a soft prototyping material just in case there is unintentional contact.

Single block is a mode of operation that causes the machine to stop at the end of every block and wait until the operator pushes the cycle start button again. Single block is very useful for determining if the program is behaving the way it was intended. The programmer can easily verify the start and finish positions of the block and decide if they make sense and if the actual position of the tool looks like it should. This method sounds simplistic in an era of high-tech simulation tools, but many mistakes are found this way, and few setup people would sign off on a setup without first running it in single block mode.

Running single block while cutting metal will result in excessive rubbing at the end of every block. It is not a good idea to let the tool spin against the workpiece without providing a feed, because this will cause the tool to heat up and become dull and may cause an uneven gouge in the material. Therefore, we usually try to use single block during a dry run while the tool is not actually in contact with a workpiece.

One of the primary tools that a setup person will use during single block is the "Distance to Go" screen (***Figure 3.11***). Most controls are equipped with this feature, which shows the distance that must be traveled in each axis to reach the end of the block. For example, the tool may be approaching the top of the workpiece, and we estimate that it has about *1 inch* before making contact.

Figure 3.11 The Distance to Go display is used to determine the actual travel of the next tool movement. This screen gives the setup person valuable information about the expected tool movements while testing a new program. In this example, the drill is about to move from point 5 to point 6—a distance of .375". Yet the Distance to Go screen is showing that the tool will move 1.375 and surely collide with the clamp!

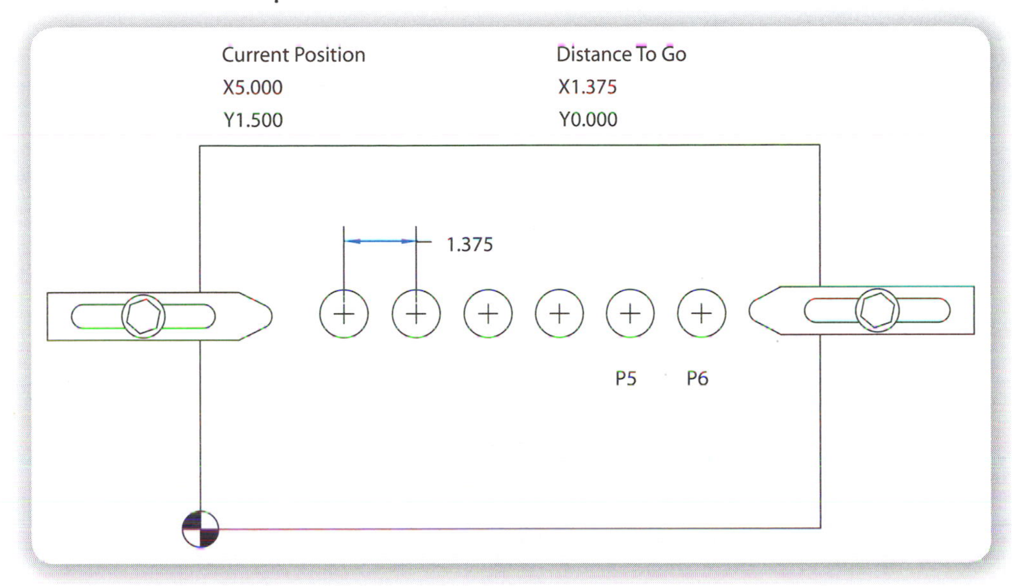

However, the Distance to Go screen indicates that there are *6 inches* left to travel in the Z direction before the end of the block. We can see that this will result in a crash, so we should stop the machine and check our numbers before continuing.

The First Cut

We are ready to cut material now that we have performed a dry run of the NC part program and single blocked through it. We still want to keep the feeds turned down to a lower level for the first cut, just in case of an unforeseen event. If possible, the first part should be cut out of a prototyping material such as machinable wax or high-density foam (see ***Figure 3.12***). These materials are readily available from industrial suppliers and are generally less expensive than a comparable volume of aluminum bar stock. They are also much more forgiving to the cutting tools if a mistake is made.

The first part can then be inspected to see if the proper dimensions have been machined. We may find that some adjustment to the diameter-offset or length-offset of the tools will be needed in order to cut the proper dimensions—this is primarily the job of the setup person. We might also have to adjust the speeds and feeds if any chatter is encountered or if the cut is too heavy to produce a consistent finish or dimension.

Figure 3.12 Machinable wax and various foam products are often used for program proving. These materials can reduce the overall cost of a program prove out as they can be cut at accelerated speeds, and they are easier on the cutting tools than metals.

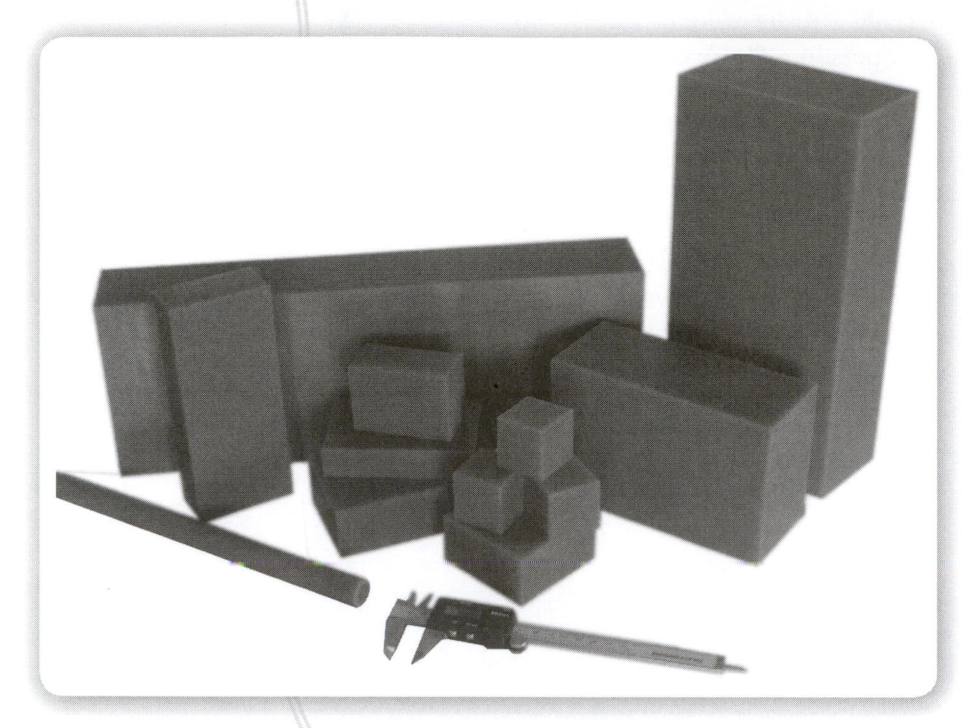

If any other mistakes are found, then the program may have to be edited and another prototype workpiece might have to be run. We have to use extreme caution anytime we make a change to a proven program. It is very easy to make a quick edit at the control and then scrap a part because you typed in the wrong number or misplaced a decimal point. Any time a program is changed, at least that section of the program should be tested before making another live run.

Program and Setup Checklist

- All rapid traverse moves are above the top of the workpiece.
- The tool numbers and offsets match.
- The tools in the machine match those listed on the setup sheet.
- All tools have been touched off, and the offset values are reasonable.
- Plunge cuts are performed with center-cutting tools.
- Speeds and feeds are reasonable.
- The spindle is rotating in the correct direction.
- Coolant is initiated prior to cutting.
- Roughing operations are performed before finishing.
- Tools are at a safe Z-level before moving to any new feature.
- Hole-making operations are performed in the proper order (i.e., spot, drill, bore or tap, etc.).
- Fixtures are bolted down tightly and squarely.
- Cutting tools are secured in their holders.

3.5 CNC SAFETY

Machining and CNC machining all have the inherent dangers associated with rapidly moving machinery, sharp edges, and hot flying chips. However, we can mitigate and minimize many of these dangers by giving some thought to our actions and following the safety rules. Many of the points that follow may seem like *common sense*, but remember that what we know as common sense is really the distillation of years of experience and the misfortunes of those who have come before us. Heed their warnings:

1. The use of CNC machine tools requires at least the same precautions as with conventional machine tools, including eye protection and standard setup procedures and operation.

2. Do not let others distract you from your work. Your concentration is critical when setting up and operating machine tools. Likewise, do not distract others while they need to concentrate.
3. Wear your safety glasses any time you are operating or are around any machine tools—even if the machine is fully enclosed. It is also a good idea to wear safety glasses when using hand tools and electric power tools.
4. CNC machine tools move automatically. They are extremely fast and powerful; therefore, special precautions should be taken when working around CNC machine tools.
5. You should never attempt to operate a CNC machine tool that you do not fully understand. Instead, you should ask your instructor or supervisor whenever you are not sure of any procedure or function. Consult the operator's manual before attempting to program or operate a CNC machine tool. Even if you are experienced with other machine tools, there may be significant and dangerous differences between makes and models.
6. You should never work inside a CNC machine tool when someone else is touching the controls. They might accidentally start the machine and cause serious injury to you. In fact, it is a good idea to put the machine in a locked mode such as "edit" before changing the tools or workpiece.
7. You should always calculate speeds and feeds for CNC machining because you will not be able to *feel* when the cutting conditions are correct.
8. Special care should be used when using rapid traverse on a CNC machine tool, including allowing an adequate distance above the workpiece and not using rapid traverse below the surface of the part.
9. Part programs should be tested by first using a computer simulation on the PC, simulating on the control, and then dry-running above the workpiece at a reduced rate and in single block mode.
10. Most programming errors can be caught by simply proofreading your code. Therefore, you should print a hard copy of your code and spend a few minutes checking for critical safety issues, including proper speeds and feeds, proper tool location when using rapid traverse, and the correct tool offsets.
11. You should never leave a CNC machine tool unattended. Tools can fail rapidly and cause damage or injury if not stopped immediately.
12. CNC machine tools should only be operated with the doors closed while in automatic operation. Do not disable or over-ride any of the safety features that were put there to protect you.
13. Never reach inside a machine tool while it is under power to force a tool changer or axis back into position. It is common to have sticky limit switches that cause a machine error and the machine to stop moving. This will sometimes cause one of the machine's moving elements to become locked into position until it can be manually moved to reset the limit switch. Never attempt this while the machine is powered on because it might start moving automatically and cause serious injury.

Your Turn

Plot coordinate points for the first initial of your name on graph paper. Record these coordinates in block form in any text program. Include other G & M codes as specified by your instructor. Transfer this document into your CNC software to verify.

CHAPTER SUMMARY

- NC part programs can be typed offline on a PC and later moved to a CNC machine tool. This is usually more efficient than standing on the shop floor and entering the program online at the machine tool.
- NC part programs can be produced on any application program that can save files in the standard ASCII text format. This includes simple text editors, word processors, and specialized NC code editors. There are some formatting differences between the ASCII and EIA/ISO formats, but they are usually transparent to the user.
- Many of the mistakes in a program can be found while the NC part program is still on the PC. Software is available that can either backplot a toolpath or show a three-dimensional solids model of the program operation. This can drastically reduce the number of errors that are sent to the machine.
- A number of techniques are employed to transfer data between the PC and the machine tool. Serial communications through an RS-232 port and cable is a common method. Other methods include floppy disks and network adapters.
- Program proving is an important part of the programming process. Many precautions should be taken to ensure that the program is first of all safe for the operator and machine tool, and, second, that it will correctly and efficiently produce the workpiece.
- CNC machine tools have all of the dangers associated with conventional machine tools plus those dangers associated with automatic movements. You should thoroughly understand the operation and behavior of the particular machine before operating or programming it.

BRING IT HOME

1. How is offline programming different from online or shop floor programming?
2. What type of program do we need in order to create an NC part program on a PC?
3. Can a word processor be used to create NC part programs? If so, is there anything different that we have to do when the program is saved?
4. Some programs have the ability to simulate the NC code. What are the advantages of simulating the code on a PC before we send it to the machine tool?
5. What are the two main classes of simulation or verification software? How are they different?
6. What are two common methods that we can use to move the program from the PC onto the machine tool?

EXTRA MILE

1. Explain how you might *prove out* an NC part program. What steps are required and what purpose do they serve?
2. Are there any precautions that should be taken if we decide to modify a working NC part program? Explain.
3. What kind of mistakes do we need to look for when preparing to set up and run a program for the first time?
4. Are there any special safety precautions that have to be taken when working with CNC machine tools?

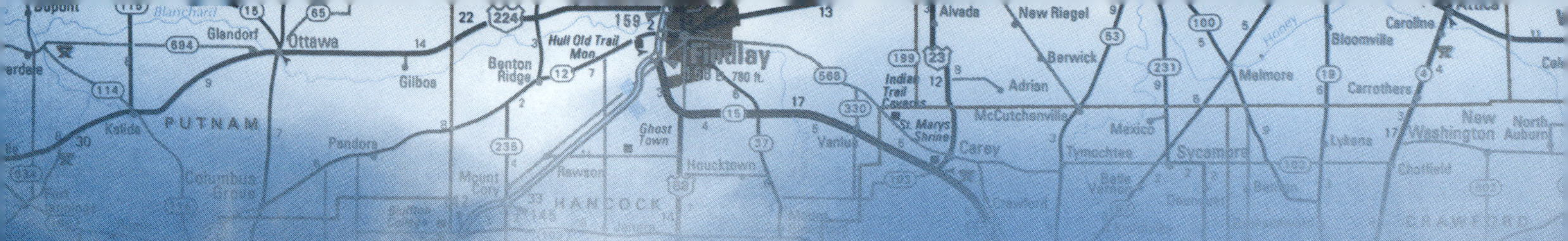

CHAPTER 4

NC Materials, Tooling and Machining Processes

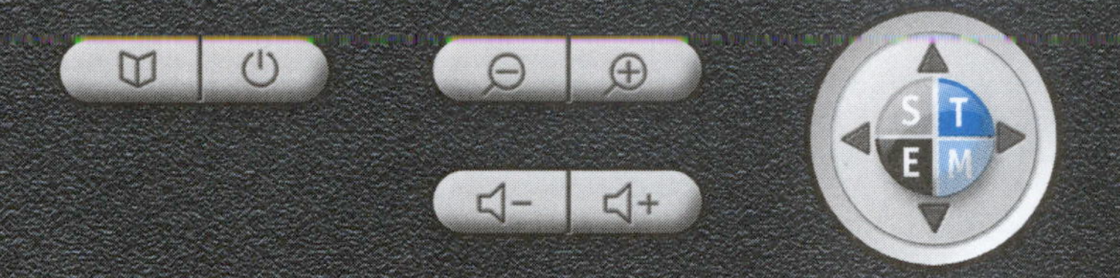

Menu

START LOCATION | DISTANCE | END LOCATION

Before You Begin

Think about these questions as you study the concepts in this chapter:

1. What are the basic machining processes?

2. What are some of the more common materials used in the machining industry today and how does their composition affect machining?
3. What types of cutting tools are used in CNC machining?
4. How do you identify what type of cutting tool to use?
5. What are some of the workholding tools used?
6. How do you select correct spindle speeds?

KEY TERMS

Fixture
Cutting Tool
Cemented Carbide
Indexable Insert
Modular Tooling
Dedicated Tooling
Spindle Tooling

The purpose of this chapter is to acquaint the reader with the basic manufacturing process and tooling that are common to CNC and conventional machining. Special consideration is given to the tools and techniques that are native to CNC machining. The information in this chapter can serve as an introduction to basic machining and should not overwhelm the novice reader. Readers with more machining experienced can also gain some insight into the differences between CNC machining and conventional machining.

4.1 CNC AND MANUFACTURING PROCESSES

CNC systems are used in various manufacturing processes, including machining, forming, and fabrication. However, machining processes are by far the most established and important use of CNC systems. Forming and fabrication processes encompass a great number of operations, including punching, shearing, bending, drawing, and cutting. CNC forming and fabricating equipment is becoming more common, but it remains outside the mainstream in discussion of CNC.

Machining is a material removal process by which a hardened cutting tool is used to remove *chips* from the workpiece (***see Figure 4.1***). The term machining is generic and refers only to this chip-making process and not to the specific operations or machine tools used to perform the process. Machining is a high-precision affair in which features are typically created to tolerances of less than 0.001". These precision standards make CNC so important to machining.

There are many different machining operations, but they all undergo the same cutting process that creates chips. Chips are formed by pushing a hardened tool into the softer workpiece until the material deforms. Soon the workpiece material can no longer resist the force and it shears. This cycle repeats itself hundreds of times every second and results in the formation of chips. If you look at a chip under a magnifying glass you will see the individual shear plates that have stacked up along the chip. This looks something like a deck of cards that has been pushed sideways.

Figure 4.1 *The chip making process.*

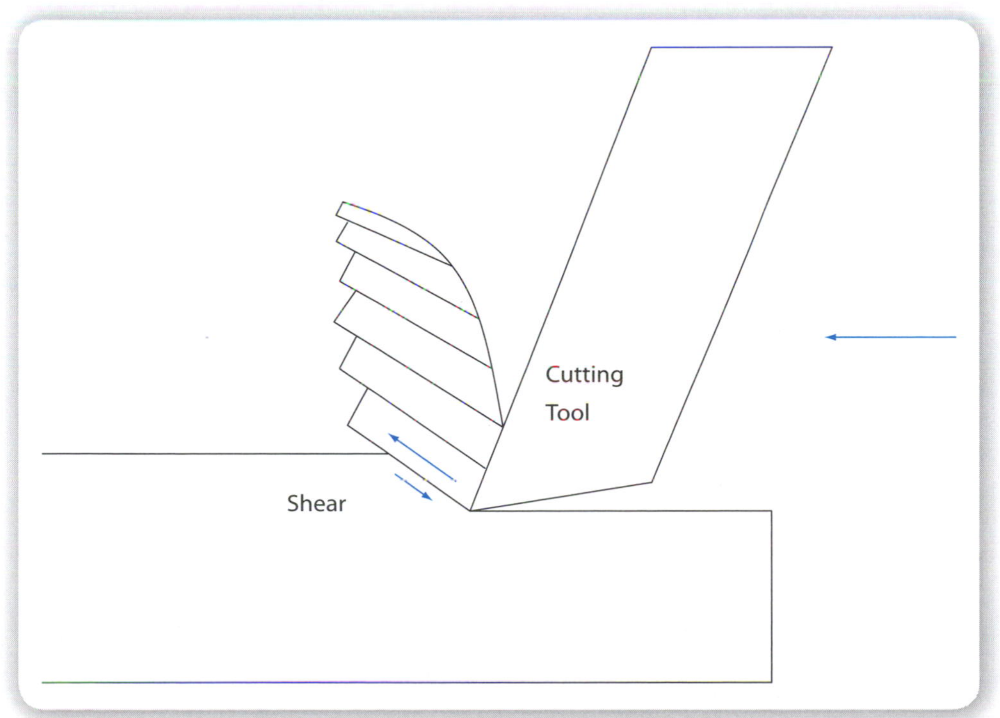

4.2 THE BASIC MACHINING OPERATIONS

Milling

Milling is a process of using a rotating tool to remove material along a contour or line. Milling has traditionally been used to create flat surfaces and straight edges on prismatic workpieces. (Prismatic is a term used to describe three-dimensional shapes that look more like cubes than cylinders.) Milling can also create curved contours. Creating curved contours on a conventional milling machine is an especially difficult task that requires a great amount of skill and specialized tooling. CNC machine tools have made curved contours much easier to create. For example, with a few simple instructions entered into the part program, a CNC machine tool can now produce a circular arc that once required a rotary table to create.

Figure 4.2 illustrates the most common milling operations. End milling and face milling are the two basic categories of milling that can be used to create common features such as contours, flat surfaces, and pockets. The only real distinction between end milling and face milling is that face milling typically uses a wide but shallow cut to create a flat surface. End milling will also produce a flat surface, but it is chiefly concerned with using the outside edges of the tool to create straight sides that are square to the surface.

Figure 4.2 Milling operations (clockwise from top) pocketing, contouring, facing, and side milling.

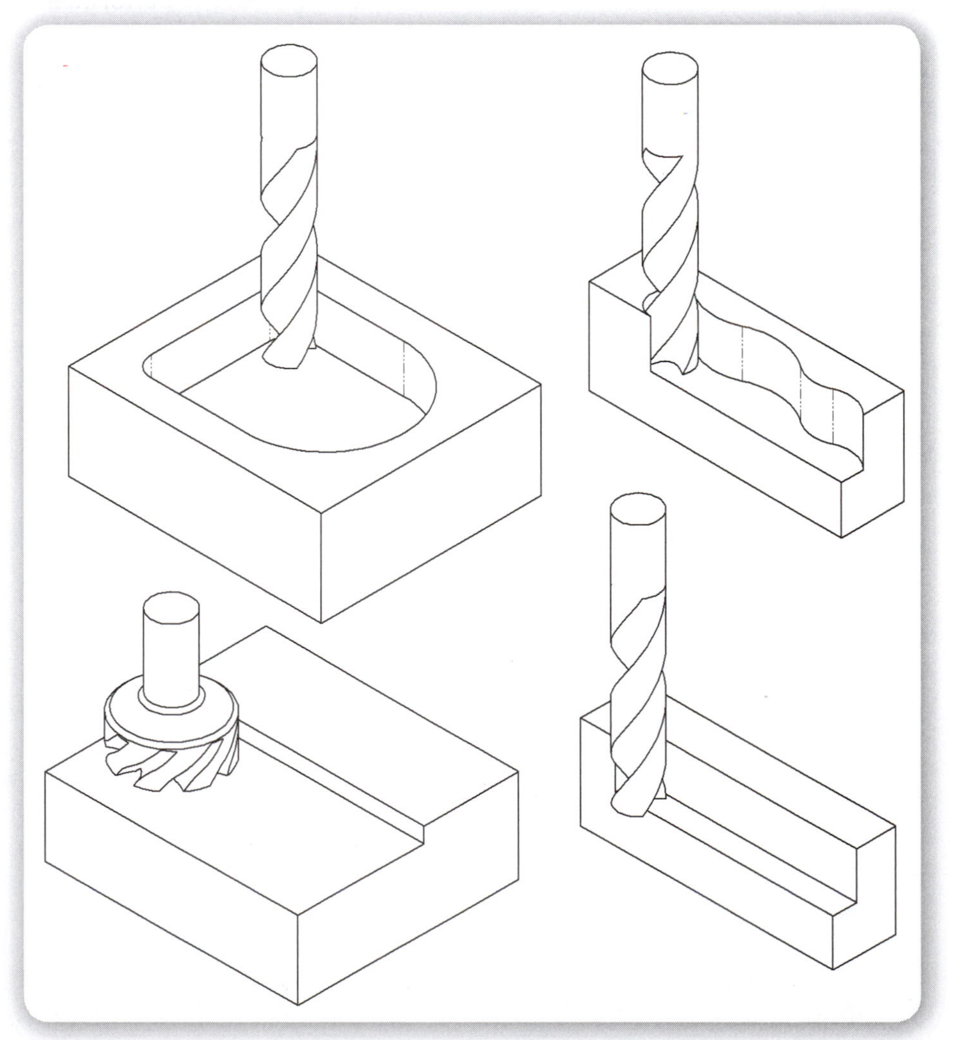

Turning

Turning is a word used to describe a number of different machining operations that are performed on a machine called a lathe. Turning is fundamentally different from milling in that the tool is held stationary while the part is rotated—the resulting shape is cylindrical. The turning process is used to create shafts, bearings, fasteners, and many other machine components that require very precise cylindrical and conical features such as outside diameters, bores, and tapers. Turning can also produce flat faces, grooves, and threads. ***Figure 4.3*** illustrates many of these turning operations.

Turning operations can be described in terms of the features they create. The most common of these operations are Outside Diameter (OD) turning and Inside Diameter (ID) turning. ID turning is also called boring, and the feature it creates is called a bore in shop-speak. OD and ID turning operations can create straight diameters, spherical shapes, and tapers. Sometimes the combination of straight, spherical, and tapered shapes on one part is referred to as profile turning.

Some other common turning operations are facing, grooving, and threading. Facing creates a flat surface that is perpendicular to the axis of the workpiece. Grooving plunges a thin, straight tool into the outside or inside diameter to create features for retaining rings, o-rings, etc. Grooving can also be performed on a face. Threads are another popular turned feature—they are commonly used on machine parts and fasteners. If you have ever had the pleasure of turning threads on a conventional lathe, then you will probably realize that it is a time-consuming, adrenaline-pumping process that often results in scrap parts. CNC machine tools have changed threading to a highly automated process that can be performed with little more than a few instructions and no operator intervention.

Drilling and Reaming

Drilling and reaming are hole-making operations (*see **Figure 4.4***) that can be performed on a variety of machine tools, including milling machines and lathes. Drilling is often considered a roughing operation because of the very nature of the

Figure 4.3 **Turning operations (clockwise from the top) profiling, ID boring, grooving or parting, OD turning, threading, and facing.**

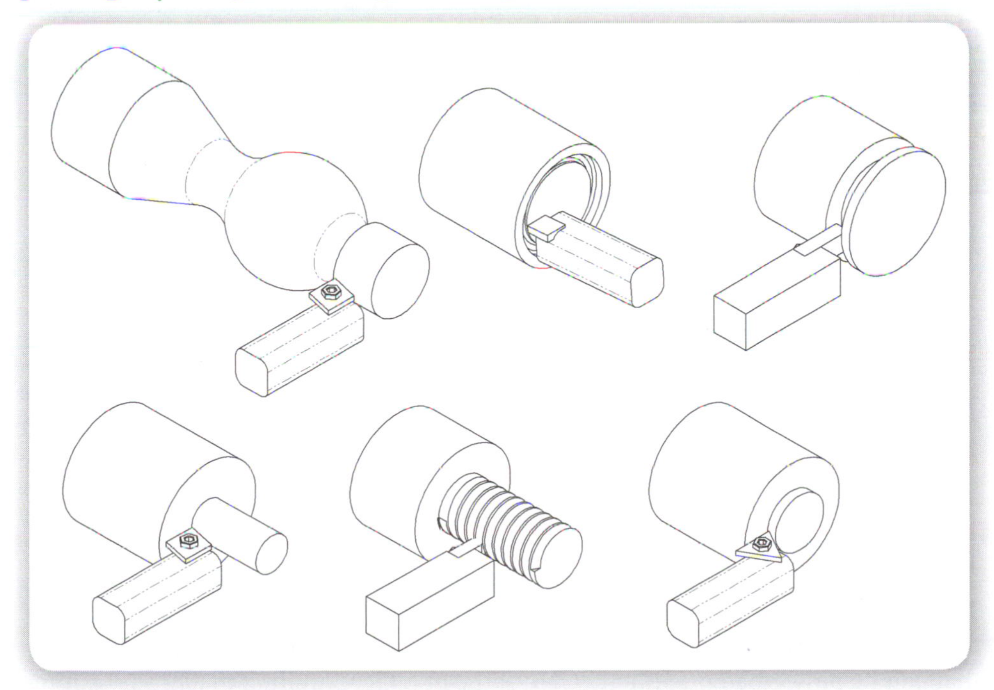

Figure 4.4 *Common hole-making operations.*

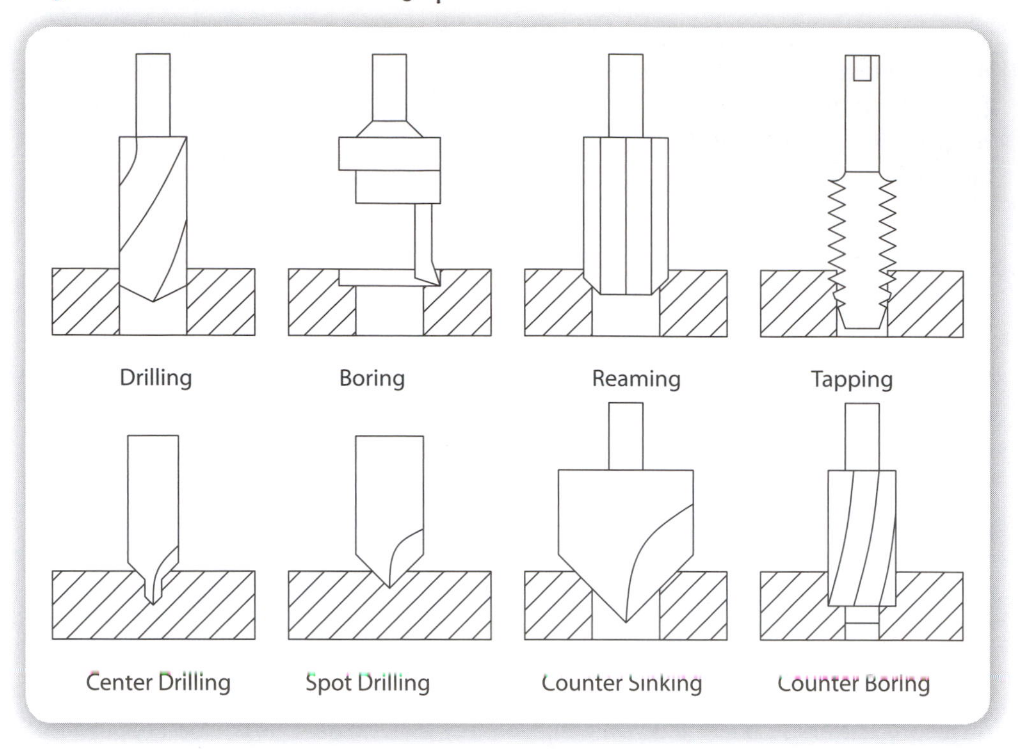

cutting tool that is used. Drilled holes are seldom round, straight, or the correct size, and it is difficult to achieve the high-precision tolerances that are required in machining. Nonetheless, drilling is used to make way for other operations whose tools cannot cut on the center (boring tools, reamers, and taps). Reaming is an operation similar to drilling, but reaming produces a higher quality hole very quickly. Reamed holes will be cylindrical and have the correct diameter. The downfall is that reaming requires a pilot hole, and reamers are expensive. Furthermore, a reamer will tend to follow the axis of the pilot hole. If the pilot hole was not straight, then the reamed hole will probably not be straight. Tight workpiece tolerances may dictate an additional boring operation to straighten the pilot hole before it is reamed.

Boring

Boring was mentioned earlier as an internal turning process that is performed on a lathe. Boring can also be performed on a milling machine with a tool called a boring head. A boring head uses a single-point cutting tool called a boring bar. The boring bar can be moved off-center with an adjusting screw. In this way, the radius of the cut can be controlled.

Boring has several advantages over other hole-making processes—it can produce very round, straight holes to tight specifications. A boring operation can also straighten a lopsided or off-center hole because the single-point tool will not be caused to follow the already crooked edge. The chief disadvantage of boring is that it is a slow machining process. Boring bars are not very rigid; therefore, relatively light cuts must be made.

Tapping

Tapping is the production of internal threads with a tool that is ground in the form of the finished thread. Tapping is an extremely fast and common operation that can be performed on either a CNC lathe or a CNC milling machine. Tapping is used primarily for small-diameter screw threads to accommodate threaded fasteners such as machine screws and socket-head cap screws. Larger threads are beyond the practical capability of tapping, so they must be produced with another process.

One of the biggest obstacles to tapping in CNC machining has been the fact that taps must be reversed to be removed from the hole. This obstacle was first overcome by the production of a self-reversing tapping head that would reverse directions when pulled out of the hole (*see Figure 4.5*). Tapping heads are expensive, and it is difficult to achieve the desired thread depth. Tapping heads also have to be held externally to prevent them from rotating, thus adding complexity to the situation. CNC systems took a while to advance to the point at which the tap could be stopped at the desired depth and then be reversed out of the hole. The problem stemmed from the fact that if even a small amount of lag or error occurred, the tap would tear itself to pieces—the spindle speed and feed must be precisely matched. Floating tap holders (*see Figure 4.6*) allow the tap to move slightly up and down to compensate for these small errors, at the expense of precision of course. Many CNC machine tools now a have *rigid* tapping ability. With rigid tapping, the tap can be held in a standard holder and the control can carefully synchronize the spindle with the feed.

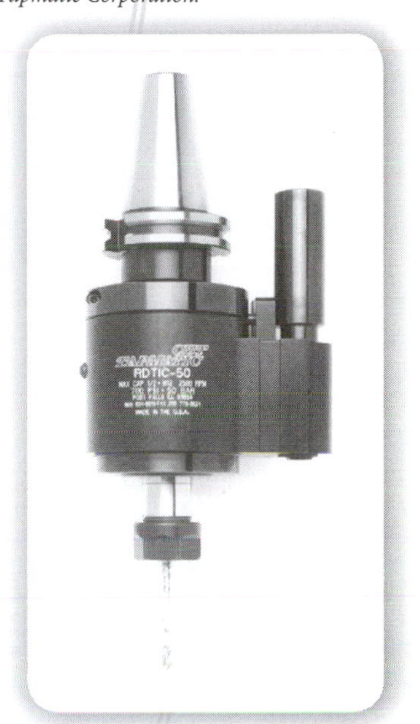

Figure 4.5 A self-reversing tapping head can be used for tapping holes on CNC machine tools. The spindle does not have to be reversed to extract the tap. *Courtesy of Tapmatic Corporation.*

Figure 4.6 Floating tap holders provide a margin of safety when using rigid tapping on a CNC machine tool. *Courtesy of Tapmatic Corporation.*

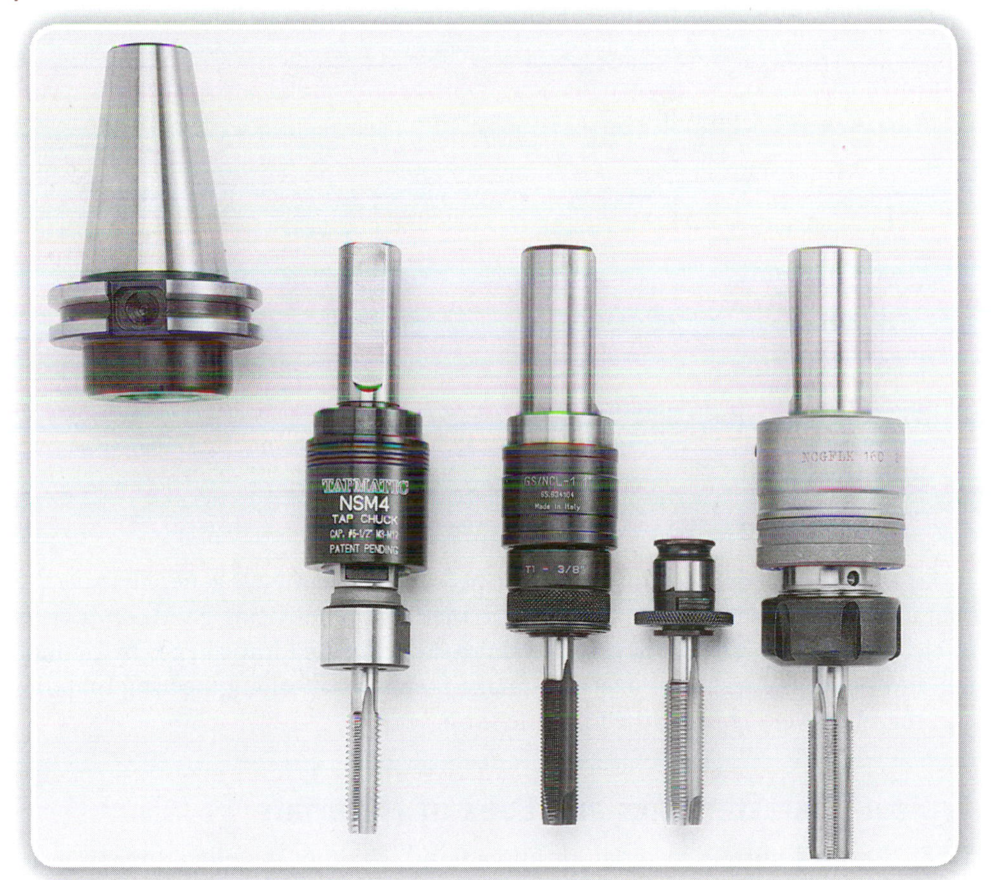

4.3 MATERIAL CONSIDERATIONS

Perhaps the first aspect that any machinist will consider when planning machining operations is the workpiece material. The specified material can range from being easy to cut with carefree performance or, on the other end of the spectrum, there are materials that are so difficult to work with that the cutting tools are always in danger of dulling and fracturing. Machine adjustments are constantly needed to maintain a good surface finish and to hold the required tolerance—this makes a machinist miserable!

So what do you know about materials? When we speak about materials in the sense of manufactured products, we normally call them *engineering materials*. These materials are suitable for consumer, commercial, structural products because they have the strength to withstand external mechanical forces. These are durable materials that are expected to last for a long time and survive a variety of environmental conditions ranging chemical exposure to extreme temperatures. Steel is an engineering material, cardboard is not. We normally categorize engineering materials based upon composition. Common engineering materials found in the machine shop include metals, plastics (polymers), and composites. Listed below are some common materials that are grouped according to their composition:

- Metals
  - Ferrous (Iron-Based)
    - Steel
    - Stainless Steel
    - Cast Iron
  - Non-Ferrous
    - Aluminum
    - Copper Alloys: Brass and Bronze
    - Magnesium
    - Nickel-Based Alloys
    - Titanium
    - Zinc
  - Non-Metals
    - Polymers (Plastics)
    - Ceramics
    - Composites

By far the majority of machining is performed on metals. Metals generally have the characteristics of being strong, having high melting points, being good conductors of electricity and heat, and having the ability to be formed into shapes by pounding and pulling. These characteristics make metals an ideal engineering material and therefore very popular the machine shop.

Typical Characteristics and Uses of Materials

STEEL Steel is a miracle material for humankind. I cannot imagine how our society could have progressed to its current state without steel. There just simply isn't another material that can replace it. It is abundant, inexpensive, and versatile.

For such a common material, very few people have any insight to the composition and properties of this incredible material. First of all, regular steel (technically carbon steel) is composed of mostly iron—usually 99% or more. The remainder is plain old carbon like we find in charcoal and living organisms. It turns out that very small amounts of carbon can radically alter the properties of pure iron. The carbon can be dissolved into the iron to form microscopic structures that exhibit various levels of strength and hardness. For example, steel can be manipulated with carbon and heat to create soft, formable steels like you might find in a "tin" can or paper-clip, or create an exceptionally hard, brittle variation that could be used as a razor blade or drill bit. Somewhere in between these extremes are steels that are strong, flexible, and impact resistant. All of these properties are possible with small adjustments to the carbon content and the proper application of heating and cooling.

Small amounts of other metals (called alloys) can further improve the properties of steel. Tiny quantities of chromium, molybdenum, and boron, for example, can be added to create steels with extreme strength. Larger amounts of alloys such as nickel and chromium so radically alter the properties that a whole new category of metal called stainless steel is created.

Steel is used for everything from structural I-beams, pipes, car bodies, and drive train components to everyday consumer products such as bicycles and stoves. Steel is also a major component in all of the concrete structure that we see in building and bridges in the form of rebar (reinforcing bar) that serves as a flexible skeleton for the brittle concrete.

STAINLESS STEEL Stainless steel is a narrow category of steels that are resistant to corrosion (rust or oxidation). Stainless contains at least 10% chromium, which is where it gets its primary resistance to corrosion. Stainless steel is five to ten times as expensive as standard carbon steels, so it is only used when absolutely necessary.

These steels are classified according to their internal microstructure and alloying elements. Chromium-based stainless steels are either ferritic or martensitic. These are specific arrangements of the crystalline structure of the steel that exhibit certain properties. Ferritic stainless steels are quite simply, soft, low-carbon steels with moderate corrosion resistance. When we add a bit more carbon, then the stainless steel is able to form a material called martensite when it is heated to an extreme temperature and then cooled quickly. The martensite is extremely hard and can be used as a cutting edge for corrosion-resistant knife blades and other tools.

Stainless steels that have the addition of nickel as well as chromium tend to have a very nice luster and high atmospheric and chemical corrosion resistance. They are unlikely to rust even when exposed to saltwater and heat. The downside is that the nickel alters the structure of the metal crystals to a material called austenite that inhibits the formation of the hard martensite. Therefore, these austenitic stainless steels are not heat-treatable, but they remain soft and very tough. Incidentally, austenitic stainless steel are relatively difficult to machine. They tend to have stringy chip and leave burrs that are troublesome to remove. These steels also have a tendency to work harden while machining. To prevent this from occurring, it is imperative to keep the cutting tools very sharp in order to shear the metal and to keep the feeds high enough to prevent any pushing or smearing of the metal that could cause the work hardening phenomenon.

CAST IRON Cast iron is a term used to describe a ferrous material with about 2% to 6% carbon content and a fair amount of silicon. When we have such large amount of carbon to iron, the material becomes brittle and is most often formed by casting rather than by rolling or drawing and we might see in the productions of carbon steels. Cast iron has a most notable property of having great compressive strength, but it is very weak in tension (tensile strength). Its brittleness is manifest by abrupt cracking. Therefore, cast iron is of limited use for structural members that are under tension or bending stresses.

Grey cast iron is the most common form of commercial iron that you will find in the machine shop. The excessive carbon tends to form as large graphite flakes. This form of cast iron has excellent vibration damping and load-bearing characteristics, which makes it ideal for machine bases. Other less common cast irons include the very hard white cast iron and a softer product made by heat treating white cast iron called malleable cast iron.

Manufacturers often select grey cast iron for machine elements for reasons of cost. Once a casting is made, the finished surfaces can be machined for a fraction of the expense of machining the part from solid steel. Grey cast iron has the property of being easy to machine. Due to the brittle nature of the iron, the chips break easily or turn completely into dust. Excellent surface finishes are also achievable. Tool life can be good when machining cast iron if the proper grade is selected. You might notice while machining cast iron that powdery residue is slippery between the fingers from the graphite—which also can provide some lubrication to the tool.

There are some recent cast iron materials on the market that claim to be an excellent substitute for low-carbon steel. The material is sold a continuously cast, cast iron product (***Figure 4.7***). These products are marketed as being able to offer improved machinability and tool life when compared to carbon steels. Cast iron barstock also has an environmental advantage over many of the "free machining" steel that use large amounts of lead achieve their machinability.

Figure 4.7 **Continuously cast iron bar stock is available in a wide variety of sizes and grades in both gray and ductile iron. Ductile iron bar stock is often used as an alternative to carbon steel because it can dramatically reduce the cost of machining through longer tool life, faster machining times and increased productivity.** *Courtesy of Dura-Bar®.*

COPPER ALLOYS: BRASS AND BRONZE Two common copper alloys are that of brass (alloy of copper and zinc) and bronze (alloy of copper and tin). The materials are used extensively for bearing surfaces, fittings and for corrosion resistant elements. The copper alloys tend to be extremely easy to machine and can therefore offset the higher material costs with increase productivity and extremely low tooling expense. Both materials are relatively soft, yet have excellent chip formation and thermal conductivity which leads to easy machining.

The misleading exception is a material called aluminum bronze (Ampco bronze) that is an alloy of copper and aluminum that is used when strength and hardness are desirable. It is a very tough material and difficult to machine—similar to cutting austenitic stainless steel. This material can turn a smile to a frown when the machinist realizes he has been fooled into thinking that this was a normal bronze machining job.

LIGHT METALS: ALUMINUM, MAGNESIUM AND ZINC Aluminum is an extremely important metal in many industries, especially transportation and packaging, because of its low density and high strength. Aluminum has good corrosion resistance and excellent machinability. One important manufacturing advantage of aluminum is tool life. Aluminum is soft and does not contain many of the hard or abrasive microstructures that we find in steel. The extremely hard cemented carbide and diamond cutting tools that are used today are a real mismatch for soft aluminum and they tend to last a very long time. It is not uncommon for a diamond finishing insert to last for weeks or months in production. Additionally, the cutting speed can exceed 3,000 surface feet per minute or about four times that which would be used for steel.

cemented carbide: Metallic carbides that have been *cemented* together with a binding metal through a sintering process. The amount and composition of the binder can dramatically affect the characteristics of the cutting tool.

Aluminums are classified according to their alloy constituents. The most common aluminum alloys you will run into in machining are 2024 (Al-Cu alloy), 6061 (Al-Si-Mg alloy), and 7075 (Al-Zn alloy). From a machining perspective, they are all pretty similar. These wrought aluminums that are specified for machined parts respond well to thermal heat treating processes. You will likely see a temper designation such as T4 or T6 that accompanies the alloy specification. The T-number is a code that can tell you how the aluminum was heat treated. The strength of the material can be increased significantly by heat treatment to a point at which the best aircraft aluminums are as strong as low-carbon steel. The hardness, however, is less significant and does not even approach the level attainable by ferrous alloys. Non–heat treated aluminum is too soft and gummy to machine with much success.

Magnesium is very similar to aluminum except that it is a bit less dense. You often find magnesium in cast transmission cases and automotive wheels. It is an extremely easy metal to machine. Just be careful not to let the chips catch on fire because they will burn very hot once ignited.

Likewise, zinc is a metal that is often cast, except that metal molds are used because of the low melting point. Zinc is a weak, brittle material that is use for non-critical consumer goods such as hardware, toys, and occasionally automotive parts. It is very easy to machine.

HIGH TEMPERATURE METALS: TITANIUM AND NICKEL ALLOYS The aerospace, defense, and energy production industries are often the consumers of metals such as titanium and nickel-based alloys that are able to handle extreme stress and temperature. These expensive metals (often $30 per pound) are used to solve very specific problems in critical systems. For example, nickel-based "super alloys" are used to manufacture the turbine blades of a jet engine. Titanium is light weight and able to handle high temperature; therefore, it is used in application such as forming the leading edges and skin of a supersonic fighter jet. These materials are not particularly hard, but they are extremely strong and tough.

Machining of high-temperature alloys can be difficult, so it is critical to select the right tool materials and cutting geometry for the job. The best source of information for such exotic machining tasks is reputable tool manufacturers. They can often provide application engineering that will result in a trouble-free job. I was once shown, by an engineer, a titanium casting about the size of a dishwasher that his company was machining. It was a complicated aerospace product that required *$8,000 worth of carbide inserts* to produce! At that kind of expense, you need to have the best tools and information available.

Material Condition

Many factors can change the mechanical properties of metals that will influence how we perform machining and the expense of the process. Metals can be made harder, stronger, weaker, softer, more brittle, and tougher, all by the application of heating, cooling, pounding, and pulling. We call these process conditioning.

THERMAL CONDITIONING Earlier, we alluded to the fact that metals can be changed with the application of heating and cooling. This process is generically known as heat treatment. It is most commonly performed on steel. Steel with more than 0.30% carbon can be hardened by heating the material to a high temperature of about 1,450° F and then quenching quickly in water or oil. This transforms the soft, workable steel into a rock-hard material suitable for use as a tool or cutting edge. The strength also increases with hardness. Therefore, harder materials are more difficult to cut. The hardness and strength increases the wear and heat at the cutting tip. Additionally, it takes a much larger amount of force to cut the chip. This will lead to rigidity problems such as deflection and chatter that can ruin a workpiece or tool.

Hardened steels can be made softer by two methods: tempering and annealing. Tempering is used when we want to adjust the hardness of a steel that has been heated and quenched to a desired level—the tempered material can still be quite hard. Tempering is performed by reheating the steel from 500° to 800°F and then cooling it slowly. An increase in time and/or temperature will cause a decrease in hardness. It is most important for a machinist to know the heat treatment condition of a workpiece before starting any machining—steels that have been quenched and then tempered can be extremely difficult or impossible to machine by traditional processes.

Annealing is the second softening process. Annealing removes all of the hardness that was imparted by the quenching process. The material is heated to above the transformation temperature again, but this time it is allowed to cool very slowly (for hours). The resulting material is as soft as it will ever get and soft enough to machine. This is the condition of tools steels that are intended to be machined to shape and then hardened. Low-carbon, hot-rolled steels are also in the annealed condition.

If you have ever overheated a high-speed steel (HSS) cutting bit or drill by causing it to turn colors, then you probably softened it to the point at which it is no longer effective for cutting.

Measuring Hardness

The Rockwell "C" Scale is a very common test that is used to measure the hardness of steel (***Figure 4.8***). The material is tested by applying a force on a small diamond point that pushes into the surface of the sample. If the material is soft, then the point makes a deep indentation. If it is very hard, then the indentation will be quite shallow. Below are some examples of hardness on the Rockwell "C" Scale (HRc):

Annealed Steel (Dead Soft)	<20
Cold Rolled Low-Carbon Steel	20–25
Lawn Mower Blade	35 (Can be Sharpened with a File)
Wrench	40–45 (Machining Becomes Difficult)
Hammer Face	55
Shear Blade	60
Twist Drill or File	62–65 (Limit of Steel Hardness)

WORK HARDENING Work hardening is a method of conditioning steel and other metals by rolling or drawing the material to change its shape. The grains of the material are compacted together and distorted, resulting in a stronger, harder, and more brittle material. This phenomenon is also called strain hardening or cold working.

Work hardening can also occur unintentionally when machining. For example, if a tool becomes dull, then it might work harden the workpiece material to a point at which the tool overheats and can no longer cut. Sharp tools and a consistent feed are essential when work hardening is a problem.

FREE CUTTING STEELS Steel is often chemically altered during the manufacturing process to make it easier to machine. Lead is one of the common elements that is added to make the steel easier to machine. Unfortunately, this means that those personnel who are working with the leaded steel risk some exposure to toxic materials. Sulfur content is also increased to make steel easier to machine. Re-sulfurized steels are popular for small steel products that are made in large numbers, such as fittings and fasteners. Sulfur can cause some brittleness at low temperature, so caution must be exercised in the design process.

A thermal process called spherodizing is sometimes employed to create free cutting steels. The process causes a normal plate-like structure in the steel to curl up into microscopic balls. The cutting tool can more easily shear through the grains with this spherodized structure, thereby making it easier to machine. The industry trend is moving toward spheroidizing and away from adding lead to steel.

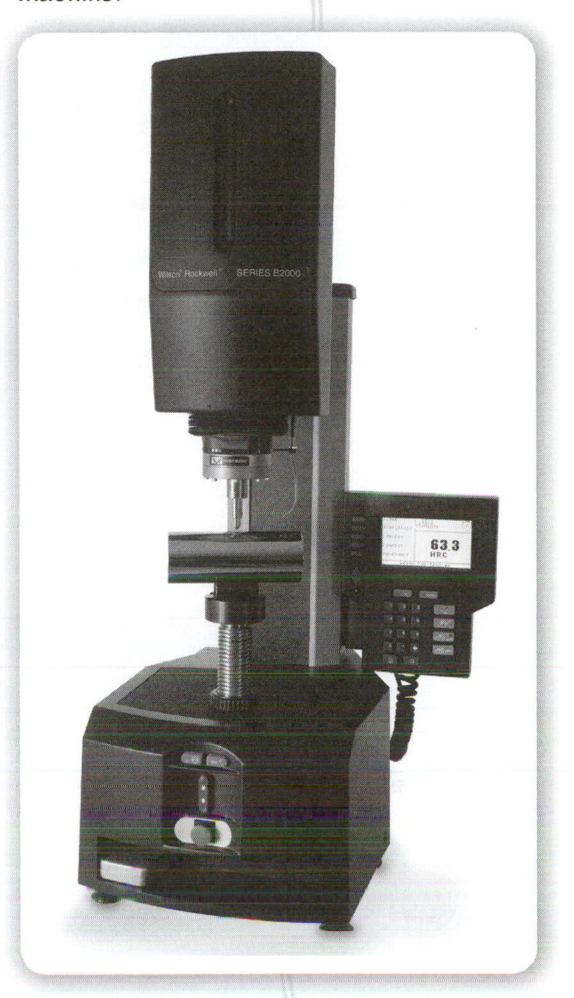

Figure 4.8 A Rockwell hardness testing machine.

Machinability

Remember that machining is a material removal process. We start with a larger block or bar of material and then use hardened cutters to whittle away the excess material until the desired shape has been achieved. Therefore, the ease at which the material can be removed is paramount to productivity and production cost.

We use the expression *machinabilty index* to compare the ease at which material can be machined. Machinability is a property that expresses the effort that is needed to remove material through a chip-making process. Perhaps the most important component of this property is simply the energy required to remove the chip. Some materials are rather easy to cut, whereas others consume a much greater amount of energy. This all translates to increased energy usage (electricity), excessive heat, undesirable surface finishes, and shorter tool life. The resulting process for a difficult-to-machine material will be slower and more expensive than for an easy-to-machine material.

There are other measures that add to the machinability index, including ability to control chips and maintain a good surface finish. Take, for example, two relatively soft materials such as brass and austenitic stainless steel. Brass is a material that forms chips that readily curl and break off. This makes it easy remove the excess material from the machine. On the other hand, the stainless steel chips tend to be long and stringy as in ***Figure 4.9***. This can create dangerous conditions and serious difficulties for the machinist as well as slowing production.

The machinability index is specifically a relative percentage that expresses how easy or difficult it is to machine a material compared to standard alloy called AISI/SAE 1112 steel. This is a low-carbon, re-sulfurized steel commonly called "screw stock" because it is used for making small screws and similar products. The relative machinability of some common materials is shown in Table 4.1

4.4 CUTTING TOOLS

Cutting tools are the instruments that actually make contact with the workpiece and produce the chips. Cutting tool technology has advanced rapidly in the past few decades, and now there are many hundreds of specialized cutting tools made from advanced materials and geometry. In the following section, we will discuss some of the more common tools and tool materials used in CNC machining.

Figure 4.9 Stringy materials can be difficult to machine. These continuous chips formed a "birds's nest" that can be dangerous to both the operator and equipment.

Table 4.1 Machinability of Common Materials

Material	Machinability Index
1112 Steel	100%
1018 Steel	77%
4130 Steel	72%
4140 Steel	62%
06 Tool Steel	48%
D2 Tool Steel	25%
304 Stainless Steel	45%
316 Stainless Steel	45%
Aluminum	175%
Brass	150%
Titanium	40%
Inconel (Nickel Superalloy)	15%

Cutting Tool Materials

Many different materials are used to make cutting tools, ranging from common steel to exotic ceramic and synthetic materials. However, two materials get most of the job done: high-speed steel and cemented carbides. The ceramics and synthetics get a lot of attention, but in reality they are only a small percentage of the cutting tools used every day.

HSS is a very common **cutting tool** material. It gets its name from its ability to maintain a cutting edge at the elevated temperatures encountered during machining. Other high-carbon steels can have the same hardness as HSS, but they lose their hardness at the elevated temperatures found at the cutting edge. HSS gets its hot hardness primarily from the addition of tungsten into the alloy. Another variation of HSS is a grade commonly referred to as M42 cobalt HSS. Its properties are superior to those of standard HSS, but it is more expensive.

cutting tool:
Any hard tool that is used to remove material. Examples include end mills, drills, saw blades, and turning tools.

HSS is inexpensive and versatile and can handle a great amount of shock. Additionally, HSS is easy to fabricate into intricate shapes. Drills, end mills, and taps are commonly made from HSS and perform well in many cutting conditions.

Cemented carbide is another popular cutting tool material. The term *carbide* refers to cutting tools made from carbides of tungsten, titanium, and tantalum. Carbides are extremely hard materials that can handle a great amount of heat and last a long time. The chemical and physical nature of carbides does not allow them to be directly melted and wrought into billets to make cutting tools. Instead the powdery carbides are mixed with another metal, such as cobalt, and sintered in an oven. The other metal will melt and act as a binder to cement the carbide particles together; hence, cemented carbide.

Cemented carbides have the ability to be used at extremely high temperatures and, therefore, high cutting speeds. In fact, the typical cutting speed for carbide tools is four to seven times that of HSS. They cost considerably more than their HSS counterparts, but when the productivity gains are measured against the cost, carbide is the clear winner.

Cemented carbides are not without flaws. First, they are difficult and expensive to fabricate. Carbides are so hard that they must be machined by either grinding or electrical discharge machining. Both are expensive propositions. Second, carbides are brittle. It is difficult to use carbide cutting tools in situations in which they will be subjected to shock and vibration. The setup must be extremely rigid or the cutting edges are likely to chip. Interrupted cuts also tend to shock carbide tools. An example of an interrupted cut is turning a shaft with a keyway already milled into it. Every time the tool hits the keyway, it will recoil heavily. This interrupted action has the tendency to break the tool. Lastly, the geometry of the cutting edge usually results in higher cutting forces. Because carbides are brittle, their cutting edges tend to be manufactured more blunt than sharp. Carbide tools are often designed to lean into the material rather than lean back as is the case with HSS cutting tools. This leaning direction is called rake; carbides most often have a negative rake angle. Negative rake angles must push across more material to create a chip, and therefore need more force. More force means that more energy (electricity) will be needed to machine the part and more wear will be put on the machine tool.

Carbides are available in several grades (composition and hardness) based upon their intended use. The grading system is a continuum from soft/tough to hard/brittle. Some machining applications require a very hard but brittle material that can faithfully hold an edge. Other applications require shock resistance and therefore must be softer and tougher. The properties of cemented carbides can be manipulated by varying the ratio of cobalt binder to carbide and by using different metallic carbides. Table 4.2 ANSI-standard C shows some of the commercially available carbide grades along with their uses. Carbide tool manufacturing has become a highly competitive industry in the last decade, and many new manufacturers have appeared on the scene. Of course, each manufacturer has its own grading system and has a grade of carbide for every conceivable machining problem (each of which is better than the competition's). The proprietary grading systems can be confusing, but they usually try to specify an equivalent in their catalogs (e.g., X335 = C3).

Table 4.2 ANSI-Standard Carbide Grades

	Grade	Primary Use	Toughness	Hardness
Grades for Aluminum, Cast Iron, and Non-metallics (≈95% Tungsten Carbide)	C1	Heavy roughing	Tough	Soft
	C2	General purpose	|	|
	C3	Finishing	|	|
	C4	Fine finishing and boring	Brittle	Hard
Grades for Steel (≈75% Tungsten Carbide, 5–10% Titanium Carbide, and 5%–10% Tantalum Carbide)	C5	Heavy roughing	Tough	Soft
	C6	General purpose	|	|
	C7	Finishing	|	|
	C8	Fine finishing and boring	Brittle	Hard

Cutting tool grades based upon the ISO system have become more prevalent in recent years (see Table 4.3). The ISO system more broadly defines the intended use of a carbide tool material that we find in the older ANSI system of industry grades (C1, C2. . .). The system does not attempt to identify the composition of the substrate or the coating—only the material or conditions within which it is intended to operate. For example, the ISO has a grade of P10. A cutting tool with this grade designation must be designed for finishing cuts in soft carbon steel. It will be a relatively hard and brittle material when compared to, for example, a P40 that has to be able to take heavy, interrupted roughing cuts in steel.

Table 4.3 ISO Carbide Grades

Grade Class	Intended Workpiece Material	Grade Number
P	Steel	01 10 20 30 40 50
M	Stainless Steel	10 20 30 40
K	Cast Iron	01 10 20 30
N	Aluminum	01 10 20 30
S	Super Alloys (Nickel and Titanium)	01 10 20 30
H	Hardened Steel	01 10 20 30
		← Increasing Wear Resistance
		Increasing Toughness →

indexable insert:

Disposable, multi-tipped cutting tools that are designed to be held in a durable tool holder. The insert can be *indexed* to a fresh corner whenever the current corner becomes dull.

The majority of carbides used today are manufactured as throwaway, indexable inserts. The inserts are held in a shank or tool body that is usually made from a good grade of steel. As soon as the inserts become dull, they are simply indexed to a sharp corner or replaced. These inserts are available in many shapes and sizes, some of which are standardized and others are the proprietary designs of the tool manufacturer. You will find it useful to be familiar with the styles that are common without regard to any particular manufacturer. ***Figure 4.10*** shows the standardized insert shapes.

It is important to select the proper insert for the job at hand. Selecting the wrong grade or geometry will lead to premature tool failure and unnecessary expense.

Figure 4.10 Some of the more common shapes of indexable insert.

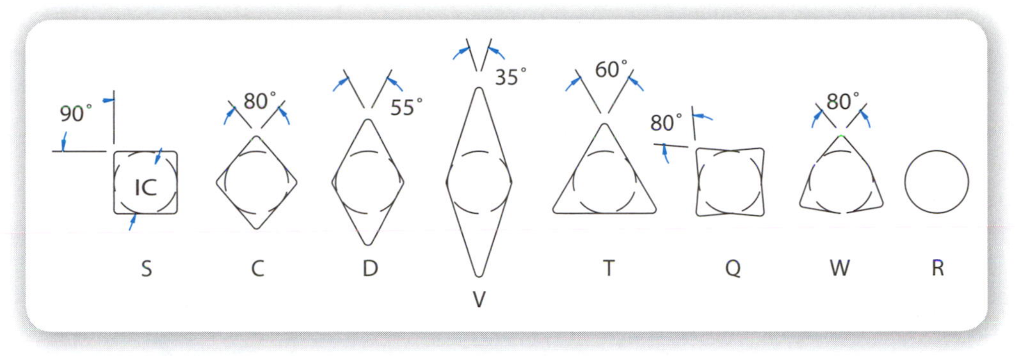

Figure 4.11 Tool holders and inserts are available in numerous styles that can be adapted to the machining job.

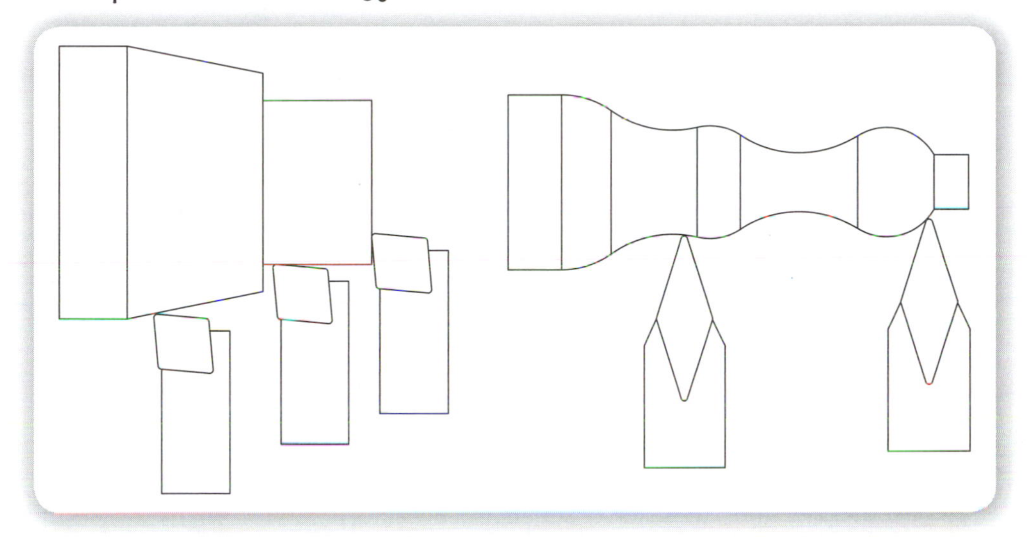

The three major factors in insert selections are the corner style, rake angle, and corner radius. Insert strength is most affected by the angle of the corner. For example, in *Figure 4.10,* the "S" and "C" shapes have relatively broad, strong corners. The "V" shape, with its sharp, 35° angle, is relatively weak. Broader corner angles will also allow the heat to be more quickly conducted away from the tip. Whenever possible, select the broadest corner that will cut the geometry without interference.

Of course, the geometry of the tool holder and of the workpiece will also dictate the style of insert that we will be able to use. For example, an intricate profile will require the use of an insert shape that has an acute angle, such as the 35° diamond (*see Figure 4.11*). Conversely, straight sections and faces can be cut with a more obtuse angle. The 80° corner is very popular geometry for general purpose turning. It can be used for straight diameters and facing. In addition, the 80° corner can be used to cut a square corner if a holder of the proper orientation is used. A straight style holder might be used for profiling.

Rake angle is another important factor in cutting performance (*see Figure 4.12*). Negative-rake cutting tools lean into the cut and tend to cause higher cutting forces and deflection. Conversely, positive-rake tools lean back out of the cut and tend to cause lower cutting forces and deflection. Most applications will use carbide inserts that are designed to use negative rake. However, positive rake is used when the setup is not very rigid and when deflection or chatter becomes a problem. For example, small boring bars are usually designed for positive rake.

Inserts designed for positive rake will have a sharp corner between the cutting surface and the edge in order to provide both rake and relief. This is not the optimal design for brittle carbide materials, and therefore the sharp corners

Figure 4.12 **Turning and milling tools with both positive and negative rake. Note the sharp corners required for positive rake.**

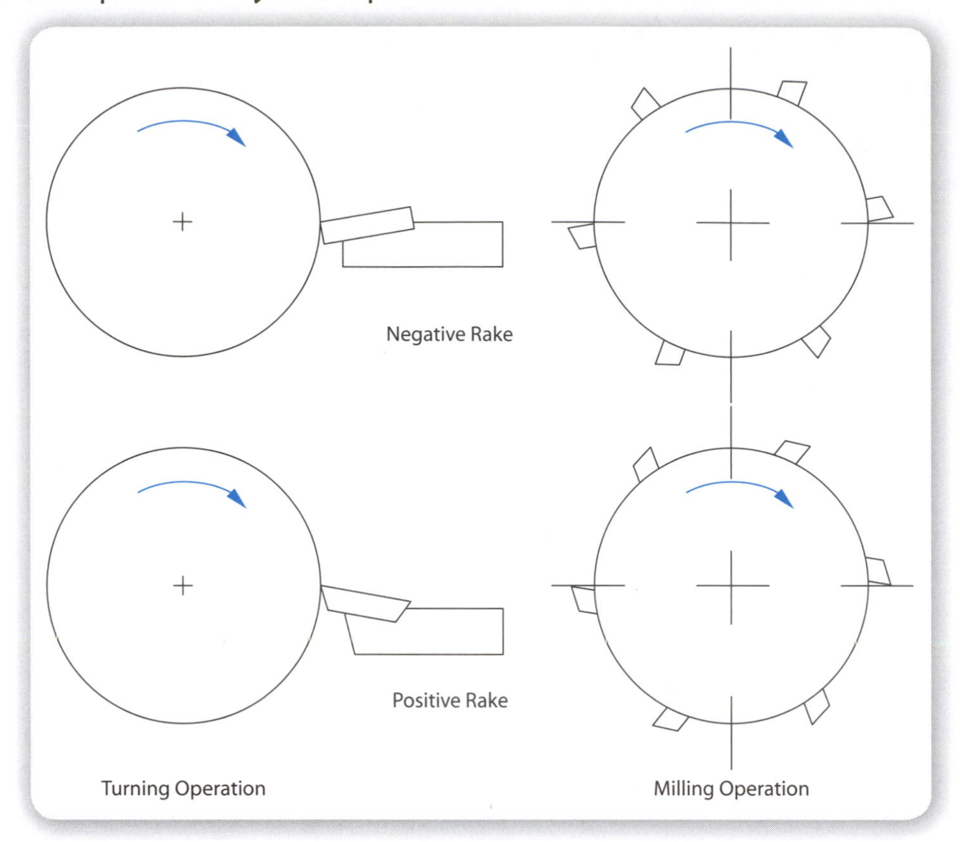

tend to fracture easily. Positive-rake inserts are not as strong as their negative-rake counterpart, but they have superior characteristics for many finishing operations.

On the other hand, inserts designed for negative rake can be constructed with strong, square corners between the cutting surface and the edge and still provide a relief angle. The square corner means that negative-rake inserts can be used on the top side and then on the back side—twice as many cutting surfaces as a positive-rake insert of the same shape. Both styles cost about the same amount, but we get twice the tool life from the negative-rake insert.

It is important to note that the tool holder and insert must be matched. You cannot use a negative-rake insert in a positive-rake holder—the underside of the insert will rub against the workpiece.

One other factor to consider when selecting an insert style is the corner radius. The corner radius affects both the finish and tool life. A relatively large corner radius leads to a stronger corner, higher feeds, and better heat removal, but at the expense of higher cutting forces. Likewise, a relatively small corner radius has lighter cutting forces, but it is weaker, less able to remove heat and requires a lower feed rate. As a rule, select the largest radius possible that will not cause chatter. In summation, each of the various insert shapes and styles has its place in CNC machining. Table 4.4 summarizes some of the uses and properties of the various insert shapes.

Insert Identification

The programmer is usually responsible for specifying the style of insert to use in a CNC machining job. Therefore, it is important to be able to accurately identify inserts according to their industry identification codes. Inserts are identified by their attributes, including shape, size, tolerance, and style. Table 4.5 illustrates the standard identification system for inserts.

Table 4.4 Common Indexable Insert Shapes

Insert Shape	Main Use/Advantage	Disadvantage
"S" – Square	Milling cutters and for rough turning. Eight corners available on negative-rake styles	Cannot be used for square shoulders
"C" – 80° Diamond	Popular turning insert for all-purpose turning. Can produce square shoulders. Sometimes used for milling	Has a maximum of four corners
"D" and "V" – 55° and 35° Diamonds	Profiling and finish passes	Lowest strength. Maximum of four corners
"T" – Equilateral triangle	A good compromise for lightly profiled or tapered surfaces. 50% more corners than diamond shapes. Positive-rake style is often used for boring bars	Weaker than the 80° corner. Does not adapt well to turning and facing with the same tool holder
"Q" and "W" – Unique polygons with 80° corners	The same uses as "C," but they are more economical because of the increased number of corners	"Q" cannot cut a square shoulder deeper than the depth of the insert
"R" – Round	Highest strength. Often used for milling cutters and for heavy turning	The large radius can cause problems with chatter

Table 4.5 ANSI Standard Identification System for Indexable Inserts

Shape	
A = Parallelogram 85°	M = Diamond 86°
B = Parallelogram 82°	O = Octagon
C = Diamond 80°	P = Pentagon
D = Diamond 55°	R = Round
E = Diamond 75°	S = Square
H = Hexagon	T = Triangle
K = Parallelogram 55°	V = Diamond 35°
L = Rectangle	W = Trigon 80°
Relief Angle	
A = 3°	**H** = 0°+11°
B = 5°	**J** = 0°+14°
C = 7°	**K** = 0°+17°
D = 15°	**L** = 0°+20°
E = 20°	**M** = 0°+14°
F = 25°	**N** = 0°
G = 30°	**P** = 11°

(continued)

Table 4.5 *(continued)*

Tolerance Class					
Letter	I.C	Thickness	Letter	I.C	Thickness
A	0.001	0.001	H	0.0005	0.001
B	0.001	0.005	J	0.002–0.005	0.001
C	0.001	0.001	K	0.002–0.005	0.001
D	0.001	0.005	L	0.002–0.005	0.001
E	0.001	0.001	M	0.002–0.004	0.005
F	.0005	0.001	U	0.005–0.010	0.005
G	0.001	0.005	N	0.002–0.004	0.001

Type	
A = With hole	B = With hole and one countersink
C = With hole and two countersinks	F = Chip groves both surfaces, no hole
G = Chip groves both surfaces, with hole	H = With hole, one countersink, and chip groove on one top surface
J = With hole, two countersinks, and chip groove on both top surfaces	M = With hole and chip groove on one top surface
N = Without hole	Q = With hole and two countersinks
R = Without hole, with chip grooves	T = With hole, one countersink, and chip groove on one top surface
U = With hole, two countersinks, and chip grooves on top surface	W = With hole and one countersink

Size I.C.
Number of 1/32nds on inserts less than ¼" IC
Number of 1/8ths on inserts ¼" IC and over.
Rectangle and Parallelogram
1st Digit = Number of 1/8ths in width
2nd Digit = Number of 1/4ths in length

Thickness
Numbers of 1/16ths on inserts ¼" and over
Numbers of 1/32nds on inserts less than ¼" IC
Use width dimension in place of IC on Rectangle and Parallelogram inserts.

Cutting Point	
0 = .000/.007 R	**A** = square insert with 45° chamfer
1 = 1/64 Radius	**D** = square insert with 30° chamfer
2 = 1/32 Radius	**E** = square insert with 15° chamfer
3 = 3/64 Radius	**F** = square insert with 3° chamfer
4 = 1/16 Radius	**K** = square insert with 3° double chamfer

Table 4.5 *(continued)*

Cutting Point *(continued)*	
5 = 5/64 Radius	**L** = square insert with 15° double chamfer
6 = 3/32 Radius	**M** = square insert with 3° double chamfer
7 = 7/64 Radius	**N** = truncated triangle insert
8 = 1/8 Radius	**P** = flatted corner triangle insert

We can identify the insert by decoding the identifier. For example, consider the insert identifier CNMG-432 (see Example of an Identification Code That Is Used to Describe One Particular Insert). From ANSI Standard Identification System for Indexable Inserts, we can determine all of the needed information about the insert. We simply look at each symbol in the code and then decode its meaning using Table 4.6.

Table 4.6 Example of an Identification Code That Is Used to Describe One Particular Insert (see Figure 4.13)

C	The insert has an 80° diamond shape.
N	The relief angle is 0°. This indicates that this insert is used at a negative rake.
M	This mid-range tolerance class is used for most applications.
G	The insert has a hole in the center for mounting and a chip breaker molded into the insert.
4	The inscribed circle is 4/8 or 0.500". An inscribed circle is the smallest circle that will fit completely within the inside shape of the insert.
3	The thickness is 3/16 or 0.1875".
2	The nose radius is 1/32 or 0.0313".

Figure 4.13 A very popular insert with the ANSI designation CNMG-432. The codes describe the shape and size of the insert geometry.

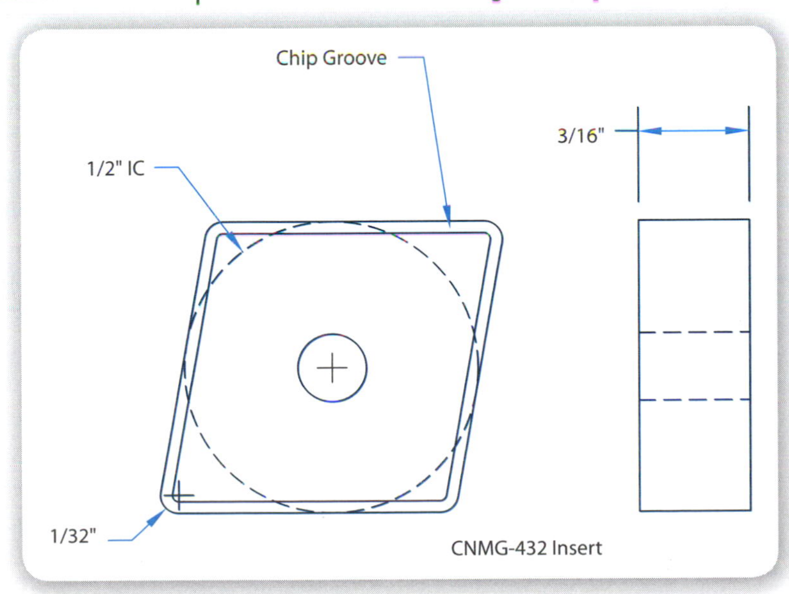

Exotic Tool Materials

New, exotic cutting tool materials have been developed for tough, specialty applications. These tool materials are extremely expensive and not very common in the average machine shop today. These applications are outside the scope of this book, but we will mention a few of the notable exotic tool materials. First is polycrystalline diamond (PCD). PCD is a synthetic diamond material that has uses in the machining of nonferrous metals. PCD can all but eliminate the need for tool changes in many finishing operations in aluminum. Second is cubic boron nitride (CBN). CBN is also a synthetic material and is the second hardest known material (the first being diamond). CBN is useful for machining hardened steel, which can sometimes eliminate the need for expensive grinding operations. Last is silicon nitride (SiN). SN nitride is a ceramic material that is used primarily to machine highly abrasive materials such as the composite materials found in the aerospace and electronics industries.

Tool Coatings

Cutting tools wear from the effects of rubbing against the workpiece material while forming a chip. The rubbing has several major effects, including abrasive action and high-heat generation. The two obvious remedies to this problem are harder tool materials and materials that are more heat resistant. Carbides enhance performance, but this is often not enough. Another way to increase tool life is to give the tool a thin-film coating of material that will enhance the tool's resistance to abrasion and heat (*see Figure 4.14*).

One of the most effective approaches has been to give the cutting tool a thin coating of an extremely hard material called titanium nitride (TiN). You can recognize a TiN coating by its gold appearance. In addition to having a high hardness, TiN coatings have a low coefficient of friction (i.e., they are slick), which results in a lower temperature at the cutting edge. Lower temperatures make the tool less likely to overheat and chemically react with the workpiece material.

These thin-film coatings are achieved with one of several processes called Physical Vapor Depositions (PVD) and Chemical Vapor Depositions (CVD). These are similar processes in which the part to be coated is placed into a controlled atmosphere along with the coating material. In the PVD process, the coating material is vaporized with a heat source and then condensed, molecule-by-molecule, onto the tool. The CVD process is similar except that the coating material is contained in a solution rather than as a solid. This results in a lower temperature process.

Figure 4.14 Insert are usually coated with materials that impart wear resistance, lower friction, and extend tool life. They are applied in layers with the PVD and CVD processes.

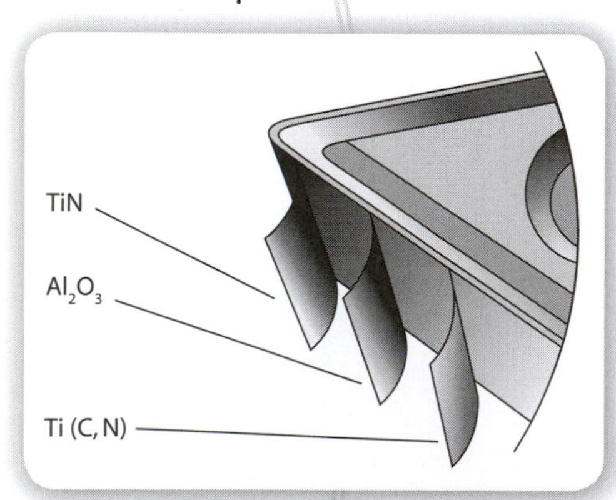

Sometimes the TiN coating is placed on top of a thin veneer of aluminum oxide (Al_2O_3). Aluminum oxide is the very hard material from which most grinding wheels are made. The aluminum oxide base will help to prolong tool life by providing good resistance to physical abrasion.

A few new tool coatings have become more popular in recent years, including titanium carbonitride (TiCN), which is silver, and titanium aluminum nitride (TiAlN), which is brown. These extremely hard substances perform similarly to TiN. Other less common tool coatings include chromium nitride, diamond, and various other proprietary coatings. Each of these offers some small advantage under the proper circumstances.

4.5 CUTTING TOOLS FOR MILLING

End Mills

An end mill is a cutting tool designed to cut on both the end and the side. Standard end mills are perhaps the most common cutting tools used in conventional and CNC milling. End mills are commercially available in various styles and materials and have become the workhorse of everyday machining applications. The most common material for end mills is HSS, but end mills are also made from solid billets of cemented carbide. HSS end mills are inexpensive and used extensively to machine ferrous and nonferrous materials. Conversely, solid-carbide end mills are four to five times as expensive as HSS and have more limited applications in nonferrous and non-metallic materials. Carbides tend to be brittle; therefore, the sharp edges found on solid-carbide end mills do not tend to hold up very well for ferrous machining applications. However, some of the newer carbide grades have been engineered for ferrous machining operations.

End mills can be described by their geometrical features, which include the number of teeth/flutes, the end-cutting style, and the edge profile (*see Figure 4.15*). The most obvious feature of a standard end mill is the outside cutting edges or teeth. The spaces between the individual cutting edges, which are called flutes, are used to remove the chips. Technically you might describe an end mill as having two teeth, but it is more common to say "two-flute." Small end mills are most commonly made with two, three, or four flutes, and each has its own advantage. Table 4.7 illustrates some of the pros and cons of the various end mill designs.

Figure 4.15 A variety of end mill styles (left to right) four-flute, two-flute with high-helix angle, roughing, and ball end mill.

Two-flute end mills usually have large flutes and therefore more room to accommodate chips. These are most often used for machining aluminum and light metals because the chips tend to be larger due to the increased feed used on soft materials. Two-flute end mills can be used on steel, but this is not the best choice. Four-flute end mills have smaller flutes, but they are stronger and stiffer than two-flute end mills. Four-flute end mills have more cutting edges to do the

Table 4.7 Comparison of End Mill Designs

Number of Flutes	Advantage and Use	Disadvantage
Two	More room for chip evacuation. Usually center-cutting. Used primarily for light metals	Weak and tends to deflect under load. Shorter tool life
Three	A good compromise between two- and four-flute. Often used for stainless steels	Less widely available
Four	More cutting edges for increased productivity and tool life. Commonly used for steel	More likely to clog-up in deep cuts. Center cutting style is less common and more expensive

work and are a better value than two-flute end mills. You should consider using a four-flute end mill when machining steel. Three-flute end mills have properties somewhere in between those of two-flute and four-flute mills; they are a good compromise for materials, particularly stainless steels, that are somewhere in between soft metals and hard steels.

The flutes of standard end mills are ground in a helix in order to distribute the cutting load to other teeth and to help pull the chips out of the cutting zone. A variation of this is the high-helix end mill. The helix angle is ground much steeper than normal to help pull the chip up and out of the flutes when machining nonferrous metals—particularly aluminum. The high cutting speeds used for CNC machining of aluminum allow for very few errors. If the flutes become clogged with a chip for even a moment, the cutter will break before the operator even realizes that something is wrong. Consider using a high-helix end mill anytime you are cutting a deep or enclosed pocket where chip evacuation is difficult.

End mills are also classified by the ability to perform plunge cutting. Plunge cutting is an operation similar to drilling, where the end mill is pushed straight down into the workpiece. Not all end mills have this ability; some are *center cutting* and the rest are *non-center cutting,* as shown in ***Figure 4.16***. Standard two-flute end mills are usually center cutting. Conversely, standard four-flute end mills are not. It is much more difficult to grind the end of a four-flute end mill into a center cutting geometry, and, therefore, more expensive. Center cutting, four-flute end mills are available from supply houses and are commonly used in CNC machining, but the machinist must be careful to select this style only when absolutely needed because of the higher cost.

If a plunge operation is needed and the only tool available is a four-flute end mill, then other machining techniques have to be employed. The simplest solution is to use a drill to create a pilot hole. Just make sure that the drill is larger than the *dead zone* on the end mill. A more elegant solution is to use a ramping or a spiral entry. These techniques will be further discussed in later chapters.

End mills may also have various different profiles. A standard end mill has a square profile. Therefore, an end mill can create flat surfaces and straight sides simultaneously and will leave a sharp inside corner. The sides can also be ground to a profile other than square. For example, it is common to create a radius on the corner in order to cut an inside fillet on the workpiece. An end mill with a full radius on the profile is called a *ball* end mill. Alternatively, a smaller radius might be ground on the corner to create a *bull-nose* end mill. Ball end mills are readily available from industrial suppliers, but bull-nose end mills are often a special order.

Figure 4.16 **Center cutting (right) and non-center cutting (left) end mill designs. Center cutting end mills are needed for plunge cutting.**

End mills with a corner radius are important in CNC machining because the radius can be used to create surfaces that would not be possible under ordinary circumstances. For example, a ball end mill can be used to create a curved or angled surface on a workpiece by taking cuts that estimate the true shape of the desired surface, as illustrated in ***Figure 4.17***. It is very common to use this technique in mold and pattern making. The resulting surface will have a slight scalloped appearance, but this can be improved by programming finer steps.

One other notable end mill profile is the roughing end mill. Roughing end mills have a serrated profile ground into the outside diameter. The serration has the effect of lowering the cutting forces while machining at an unusually high feed rate. A roughing end mill should be used anytime there are large amounts of material to remove. They cost a little more than a comparably sized standard end mill, but the increased productivity will more than make up for

Figure 4.17 Surface estimation with a ball end mill. The resulting surface will have a slight scalloped appearance, but this can be improved by programming finer steps.

the additional cost. The downside of roughing end mills is that the serration leaves a rough surface on any side that was milled. Another finishing operation will be required to smooth the surface.

Insert-tooth Cutting Tools

Although the previously mentioned HSS end mill is probably the most common and versatile cutting tool, HSS simply does not have the durability of carbide. It might seem that the obvious solution would be to fabricate end mills from carbide. Solid-carbide end mills are very expensive, and the brittleness of carbide makes them unlikely candidates for many machining applications. A more robust solution is to make a tough steel body with hard carbide cutting edges. Such cutting tools are called insert-tooth cutting tools (*see Figure 4.18*). These cutting tools are very economical because the carbide inserts are designed to be thrown away. The inserts are held in with a clamping device or screw, and when they get dull they are

Figure 4.18 Insert-tooth face and end mills.

simply removed and replaced with sharp inserts. The insert styles vary, but most are made in the indexable style—they have several corners per insert. When the insert becomes dull, it is *indexed* to a fresh corner.

A popular application of the insert-tooth concept is the face mill. Face mills are also made from HSS, but the high performance of carbides makes the insert-tooth variety more economical. Face mills are used to create flat and wide surfaces. Face mills are usually designed to take a shallow cut, and most are not intended to plunge cut or to mill pockets like an end mill. The inserts tend to have either a large radius or leading angle on them to produce a smooth surface. Consequently they are not usually used for machining shoulders. There are face mills on the market that can produce a square shoulder, but this is achieved at the expense of the number of available corners per insert.

Insert-tooth end mills are less common but have been gaining in popularity in the past few years. Their main application has been in the machining of hard steels used in mold making and in the machining of tough aerospace alloys. They often use multiple inserts along each flute and sometimes have a different insert style on their corners from their sides. In the right applications, insert-tooth end mills may be a very productive solution.

In everyday applications, small, insert-tooth end mills seem to offer fewer applications and less of an economic advantage over small HSS or solid-carbide end mills. It is common for one insert to cost nearly as much as a standard HSS end mill. This is probably true because end mills may require a more sophisticated geometry and the inserts tend to be of a proprietary design.

Spindle Style

Regardless of the cutting tool you select for milling, the tool must somehow be mounted in the spindle of the CNC milling machine. No discussion of tool mounting can take place until we have a common knowledge of the basic spindle styles that are used on machine tools. The spindle is the part of the machine that holds the cutting tool (in the case of a milling machine) and rotates by means of an electric motor. The spindle is usually hollow and is supported by at least two bearings—one at the top and another near the bottom. The spindle bearings are anti-friction bearings mounted in a cartridge containing both sets of bearing races and the balls or rollers. They are some of the highest grade precision bearings available, and they have a price to match.

The end of the spindle is precisely ground to an industry-standard shape to accommodate some type of tool holder. You may be familiar with conventional machine tools that often use an R-8 collet system or the NMTB steep taper. The cutting tools are then mounted in a collet or an adapter that is held in the spindle with a threaded drawbar. CNC machine tools also use a taper and a spindle adapter to mount the tools, but the style is slightly different, and provisions are made to automatically hold or release the adapter. ***Figure 4.19*** illustrates end mill adapters for two different spindle styles.

The two most popular spindle nose styles are the Caterpillar (CAT), or V-flange, and the BT. These steep tapers can carefully locate the tool concentric to the spindle and provide a fair deal of resistance to being deflected side-to-side. The taper is not so shallow as to be self-holding like you would find in a Morse taper—we want the adapter to easily fall out of the spindle when the clamping force is released. The CAT and BT adapters look nearly identical, but they have a slightly different configuration for mounting in an automatic tool changer. The adapter is held into the spindle by a pull stud that is mounted to its end and is gripped by a mechanical device inside the spindle.

Figure 4.19 Tool adapters for the most common spindle styles.

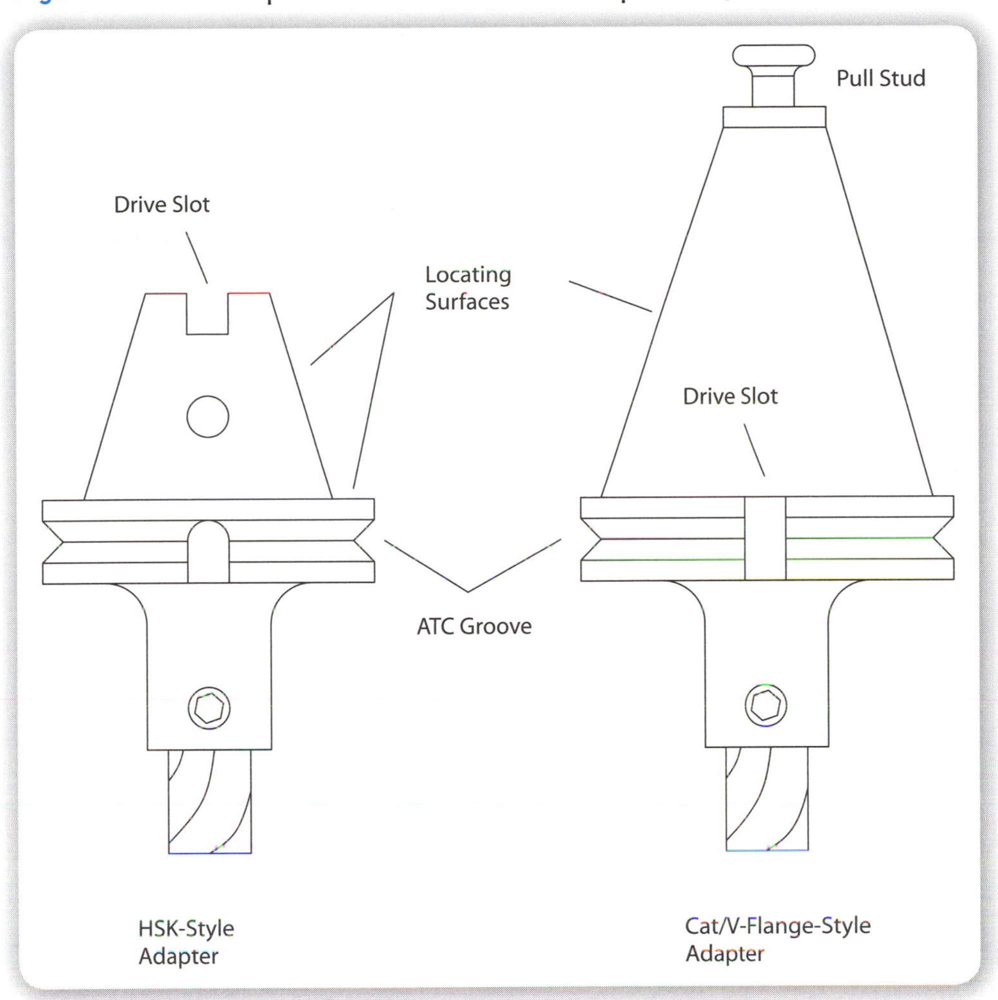

The tapers are available in several sizes, including #30, #40, and #50 tapers. The #40 taper is probably the most common and is found on many small vertical machining centers. The #50 taper is used on larger vertical and horizontal machining centers that have the power to drive larger cutting tools.

A recent addition to the spindle taper family is HSK. HSK uses a relatively short taper and one flat surface for locating. The chief advantage of HSK is that it was designed specifically to be hollow, thereby allowing coolant to be piped directly through the center of the tool. HSK is not very common in the United States, but we can expect it to become more popular as its advantages become known and more machine tool builders start offering HSK as an alternative.

Spindle Tooling

spindle tooling: Tooling that adapts cutting tools to the rotating spindle of a machine tool. Examples include end mill adapters, collet chucks, and boring heads.

The most common spindle adapter is the end mill adapter (*see Figure 4.20*).

End mill adapters are available in standard sizes and come in several lengths. The end mill is secured in the holder with a set screw that is aligned with a groove on the cutting tool—the groove will prevent the tool from rotating or pulling out. It is important to select the shortest adapter that will allow the tool to reach the cutting zone without any interference. Longer adapters will give a better reach, but they will cause a higher torque to be exerted on the spindle, and their lack of rigidity can cause damaging chatter.

Closely related to the end mill adapter is the shell mill adapter. Some larger milling tools—specifically face mills—do not have a shank. They are manufactured as a *shell* that is intended to be mounted on a rigid adapter. The adapter has

Figure 4.20 **Spindle tooling: drill chuck, collet chuck and collets, end mill adapter, and boring head.** *Courtesy of Command Tooling Systems.*

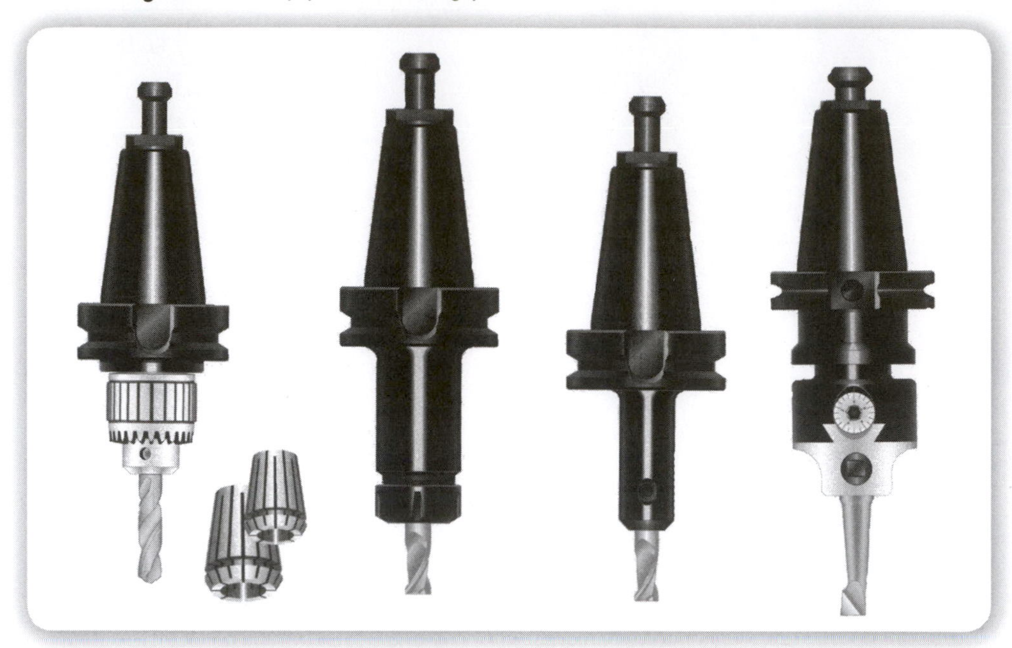

a cylindrical locating surface to maintain concentricity and two drive tangs to keep the tool from slipping as it cuts. The milling tool is usually fastened to the adapter with a socket-head cap screw. The chief advantage of the shell configuration is that the cutting tool is held very rigidly, and the tool is unlikely to slip under heavy cutting conditions (a real problem for round shanks).

Hole-making tools are often held in a drill chuck adapter. These adapters work well for holes that do not require a high degree of accuracy. Drill chucks can accommodate a wide range of diameters, usually from 0.093 to 0.500" in a single chuck. This feature makes the drill chuck a very economical tool to own. It is common to perform spot drilling, drilling, counter-boring, countersinking, and sometimes reaming operations with the tool held in a drill chuck. However, drill chucks tend to have an unacceptable amount of circular runout for precision operation; therefore, reamers and taps are usually held in a spring collet.

Drill chucks are only used for cutting tools that are plunged downward into the workpiece—never attempt to install an end mill into a drill chuck, not even for plunge cutting.

A collet chuck is another popular adapter that is used to hold drills, reamer, taps, and end mills—virtually any tool used for machining. It offers a great amount of flexibility with regard to the types of tools it can hold and the range of diameters it can accommodate. Collet chucks use a tapered spring collet that can deform to fit an entire range of diameters (***Figure 4.21***). The tapered design ensures that the shank of the tool will be tightly and evenly gripped and highly concentric. Collet chuck systems are relatively expensive compared to end mill and drill chuck adapters, so they are only used when a less expensive adapter is not available.

Taps can be held in a collet for rigid tapping, but there are a few specialized adapters made specifically for tapping that offer a better solution. Quick-change tap holders allow for rapid tool changes that will maintain the correct depth of the threaded hole. Tapping routines often require a certain amount of adjustment by the operator in order to meet specifications, so a tool change that maintains the depth will save time and money.

Figure 4.21 A collet chuck is used to secure drills, reamers, and end mills for performing machining operations. The tapered geometry creates a very tight and concentric fit.

4.6 CUTTING TOOLS FOR TURNING

The cutting tools used for turning are very different than those used for milling. Turning operations use single-point tools that only use one cutting edge at a time. *Figure 4.22* illustrates a variety of turning tools for external and internal work. The great demand put on this single edge makes carbide the overwhelming choice for everyday CNC turning applications—HSS is rarely used except for form tools and intricate geometry. CNC turning relies heavily on indexable insert tools for most operations, and just a few tool styles can be adapted to get most of the work done.

Outside Diameter Turning Tools

OD turning tools come in a few basic styles to hold the standard inserts that we have already discussed. The workhorse of the bunch is the 80° diamond turning tool. The most popular configuration is to set the insert in the holder with a 5° side and end angle and a negative rake angle. The popularity of this configuration comes from its ability to perform both OD turning and facing with the same tool—eliminating the need for a tool change.

Of course, OD turning tools can also use a variety of other insert styles, including the 60° triangle, 55° diamond, and 35° diamond. The holders are available with a number of different side and end cutting angles and rake angles to meet almost every turning requirement (*see Figure 4.23*).

Figure 4.22 Single-point turning tools (from left to right) boring, ID threading, ID grooving, OD profiling, OD turning, and OD threading.

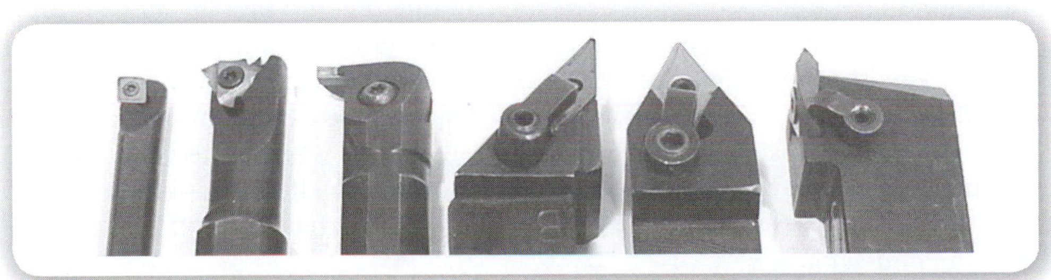

Figure 4.23 Turning tools are manufactured in a variety of orientations for turning from any direction.

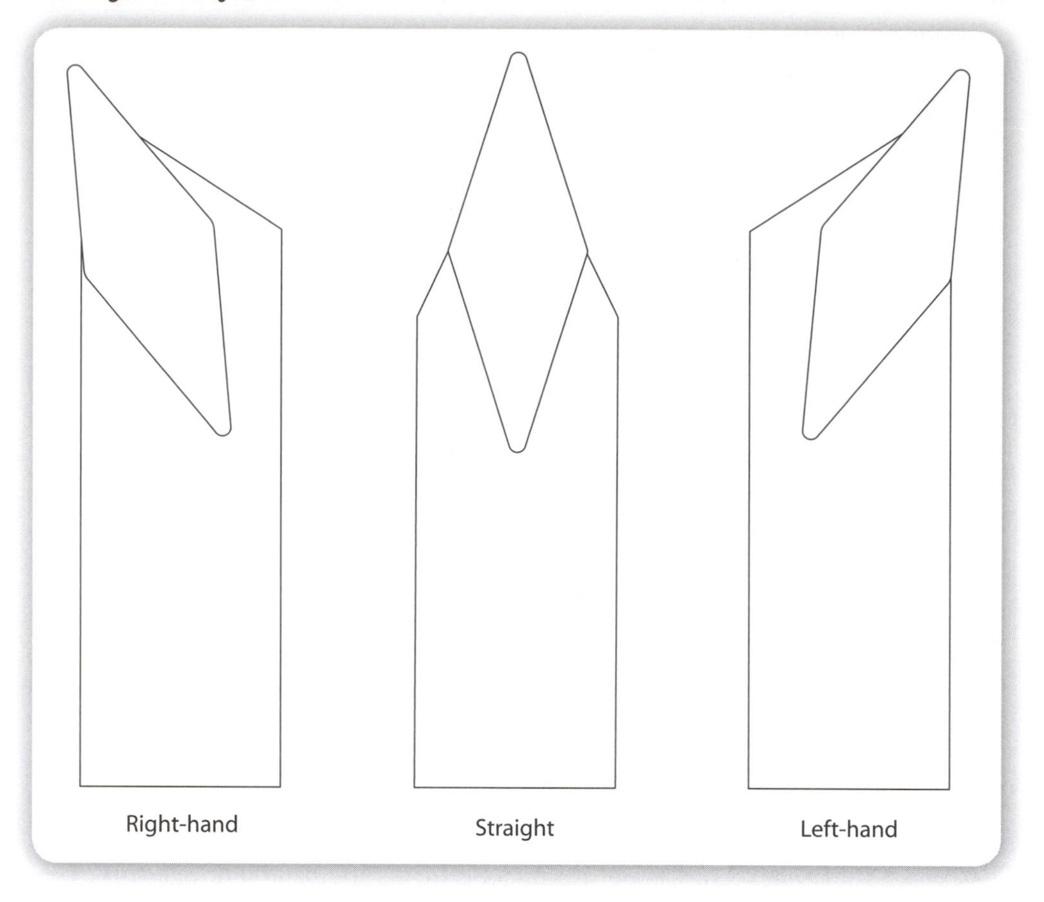

Boring Bars

Indexable insert boring bars are also available in any number of different configurations. The main difference between OD tools and boring bars is that boring bars are usually designed to use a positive-rake insert. Positive rakes will generally produce lower cutting force than negative rakes. Boring bars inherently suffer from rigidity problems, so positive-rake angles will help lower the amount of deflection. Another approach to the rigidity problem is to manufacture the boring bar shank from solid carbide. Solid cemented carbide is much stiffer than steel, and its density is about two and a half times that of steel. The combination of these two properties allows for less deflection and significant vibration damping.

Tool Holding Methods for Turning

The turning tools on a lathe are held in one of several methods. The most common method is simply to mount the tools in a turret. The turret is an indexable wheel that contains stations for square-shank tools and stations for center tools (boring bars and drills). There is nothing special about mounting square-shank tools that are used for OD turning and facing; they are the same style that we might use when using a conventional lathe. The tools are generally held in with set screws or wedges, and there is sometimes a provision for adjusting the center height. OD tools can usually be mounted either right-side up or upside down. This ability allows a right-handed tool to cut either left or right by simply reversing the direction of the spindle. Center tools are usually held in a collet chuck and then secured to the turret with a bushing and set screw.

Turning centers with high-capacity tool magazines do not use a turret. Instead the tools are mounted in individual holders, which can be held in a magazine and called up at will. The most common tool mounting system is called VDI, which allows the use of *live tooling,* but taper holders are also used.

4.7 CUTTING TOOLS FOR HOLE MAKING

Drills and Reamers

Drills and reamers are popular tools for machining small-diameter holes. (Several others are shown in *Figure 4.24.*) They are readily available in many sizes ranging from 0.020" to well over an inch, but drills under 1" are the most prevalent. HSS twist drills are used for much of the general-purpose work and short production runs. Small twist drills are inexpensive, so most shops will usually keep common sizes in stock. Furthermore, they are easily resharpened when dull, either by hand or with a specialized drill grinder.

Figure 4.24 Hole-making cutting tools (from left to right) HSS twist drill, reamer, center drill, spotting drill, single-flute and zero flute countersink, and counterbore.

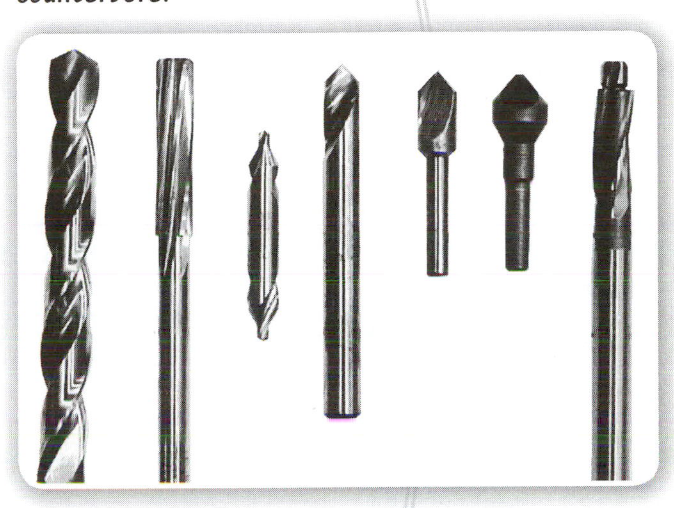

Solid-carbide or carbide-tipped drills are often used for tougher materials and higher volume production. They are significantly more expensive but are often the preferred choice for many applications. Solid-carbide drills are often supplied with TiN coating and through-the-tool coolant holes to increase their tool life. Tool manufacturers are also now marketing small twist drills with throwaway carbide tips—a very economical solution.

Twist drills are generally not rigid enough to start a hole on their own; the point of the drill is somewhat blunt and tends to cause the drill to *walk* off the centerline when starting. Consequently, most drilling operations have to be preceded by a spot drilling or center drilling operation. Spotting drills and center drills have a pointy end and a stiff shank that are able to produce a small divot (spot) for the twist drill to start in without walking. Center drills were designed to produce a 60° bearing area for lathe centers, but they also work well for spotting. Spotting drills have a slightly different design that is made specifically for starting drills; they are less common than center drills, but they are the superior tools for the job.

Figure 4.25 Carbide tipped drill with through-the-tool coolant.
Courtesy of Clackamas Community College.

Larger drill sizes are often constructed to use indexable carbide inserts, and they are often manufactured with through holes to deliver coolant to the cutting zone (*see Figure 4.25*).

Insert-tooth drills are commonly available in sizes from ¾ to 2", and they are the preferred cutting tools for CNC machining of larger diameter holes. Some hybrid designs can even act as a drill and boring bar all in one tool. These are great for lathe work, and they can save a tool change.

Reamers are used to enlarge a predrilled hole to very precise size and roundness. Reamers are often constructed from HSS or carbide and are available from industrial suppliers in standard sizes or by special order in any size needed. However, large reamers can be expensive and are only purchased if the particular job will justify the expense.

Boring Head

Odd-sized and precision holes are often produced with the boring process. Lathes excel at boring, but boring is a little more difficult on a machining center. Boring on a milling center is accomplished by changing the radius of the cutting tool while the spindle remains stationary over the centerline of the hole. A tool called a boring head uses a precise adjusting screw to move the boring bar off-center and thus produces the desired radius in the workpiece (***see Figure 4.26***). This is often a haphazard affair, so the part program may be written to allow the operator to stop the tool and make adjustments.

4.8 WORKHOLDING TOOLING

Workholding is the act of securing the workpiece while machining. Its two primary purposes are to hold the workpiece in place while cutting and to provide an accurate reference surface to align the workpiece. The tools used to perform workholding are usually referred to as simply *tooling* (not to be confused with cutting tools).

Figure 4.26 Eccentric boring head. The boring bar is moved off center to control the hole size.

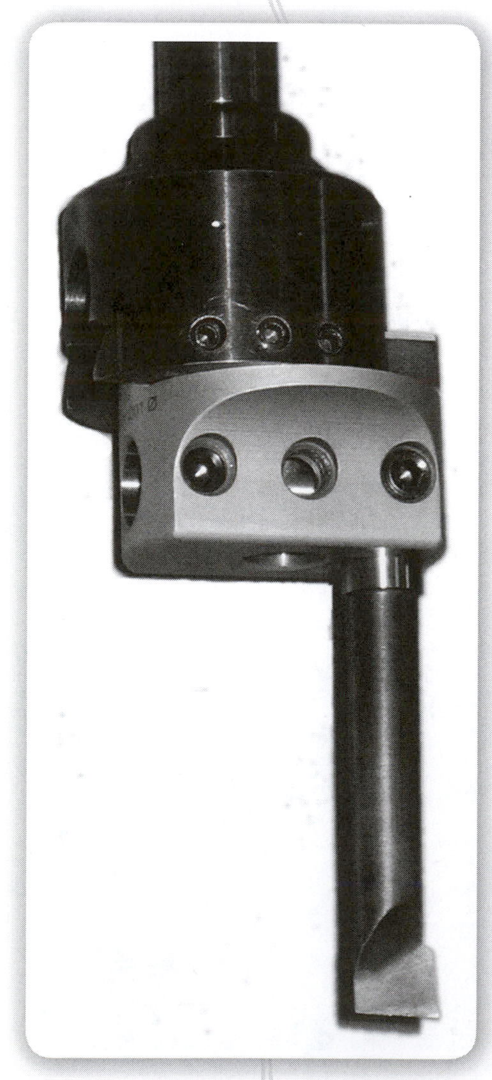

The workpiece must be held rigidly so that it will not slip or deflect during machining. If the workpiece comes loose during machining, it can be quite catastrophic and may cause injury to the operator and damage to the tooling and workpiece. CNC machining offers some special challenges to the machinist because much of the immediate feedback has been removed from the machinist. You cannot simply stop turning the feed wheel if a problem arises as you might on a conventional machine tool. Therefore, it is important to understand the limitations of the various devices used to hold the workpiece.

The second function of providing an accurate reference surface is also important. The surfaces of some workholding tools are expected to provide the alignment for the workpiece; therefore, they must be flat, square, and parallel, or the workpiece will not be aligned correctly. Because of these requirements, tooling is usually made to very high quality standards and exceptional tolerances. After all, the quality of the workholding tooling will directly affect the accuracy of the finished dimensions.

To provide for these two functions, tooling falls into two main categories: alignment tools and clamping tools. Alignment tools provide the reference surfaces, and clamping tools provide the force needed to resist the cutting forces. Of course, these functions are sometimes found in a single tool such as a milling vise or a lathe chuck.

One important concept that you must understand is that clamping operations work on the principle of friction. Friction is the resistance you feel as you try to slide a heavy box across the floor, and it is the force that keeps your car on the road when you drive around a curve. Friction is caused by the microscopic roughness and adhesion between two surfaces that are in contact with each other. Some surfaces are rough and will cause a great deal of friction, whereas others are smooth and cause very little resistance. For example, rubber on asphalt creates a lot of friction, but hardened steel against steel creates very little.

Friction in machining is used to keep the workpiece from moving as it is being cut (***see Figure 4.27***). However, smooth metal surfaces tend to produce relatively small amounts of friction and are consequently slippery. Slippery surfaces require a great amount of clamping force to keep the parts from moving as a cut is taken. The heavy clamping forces can potentially cause the tooling or the workpiece to deflect. Consequently, the tooling must be very rigid to avoid deflecting to an unacceptable degree. Heavy clamping can also distort a fragile workpiece enough to cause dimensional errors. Sometimes a workpiece is distorted during clamping and will not show any visible signs of damage. Nonetheless, as soon as the clamping is removed the workpiece will tend to spring back and cause the feature dimensions to change, as illustrated in *Figure 4.28*.

Figure 4.27 **Clamping forces create enough friction to resist the cutting forces and hold the workpiece in place.**

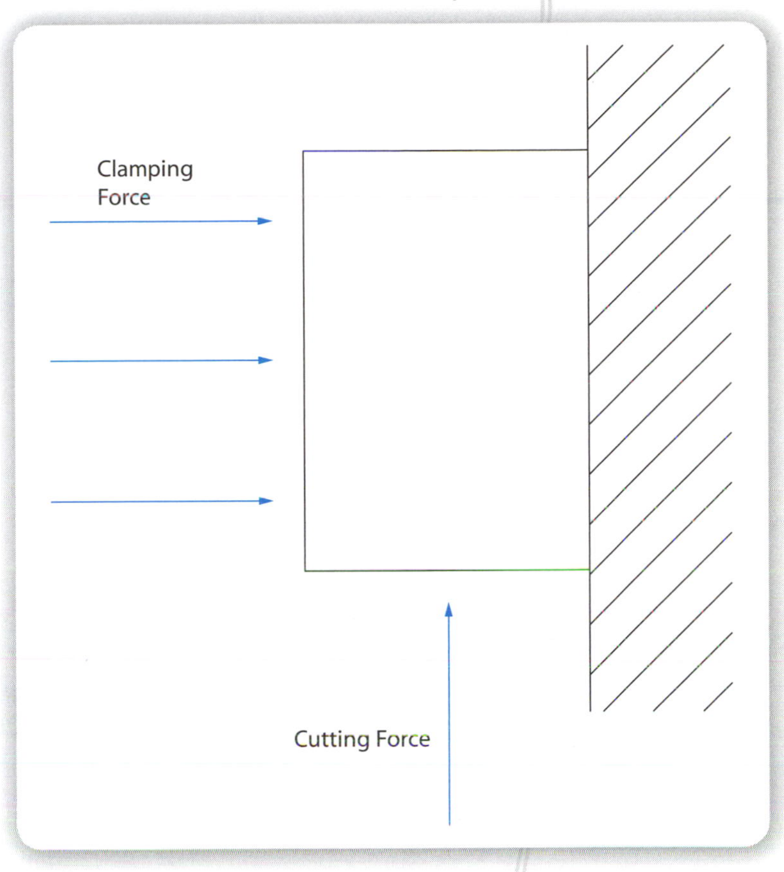

The Milling Vise

The milling vise is perhaps the most common workholding tool in use today (*see Figure 4.29*). Milling vises are available in many different sizes and styles to fit a variety of machining chores. They provide the clamping force needed to hold the workpiece in place and they provide the precise reference surfaces that we use to locate the workpiece—all in one package.

The key to the milling vise's success is its flexibility to hold just about any small workpiece. This is accomplished by the addition of many accessories for holding and locating work. The principal accessory is the vise's removable jaw (*see Figure 4.30*). These jaws can be constructed of hardened steel and precisely ground for holding workpieces at the proper orientation. Another popular variation is the step jaw, which can eliminate the need for troublesome parallels to align the workpiece.

Figure 4.28 **Care must be taken when clamping fragile workpieces. Clamping forces can distort a fragile workpiece.**

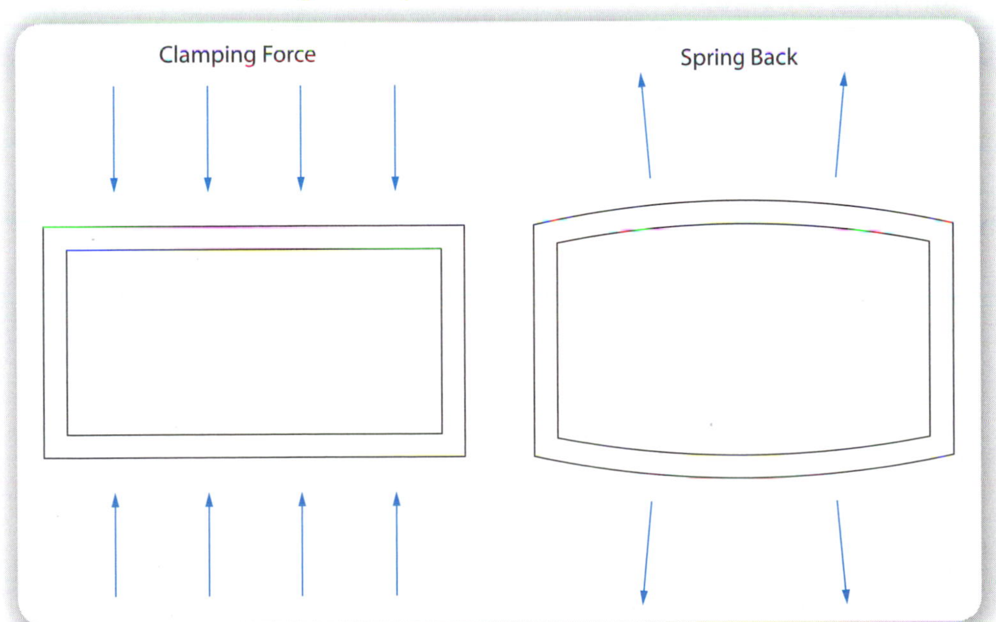

Figure 4.29 A milling vise is the versatile.

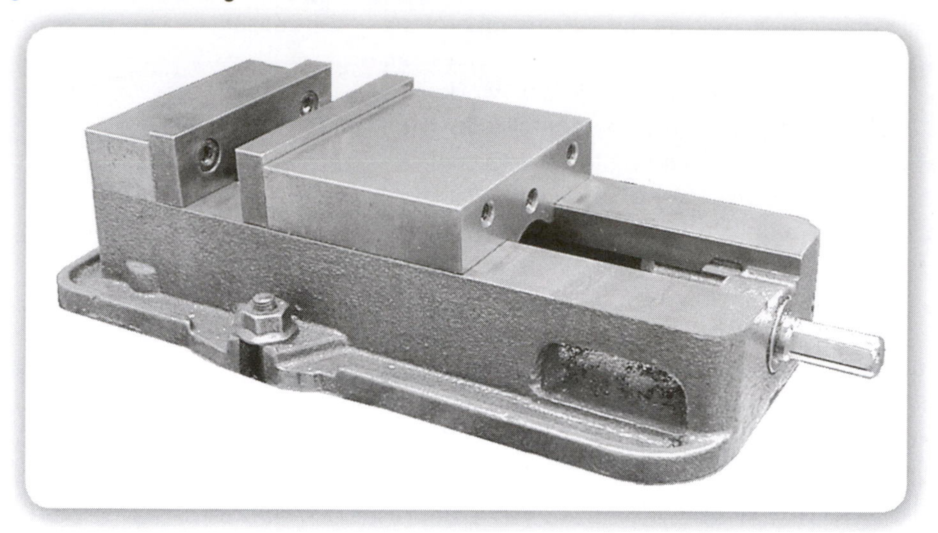

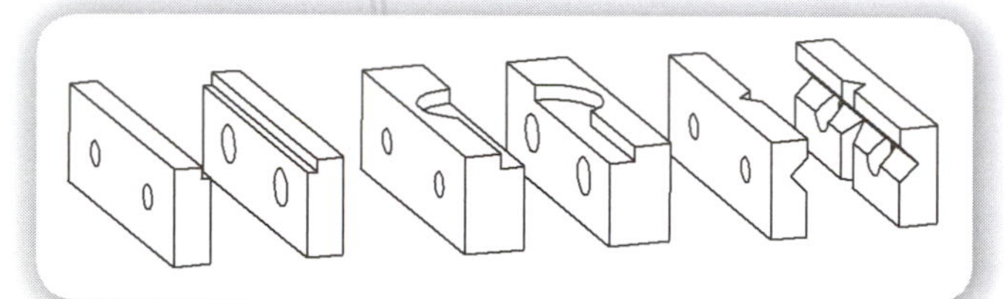

Figure 4.30 A variety of interchangeable vise jaws. Step jaws, soft jaws milled to accommodate an odd shape, and V-groove.

When odd shapes need to be held, the jaws can be made from soft steel or aluminum. Soft jaws, as they are known, can then be machined to accommodate the unique shape of the workpiece—the programmer will often write a part program specifically for this purpose. Many shops will keep a large stock of soft jaws on hand because they are so versatile. The only downside is that the soft material does not have the wear resistance or stiffness that hardened jaws possess, so they will have to be periodically re-machined to the original accuracy.

Workstops are another important vise accessory. Workstops provide a single point of contact to help align the workpiece. They are available in many styles ranging from spider-like arms to simple clamps that attach to the vise jaw (***see Figure 4.31***). Regardless of the workstop style, workstops are never used to keep the workpiece from moving while it is being cut; their purpose is only for alignment.

Clamping Devices

Larger workpieces are not readily held in a milling vise, so they are often clamped to the milling table or other fixture with a strap clamp (***see Figure 4.32***). Strap clamps are oblong billets of steel that rest like a lever with one end on the workpiece and the other on a riser block. They have a slot in the middle to accommodate a threaded stud, which can then fit into a nut or threaded holes to be tightened with a wrench. This is a very flexible clamping system and is usually purchased in a set containing many different sizes of clamps, riser blocks, and studs. However, they can be awkward to use in a production environment. The programmer must also use great care not to crash the cutting tool into the clamps, because they tend to protrude above the surface.

Edge clamps are another style of clamp that is used to secure the workpiece. Edge clamps work by pushing against the surface at a slight angle, which produces both forward and downward thrust. This configuration keeps the clamps below the surface of the workpiece and away from cutting tools. However, the clamping force is usually not as high as could be attained with strap clamps. A notable variation on this principle is miniature edge clamps that are activated with an eccentric mechanism. These are an ideal solution for securing small workpieces on a fixture plate.

Figure 4.31 Different workstop configurations used to locate the workpiece on a single a reference point.

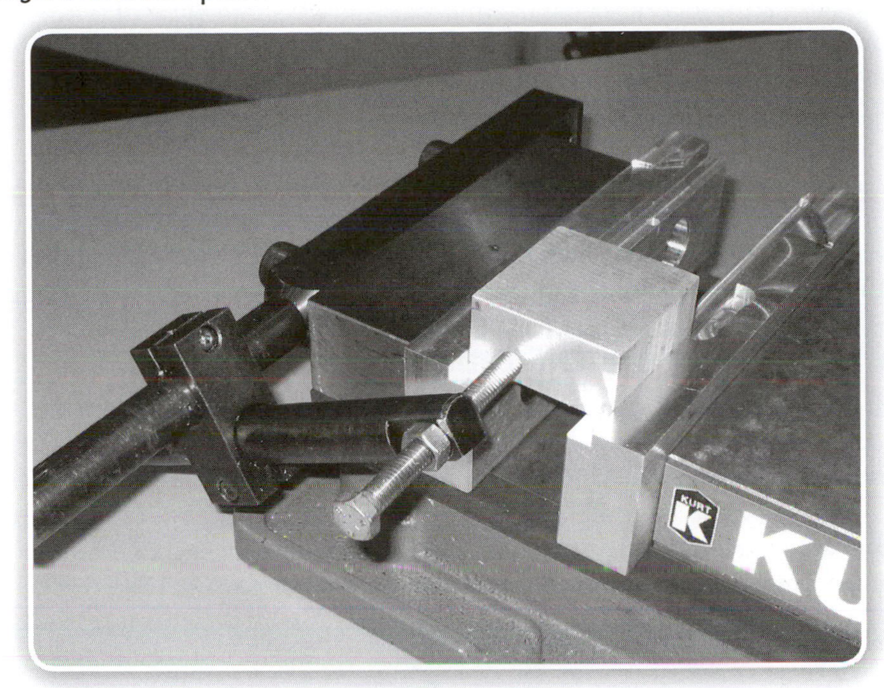

Figure 4.32 Strap clamps in use. The slight inclination of the clamp prevents the workpiece from moving when the clamp is tightened. A spring can also be used under the clamp to keep it from falling during part changes.

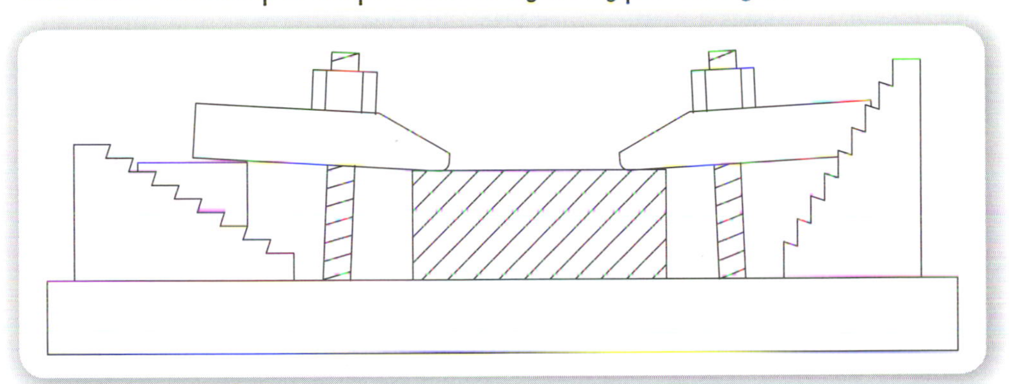

For higher production environments, the use of hydraulic power clamping is quite common. Advantages of hydraulics is their ability to provide high clamping forces in a very small package and to relieve the machine operator of repetitive manual operations that are known to cause injury. Most CNC machine tools also have the ability to automatically clamp or unclamp a hydraulic device by simply giving it an instruction in the program.

Fixturing

A **fixture** is technically any device used to hold and support a workpiece machining. However, the term fixture is usually used to describe any specially built tooling that is designed to be used for a particular workpiece or family of workpieces. Fixtures are generally a collection of clamping and locating tools that are mounted on top of a steel base. They may contain work supports, locating pins, angle brackets, and any number of clamping mechanisms. Fixtures that have been custom designed are sometimes called *dedicated tooling* because they have been built for a specific purpose and cannot be easily reconfigured for any other purpose. Dedicated tooling is avoided unless the quantity can justify the expense.

fixture:
Workholding tools that are designed to clamp and position a workpiece for machining.

dedicated tooling:
Workholding tools that are designed for one particular job and are not intended to be used elsewhere.

A better solution to tooling for smaller production quantities is modular fixtures. Modular fixturing is a system of tooling components that are designed to be interchangeable and fit together in many different configurations (like a big Lego set). The system starts with a fixture base that has been drilled with an accurate pattern of locating and mounting holes. ***Figure 4.33*** shows a variety of modular fixturing components that can be configured for any number of machining jobs. All of the other components in the system are then manufactured with the same bolt hole pattern so that any component is capable of being mounted at any location on the base. Modular tooling provides the best of both worlds—we can design custom fixtures to hold unique workpieces and then turn around and reuse the components on another job. This can eliminate the need to have rooms full of dedicated tooling.

modular tooling:
Workholding tooling that is manufactured in standard mounting units and designed to be quickly reconfigured for other setups.

Workholding for the CNC Lathe

A chuck is a common workholding tool on both conventional and CNC lathes. The chuck is to the lathe what the milling vise is to the milling machine (i.e., its extreme versatility makes it the workhorse for lathe work). In addition, it has interchangeable jaws much like the milling vise. The typical lathe chucks for CNC turning centers are hydraulically activated and much different from the mechanical-style scroll and independent chucks that we find on conventional lathes—mechanical chucks are rarely used on CNC turning centers. Hydraulic control makes it easy to open and close the chuck automatically and allows a fine adjustment of the clamping force.

CNC lathe chucks are used almost exclusively with soft jaws (***see Figure 4.34***). For each new job, the jaws are adjusted into position and then bored to a diameter that is very close to the size of the workpiece. This may seem time consuming, but the close bore creates a highly concentric and close-fitting locating surface that is necessary to maximize the power of the machine tool and cutting tools. CNC

Figure 4.33 Modular fixturing components showing a variety of locating tooling and clamping tooling. *Courtesy of Carr Lane.*

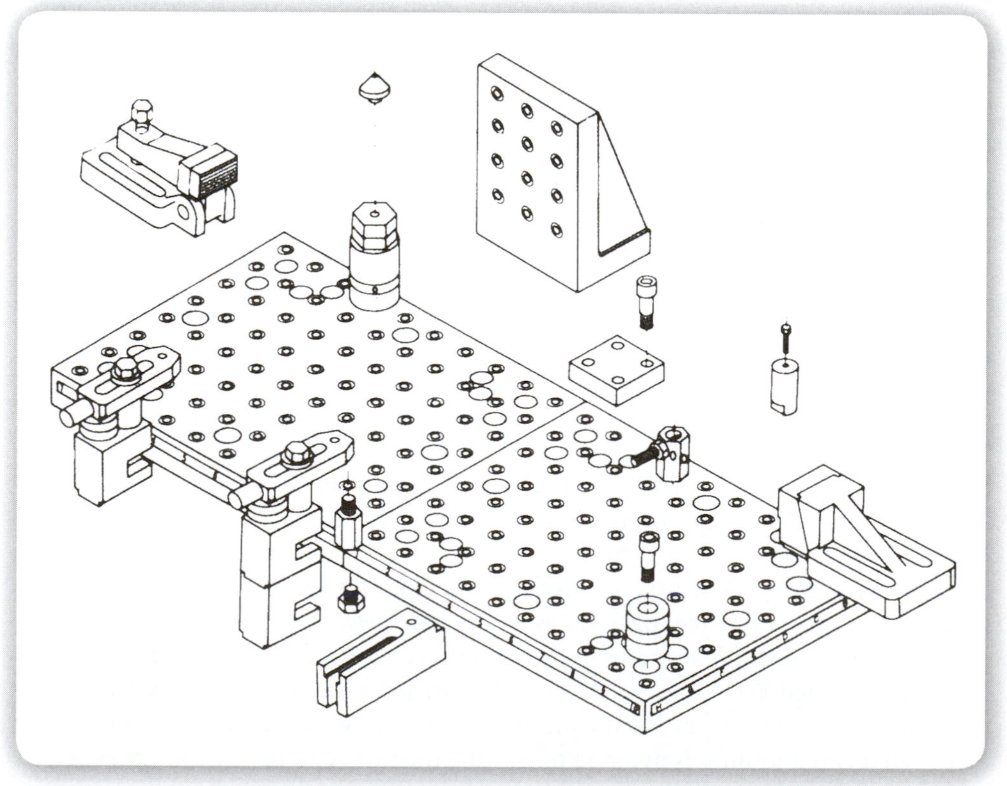

lathes are capable of high spindle speeds and heavy feeds, so good workholding is extremely important or the workpiece can slip—or worse—come out of the chuck.

CNC turning presents some dangers that are not as obvious in CNC milling. A large workpiece will develop a tremendous amount of kinetic energy when it is held in a chuck spun at several thousand RPMs—particularly if the workpiece is unbalanced. So much energy is generated that a workpiece coming out of the chuck could cause serious injury to the operator. Consequently, most hydraulic chucks will have a maximum RPM rating to prevent damage or injury resulting from failure from high centripetal forces. The bottom line is that you should make sure the workpiece is properly secured and that the guards are in place before machining.

Figure 4.34 **A hydraulically actuated lathe chuck with soft jaws bored to accommodate workpiece.** *Courtesy Clackamas Community College.*

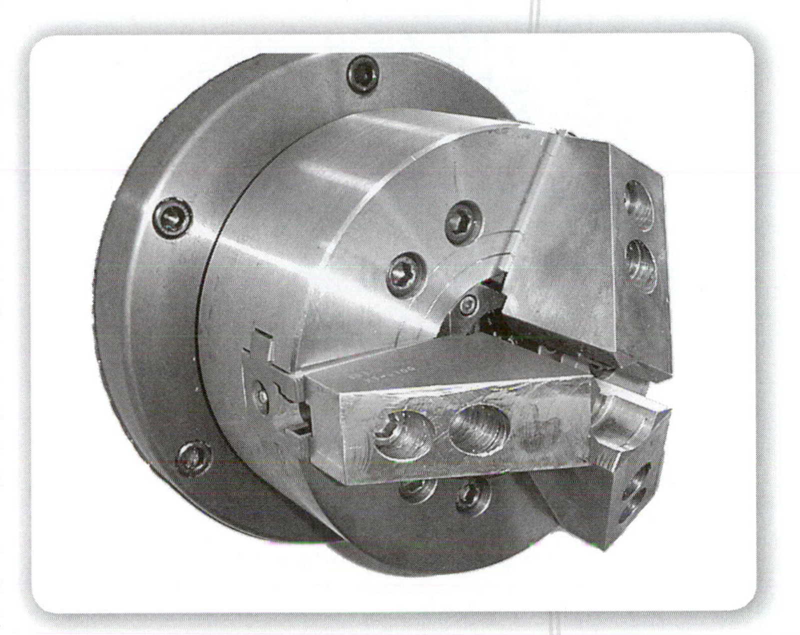

Collet chucks are another important workholding tool for CNC turning centers. Collets are used for smaller diameter workpieces that must be held to close concentricity. The chief advantage of collet systems is that they do not have to be bored like chuck jaws. Furthermore, the smaller diameter chuck can be safely rotated at much higher speeds to produce a good surface finish on small diameters.

Your Turn

1. Report to the class the advantages and disadvantages of the readily available types of material you have to machine within your classroom setting.
2. Create a presentation on the different types of cutting tools available and their purpose. List all pros and cons.

CHAPTER SUMMARY

- Some of the basic machining processes performed on CNC machine tools are turning, milling, boring, drilling, reaming, and tapping.
- The milling processes can produce flat surfaces, pockets, contours, and three-dimensional surfaces.
- Turning describes many operations that are performed on a lathe to produce cylindrical workpieces. These operations include OD and ID turning, profiling, facing, grooving, and threading.
- Metals are the most common materials that are encountered in machining. Ferrous metals such as steel, stainless steel, and cast iron contain large amounts of iron. Nonferrous metals include aluminum, brass, bronze magnesium, nickel, titanium, and zinc.
- Materials all have properties such as strength and hardness that allow us to predict the ease at which they will machine. We can also use the index of machinability to compare the relative values of metals.
- HSS is a popular tool material for end mills and drills. It is a very tough material but can handle only a moderate amount of heat.
- HSS end mills are very popular tools for milling. They are available in various styles, and it is important to select the proper geometry for the cutting conditions.
- Carbide cutting tools are becoming increasingly important for turning, milling, and drilling operations. Carbides can withstand great amounts of heat and can substantially increase productivity. However, carbides are brittle and must be used with care.
- Carbides are available in many grades and coatings. The performance of carbide depends largely on selecting the proper compositions and coating for the workpiece material and process.
- Many different types of tooling are available for locating and clamping the workpiece. CNC machining brings on some special challenges, so it is important to understand the qualities and limitations of tooling in order to maximize the machining efficiency.

BRING IT HOME

1. What is the difference between iron and steel?
2. How does the hardness of a material affect machining?
3. Describe the conditioning process that we might use to soften a block of previously hardened steel in order to perform machining operations.

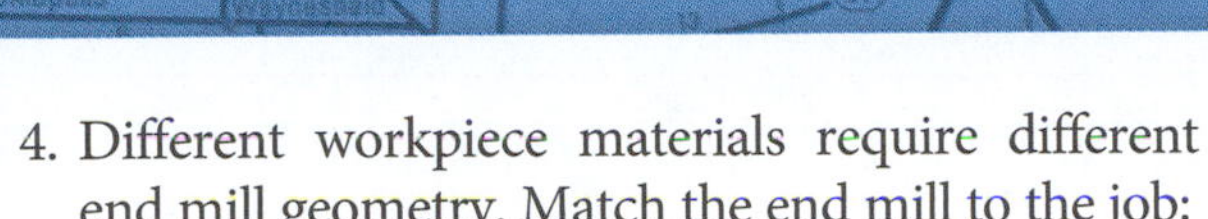

4. Different workpiece materials require different end mill geometry. Match the end mill to the job:
   A. Finish milling of steel
   B. Deep pocket milling in aluminum
   C. Milling of soft stainless steel
   D. Heavy material removal in a variety of materials
   E. Two-flute, high helix
   F. Three-flute end mill
   G. Roughing end mill
   H. Four-flute end mill
5. What is the difference between center cutting and non-center cutting end mills? What will happen if you attempt to plunge cut with an end mill that is not center cutting?
6. What applications would call for exotic tool materials such as CBN or ceramics?
7. What is the primary tool material used for CNC turning?
8. What is the purpose of tool coating such as TiN and TiCN?
9. How is insert strength related to the shape of the corner?
10. Is it necessary to spot drill before drilling a hole with a twist drill? Why or why not?
11. Is it acceptable to mount an end mill in a drill chuck? What might happen if you try this?
12. Under what circumstances would you need to bore a hole? Why not just drill it?
13. What are the two primary purposes of work-holding tooling?
14. Select the proper ANSI and ISO grades of carbide for the following machining conditions:
    A. Finish turning of aluminum
    B. Interrupted cut and heavy roughing of steel
    C. General purpose cutting of cast iron
    D. General purpose cutting of steels
15. Select the correct insert description based on the identification code TNMG-322:
    A. Square, positive rake, 1/64 nose radius
    B. 80° diamond, negative rake, 0.375 inscribed circle
    C. Triangle, negative rake, 0.375 inscribed circle, 1/32 nose radius

EXTRA MILE

1. How is modular tooling different from dedicated tooling?
2. What are some of the dangers associated with workholding while milling or turning?
3. Describe one method that you might use to hold an oddly shaped workpiece in a milling vise.
4. Why do we have to be careful when clamping thin or fragile workpieces? Can you think of any clamping techniques that might eliminate some of these problems?
5. What is the primary tool for holding larger workpieces in a CNC turning center? What are some of the dangers associated with turning that are not as prevalent in milling?

CHAPTER 5 Tool and Workpiece Setup

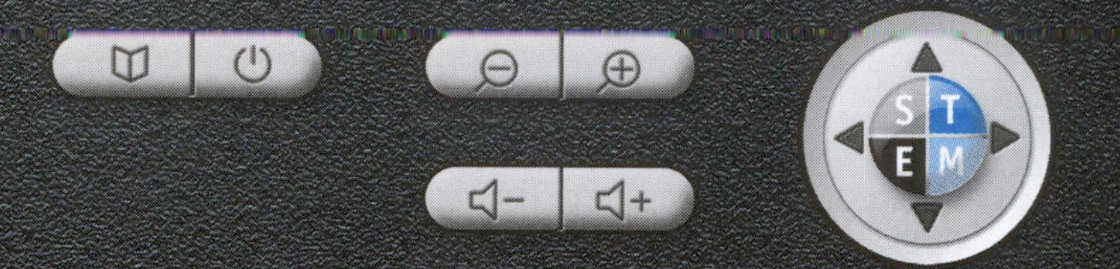

Menu | START LOCATION | DISTANCE | END LOCATION

Before You Begin

Think about these questions as you study the concepts in this chapter:

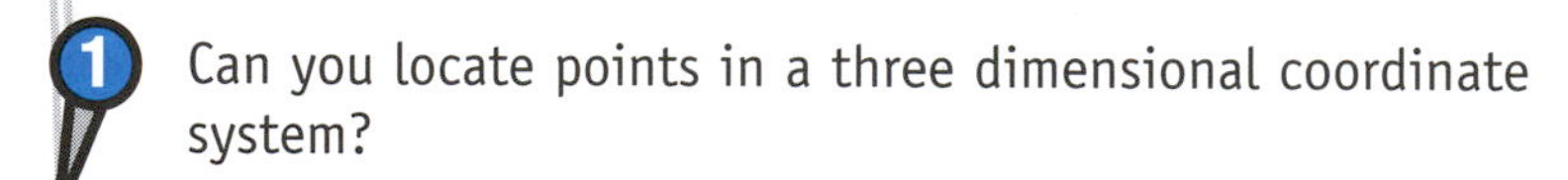

1. Can you locate points in a three dimensional coordinate system?
2. What are the differences between absolute and incremental coordinates?
3. What does an offset registry do?
4. What are the differences between the two methods of setting the Z-offset and length offset?
5. How are workplanes defined to allow multiple setups?
6. Why do we preset tools in production machining?

Key Terms

Axis
Cartesian Coordinate System
Compensation
Coordinate
Diameter Offset
Fixture Offset
Height Offset
Machine Home
Offset Register
Origin
Part Zero
Reference Position
Work Offset
Workplane

5.1 THE CARTESIAN COORDINATE SYSTEM

In order to describe the location of any point on a two-dimensional plane or in three-dimensional space, we must first develop a coordinate system to define our directions and relative position. This coordinate system is often called the **Cartesian coordinate system**.

The simplest coordinate system related to machining has just two dimensions and forms a plane much like a flat sheet of paper. We can also think of a milling machine table as a flat plane with only two dimensions. The table can move in two directions, left to right and in and out. These two directions are perpendicular to each other, and each direction is called an **axis**. The axes establish the X direction and the Y direction.

The point at which the two axes cross is called the **origin**, as shown in *Figure 5.1*. The origin will distinguish the positive direction from the negative direction. Another function of the origin is to give us a natural reference point. We can describe any point on the plane by giving the point's distance from the origin in each direction. Of course, that distance can be in either the positive or the negative direction.

Take for example the points drawn in *Figure 5.2*. Each point is defined by looking at its distance along the X- and Y-axes from the origin. These distances are called absolute coordinates and are often written in parentheses in alphabetical order. For example, there is a point in the upper right with the coordinates (1.25, 1.0). We can see that this point is 1.25" to the right of the origin and 1.0" above the origin.

We mentioned before that it was also possible to have points that are in a negative direction from the origin. Again in Figure 5.2, there is a point in the lower left with the coordinates (−0.75, −0.75). The negative values indicate that the point is to the left

cartesian coordinate system: A system used to define points in space by establishing directions (axes) and a reference position (origin). Coordinate systems can be rectangular or polar.

axis: A single linear direction used in the Cartesian coordinate system to locate points in space (e.g. left-right, in-out up-down).

origin: The point where two or more axes of a coordinate system intersect. The origin will have the coordinates of zero.

coordinate: A distance along an axis or axes from the origin of a Cartesian coordinate system. Any point in space can be located relative to a known position with the use of coordinates.

Figure 5.1 A flat, two-dimensional coordinate system. The intersection of the perpendicular axes X and Y defines the origin. Any point on this plane can be described by giving its distance from the origin along each axis—the coordinates.

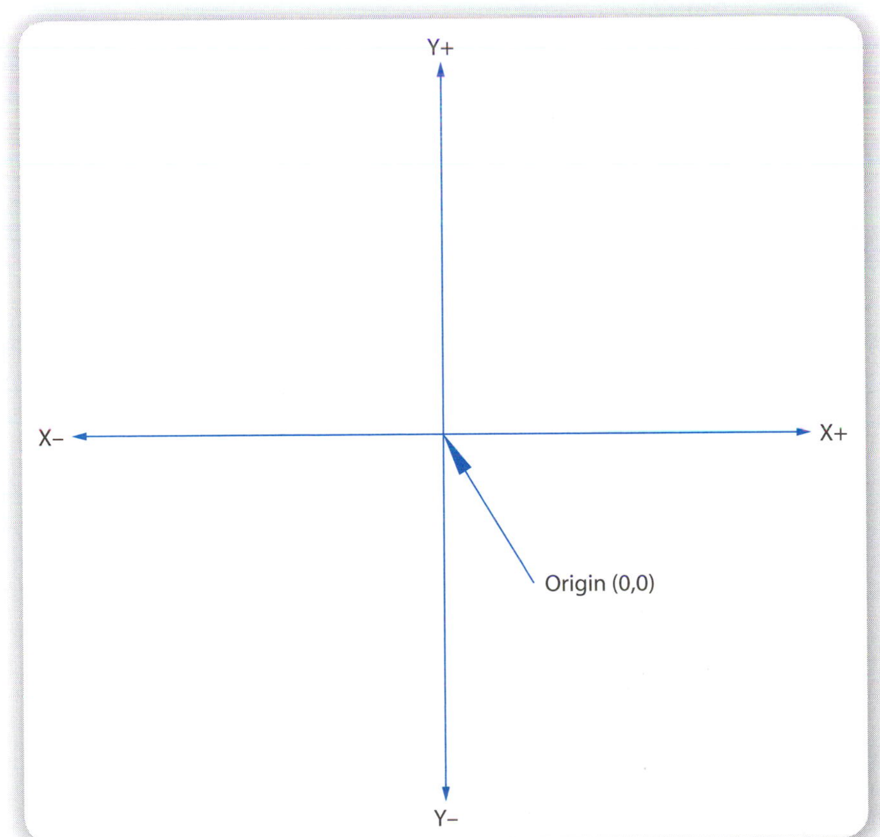

Figure 5.2 Points in a two-dimensional coordinate system. The absolute coordinates are specified by first giving the distance along the X-axis and then the distance along the Y-axis. Note that a point located directly on an axis will have a value of zero for the opposite axis coordinate.

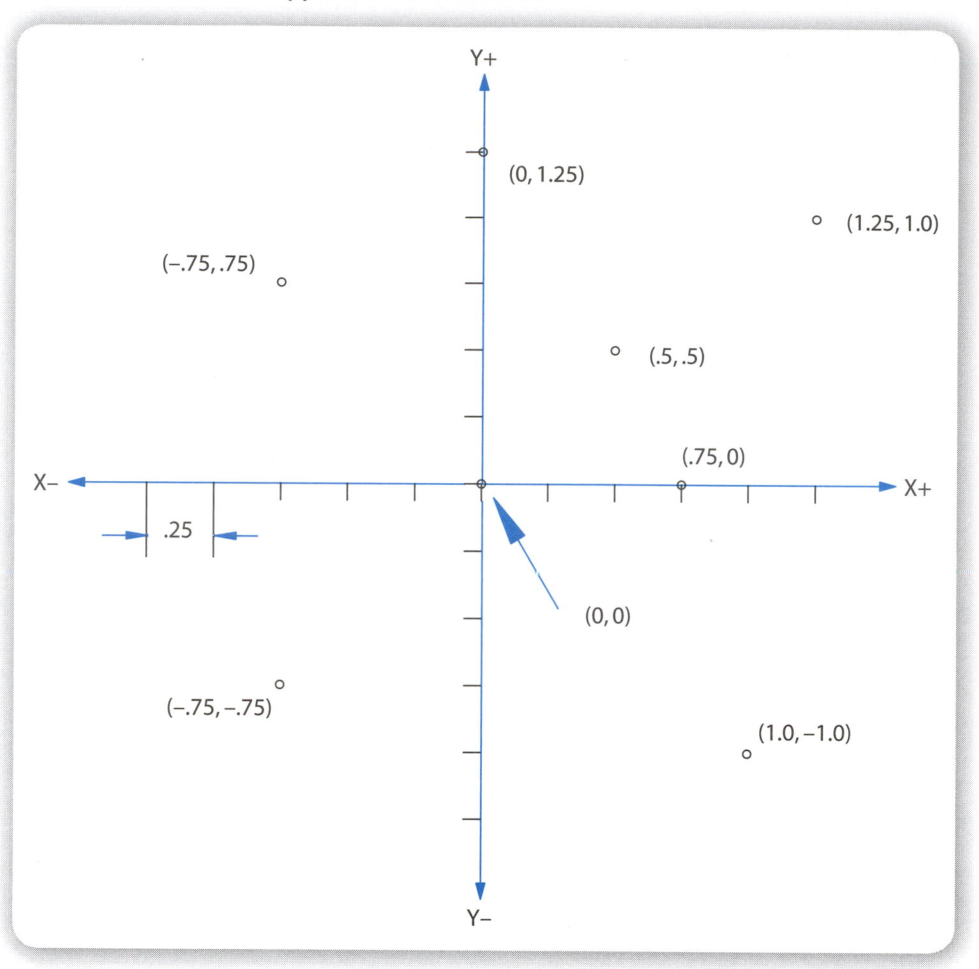

of the origin (a negative X value) and below the origin (a negative Y value). Negative coordinates are no more difficult to work with than positive values, but we must diligently remember to include the minus sign or an error will occur in our calculation.

The last variation of coordinates occurs when a point falls directly on one or more of the axes. For example, if a point is drawn directly at the origin, it will be zero inches from the origin along both axes and will consequently have the coordinates (0, 0). Or if the point falls directly on the X-axis, such as the point on the right side of the origin with the coordinates (0.75, 0), the distance from the origin will be zero inches in the Y direction.

It is often most useful in CNC machining to work in a three-dimensional coordinate system (*see Figure 5.3*) rather than in only two dimensions. A three-dimensional coordinate system will have an additional vertical axis (up and down) that we will call the Z-axis. The previous examples used a two-dimensional coordinate system. We had to use only two dimensions to describe exactly where the points were located. Two dimensions are a simplified way to find hole locations and tool center points for writing NC code. However, three dimensions are actually needed to produce real-world parts. (*Figure 5.4* shows the axes designation for most common machine tools.)

Points in a three-dimensional system are described in a manner similar to those in a two-dimensional system. The coordinates are given within parentheses in alphabetical order: (X, Y, Z).

Figure 5.3 A three-dimensional coordinate system has an additional Z-axis that can describe a vertical distance to from the origin.

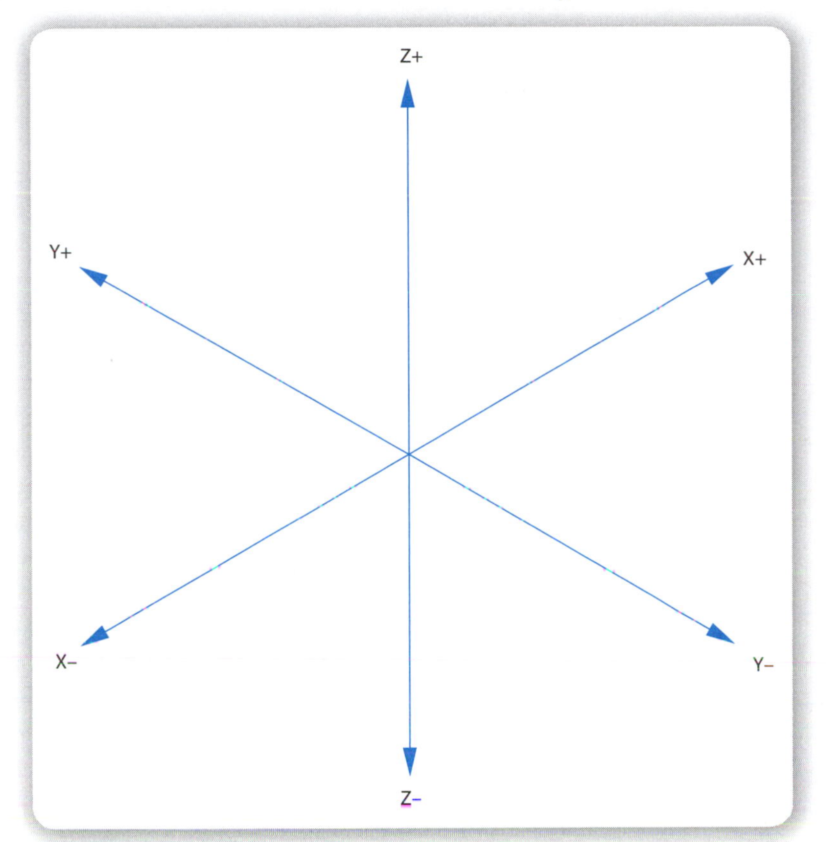

Figure 5.4 (Clockwise From the Top Left) The major axes of a vertical machining center (VMC), horizontal machining center (HMC), and slant-bed lathe.

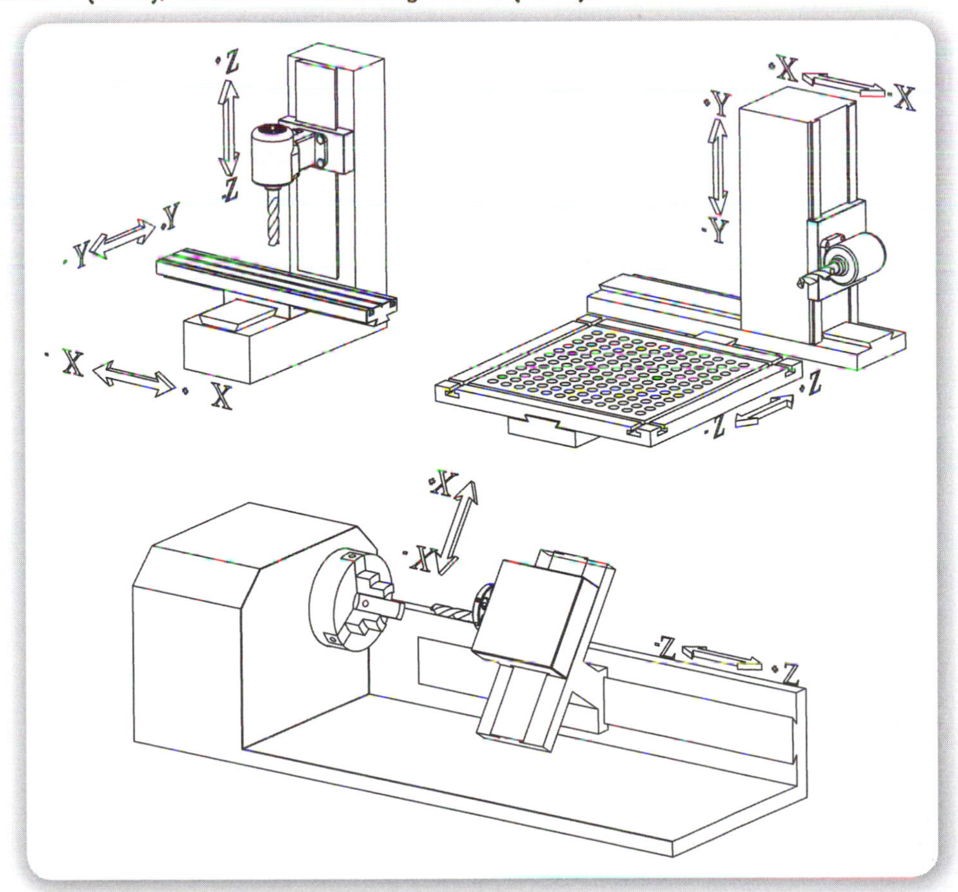

Coordinate System on the Workpiece

In order to machine a workpiece on a CNC machine tool, we need to establish a coordinate system in relation to the workpiece to be machined. Fortunately, the electronic nature of CNC allows us to place the coordinate system any place that we find convenient by simply pressing the right keys on the control. One common practice is to set the origin of the coordinate system on a corner of the workpiece so that the top of the finished workpiece forms an X-Y plane with the Z-coordinate of zero (*see* ***Figure 5.5***). This makes it easy for the programmer to quickly determine when the tool is below the surface of the workpiece, and makes it easy for the operator to touch off the tools. The specifics of setting the coordinate system will be discussed later in this chapter.

Figure 5.5 The coordinate system can be electronically placed at any convenient location on the workpiece in CNC machining. However, it is common to define the top of the finished surface as Z0.0.

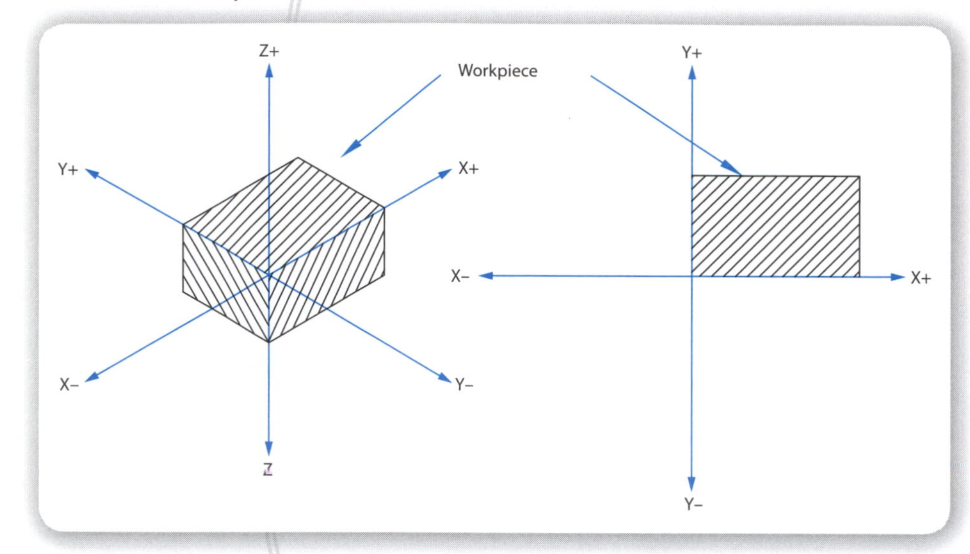

The Role of Coordinates in NC Programming

The role of coordinates in NC programming is quite simply to establish the position of the cutting tools in relation to the workpiece so that the proper dimensions can be produced. The NC programmer will have to plan and program the individual hole locations and toolpaths before the workpiece can be produced. Therefore, the coordinate system must be established prior to any programming and prior to any machining. Furthermore, the shape of any workpiece can be described as a collection of geometrical features that will form the boundaries of the workpiece. If we look at a typical workpiece, such as in ***Figure 5.6***, we can see that it is made up of lines and arcs. Some of these lines and arcs will form contours to represent the exterior boundary, whereas others will define inside pockets or even drilled holes.

To create this workpiece on a CNC machine tool, we have to first find every point that contains a change of geometry, including changes in direction or changes from one type of feature to another. For example, the intersection

Figure 5.6 Any workpiece will consist of geometrical shapes such as lines, arcs, and holes. The programmer must first determine the coordinates of any intersections or centerlines before writing the NC code.

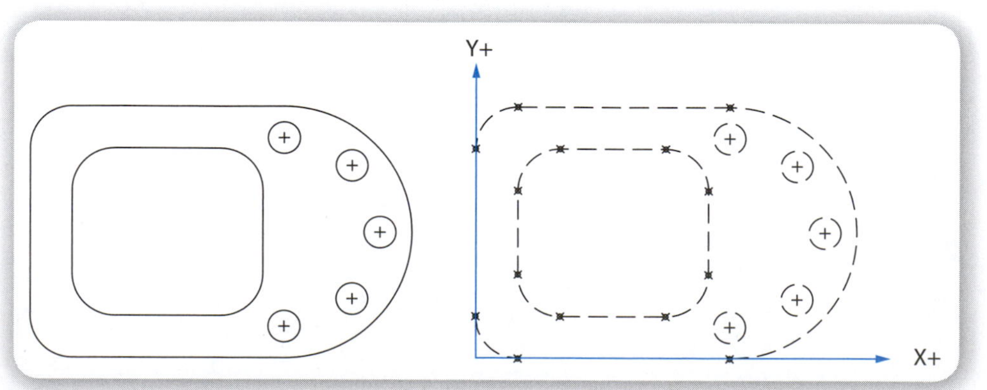

of two lines is a change in geometry. A horizontal line might intersect with a vertical line; therefore, a change of direction would occur. To machine such a contour you will have to program the machine tool to change directions at this point. The geometry might also transition from one type of geometry to another. For example, a line might become tangent to an arc. This point will have to be known before the proper instruction can be written to produce such a contour.

The first thing that an NC programmer will usually do after receiving a blueprint is to locate all of the key positions needed to plan the toolpaths and to write a part program for the job. This will make it a lot easier to write the code because all of the endpoints and center points have been located.

5.2 ABSOLUTE VERSUS INCREMENTAL COORDINATES

So far the coordinate systems we have described have all been based on the idea of absolute coordinates. That is, all coordinates were given relative to a fixed point called the origin. Another way to look at coordinates is as incremental coordinates. Incremental coordinates are not given from a fixed reference point, but rather from the previous point.

Incremental coordinates are referenced from the previous point as though the previous point has become a new origin. The reference point is essentially moving with every new coordinate. Incremental coordinates were once required on older NC machine tools, but modern CNC controls allow the programmer to use either absolute or incremental systems. There are still some situations in which it is still advantageous to use incremental coordinates, but they are rare.

Take, for example, the points in *Figure 5.7*. Imagine that it is our job to drill a series of holes in a part at points 2 through 5 and that the tool was currently located at point number 1. Unfortunately, the machine tool we are to use only allows incremental coordinates. We could start by making a table that describes the absolute coordinates that we already understand, and then calculate the incremental coordinates. This has been accomplished in Table 5.1.

One final point before we begin is that when using only incremental position, we must be careful to return to the starting position when we are finished. Failure to do so will cause next workpiece to be machined incorrectly.

Figure 5.7 Incremental coordinates are referenced from the previous point rather than the origin.

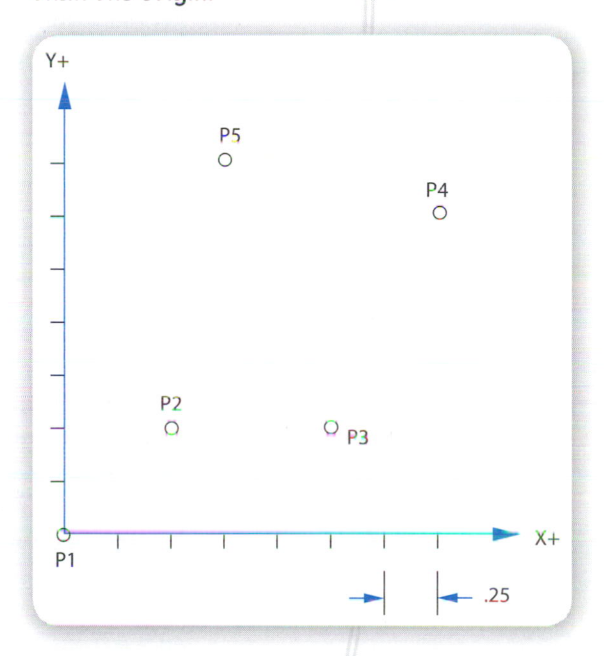

Table 5.1 Absolute and Incremental Coordinates

Point	Absolute		Incremental	
	X	Y	X	Y
P1	0	0	0	0
P2	0.5	0.5	0.5	0.5
P3	1.25	0.5	0.75	0
P4	1.75	1.5	0.5	1.0
P5	0.75	1.75	−1.0	0.25
P1	0	0	−0.75	−1.75
	Sum		0.0	0.0

You can see that the incremental coordinates are the new absolute coordinate minus the previous coordinate. For example, take the tool movement between point 2 and 3. The absolute X-axis coordinates of points 2 and 3 are 0.500 and 1.25, respectively. We can then find the incremental coordinate of point 3 by subtracting:

$$1.25 - 0.500 = 0.750$$

Therefore, the incremental X-axis coordinate of point 3 is 0.750". We could also find the incremental Y-axis coordinate in a similar manner. This time both absolute values are 0.500 and will yield an incremental value of zero:

$$0.500 - 0.500 = 0$$

Incremental coordinates are intrinsically prone to mistakes because of all the calculations. One simple method that we can use to verify that no mistakes have been made is to add up all the X-coordinates and then add up all the Y-coordinates. The values must sum to zero or else a mistake has been made. The values have been summed in the bottom row of Table 5.1 to confirm that our coordinates are correct.

5.3 POLAR COORDINATES AND ROTARY AXES

Polar coordinates are not often encountered directly in writing NC code, but they do crop up in the calculations needed for programming and in general technical problems in manufacturing.

A polar coordinate is given by stating the radius from the origin and the angle of rotation. The convention for referencing the angle of rotation is to set the 3-o'clock position equal to zero degrees, as shown in ***Figure 5.8***. The counterclockwise direction will then define the positive direction, and the clockwise direction will indicate the negative direction. A full circle is also divided into quarters—called quadrants—that are 90° of arc and are bounded by the horizontal and vertical axes.

CNC machine tools can also have polar (or rotary) axes. The most common polar axis is the rotating spindle. However, many machines tools will have one or two additional axes. These axes can be used to swivel the milling head or to rotate the workpiece on a rotary table or in an indexing head.

Rotary axes are designated as rotation around a linear axis and they are paired in alphabetical order as A, B, and C to rotate around axes X, Y, and Z, respectively (***see Figure 5.9***). Determining the sign of a rotary axis can be accomplished by applying the right-hand rule. When your thumb is pointing toward the positive linear direction, your finger curls around in the positive angular direction. The

Figure 5.8 (Left) Rotation around the origin in the positive direction is referenced from the 3-o'clock position (0°). Each quarter of the circle forms a quadrant of 90°. (Right) Polar coordinates are simply the length of the vector and the angle of rotation.

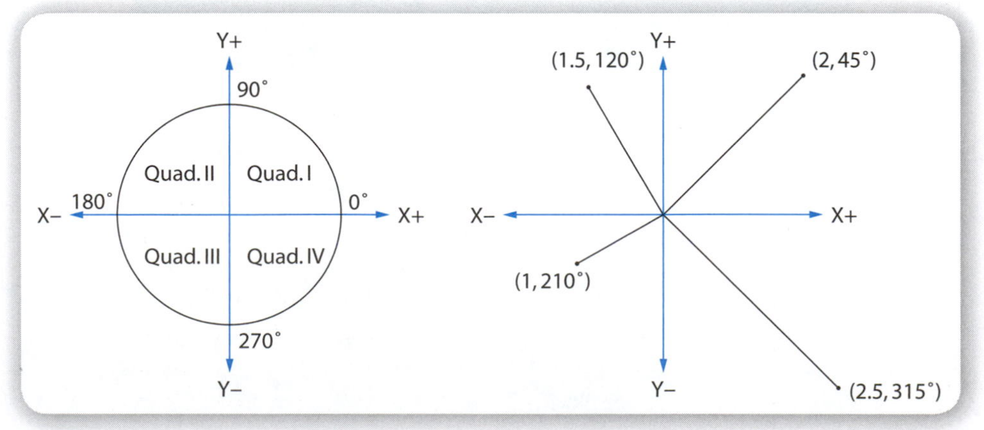

Figure 5.9 Rotary axes are labeled A, B, and C respective to linear axes X, Y, and Z. You can always find the angular direction by applying the right-hand rule. When your thumb is pointing toward the positive linear direction, your finger curls around in the positive angular direction.

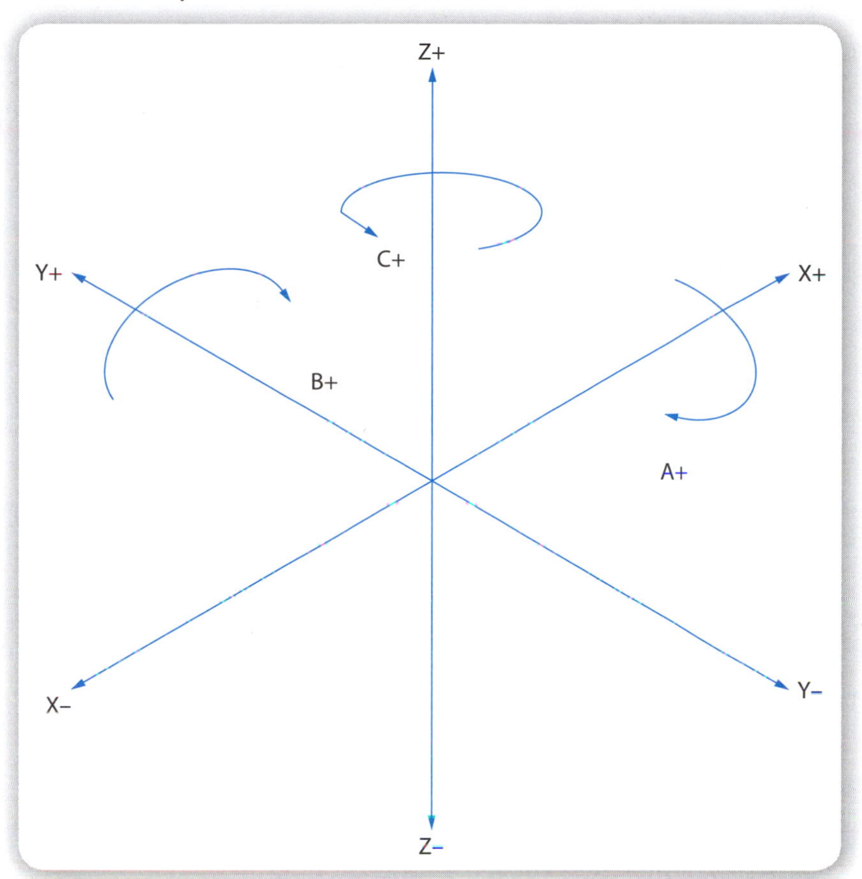

Z-axis is aligned with the spindle, so it is a given that the C-axis is part of most machine tools. A and B axes only have meaning when the machine tool has been built as a four- or five-axis machine or when a rotary table is used.

5.4 WORKPIECE AND MACHINE OFFSETS

Positioning in NC programming is based on Cartesian coordinates; therefore, the machinist must establish the position of the coordinate system before machining the part. In conventional machining it is a common operation to use an edge finder to locate the edges of the workpiece. The machinist will touch the edges of the part with the edge finder and then zero-out either the dials or the Digital Readout (DRO). All positions are then referenced from this point—the origin of the coordinate system.

CNC machine tools work much the same way: the part must be located on the table with an edge finder or indicator before it can be machined. The operator will then press a few buttons to establish the origin of the coordinate system (better known as work zero).

There are several "zeros" that we need to understand to set up a job in a CNC machine tool. First, each machine will have a reference position that is also known as the **machine zero** or the machine home position. This is a position that is designated by the machine tool builder and is established by limit switches or other sensors. For all practical purposes, the machine zero will remain in the same physical place, which is often the upper right-hand corner of the machine. At machine start-up, a routine will usually cause the machine to physically move to the machine zero and then electronically store the position in the machine control unit (MCU).

reference position:
A known position or datum from which other measurements can be taken.

machine home:
A physical location established by the machine tool builder from which all other measurements are referenced. Also called machine zero.

Figure 5.10 Each axis has an associated offset—the distance from the machine zero to the work zero.

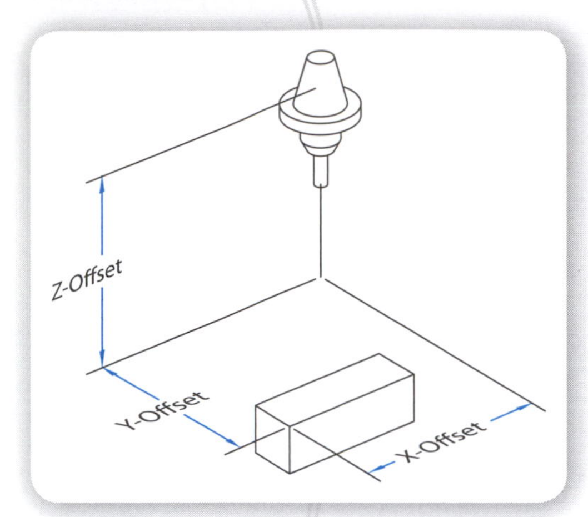

part zero:
The location on the workpiece that serves as the origin for the programmed coordinates in the part program.

work offset:
The distance from the machine zero to the work zero along each axis.

workplane:
A virtual coordinate system that allows the definition of more than one series of work offsets. Workplanes are called from within a part program with G54, G55, G56, etc.

The second zero to be concerned with is the workpiece zero (also called the part zero or work zero). The work zero is the origin of our coordinate system from which our part program is written; it is established as previously discussed by edge finding or indicating the workpiece.

The distance from the machine zero to the work zero is called the work offset. ***Figure 5.10*** illustrates these offsets. Each axis will have an offset value that is stored in the offset registry in the MCU. The MCU keeps track of these values and uses them to move the tool to the proper position for machining. It is possible to adjust these values by editing them in the MCU. For example, adding 0.005" to the X-offset will move the entire coordinate system to the right by 0.005". This is a common adjustment used to control the quality of the finished product.

Many modern machine tools can establish multiple work offsets to define multiple work zeros. In fact, you could have several jobs set up at the same time without having to reestablish the work zero. Multiple work offsets are accomplished by using G-codes designated for work offsets. The designation for the first offset is usually G54, and other offsets are designated G55, G56, etc. For example, ***Figure 5.11*** illustrates the X- and Y-offsets for the G54 workplane. Additional workplanes, such as G55, can coexist in the same work area, as illustrated in ***Figure 5.12***.

It is up to the machine tool builder to determine the G-codes for the offsets. You must call out the offset number in your part program. For example, it is common to call a work offset in the safe line:

```
N010 G20 G54 G90 G98
```

Or, a work offset can be called after a tool change:

```
N200 M03 T06
N205 G43 H03
N210 G54 G00 X1.0 Y1.0
```

Figure 5.11 Offsets are entered in the offset registry.

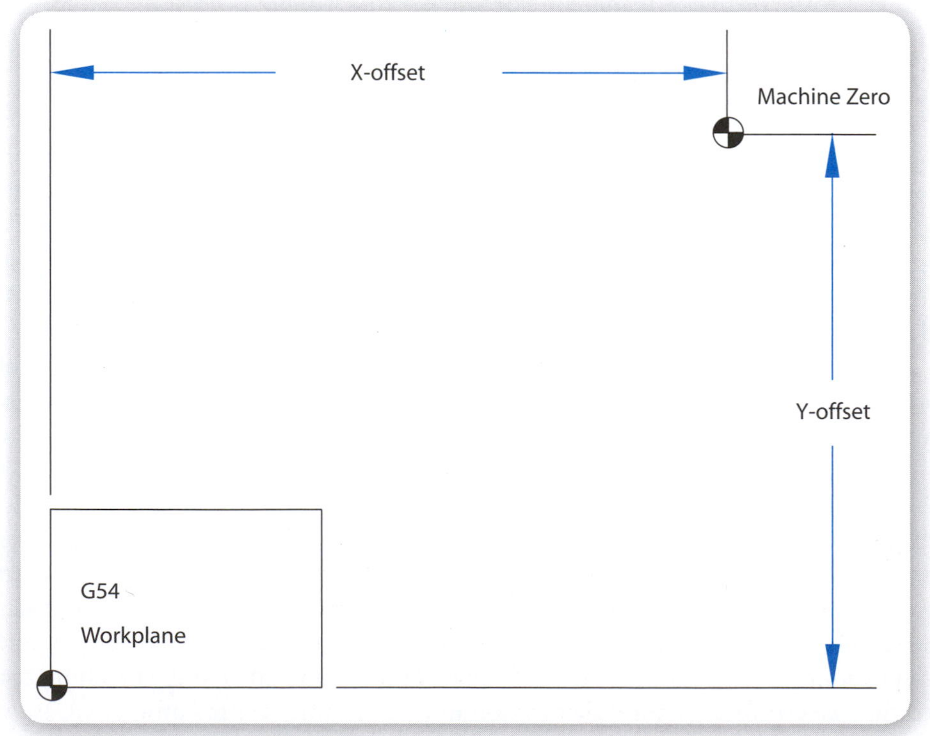

Figure 5.12 Many machines support multiple work offsets. Each offset will have a separate entry in the offsets registry of the control.

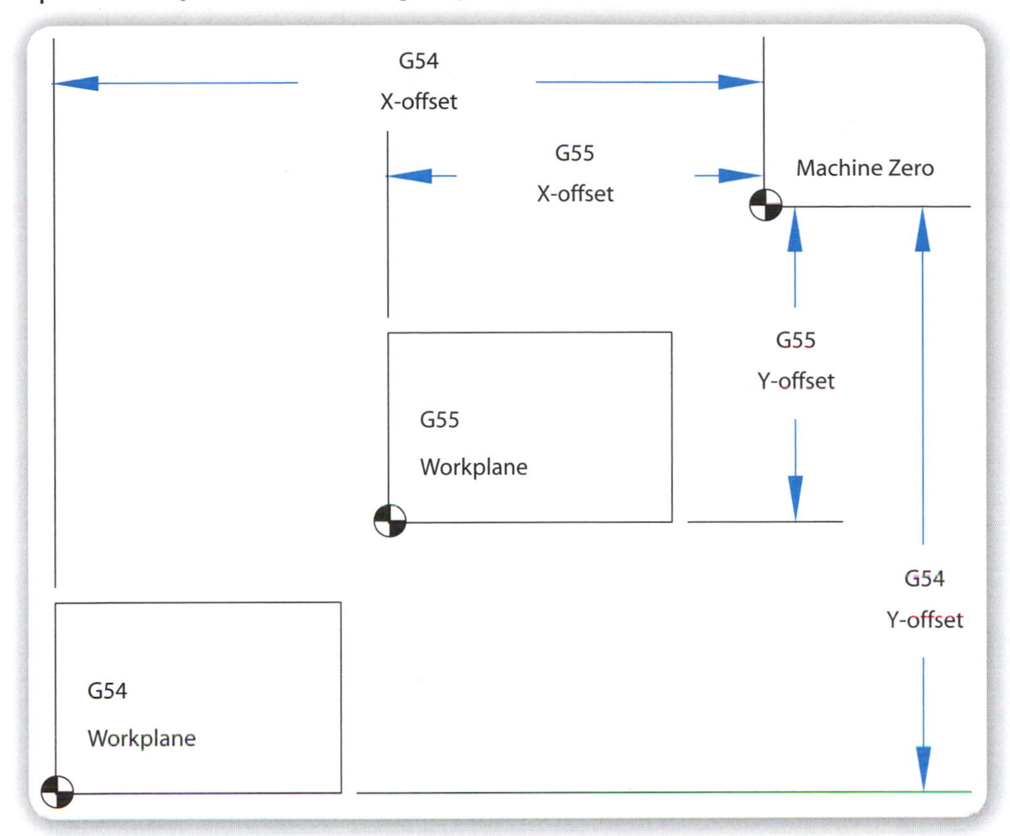

Table 5.2 Possible Offset Values for Different Workplanes on the Same Machine

Designation	X-offset	Y-offset	Z-offset
G54	−12.3450	−6.3300	0.0000
G55	−5.500	−2.3000	0.0000
G56	−6.2322	−5.0987	−2.0

Each work offset must be established by touching off the part with an edge finder or other method. The X, Y, and Z values for each work offset are then stored in the MCU in the offset registry. If we looked in the offset registry of the control, we might see something like the numbers in Table 5.2.

5.5 THE Z-OFFSET AND TOOL LENGTHS

The Z-axis will also have an offset value. However, having a tool in the spindle will complicate setting the Z-axis offset. Recall that the offset value is the distance that the machine will have to travel from the machine zero to get to the edge of the part. The X- and Y-offsets are easy to visualize because we are trying to line up the center of the spindle with the edge of the part. The Z-offset is a little different because of the length of the tool in the spindle. The Z-offset will have to be coordinated with the tools, which also have an offset called the tool **height offset**.

height offset: The difference between the length of a cutting tool and a reference value. Tools that are different lengths will have different height offset values.

When the control reads an instruction in the program to move to a certain Z-level, what it really does is add the programmed coordinate to the Z-offset and then add in the tool length offset. For example, the control reads the instructions

offset register:

The storage location in the control that contains the values for the tools and fixture offsets. These values are established during setup and are used by the control during automatic operation.

G01 Z−.500 while the Z-offset is −5.32 and the length offset is 0.750. The sum of these three values is the actual distance from machine zero to reach the programmed value (Z−.500):

Programmed Value	*−.500*
Tool Length Offset	*.750*
Z-Axis Work Offset	*−5.320*
Sum (Machine Z-Position)	*−5.070*

The control immediately does the math and moves to the machine position:

$$\text{machine Z-position} = -0.500 + (-5.32) + 0.750\text{m}$$
$$\text{machine Z-position} = -5.070$$

The absolute, real machine position is the only position that the control is concerned with from a motion control point of view. All of the other values are contrived to make it easy to program and set up the machine tool.

The Z-offset can be established by two different methods. The first method is the large offsets method; the second is the reference tool method.

The large offsets method leaves the Z-offset in the same location as the machine zero (***see Figure 5.13***). The machine is positioned to machine zero and then the Z-offset is set. The resulting value in the registry will be 0.0000" for the selected workplane. Next, each tool is inserted in the spindle and carefully jogged down to the surface of the part (assuming that the part program uses the top surface as zero), and the tool length offset is set. The resulting offsets will be relatively large negative numbers. (***Figure 5.14*** illustrates the relationship between the actual length of the tool and its length offset value.)

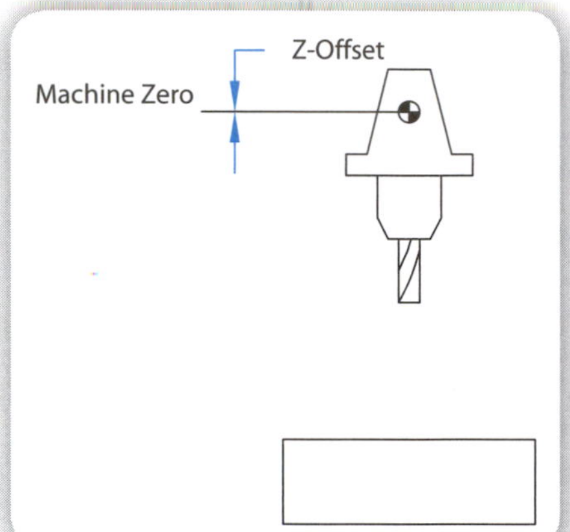

Figure 5.13 When the large offsets method is used the Z-axis offset will be zero because it is set at the machine zero.

When the tool is called in the program, the machine will move the Z-axis the value of the tool length offset plus the value specified in the program. For example, imagine that the Z-offset has been set to zero and the length offset of a drill is −10.505". If the program instructed the drill to create a hole −2.0" deep, then the tool would have to move −12.505" before reaching the bottom of the hole.

The second method for establishing Z-offsets is called the reference tool method. The reference tool method uses a standard tool mounted in a tool holder to establish the Z-offset. The set up person will jog the reference tool down to the surface of the part until contact is made. Then the Z-offset and the tool

Figure 5.14 (Left) Tools of different lengths are positioned above the workpiece at the machine zero. (Right) When using the large offsets method, shorter tools will have larger, negative compensation values than long tools. Short tools must move further to reach the top of the part.

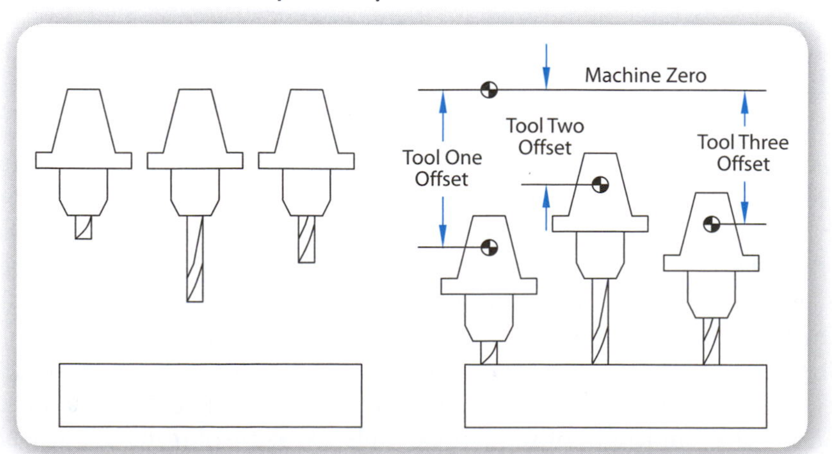

length offset are set at the same time. The results of this method will be a Z-offset value (for the selected workplane, such as G54) of some negative number and a length offset value of zero for the reference tool, as illustrated in ***Figure 5.15***. The reference tool will have a value of zero because the control will not have to add or subtract any number to reach the top of the workpiece.

Subsequent tools are also touched off on the top of the workpiece. Any tool that is longer than the reference tool will have a positive length offset value, and any tool that is shorter than the reference tool will have a negative value, as illustrated in ***Figure 5.16***.

The chief advantage of the reference tool method is that it makes it easy to set up subsequent jobs in the machine without having to touch off each tool again. The operator can simply load the reference tool into the spindle and touch the top of the workpiece. At this point, the Z-offset is set and machining can begin. The other tools do not have to be set again because they are already set relative to the reference tool.

On the other hand, if the large offsets method had been used, each tool would have to be touched off again. This can be a slow process, but it does tend to prevent the operator from becoming complacent. It is generally not a good idea to assume that the tools are already set to the proper length unless some methods are in place to ensure that they have been.

compensation: The act of moving the tool centerline away from the part edge in order to maintain the correct dimensions.

Figure 5.15 The reference tool will have a height offset value of zero. The Z-offset and the length offset are set while the reference tool is touching the top of the workpiece. Subsequent setups require only the Z-offset to be set because the length offsets of the other tools are relative to the reference tool and will remain unchanged.

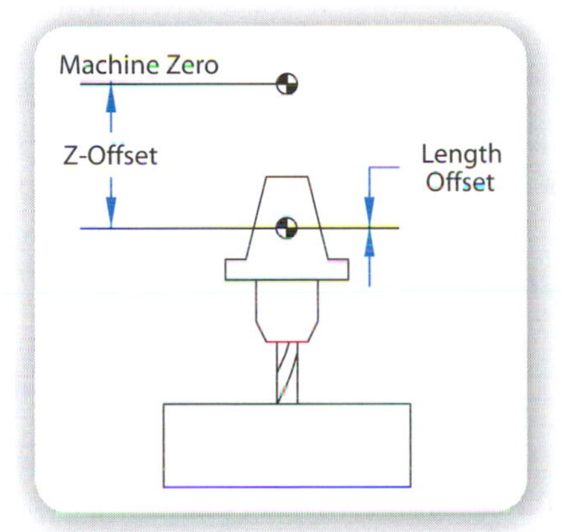

Figure 5.16 Tools that are shorter than the reference tool will have a negative height offset (left), whereas longer tools will have a positive length offset (right). However, the Z-offset will not change heights—it was set while the reference tool was positioned on the top of the workpiece.

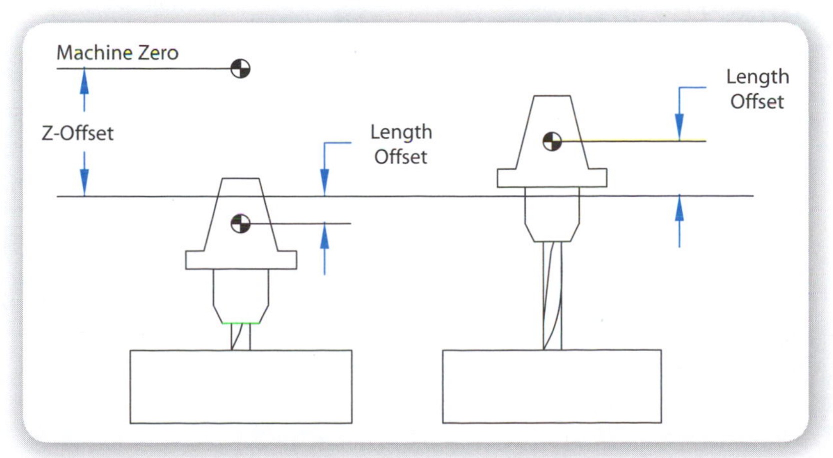

Summary of Tool Touch-Off Methods

Reference Tool Method

1. Install all tools in their adaptors and load them into the tool magazine.
2. We are going to use T01 as the reference tool for this example. All other tool-length offsets will be measured relative to T01.
3. Manually jog the tool over the workpiece and then move it down until it just touches the top. You can use a piece of paper to feel the contact between the part and tool.
4. Without moving the tool, enter the workpiece setup area of the control and set the Z-axis work offset. It should now be a fairly large negative number. This is where we tell the machine that we need to move the tool down by some distance to reach the top of the workpiece.
5. Again without moving the tool, enter the tool setup area of the control and find the T01 length offset. When set, it should read "0.0000."
6. Next, change to the second tool (T02) in the program and bring it down until it touches the top of the workpiece. Without moving the tool, enter the tool setup area of the control and find the T02 length offset. When set, the offset value should show the difference in length between T01 (the reference tool) and T02. This is usually minimal—plus or minus a couple of inches unless there is an exceptional difference between the tool lengths.
7. Repeat step 6 for each additional tool. ***See Figures 5-17, 18,*** and ***19*** for the sequence.

Figure 5.17 In a typical setup operation, the tool is manually jogged into position until it just slightly makes contact with the top of the workpiece.

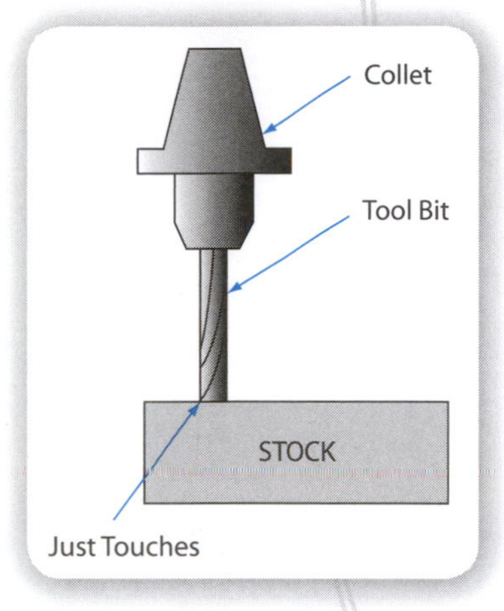

Large Offsets Method

1. Enter the workpiece setup area of the control and manually set the Z-axis work offset to the value "0.0000."
2. Install all tools in their adaptors and load them into the tool magazine.
3. Manually jog the first tool (T01) over the workpiece and then move it down until it just touches the top. You can use a piece of paper to feel the contact between the part and tool. We are not using a reference tool so there is nothing special about using T01 first. We could have picked any tool.

Figure 5.18 In the common reference tool setup method, the Z-axis work offset is calibrated while the tool is even with the finished work surface. This establishes the top datum of the part.

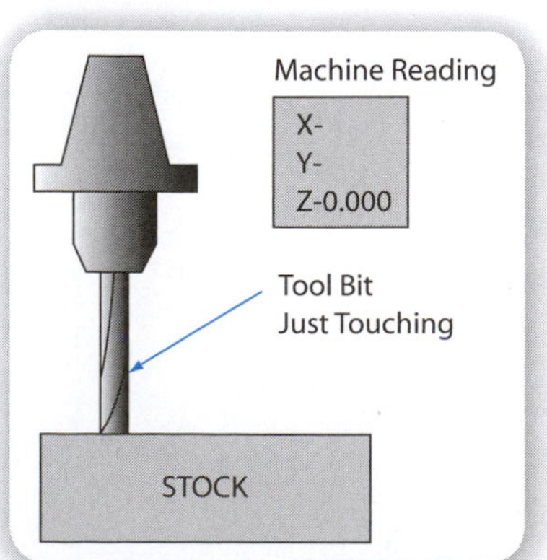

Figure 5.19 Next, the tool length offsets are set for every tool by touching off the tools to the finished surface and then calibrating the tool length offsets individually. The reference tool will have an offset value of 0.0000.

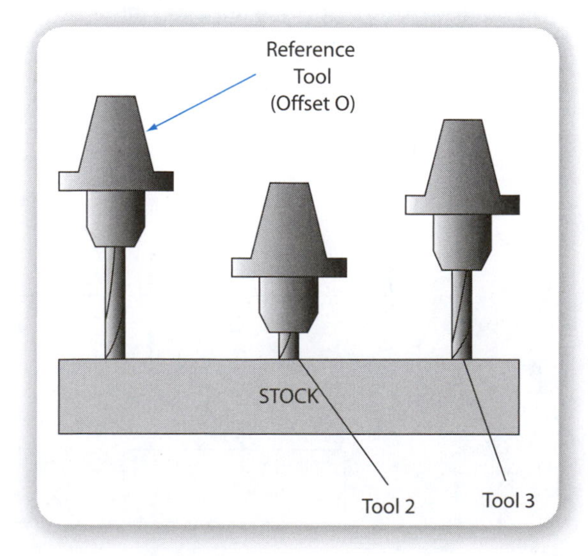

4. Without moving the tool, enter the tool setup area of the control and find the T01 length offset. When set, it should be a fairly large negative number.
5. Repeat steps 3 and 4 for each additional tool.

The manual tool setting process of touching the tool to a solid workpiece can be an inaccurate process. Many CNC machine tools (particularly lathes) are now equipped with "tool probes" that use an electronic measuring device to establish the tool offsets (***Figure 5.20***). The process is much the same in that the setup person will manually jog the tool, but instead of touching the workpiece, the tool will touch a pad that reads the measurement very accurately.

Figure 5.20 An electronic tool setting probe is shown in a CNC lathe. The operator will maneuver each tool to make contact with the sensing pad. The process is more accurate and reliable than physically making contact with the workpiece.

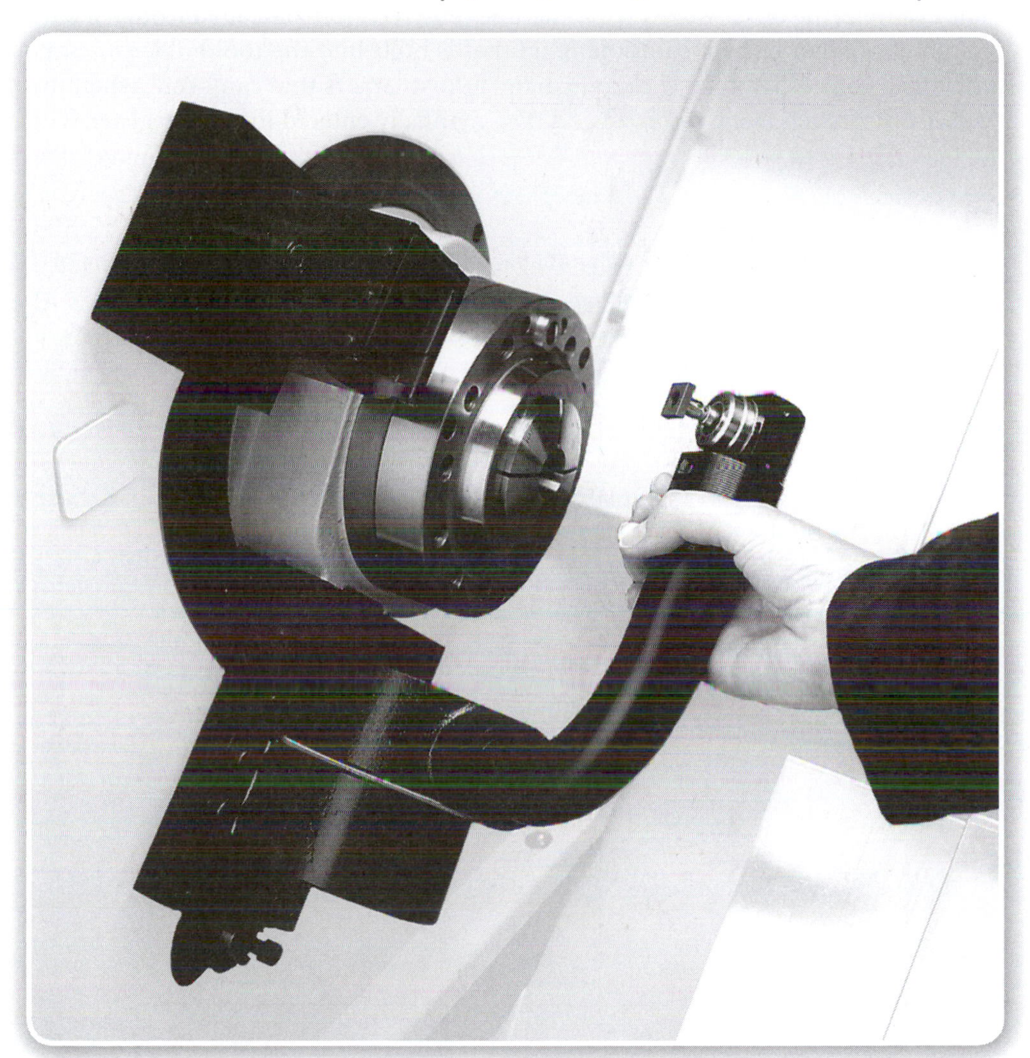

5.6 TOOL PRESETTING

Tool presetting is a method commonly used to establish the length offset and diameter of a newly installed cutting tool. Rather than stopping the machine tool to physically touch off the tool to the workpiece, a special device called a tool presetter is used. The tool presetter is a precision measuring device that can establish the length and diameter of the cutting tool before it is inserted in the CNC machine tool.

diameter offset:

A value stored in the offsets register in the control that will be used for automatic cutter compensation. Each tool will have its own diameter offset value.

fixture offset:

Work offset.

The tool presetter has a bearing-mounted, tapered spindle that is similar to the spindle of the CNC machine tool. The tool and adapter are mounted in the presetter, and then a contact probe is adjusted to the end and side of the cutting tool to establish the length and diameter offsets, as illustrated in ***Figure 5.21***. These values can then be manually entered into the offsets registry, and cutting can begin with extreme precision.

Tool presetters tend to be more accurate than manually touching off the tool to the workpiece. Manual touch-offs tend to have some variation due to the limitations of gauging when the actual contact between the tool and workpiece occurs. Similarly, the diameter offsets tend to be more accurate than when established by simply measuring the diameter of the tool with a micrometer. The spindle of the presetter can rotate and therefore take any eccentricity into account (off-center tools will produce an oversized cut).

Some tool presetting systems are more sophisticated than described earlier. For example, the offset values are sometimes entered electronically through a data exchange system or by a computer chip that is built into the tool holder to carry the offset values. This can eliminate many human errors that can occur when the offset data are manually recorded and then manually entered into the register. The

Figure 5.21 **A tool presetter can establish the length and diameter offsets of the tool without costly downtime.**

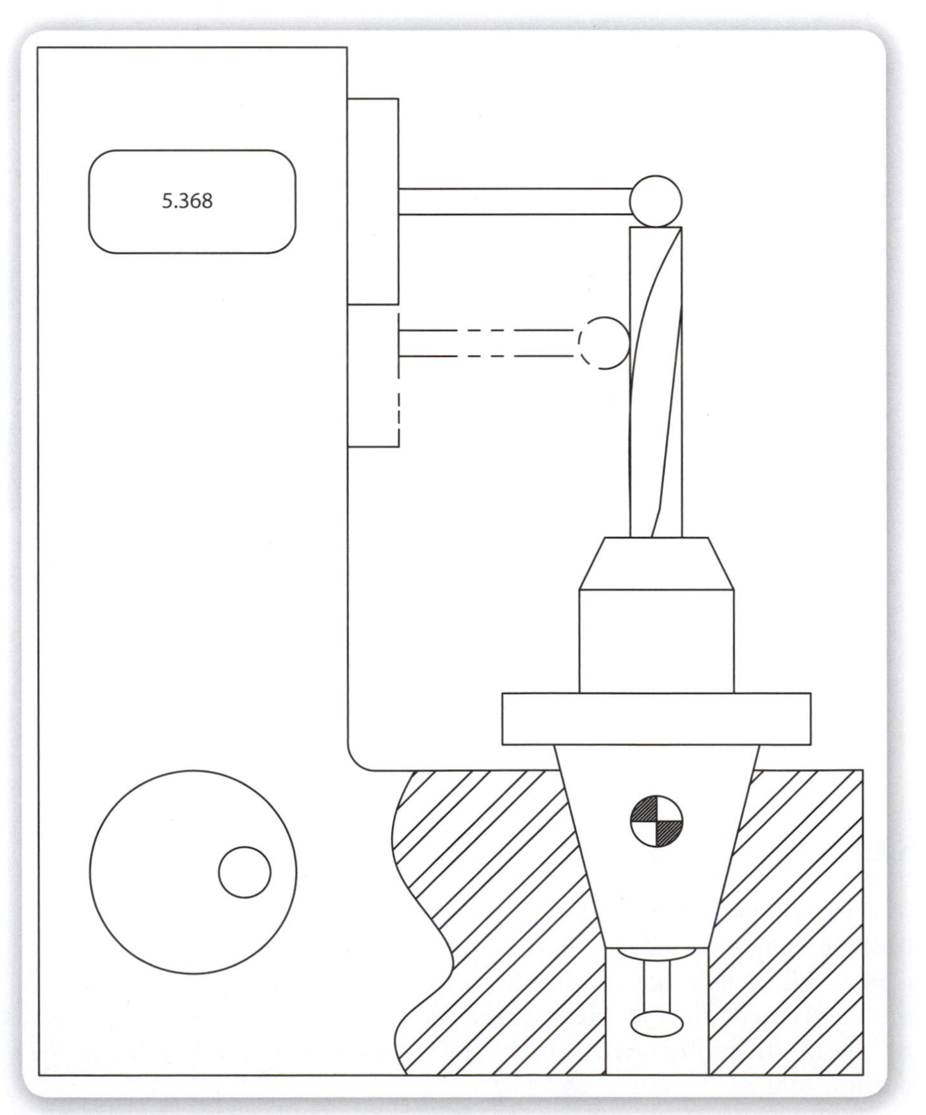

tool presetters may also be more sophisticated. The latest technology uses a non-contact system to optically measure the tool. This is often coupled with a profile projector that allows the technician to visually inspect the tool at an increased magnification.

The primary use of tool presetting is found in production machining environments where the cutting tools are fairly constant compared to job shop environments. The same tools become dull over and over again; therefore, it is common to perform many dozens of tool changes during a production day. A technician can mount cutting tools in their holders and establish the preset values before the current cutting tool becomes dull. The machine operator can then load the sharp tool into the magazine and load the offset values into the register with zero or minimal downtime. Tool presetting saves time and increases quality by reducing downtime and ensuring that the tool cuts the proper dimension the first time. Conversely, manual touch-offs are slow and will usually require some readjustment after the first cut to bring the workpiece to the correct dimension.

Your Turn

Plot your initials on the Cartesian coordinate system using both the absolute and incremental plotting systems.

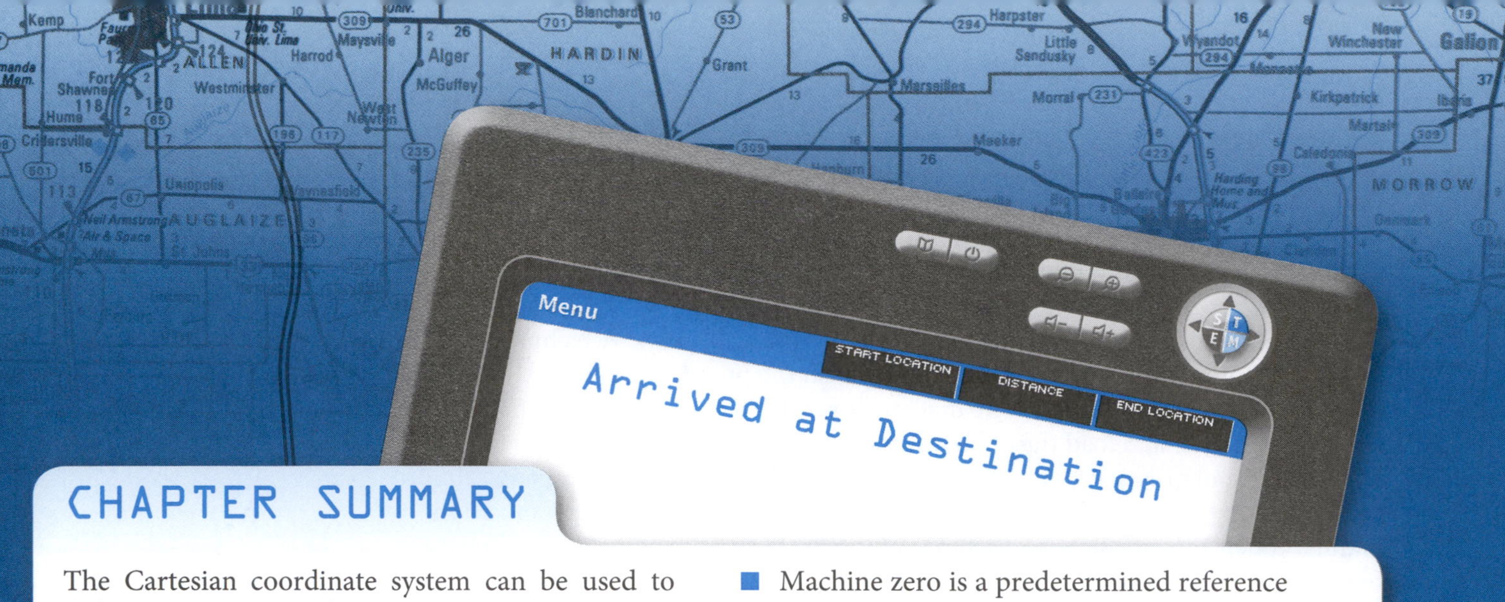

CHAPTER SUMMARY

The Cartesian coordinate system can be used to define a point on a plane or in three-dimensional space.

- NC programming requires the use of the Cartesian coordinate system to define toolpaths to create an accurate workpiece.
- Absolute coordinates are always given as a distance from the origin of the coordinate system. Most NC programming is done in absolute coordinates.
- Incremental coordinates are always given as a distance from a previous point.
- A work zero is established by electronically setting the origin of the coordinate system at some point on the workpiece. Also, this origin is used by the NC part program.
- Machine zero is a predetermined reference position that remains static. Sometimes this is called the machine home position.
- The work offset is the distance from the machine zero to the workpiece zero. Some machines support multiple work offsets.
- Tool length offsets are a compensation value that the control will use to accommodate tools of differing lengths. There are at least two methods used to establish this value; each has pros and cons.
- Tool presetting uses a precision measuring device to determine the length and diameter of a tool before it is installed in the machine tool. Presetting generally provides a more accurate offset value and reduces downtime.

BRING IT HOME

1. What are the three linear axes used in the Cartesian coordinate system?
2. What are the three rotary axes associated with each axis, X, Y, and Z?
3. What is the difference between absolute and incremental coordinates?
4. What is the work zero? Machine zero?
5. Explain the concept of a work offset. What distance is it actually measuring?
6. Where are all of the offset values stored in the control?
7. *Figure 5.22* shows a tool positioned at machine zero. In the large offsets method, what would be this tool's length offset in the register?

Figure 5.22 In the large offsets method, what would be this tool's length offset in the register?

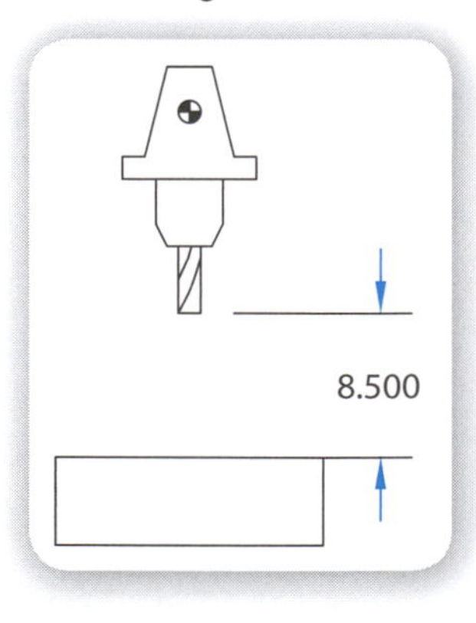

8. If the feature cut by the tool from the previous question were found to be 0.015" too tall, what action would you take?

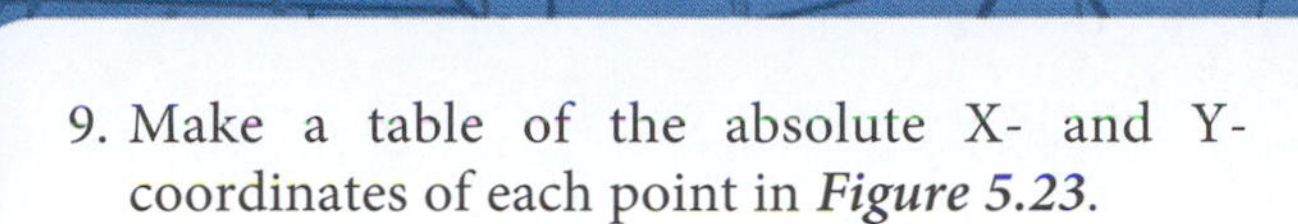

9. Make a table of the absolute X- and Y-coordinates of each point in *Figure 5.23*.

10. Make a table of the incremental X- and Y- coordinates of each point in Figure 5.23. Assume that you are starting tool movement at the origin and then return to the origin after point seven. Check your answer by summing both the X- and Y-coordinates.

Figure 5.23 Find the X and Y coordinates of each point.

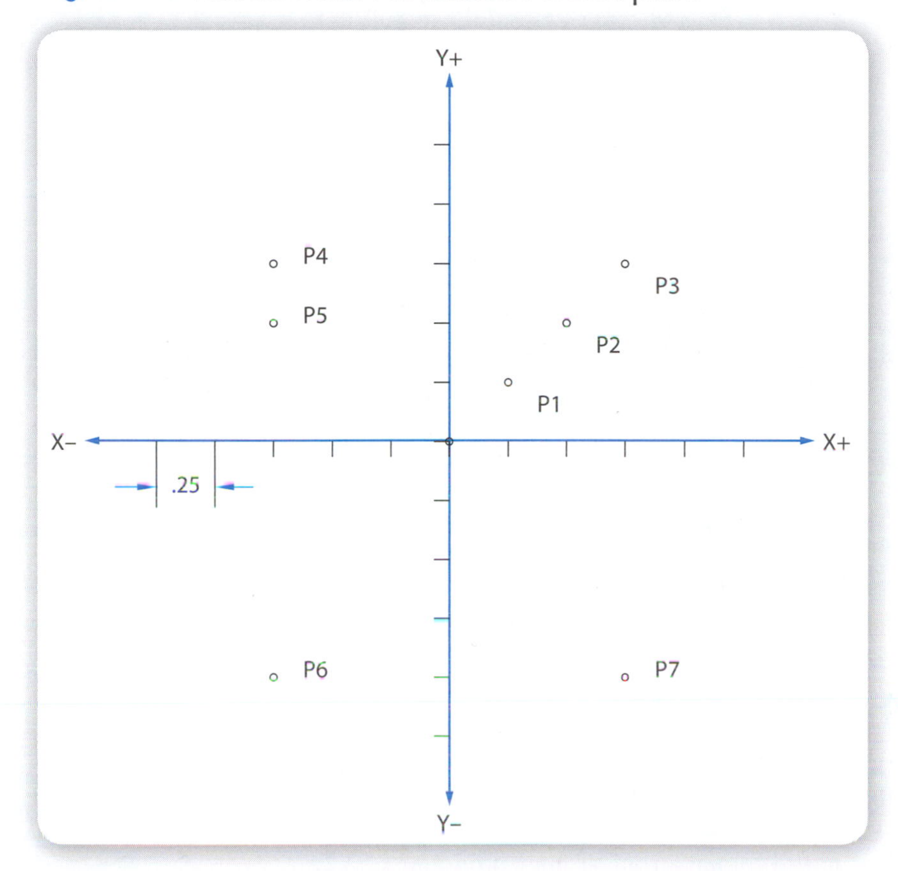

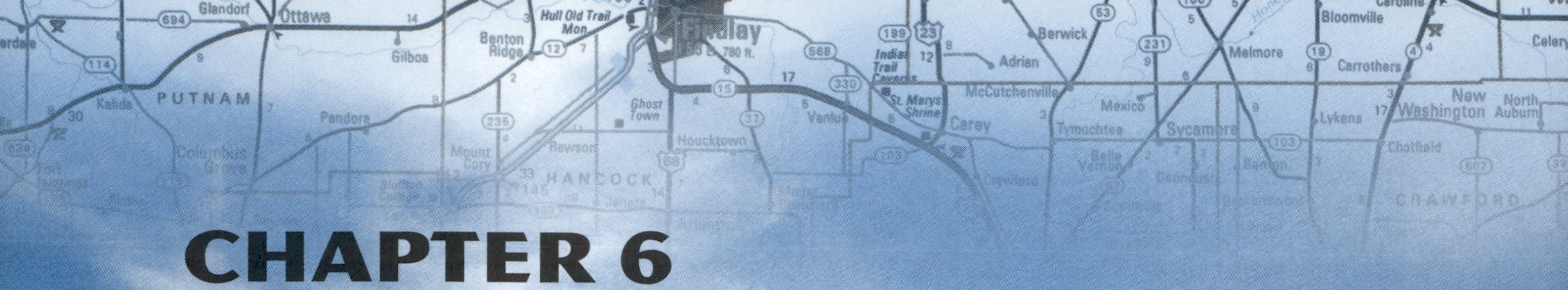

CHAPTER 6

Programming Concepts and Job Planning

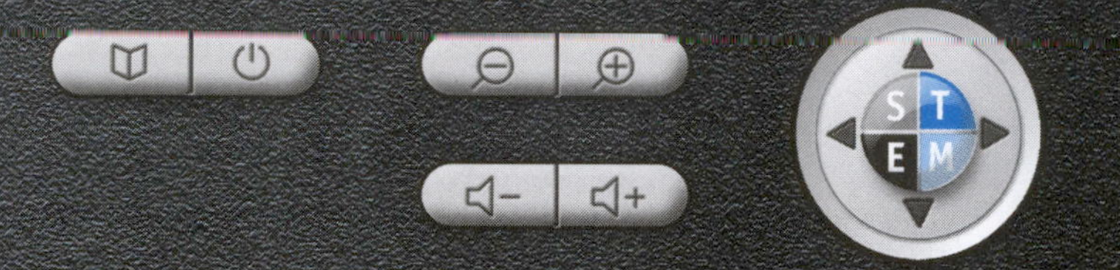

Menu

START LOCATION | DISTANCE | END LOCATION

Before You Begin

Think about these questions as you study the concepts in this chapter:

1. What is the format of an NC part program?
2. What defines the functional areas found in all NC part programs?
3. What are some of the most commonly used G & M codes and what are their functions?
4. Why are modal commands used?
5. What is the purpose of the safe line?
6. How is grammar associated with programming?

Key Terms

Address
Block
Datum
Miscellaneous Code (M-code)
Modal
Part Program
Preparatory Code (G-code)
Word

6.1 PROGRAMMING WITH G & M CODES

Modern CNC machine tools can be programmed to perform machining operations with a language commonly referred to as G & M codes. G-codes and M-codes are simple instructions that the programmer will write to make the machine behave in a certain way. For example, if the programmer wants the machine to cut a straight line, he or she will use the G01 code. Or if the programmer wishes to turn the coolant on, the M08 code can be used.

There are dozens of G-codes and M-codes used to perform everything from complex machining operations to mundane tool changes. However, just a few codes, if learned, will get most of the work done. The others are just enhancements to make programming more powerful and provide advanced machine functions.

In Table 6.1, you will find a list of the fundamental G & M codes that we will concentrate on in the next few chapters. It will be worth your time to look over these codes and become familiar with their meaning and format. Do not be concerned about every detail at this point—each code will be explained in further detail in the next few chapters. A more comprehensive reference can also be found in Appendix A when you are ready.

address:

A letter used in G & M code programming to designate a class of functions. Examples include G, M, X, Y, and Z. Letters are never used alone, but instead are combined with numbers to form words.

modal:

Describes codes or values that stay active until changed by another code or value. Most G & M codes are modal.

part program:

The instructions written by the programmer to produce a workpiece.

word:

The programming expression formed when a letter (address) is combined with a number. Examples: G01, X3.500, M30.

Table 6.1 *Summary of Common Programming Codes*

Task	Definition and Example of the Code in Use
Axis Movement	
G00	Rapid traverse—Moves the machine at a very fast rate to a specified point G00 X1.5 Y2.5
G01	Linear interpolation—Mills a straight-line at a controlled rate of feed G01 X1.5 Y.2.5 F10.0
G02	Circular interpolation—Mills in a clockwise circular arc G02 X1.5 Y2.5 I.5 J0 F10.0
G03	Circular interpolation—Mills in a counterclockwise circular arc G03 X1.5 Y2.5 I.5 J0 F10.0
G28	Return to reference point G91 G28 Z0
Machine Setup	
G20	Inch programming units G20 G00 X1.0 Y2.0
G21	Millimeter programming units G21 G00 X25.4 Y50.8
G90	Absolute positioning—All coordinates are relative to the reference position G90 G00 X1.0 Y1.0
G91	Incremental positioning—All coordinates are given from the current position G91 G00 X1.0 Y1.0
Hole Making	
G81	Drilling cycle G81 X1.0 Y1.0 Z-.75 F4.0 (DRILL A HOLE .75 DEEP) X2.0 Y1.0 (SUBSEQUENT HOLE LOCATIONS) X3.0 Y1.0
G82	Drilling cycle with dwell G82 X1.0 Y1.0 Z-.75 R.2 P.5 F4.0

(continued)

Table 6.1 (*continued*)

Task	Definition and Example of the Code in Use
G83	Peck drilling cycle G83 X1.0 Y1.0 Z-.75 Q.25 R.2 F4.0
G84	Tapping cycle M03 S100 G84 X1.0 Y1.0 Z-.75 F5.00
G85	Boring cycle G85 X1.0 Y1.0 Z-.75 F4.0
Machine Functions	
M00	Program stop—The program execution will pause at the current instruction until restarted by the operator M05 (STOP SPINDLE) M00 (CLEAR CHIPS FROM CUT) M03 S1000 (START SPINDLE) G00 X1.0 Y1.0
M01	Optional program stop—The program will pause only if the optional stop button on the control is activated M05 M01 (CLEAR CHIPS FROM CUT) M03 S1000 G00 X1.0 Y1.0
M03	Spindle on clockwise M06 T05 (TOOL CHANGE) M03 S1200 (TURN SPINDLE ON CW)
M04	Spindle on counterclockwise M06 T05 (TOOL CHANGE) M04 S1200 (TURN SPINDLE ON CCW)
M05	Spindle off M05 (TURN SPINDLE OFF) M06 T05 (CHANGE TOOLS)
M06	Tool change M06 T05 (CHANGE TOOLS) G43 H05 (USE HEIGHT OFFSET #5)
M08	Coolant on G01 Z-.75 M08
M09	Coolant off G00 Z0. M09
M30	Program reset—Rewinds to the beginning of the program G00 Z6.0 G00 X10. Y10. M09 M30 (RESET TO TOP OF PROGRAM) % (END OF PROGRAM)

6.2 STRUCTURE OF AN NC PART PROGRAM

NC part programs are made up of codes that are read by the control to perform various functions. There are two fundamentally different types of codes found in an NC part program. The first type of code controls machine functions and settings. The second type of code is used to modify and perform machining functions. The arrangement and use of these codes will form the fundamental structure of the part program.

We can start to understand NC programming by looking at sample programs. An NC part program will have to perform the same functions that we would to produce a workpiece conventionally. We would have to decide how to interpret the drawing, load our tools and turn on the spindle, locate the corner of the part, and then touch off the tool. Then, we would be ready to machine the profile according to the print. Let's look at the short NC program in Listing 6-1 that will be used to machine the square profile in ***Figure 6.1***. We will dissect the sample program by looking at the various elements and formats that are present in the code and then discussing the function of each section.

Figure 6.1 A simple toolpath to machine a square shape.

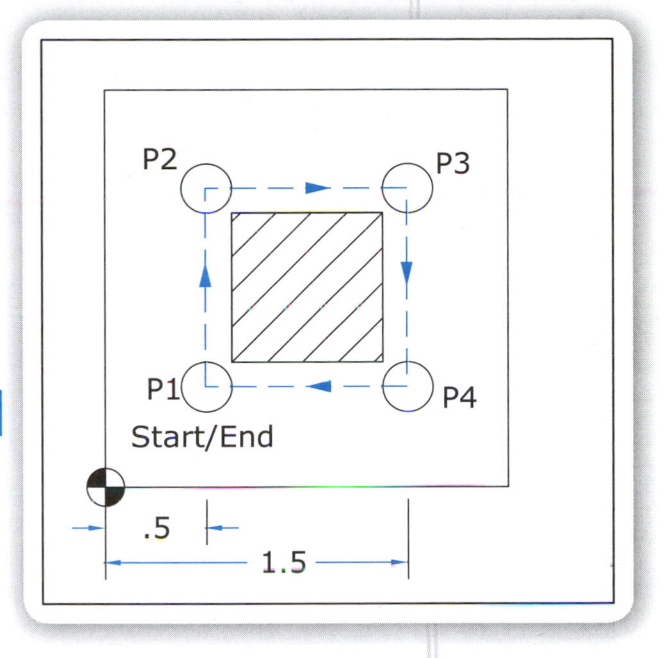

Listing 6-1 Program Code

```
O0001 (Program 6.1)
N10 G20 G40 G49 G54 G80 G90 G98
N20 M06 T03 (.25 EM)
N30 G43 H03
N40 M03 S2000
N50 G00 X.5 Y.5 (P1)
N60 G00 Z.2
N80 G01 Z-.1 F5.0
N90 G01 X.5 Y1.5 F10. (P2)
N100 G01 X1.5 Y1.5 (P3)
N110 G01 X1.5 Y.5 (P4)
N120 G01 X.5 Y.5 (P1)
N130 G01 Z.2
N140 G00 Z5.0
N150 G00 X0. Y6.0
N150 M05
N160 M30
%
```

The NC part program is made up of distinct units of instruction called **blocks** (or just plain lines). The machine control unit (MCU) will read and execute the blocks sequentially (i.e., one line after the other). Sometimes the blocks are numbered using N10, N20, etc., but these sequence numbers are not required. The MCU will read several blocks at once but will execute only one block at a time. There are some cases in which the MCU must know what is coming up in the next block in order to operate properly.

block: A single line of code in an NC part program.

First the program tells the MCU where the program begins (%) and what its name is (O0001). These first two blocks have no effect on the actual machining, but are necessary for the MCU to distinguish one program from another. You must not put line numbers on these blocks:

```
%
O0001 (Program 6.1)
```

The next block tells the machine how to set itself up and how to interpret the instructions in the remainder of the program. For example, the G20 code tells the MCU that we intend this program to be in inches rather than millimeters. Sometimes this is called the safe line because it will also load a number of instructions into memory that will keep an errant programmer out of trouble (more on this later):

```
N10 G20 G40 G49 G54 G80 G90 G98
```

The next few blocks will tell the machine to get ready for machining by putting a tool into the spindle and turning the spindle on in the proper direction at a specified speed. You also see the use of comment characters. The MCU will ignore any code that is placed within the ( ) characters, which are used in this case to describe the tool (Ø 0.250" end mill):

```
N20 M06 T03 (.25 EM)
N30 G43 H03
N40 M03 S2000
```

preparatory code (G-code): Codes that carry out machining operations or establish machine settings; G-codes.

Now we are ready for the meat of the program. The blocks of code numbered N50 through N150 are the blocks that actually perform the machining operations. They instruct the machine to move to the specified end points in one manner or another. You will notice that there are only two different **G-codes** used in this section. The G00 and G01 codes tell the machine to move from wherever it is currently positioned to the point specified by coordinates that follow the G-code. For example, line N110 tells the machine to move from its current position to X1.5 and Y.5:

```
N50 G00 X.5 Y.5 (P1)
N60 G00 Z.2
N80 G01 Z-.1 F5.0
N90 G01 X.5 Y1.5 F10. (P2)
N100 G01 X1.5 Y1.5 (P3)
N110 G01 X1.5 Y.5 (P4)
N120 G01 X.5 Y.5 (P1)
N130 G01 Z.2
N140 G00 Z5.0
N150 G00 X0. Y6.0
```

The last few lines of the program control a few ordinary machine functions and administrative duties such as shutting off the spindle, resetting the program, and telling the MCU where the program ends:

```
N150 M05
N160 M30
%
```

All NC programs will follow the same basic pattern—code that gets the machine ready mixed with code that will perform the machining. We will concentrate on the basic structure and format of NC part programs for the remainder of this chapter and then take a closer look the codes in the following chapters.

6.3 WORDS, ADDRESSES, AND NUMBERS

The programming format that we studied in Listing 6-1 is known as word address programming. Some sources will refer to this format as letter address programming. A block in an NC program is a line of code that is constructed from words. Each word is made up of letters (or addresses) and numbers that are related to the letter. For example, the letter X refers to the X-axis. The number that follows it indicates a coordinate along the X-axis. Simply put, a letter is followed by a number and these will have some important meaning to the MCU. In summary,

- X is an address (or letter)
- 1.5 is a number
- X1.5 is a word

Likewise,

- G is an address
- 01 is a number
- G01 is a word

Put the two words together and you have a valid block:

```
G01 X1.5
```

Incidentally, the number associated with a G- or **M-code** does not have to include the leading zero, as shown in single-digit numbers such as G00, G01, G02, M01, etc. These could be written as G0, G1, G2, M1, etc. However, many programmers will include the leading zero out of style or habit.

miscellaneous code (M-code):
Codes that control machine functions other than tool movements. For example, M-codes are used to control coolant and spindle rotation.

6.4 MODAL AND NON-MODAL CODES

Modal and non-modal are the terms used to describe the two classes of codes used in NC programming regarding their ability to remain in the MCU's memory. Non-modal codes are used just once and then discarded. Modal commands, on the other hand, remain in the active memory until they are replaced by another code (superseded).

Programming codes are grouped into different categories of related functions, as listed in Table 6.2. Only one code from each group can be active at any one time. For example, G00 and G01 are in the same group. Therefore, you cannot invoke G00 and G01 at the same time. It would be analogous to putting your car in reverse and drive at the same time. However, you can have codes from different groups active simultaneously. For example, G90 and G01 are in different groups and can be active at the same time. This is like having your car in drive and turning on the radio at the same time—they are not related, so there is no conflict.

The advantage of modal codes is that you do not have to enter the code in the subsequent blocks if the code has not changed. For example, G01 is the modal code used to cut a straight line. Therefore, if you have a series of straight-line moves to make, you do not have to reenter G01 in every single block. However, if G01 is superseded by another code in the same group, then that new code will remain in effect until changed again.

There are many different codes used in NC programming, but the vast majority of codes are modal. It is important to know which codes belong to which group, but fortunately it is fairly easy to think about in a logical manner.

Most other address words are modal in the sense that they have a place in the memory reserved for only one value at a time. For example, if we wish to position the X-axis to 5.500", then we might use the code G00 X5.500. When the control reads this code, the value 5.500 will be placed into the X-register and held there until it is replaced with another value.

M-codes have also been included in the list, although they are not traditionally thought of as modal or non-modal. A few M-codes can be thought of as modal, such as those that are used to control the spindle or coolant. They simply throw a

Table 6.2 *G & M Code Functional Groups*

Functional Group	Codes and Addresses
Axis movement	G00, G01, G02, G03
Positioning system	G90, G91
Units	G20, G21 (G70, G71 on some systems)
Canned cycles	G80, G81, G82, G83, G84, G85,...
Coordinate systems	G54, G55, G56,...
Tool radius compensation	G40, G41, G42
Tool height	G43, G44, G49
Return plane for canned cycles	G98, G99
Plane selection for arcs	G17, G18, G19
Axis position addresses	X, Y, Z, A, B, C, U, V, W
Additional addresses	D, F, H, I, J, K, Q, R, S
Coolant	M07, M08, M09
Spindle control	M03, M04, M05
Non-modal	
Miscellaneous G-codes	G04, G28, G29, G92
Auxiliary addresses	L, P
Miscellaneous M-codes	M00, M01, M02, M06, M30, M97, M98, M99

switch that will remain on until another code is used to turn it off. However, most M-codes function non-modally. For example, M06 is used to change tools, but as soon as the tool change is complete, M06 is discarded from memory. Regardless of a code's function, most controls only allow one M-code per block.

6.5 PROGRAMMING GRAMMAR

It is possible for programs that look very different to perform exactly the same machining operation. This occurs because the MCU will remember the previous commands and settings and run the program even though it appears to have parts missing. The MCU treats most commands as modal commands; each remains constant until another command is called to replace it. In the following three blocks, the command to cut a straight line is G01, and the tool will move to the X- and Y-coordinates that follow.

```
N01 G01 X1.5 Y3.0
N02 G01 X6.0 Y3.0
N03 G01 X6.0 Y6.0
```

However, in the previous section we learned that the G01 code is modal, so the same code could be rewritten as

```
N01 G01 X1.5 Y3.0
N02 X6.0 Y3.0
N03 X6.0 Y6.0
```

The MCU reads the first line where it is instructed to cut a straight line to X1.5 and Y3.0. Then the next line is read, but there is no G-code given. This is not a problem because the MCU has stored the G01 in memory and knows that it should perform a G01 until told otherwise. The X and Y words are also modal; they will remain in memory until replaced or cancelled. This code can now be rewritten as

```
N01 G01 X1.5 Y3.0
N02 X6.0
N03 Y6.0
```

The intention of the second line is to cut to X6.0 and Y3.0, but the MCU knows that it is already at Y3.0 and, unless told otherwise, will remain at Y3.0. The same principle is true in the third line. The X position does not change, so this does not have to be updated.

The MCU processes the program one block at a time and reads each entire block before processing it. Therefore, it is not usually important to the MCU that the codes are in any particular order. Some older machines with primitive control systems had funny rules about what should go where and even where the spaces should be located. Of course, these rules were difficult to follow, so the MCUs eventually evolved to a more human-friendly design. Modern MCUs are robust enough to handle most of the grammar variations without getting too upset. To the MCU, the two blocks below are identical.

```
N01 G54 G01 X1.5 Y3.0 M08
N01 M08 Y3.0 G54 X1.5 G01
```

However, people are not so flexible, and many prefer that a particular order be followed. This is less out of habit, and more for readability. The second line in the previous example follows no logical order and is somewhat difficult to read.

We could look at a list that specifies the exact order in which the addresses should occur, but that would probably not serve any purpose unless you really enjoy memorizing lists. Another way of looking at address order, without going into too much detail, is to group the codes into functional patterns that place the addresses in a logical order according to the intended function. In this case, the order is

1. line number
2. G-codes
3. axis coordinates (alphabetical order by group X, Y, Z; I, J, K; A, B, C)
4. all other addresses (somewhat in alphabetic order)
5. M-codes

After line numbers, the G0 codes should come next. G-codes are like the verb in a sentence—they immediately tell us what action the block is supposed to perform. In some cases, the G-codes will have axis addresses, so they should be next. Any other addresses related to the G-codes should then be included.

M-codes are usually the last code in the block. However, they are kind of a special case. They are usually placed at the end of a block when the block also includes a G-code. However, when M-codes are used alone, some people prefer to place them at the beginning of the block because they imply some action. The blocks below illustrate this point.

```
N100 G54 G01 X1.5 Y3.0 M08
N100 M03 S1000
```

There are a few other loose ends in the grammar department, such as spaces and line numbers. Spaces must not occur between a letter and its number or within a G- or M-code. In the line below, three mistaken spaces will cause the MCU to error.

```
N02 G 01 X 6.0 Y 3.0
```

Other errors occur when a tab or other hidden characters are accidentally typed in on a word processor. If you consistently get an error and you believe the code is correct, try displaying the program on a word processor with the hidden characters switch turned on. You might find some funny characters such as →, •••, or ¶ in places where they do not belong. These will cause problems in the control and they should be deleted.

Spaces between words are not usually needed by the MCU, so they are sometimes left out to conserve memory. However, it can be difficult for a human to read code that runs together without spaces between the words, and this may lead to frustration and mistakes. Note that the blocks of code in A) and B) below are identical, but A) does not contain any spaces and is difficult to read.

A) No spaces:

```
N01G01X1.5Y3.0M08
```

B) Rewritten with spaces:

```
N01 G01 X1.5 Y3.0 M08
```

Sequence numbers (also known as line numbers) are not needed by most controls. They are useful for troubleshooting, and some specialized programming techniques require sequence numbers, but they are not needed for most programs. Therefore, the code blocks

```
N01 G01 X1.5 Y3.0
N02 X6.0
N03 Y6.0
```

could be rewritten as

```
G01 X1.5 Y3.0
X6.0
Y6.0
```

The machine tool manufacturers will also vary the number format that is required. The typical control will allow three places before the decimal place and four places after (e.g., 100.1234). Most support a standard decimal format. This means that numbers can be entered as they would be entered on a calculator—this should seem very natural. The format is as follows:

```
X.500
```

Other suitable variations of the decimal format can include or not include leading and trailing zeros:

```
X.5, X0.50 or X0.5000
```

The same controller may also allow common decimal numbers to be written as integers representing 1/10,000 of an inch (or sometimes 1/1000 of an inch). This format, called leading zero suppression, will seem a little awkward. The word below indicates 0.500 on the X-axis.

```
X5000
```

Larger numbers in this format can be difficult to read:

```
N100 G01 X155124 Y105001
```

On controls that support both formats, you must be careful not to forget the decimal point, or else numbers such as 3" will be read as 0.0003". For example, a common programming error occurs when a feed is entered as F10 and the programmer expects the machine to move along at 10" per minute, but instead the feed is only 0.0010" per minute—a little slow.

Grammar and Readability

The points made earlier about grammar and the various formats were meant only to inform you of the specifications of many machine tool controllers. The reality of writing G & M code programs is that eventually a human will have to read through the code to find mistakes or to make modifications. It is generally much easier to read a program if all the information is provided and the code is presented in a structured and consistent format.

Not too many years ago, writing a program was quite labor intensive. Inexpensive computers were not readily available, and much of the programming was completed on the MCU at the machine itself or typed on a machine that punched a paper tape. This was a very time-consuming practice, and programmers would do anything to save a few keystrokes. Most programs today are written on modern, PC-based computers with a text editor that makes it very easy to copy and paste redundant code and to make changes at will. Furthermore, computer memory is inexpensive today, so saving an extra 100 bytes in the MCU's memory is not very meaningful. Look at the two programs in Listing 6-2 and decide for yourself which is more readable. Both are identical in function and have correct syntax, but the program on the right has been shortened using the rules that we discussed earlier in this section.

Listing 6-2 Long Version	Condensed Version
%	%
O0001 (Mill a Square)	O0001
N10 G20 G40 G54 G80 G90 G98	G20
N20 M06 T03 (.25 EM)	G54G80
N30 G43 H03	T3M6
N40 M03 S2000	G40
N50 G00 X.5 Y.5	M3G90G98S2000
N60 G00 Z.2	G43G00Y.5X.5H3
N80 G01 Z-.1 F5.0	G0Z.2
N90 G01 X.5 Y1.5 F10.	G1F5.0Z.1
N100 G01 X1.5 Y1.5	X.5Y.5F10.
N110 G01 X1.5 Y.5	X1.5
N120 G01 X.5 Y.5	Y.5
N130 G01 Z.2	X.5
N140 G00 Z5.0	Z.2
N150 G00 X0. Y6.0	G0Z5.0
N150 M05	X0.Y6.0
N160 M30	M05
%	M30
%	

Clearly the programming style can affect the readability of the code. Spaces, line numbers, and a logical word order have a great impact on readability during manual programming and therefore affect the reliability and accuracy of the code. For example, did you find the two mistakes? Your eyes probably moved right over them. Try finding the first mistake in block number . . . I mean in the first linear interpolation code, and the second irregularity in the following block. By now, you get the point.

6.6 MORE ON THE SAFE LINE

Earlier we mentioned that the so-called safe line is used to set machining parameters and provide a degree of safety. The functions performed by the safe line fall into just a couple of categories, such as safety resets and setup parameters.

Safety Resets

There are four codes in the following safe line that will set the control back to a safe state before starting again.

```
N10 G20 G40 G49 G54 G80 G90 G98
```

Many codes are modal and will remain in effect until they are cancelled or superseded by another code from the same group. It is easy to forget to call the new code and unintentionally invoke the previous modal code.

G40 cancels automatic tool radius compensation. Tool radius compensation automatically offsets the tool from the programmed path. The compensation may still be active if you are using cutter compensation at the end of a program and then forget to turn it off. Therefore, the next toolpath may not be produced where you expected it to be, and the result could be a scrapped workpiece.

G49 cancels any length compensation values from the previous tool. This is important because if you forget to call a new length after a tool change, a crash may result.

G80 cancels any canned drilling cycles and their associated parameters. This is important when multiple drilling cycles are used in the same program. Forgetting to cancel a drilling cycle will leave the previous values active in the memory, which could result in a crash, excess tool wear, or the wrong geometry.

G98 is related to how drilling cycles retract between holes. G98 forces the tool back to a position that is normally higher above the part and therefore helps to avoid unintended collisions.

Setup Parameters

Three codes in the following safe line are used to set the machining and axis position parameters. However, many controls will have an automatic default value that will be activated at startup.

```
N10 G20 G40 G49 G54 G80 G90 G98
```

G20 tells the control that all coordinates and feeds are to be interpreted in inches. (This could also have been G21 for metric units.) It is possible to switch between English and metric units within the program, so it is a good idea to include G20 or G21 in the safe line.

G90 is used for absolute positioning, and G91 is used for incremental positioning. Most programs are written completely in absolute units, but there are situations in which incremental is advantageous. It is common to switch between incremental and absolute coordinates within a program, so the programmer must be sure to set the proper system before machining or else all of the geometry will be incorrect and a crash could result. Both of these codes are modal and will remain in effect until superseded by the other.

G54 is one of many preset coordinate systems available on most modern machine tools. G54 could also have been G55, G56, G57, etc. It is meant only to represent the primary coordinate system that the setup person has selected for the job. This is included because sometimes the coordinate system will be changed within the program to make it easier to program a particular feature (e.g., bolt hole patterns). Like most other G-codes, the coordinate systems are modal and, once changed, will remain in effect until superseded.

6.7 PROGRAM ANNOTATION AND SETUP SHEETS

In many shops, it is common to have hundreds or thousands of programs contained in files for various jobs. It is simply not practical to believe that the programmer or setup person will be able to remember all of the details of a program. Furthermore, the programmer is usually not the person who will set up and operate the machine tool. Therefore, it is critical that any important information be communicated to the personnel who will need it. This information is usually conveyed via two separate methods: setup sheets and program annotation (comments).

Programs are often annotated with comments to make it easier for the other people who will have to use the program to understand. Comments make it much easier to set up the job and will lower the chances of mistakes (not to mention the stress level of the setup person). Some of the critical information to be included in the program might include the following:

- Location of the work offset
- Style, length, and diameter of cutting tools
- Insert specifications
- Work-holding information
- Warnings of past mistakes
- Customer part number and material information
- Individual processes within the program
- Program information such as date, file name, programmer, etc.

Take the program in Listing 6-3, for example. On the left is the original program, and on the right the annotated version.

Listing 6-3	Original Program	Annotated Version
	%	%
	O0001 (Program 6.1)	O0001 (Program 6.1)
	N10 G20 G40 G49 G54 G80 G90 G98	(FILE NAME C:\ACME51.NC)
	N20 M06 T03 (.25 EM)	(PROGRAM DATE 7/5/99)
	N30 G43 H03	(CUSTOMER: ACME MFG)
	N40 M03 S2000	(PART # 0123)

(continued)

Listing 6-3 *continued*

N50 G00 X.5 Y.5	(MATERIAL: 2024 T5 ALUMINUM)
N60 G00 Z.2	(WORK ZERO, LOWER LEFT, Z0 TOP)
N80 G01 Z-.1 F5.0	(TOOL #3, .25 DIA, 2-FLUTE)
N90 G01 X.5 Y1.5 F10.	(TOOL MATERIAL: HSS, TICN)
N100 G01 X1.5 Y1.5	N10 G20 G40 G49 G54 G80 G90 G98
N110 G01 X1.5 Y.5	N20 M06 T03
N120 G01 X.5 Y.5	N30 G43 H03
N130 G01 Z.2	N40 M03 S2000
N140 G00 Z5.0	N50 G00 X.5 Y.5
N150 G00 X0. Y6.0	N60 G00 Z.2
N150 M05	N80 G01 Z-.1 F5.0
N160 M30	(BEGIN MILLING SLOT)
%	N90 G01 X.5 Y1.5 F10.
	N100 G01 X1.5 Y1.5
	N110 G01 X1.5 Y.5
	N120 G01 X.5 Y.5
	(FINISHED WITH SLOT)
	N130 G01 Z.2
	N140 G00 Z5.0
	N150 G00 X0. Y6.0
	N150 M05
	N160 M30
	%

You can see that there is almost no limit to information that you can include with comments. However, you probably do not want to spend all of your time typing in comments, so just give the essential information that will aid the setup person and operator in performing their jobs.

Another important tool that is used to communicate information about the job is the setup sheet. Setup sheets contain all the information needed to set up the job and usually include a sketch of the part and the fixture. Setup sheets contain much of the same information that can be found in the program annotation. However, program comments aid in following the structure of the program, whereas setup sheets contain generic information that is not related to the actual code.

Figure 6.2 A typical setup sheet. The setup sheet is a quick way to communicate the essential information about a CNC machining job.

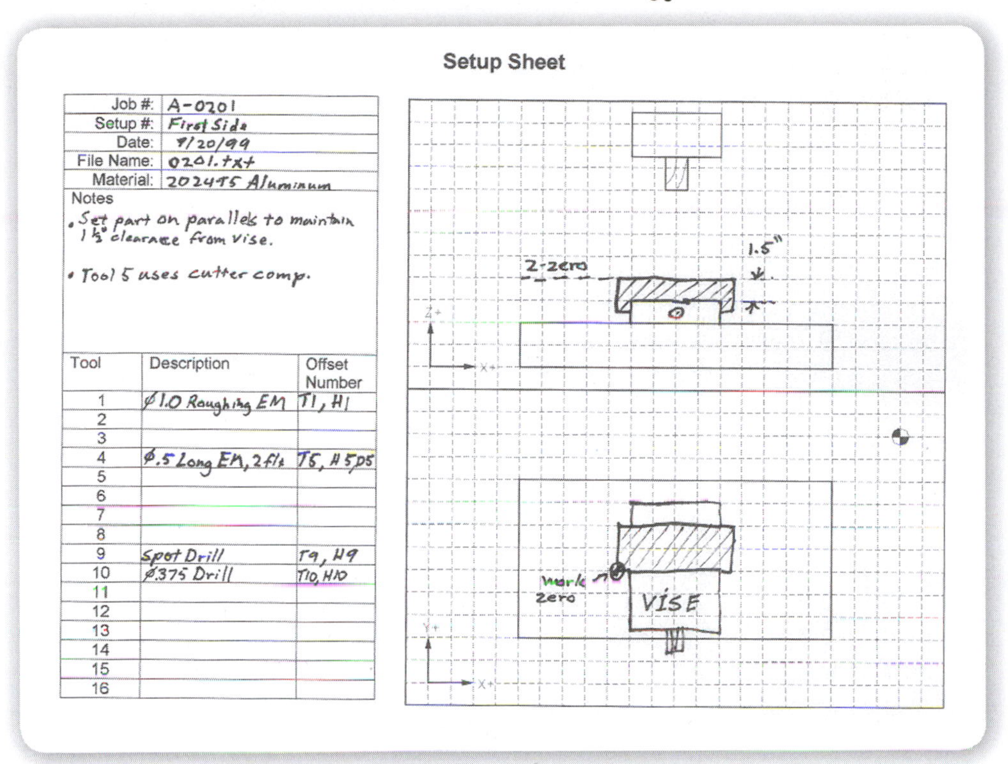

Setup Sheet

Job #:	A-0201
Setup #:	First Side
Date:	9/20/99
File Name:	0201.txt
Material:	2024T5 Aluminum

Notes

- Set part on parallels to maintain 1½" clearance from vise.
- Tool 5 uses cutter comp.

Tool	Description	Offset Number
1	Ø1.0 Roughing EM	T1, H1
2		
3		
4	Ø.5 Long EM, 2 fl	T5, H5 D5
5		
6		
7		
8		
9	Spot Drill	T9, H9
10	Ø.375 Drill	T10, H10
11		
12		
13		
14		
15		
16		

The main advantage of a setup sheet is that it serves as a quick reference to gather the tools and plan the machine setup. Part programs are usually kept as electronic files and are not very convenient to print out just to see which tools are needed. On the other hand, a setup sheet is usually a single piece of paper that can be quickly scanned to find the needed information. This is not to say that setup sheets are a replacement for program annotation, but they can be used in conjunction with each other.

Figure 6.2 shows an example of a completed setup sheet. You can make additional copies from the blank setup sheet in Appendix B.

6.8 PLANNING A CNC MACHINING JOB

It can be a little overwhelming to receive a part drawing and then be told to *make it*. Your mind may start racing to understand the geometry and plan the machining operations. How will you hold onto the part? Which tools should you use? What operation will be the first or second or last? Will this job require more than one setup?

All of these questions have to be answered before you can even consider writing one line of code. Nevertheless, do not be overwhelmed by trying to answer all the questions simultaneously. There are some tools and guidelines to help you organize and plan the job in a systematic manner. The following steps form the usual path toward planning and executing a CNC machining job:

1. Study the workpiece drawing.
2. Plan the operations and setups.
3. Determine which tools are available.
4. Write the part program.
5. Test and revise the program.

Study the Workpiece Drawing

This is really the first step to understanding the job ahead of you. Do not be rushed into planning operations and writing code before you understand the geometry. After all, the geometry of the finished workpiece will be the ultimate measure of the success or failure of your programming. ("Oops, I didn't mean to put that there.")

Complicated drawings can take hours to fully comprehend. You need to understand the form of each feature, its relationship to other features, and the implication of the tolerances before planning any machining operations. For example, you might notice a hole in the center of the workpiece. What kind of hole should this be? Should you drill the hole, ream it, or bore it? Can this hole be milled with an end mill or is a counterbore the proper tool to use? All of these questions are based on the specified tolerances and the required quality of the finished feature. Failure to understand these relationships will lead to wasted time and a scrapped part.

Plan the Operations and Setups

After becoming familiar with the features of the workpiece, it is time to start formulating a plan for making it. Again, this leads to many more questions that you will have to answer. Where will the work zero be located? Do any surfaces need to be qualified? How will you hold onto the workpiece? Will more than one setup be required? Are there any operations that must be done before the others?

Let's take as an example the workpiece shown in ***Figure 6.3*** and develop a plan to manufacture the workpiece. First, we need to become familiar with the features, tolerances, and any other information given in the drawing. A quick review of the drawing shows four bolt holes, two dowel pin holes, and a mounting hole for a bearing. The drawing also reveals that the designer intends most of the surfaces to be left as they are from the mill—a common practice on noncritical surfaces. The stock size is 4" by ¾" and will more than likely be saw-cut slightly longer than 6" to allow the ends of the workpiece to be milled square.

Figure 6.3 **Bearing support.**

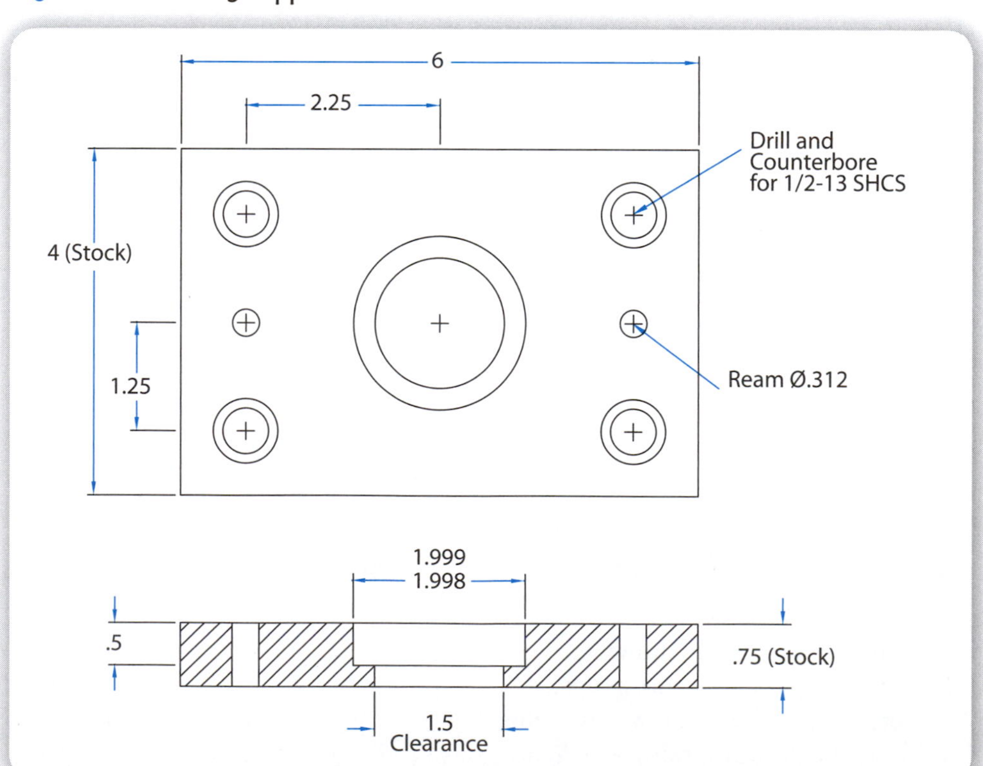

We might also notice holes in the part, and each hole will have very different requirements. The four corner holes are to be made to accommodate a socket-head cap screw, possibly to bolt the plate to a mating part. These do not have to be made to particularly close tolerances, so you would probably drill the through-holes and use a counterboring tool to produce the counterbores. The two smaller holes are specified as dowel pin holes. You may know that dowel pins are used to accurately locate two mating parts and therefore must be very accurate and perpendicular. Small dowel pin holes can usually be drilled and then reamed; larger holes will have to be bored first.

What about the center hole? This part comprises two concentric holes with very different requirements. The through hole is only for clearance and therefore does not need to be manufactured to close specifications. We might simply drill this hole and then mill the final size with an end mill if you do not have a Ø 1.5" drill. Conversely, the counterbored portion of the hole is for a bearing fit and has a very tight tolerance. A rational approach to making the bearing seat might be to mill a circle with an end mill to establish the rough size and then use a boring head to produce the final diameter.

Our final list of operations can then be listed in logical order for machining. Of course, the order of some operations is important, whereas others are strictly arbitrary. For example, we have to spot drill the holes before drilling them and we must rough the bearing hole before boring the final diameter. Yet, it does not really matter if the counterbore is performed before reaming. For our bearing support, the operations might look like the following:

1. Spot drill all hole locations.
   a. Spot drilling is a method used to create a small divot in the work material in order to later drill a hole. Without this starter hole, it is difficult for the twist drill to get started in the correct location. Twist drills have a blunt point and will often "walk off" the center location, which can lead to a scrapped part. Center drills are also commonly used for this operation. They perform a similar function—just not as well because of their blunt, thick web. The idea behind this operation is that the spot or center drill is a short, stiff tool that creates a small hole without being deflected. A twist drill, on the other hand, is usually quite long when compared to the diameter. They are flexible enough to deflect wildly.
2. Drill Ø.531 bolt holes and pilot for center holes.
   a. These holes are just clearance for a socket-head cap screw. The hole is specified to be Ø.531", so there is lots of room for error when the Ø½" screw is installed. This is well within the twist drill's crude capability of perhaps ± .003" for position and diameter.
   b. We will later mill out the rough stock for the bearing and clearance hole. It is much, much easier on the end mill if it can plunge into a little bit of material rather that having to rely on its inefficient center cutting capability to drill through the material. Therefore, it is better to use a twist drill to create a quick pilot hole first.
3. Counterbore bolt holes.
   a. Special cutting tools called counterbores are made to create the recessed holes for screw heads. They are available in all the standard sizes. The counterbores rely on a noncutting "pilot" stud to guide the cutting tool into the material and to keep it concentric with the clearance hole for the screw. The pilot is usually a fixed diameter, so be sure that it is the same diameter as the hole you drilled previously. It can be very messy to shove a counterbore into an undersized hole at 1,000 RPM.
4. Drill Ø.296 pilot hole for reamer.
   a. Reamers have a very small cutting edge at the periphery and are not capable of center cutting. So we must always drill a pilot hole that is just slightly smaller than the reamer. This must be straight hole that is on the correct

location; otherwise, the reamer will tend to drift off center and follow the pilot. Sometimes a boring head is used to create a shallow, but very round and concentric starting hole for the reamer after the pilot hole is drilled.

5. Ream dowel holes.
   a. Dowel pins are cylindrical pins that have been ground slightly oversized (about +.0002"). They are used to precisely locate parts when they are assembled. Bolts and screws are not used for this purpose in precision machinery. It is important that the dowel pin holes are reamed to a size that will allow the pins to be pressed into the workpiece. Reamers are used when this kind of precision is needed with small holes.
6. Drill clearance hole.
   a. This is a low-precision hole that is used for the clearance of some moving part—a rotating shaft in this case. Large drills, like the Ø1.5" drill called for here, are expensive and prone to chatter. One solution is to use a smaller drill to remove the bulk of the material and then use an end mill to mill a circle. Milling of circles and curves is a process that is very difficult to perform manually. CNC controls make it possible to easily create holes of any size.
7. Mill the bearing hole and square ends.
   a. The next step is to finish mill the clearance hole and the rough-out the step for the bearing hole (Ø1.999"). Milling circles with an end mill is easy to do, but there are some drawbacks. The hole will be much less perfect than those produced by boring and reaming. The actual roundness may suffer, and it can be difficult to mill circles to close tolerances. Therefore, we will use milling to produce only second-class holes such as the clearance hole and the rough hole for the bearing, which will be finished later.
   b. While we have the end mill installed, we can also finish mill the ends of the workpiece.
8. Bore bearing hole.
   a. An offset boring head is a good choice for creating the high-precision bearing hole. Our total tolerance is only .001" so special steps are needed to make sure that we get it right. We previously milled the hole with an end mill. The size we would shoot for is about .007 to .015" less than the finished size. If we leave more material, then it can be difficult to hit the finished size in one step. If we leave too little material, then there is a good chance that some of the rough hole will not be cut at all or that there will not be enough material to get a good surface finish. The boring head is manually adjusted by the setup person and will not be moved again unless the hole is out of tolerance.

Now that we have the order of operations established, we can start putting together an operations sheet to document the manufacturing process. Operations sheets combine information about the operations, tools, speeds, and feeds into one location that is easy to read at a glance. Sometimes they will also include a graphical sketch of the setup so that the communication is clear between the programmer, setup person, and operator. Operations sheets are a common tool used in industry to plan manufacturing processes, and you will find that the few moments needed to complete an operations sheet will make it easier to write the part program. ***Figure 6.4*** shows the operations sheet for our bearing support. You can make copies of the setup sheet from Appendix B.

6.9 WORKHOLDING AND LOCATING PRINCIPLES

An integral part of processes planning is determining how we will hold on to the workpiece while machining. Some work-holding situations are easy to accomplish, whereas others require some creativity. For example, the bearing support from the previous example was easy to hold in a vise. The material was

Case Study

CNC in the Kitchen

What started out as a hobby for Bob Opsal eventually led to his current career as a stone fabricator and owner of Ancient Stone located in Brentwood in Northern California. Utilizing a CNC machine, the company can create an average of eight kitchens a week, and handles all aspects of natural stone fabrication, including flooring, countertops and full shower enclosures.

While working for his father, who owned a semiconductor business in Silicon Valley, Opsal and his childhood friend began fabricating granite countertops for their friends' parents, just for fun. As the jobs starting accumulating, he no longer had the time to work at his dad's company, and in 2003, he opened his own stone fabrication shop. The 3,000-square-foot facility houses a variety of state-of-the-art equipment, including a Haas GR-710 CNC stone working center, to which Opsal attributes much of the company's success.

Courtesy of Ancient Stone.

"I am the only one that operates it," said the owner, adding that he took a 4-day training course to learn how to use the machine. "After the training, I could run it on my own, but it really is an ongoing process. I have had it for two years now, and I am always learning new things on it."

Ancient Stone uses the CNC machine for all aspects of stone fabrication, including sink cutouts, edging and drain boards, among other functions. Opsal knew he wanted a CNC since before opening his shop, as he feels it is key for quality production. "It is a huge advantage for granite shops to have. It opens up every aspect when you have a CNC running. We can push three kitchens out of it a day. We can edge pieces and do sink cut-outs at the same time without even taking the pieces out."

Opsal said that when a slab comes off of the CNC, it is about an 80% finished product, and then employees do the final polishing by hand to get the perfect shine.

Ancient Stone's production is 80% residential and 20% commercial, and the company recently completed a large-scale project for a Gold's Gym, "All done on the CNC," said Opsal, adding that the project was a great success. "We also recently did some work for a local steak restaurant. We did all the countertops in Black Galaxy granite, and some walls feature granite as well."

Opsal has come a long way since he first began fabricating stone by hand, averaging one kitchen a week. He does research on the Internet to learn about new machines and new material. "I didn't try to make this business happen too fast," he said. "I try to run as lean as I can. My main focus is getting equipment that will do the quality that I want to put out."

already square, and the locating surfaces (backside and bottom) did not need to be finished. However, many machining situations requiring closer tolerances are not this simple.

In the remainder of this section, we will discuss some of the concepts that you will need to understand in order to plan the setups for your CNC machining jobs.

The 3-2-1 Locating Principle in Theory

Locating is the act of positioning the workpiece against a reference surface on the fixture in such a way as to produce accurate and repeatable results. The key to properly locating a workpiece is a principle called 3-2-1. The 3-2-1 locating principle is an idea used to locate a workpiece by making contact on three surfaces

Figure 6.4 Operations sheet for the bearing support. The operations sheet specifies the order of each manufacturing process, the cutting tools, and the speeds and feeds.

Manufacturing Operations

Name: Joe Date: 7-99 Project Name: Bearing Support Material: 1018 CRS

Op. #	Description	Cutting Tools	Velocity	D.O.C	Feed	Surface Finish	MRR or Cutting Time	Graphic Description
1	Spot Drill Holes	90° S.Drill	120	—	.003	—	—	Hold in Vise
2	Drill Bolt Holes	17/32 HSS	110	—	.004	—	—	
3	C'bore Bolt Holes	5/8 C'Bore	40	—	.006	—	—	
4	Pilot for Dowel	19/64 Drill	110	—	.0019	—	—	
5	Ream Holes	5/16 Reamer	40	—	.004	—	—	
6	Clearance Hole	Ø1.0 Carbide	400	—	.008	—	—	
7	Mill Ends and Holes	Ø3/4 EM	110	.375	.006	63	—	
8	Bore for Bearing	Solid Carbide	400	.02	.0015	63	—	Mill Ends and Center Hole

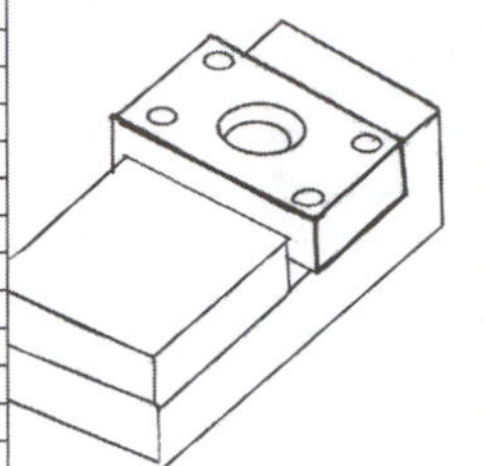

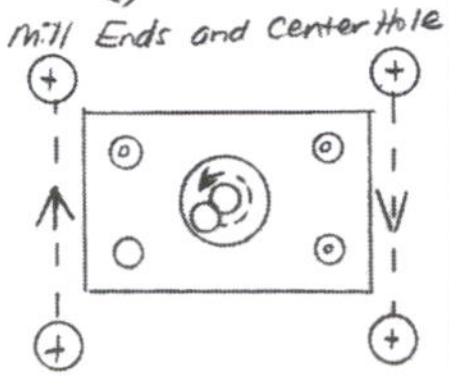

simultaneously (see *Figure 6.5*). The 3-2-1 locating principle takes into account that surfaces will be imperfect; that is, under most circumstances, you cannot count on physical surfaces to be flat, square, and parallel to each other with any degree of certainty.

datum:

A reference plane or point that is defined on the engineering drawing from which other dimensions are established.

The 3-2-1 locating principle starts by taking a primary surface (**datum**) of the workpiece and locating it on three points. From geometry, we know that three points will form a plane. Therefore, if we place a workpiece on three points of contact, it will be sitting on a plane. This first plane will form our first reference surface. The only practical way to form a point is to use a spherical radius as a contact. When a spherical radius is tangent to a plane, it will have only one point of contact. Imagine a ball bearing sitting on top of a table—it makes only one point of contact with the table. Now place three ball bearings on the table and place a book on top of the bearings. The book will sit flat on the ball bearings in a (nearly) perfect plane.

The next step is to locate the adjacent side of the workpiece on two points of contact. The two points of contact form a line rather than a plane, so the surface will

Figure 6.5 The 3-2-1 locating principle starts with the concept that a sphere will contact another solid surface at only one point. Three points contacting (tangent to) the surface will create a flat plane. In this illustration, three ball bearings are resting in the face of a workpiece creating a plane.

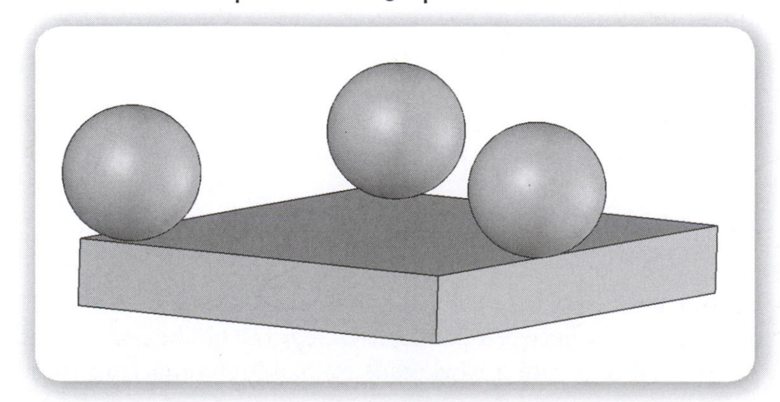

Figure 6.6 A workpiece that is not square cannot be held flat against two perpendicular planes at the same time. A common setup method to deal with this situation is to use a round bar to create a single line of contact with the workpiece. The bar allows the workpiece to rotate into position without interference from the non-square faces. The clamping forces from the vise will ensure that the flat face of the stock is held flush with the upright solid jaw (forming a datum surface). The part can then be machined for excellent perpendicularity between the top and left side.

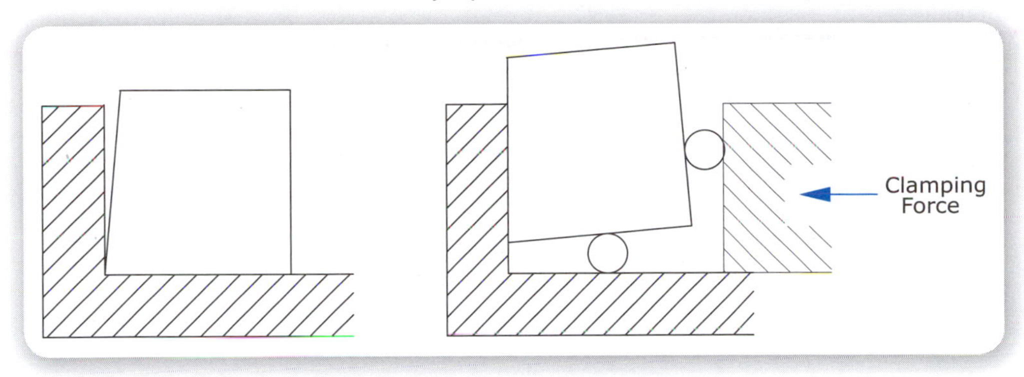

be free to revolve around it. Imagine two ball bearings on a table with a book balanced on top of them. The book can rotate in one axis but not the other. The reason we use two points rather than three is to allow this free rotation. With two points of contact, the adjacent side of the workpiece can be out of square without disturbing the first locating surface. However, if we use two perfectly square planes to locate an imperfect workpiece, then we cannot be certain how it will be located—maybe it is flat against the back or maybe against the bottom. The only thing we can be certain of is that an out-of-square workpiece cannot be held simultaneously flat on two planes, as illustrated in *Figure 6.6*.

The last step is to locate the surface that is adjacent to our primary and secondary surfaces with just one point. The single point of contact will not disturb either the first or the second locating surfaces. Again, this is done to account for any imperfect surfaces. This locating principle is illustrated in *Figure 6.7*.

Figure 6.7 The 3-2-1 locating principle. The bottom is located on 3 points, the back on 2 points, and the left side on only 1 point.

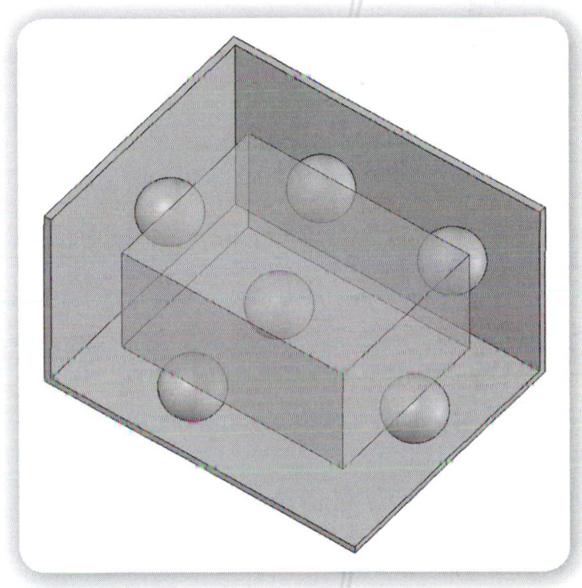

The 3-2-1 Principle in Practice

In practice, the 3-2-1 principle is modified slightly. On most raw materials, the stock faces are relatively flat and square from the mill. For fixturing purposes, these slight variations will not pose a problem—the same cannot be said about cast, saw-cut, and flame-cut faces.

Rather than having only three points of contact, the primary surface is located against a very flat surface, such as the solid jaw of a vise. The secondary reference surface can be formed by a thin parallel bar or pin without disturbing the primary surface. Finally, the third point of contact is constructed from a work-stop such as a rod or screw. *Figure 6.8* shows a typical setup in a milling vise. (The sliding jaw is removed for clarity.) The solid jaw forms the primary surface, a thin parallel forms the second, and a work-stop provides the final point of contact.

Work-holding fixtures, including vises, usually have one or more reference planes on which the workpiece will be located, but only one surface serves as the primary plane for a setup. On a vise, the solid jaw and base are generally flat, square, and parallel and can serve as reference surfaces. On the other hand, the sliding jaw can slightly twist and move, so it cannot be used as a reference surface.

Figure 6.8 The 3-2-1 principle in practice. The primary datum of the workpiece (translucent) is located against the solid jaw of a machining vise (sliding jaw removed for clarity). The secondary datum rests on a parallel bar that will allow it some degree of rotation. Finally, the tertiary datum is formed by the bar stop that is mounted to the vise.

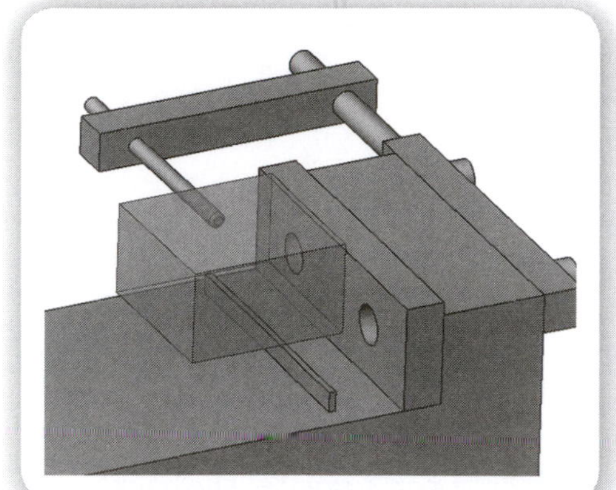

Preparation and Mounting

In order for locating to be effective, the locating surface of the workpiece must be in a finished condition. A saw-cut surface, for example, does not make a suitable locating surface because we cannot be certain that it is flat, square, or parallel. Rough, cast, and saw-cut surfaces must usually be *qualified* before they can be used as locating surfaces. This usually involves machining the surface flat in an earlier operation. However, these extra setups and operations are expensive and should be eliminated whenever possible. For example, ***Figure 6.9*** shows a raw material blank that is rough cut stock. The back surface of this workpiece will have to be machined flat before the part can be mounted in a vise to machining the five holes.

An important concept in CNC machining is to eliminate the need for operations to *qualify* the workpiece. The programmer will sometimes find a way to combine operations to qualify the surfaces and create the features simultaneously. That is, rather than starting with a workpiece that has been squared-up with preliminary operations, every attempt is made to machine the workpiece from a rough piece of stock whose edges are neither square nor flat. The programmer will use the end mill to create the straight and square edges in one setup without the need for premachined reference surfaces.

This is a particularly effective technique if there is enough excess material to hold the part while creating the finished surfaces. The excess material can then be milled away in subsequent operations. For example, if the raw material is large enough, the excess material can be used for clamping and then removed in a second operation. ***Figure 6.10*** shows this situation. The stock is oversized, and the lower portion is held in a vise while the finished features are milled into the top. In this case, the raw material can have slightly rough and uneven surfaces with no ill effects.

Figure 6.9 Uneven surfaces will have to be machined before the surface can be used for locating.

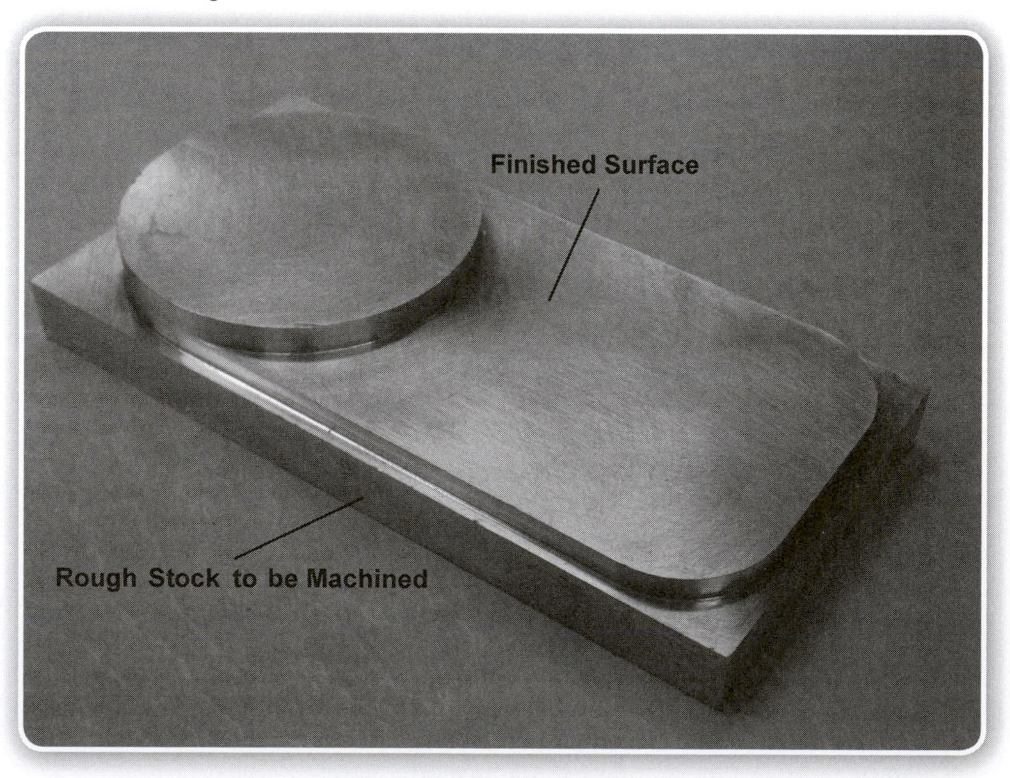

Figure 6.10 A portion of the raw stock can sometimes be sacrificed for work-holding. This can eliminate machining operations needed to "qualify" the edges (i.e. the end mill and machining operations will create the straight and square edges). The excess material will later be removed.

You must be careful when using the above technique, because it is not always practical. If the workpiece is taller than it is wide, then you risk pulling it out of the vise during machining. Make sure the material is inserted in the vise by at least ⅓ of the total height, or deeper if the material is tough.

Other workpieces are too large or impractical to mount in a vise. In many cases, the workpiece can then be mounted directly to the table or a dedicated fixture can be built. This often requires that the programmer develop creative work-holding techniques to machine the workpiece in as few setups as possible. One such situation occurs when a relatively thin and flat workpiece must be milled around the entire outside of the material. In this case, the material can be clamped to a flat plate (see ***Figure 6.11***). The clamps will obviously interfere with the machining, so the periphery can be partially machined and then the clamps moved so the remainder of the machining can take place. Of course, the second set of clamps should be installed before removing the first set or the workpiece will move. This technique can drastically reduce the need for material preparation. Furthermore, the concept can be extended to other clamping methods such as socket-head cap screws and edge clamps.

For other odd and large workpieces, you may have to design a fixture. Fixtures can be as simple as a plate with a couple of locating pins or as complicated as a fully automated system to hold multiple workpieces with hydraulic clamps. Fixture design is beyond the scope of this book, but in general, you should only make what you cannot buy. There are numerous tooling components that are available off-the-shelf, and it is usually more economical to buy something rather than to build it yourself. There are also modular tooling systems that allow you to reuse the components for other jobs—a much better solution than having dedicated tooling sitting around and taking up space.

Figure 6.11 The workpiece geometry may require several clamping operations.

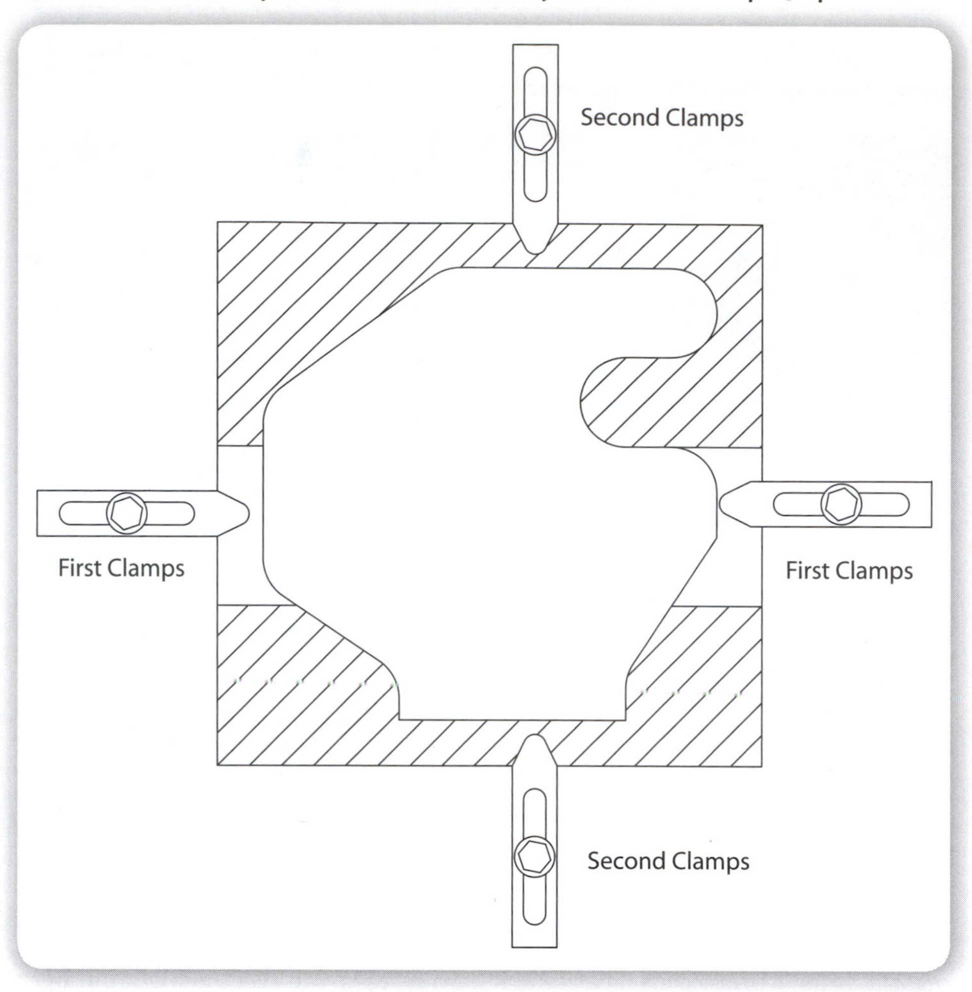

In summary, the following guidelines will help you to develop setups that are more efficient:

- Follow the 3-2-1 locating principle when applicable.
- Use as few setups as possible.
- Eliminate preparatory and secondary operations if possible.
- Use off-the-shelf tooling when possible and build dedicated fixtures only as a last resort.
- Coordinate the raw material purchase with the process plan.

6.10 SELECTING THE WORK ZERO

At this point, we should be able to decide where our workpiece zero will be located. Work zeros are selected for two important reasons: convenience and repeatability. First, let's look at work zero selection based upon convenience.

We know from our study of the Cartesian coordinate system that any location in the first quadrant will have positive coordinates. Most people find it easier to program with positive coordinates, so the first quadrant is a natural choice. For example, ***Figure 6.12*** shows a workpiece that will be set up to use the lower, left-hand corner as the work zero. The result of this is that most of the tool locations will have positive values.

Figure 6.12 The work zero can be set at one of many locations. Each location will offer some advantage.

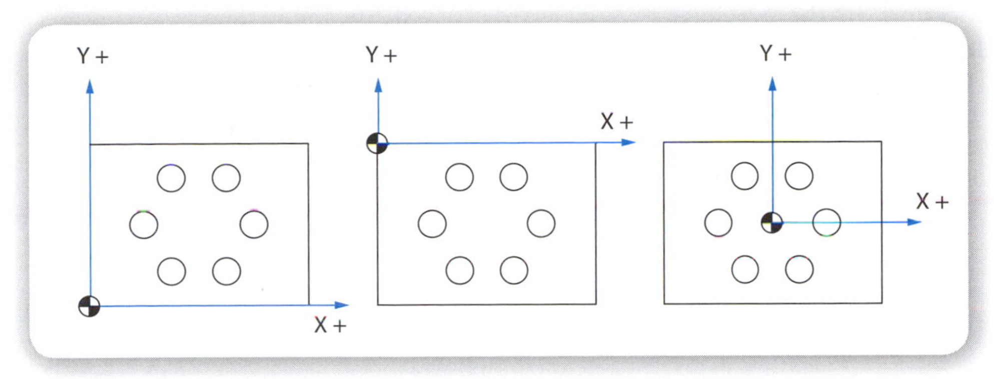

Figure 6.13 Workpiece widths can vary; therefore, we should only locate the workpiece from a reference surface. The locating surface should not be confused with the work zero. The two concepts are independent of one another.

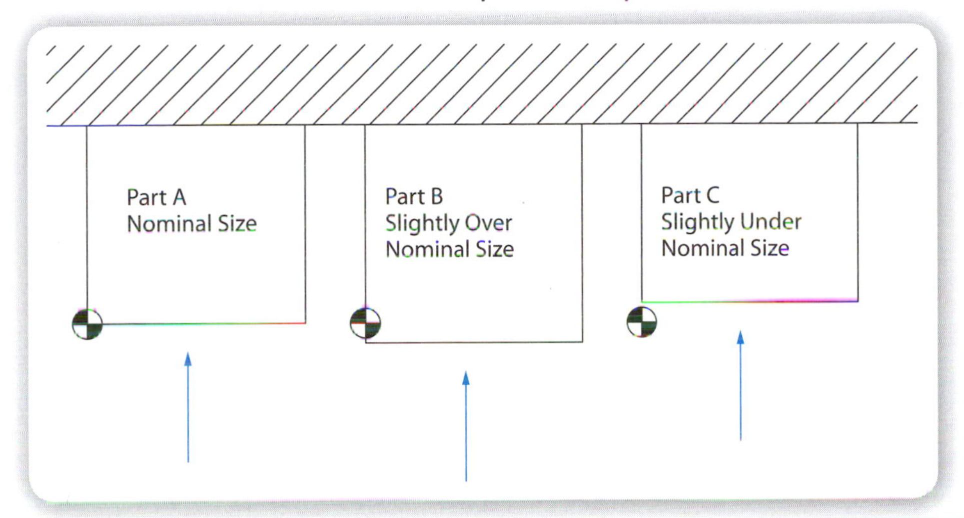

Common shop wisdom may lead you to believe that the surface against the sliding vise jaw cannot be used as the zero because the width of the part may vary. The variation will cause the dimensions of the feature to *float around* as the width changes and therefore cause errors in the dimensions. Although this is true, it can be misleading. We are not actually going to use the side against the sliding jaw as a reference—that would be foolish. We will use the side against the solid jaw (or other reference surface) to edge-find or probe the workpiece. Then it is just a matter of moving the grid electronically by the nominal width to the lower, left-hand corner. ***Figure 6.13*** illustrates this point. The exaggerated width will change from part to part, but the work zero will remain in the same—and correct—place in relation to the reference surface.

We could also put the work zero in the upper left corner to coincide with the two reference surfaces. This is actually easier to set up because we would not move the grid after edge-finding the workpiece. Yet, it is more difficult to program because most of the Y-axis coordinates will be negative.

A third possibility with our workpiece in ***Figure 6.13*** is that we could use the center of the bolt hole circle as the work zero. Bolt hole circles require a number of manual calculations. If we were to select any of the corners as the work zero, then we would have to add a value to each location in our bolt hole calculations. Again, this can be time consuming and is prone to errors. However, if there are numerous other features besides the bolt hole circle, then the center might not be a good choice for the work zero.

The second reason for selecting a particular work zero is repeatability. This is actually related more to workpiece location than to an arbitrary work zero location. In the previous section, we discussed the importance of workpiece location and some of the methods used to establish the proper location against a reference surface. However, we never discussed why a particular feature should serve as a locating surface. The safest way to select a locating surface is to use the datums that are specified in the drawing. For example, in ***Figure 6.14*** the same workpiece has the datum in a different location. If the workpiece arrives only to have the hole bored in it, the only choice we have in either case is to locate the workpiece against the datum. We could also use the datum as the work zero for convenience, but this is an arbitrary decision as long as we edge find from the datum.

Figure 6.14 **In some cases, the workpiece datum will dictate the locating surfaces.**

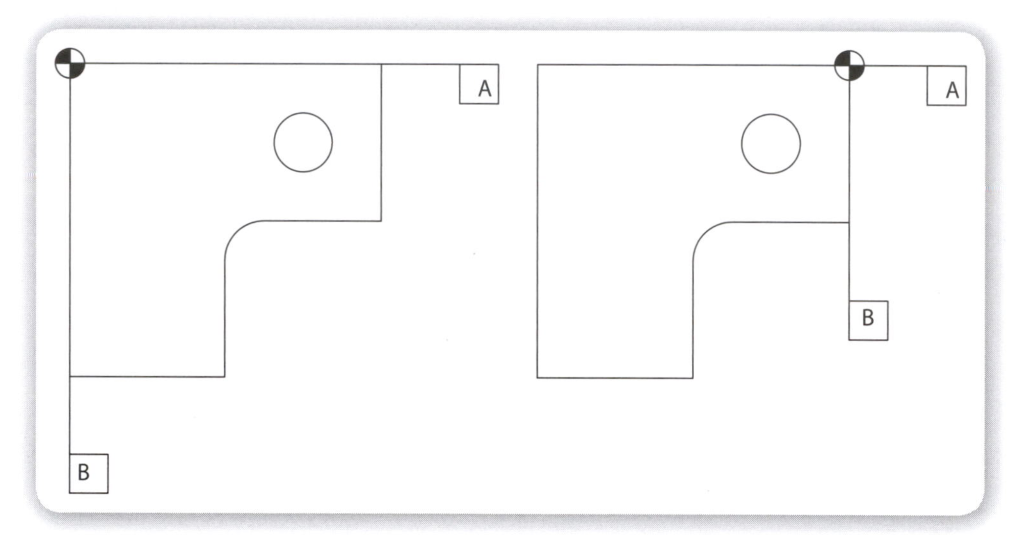

If datums are not specified, then the water becomes a bit murkier. In this case, we will want to find the surface features that are most likely to be stable and use only those features for location. This method will at least keep the finished dimensions consistent. For example, in ***Figure 6.15*** the dimensions are all given from the centerline. Because we cannot actually use the centerline as a locating surface, we will select the top and left faces for locating, and the upper left corner for the work zero—assuming that the surfaces are already machined to size.

This workpiece will need a second setup to machine the backside. Therefore, we must be sure to use the same faces for locating on the second side or the finished dimensions are likely to suffer. You might think that because this part is square, the second side could be located on any corner. However, you must remember that the size of the workpiece will vary and this error can add to any previous error to create a scrap workpiece.

In the end, it does not really matter where you put the work zero as long as it is kept in the proper relation to the locating surfaces. However, we will usually try to select a location that will minimize the number of calculations we will have to make and thereby minimize the possibility of human errors. Perhaps of more importance is that regardless of where we place the work zero, we should never establish a work zero by edge finding from a surface that is free to move as the workpiece changes.

Figure 6.15 Dimensions given from the centerline indicate no clear locating surface.

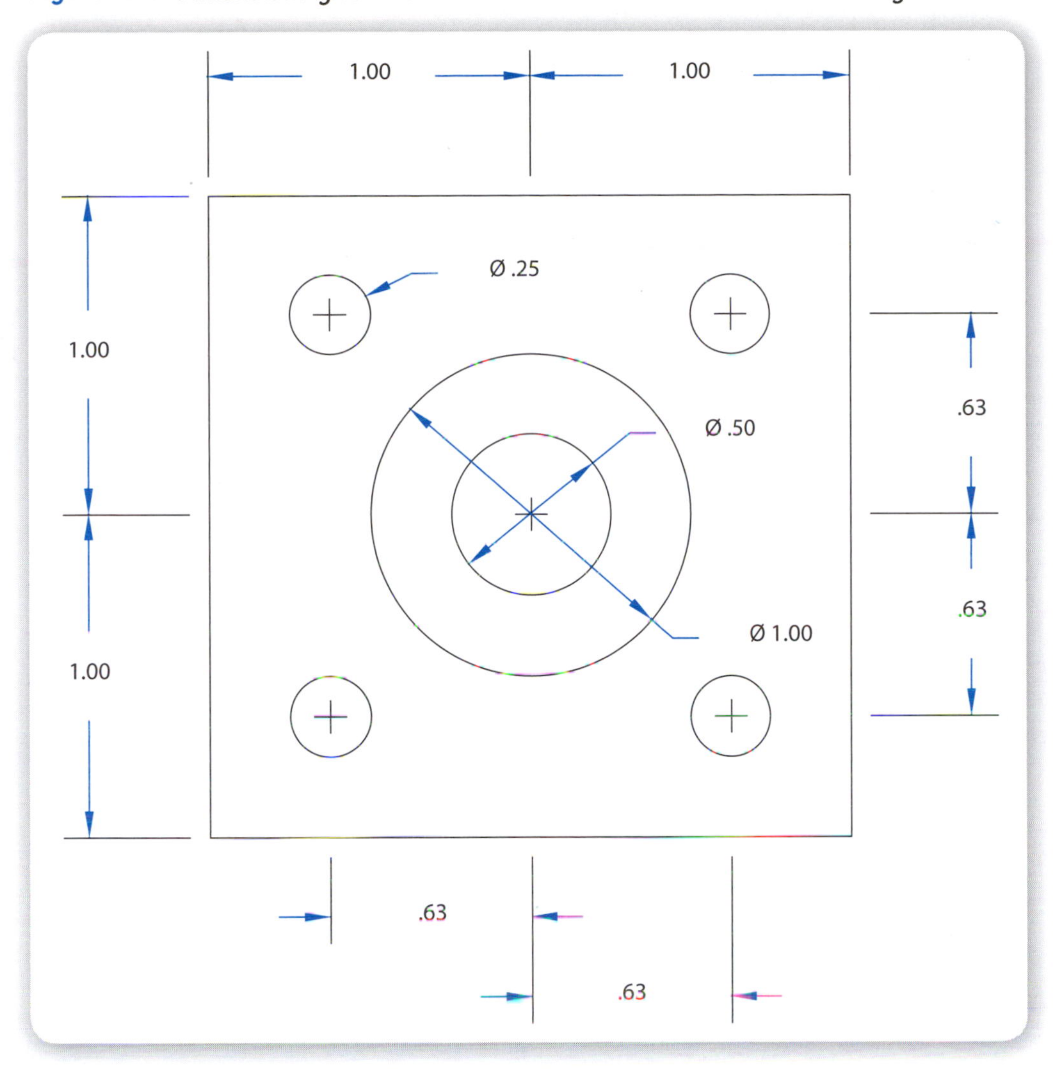

6.11 PLANNING THE TOOLPATH

Now that we have planned the operation and established the locating surfaces and work zero, we are almost ready to the write the part program. The next step is to plan the toolpaths. Essentially, this involves plotting out the center locations of the tools as they make their way around the workpiece. We could do this while we write the code, but it can be disruptive to stop in the middle of coding to calculate a tool position.

Center operations such as drilling are easy to plot because we already know where the center of the tool should be located. Toolpaths are a little more difficult because we have to manually compensate for the radius of the tool. In other words, the part drawing will show us where the finished surface is located, but we will have to move the tool-center over to cut the surface in the proper place. Therefore, we must already have the cutting tools, work-holding method, and raw material size before we can plan the toolpath.

As an example, let's look at the workpiece shown in *Figure 6.16*. The material for the job is just slightly bigger than the finished features, and it does not appear that we will need to qualify any surfaces. We can simply mount the stock in a vise and start machining. However, the part is longer than it is wide, so we should turn the workpiece 90° before mounting it in the vise. As for the work zero, there is no clearly defined datum, but the lower left corner (of the original drawing) looks like a good candidate.

Figure 6.16 A finished workpiece to be machined from a larger block of raw stock.

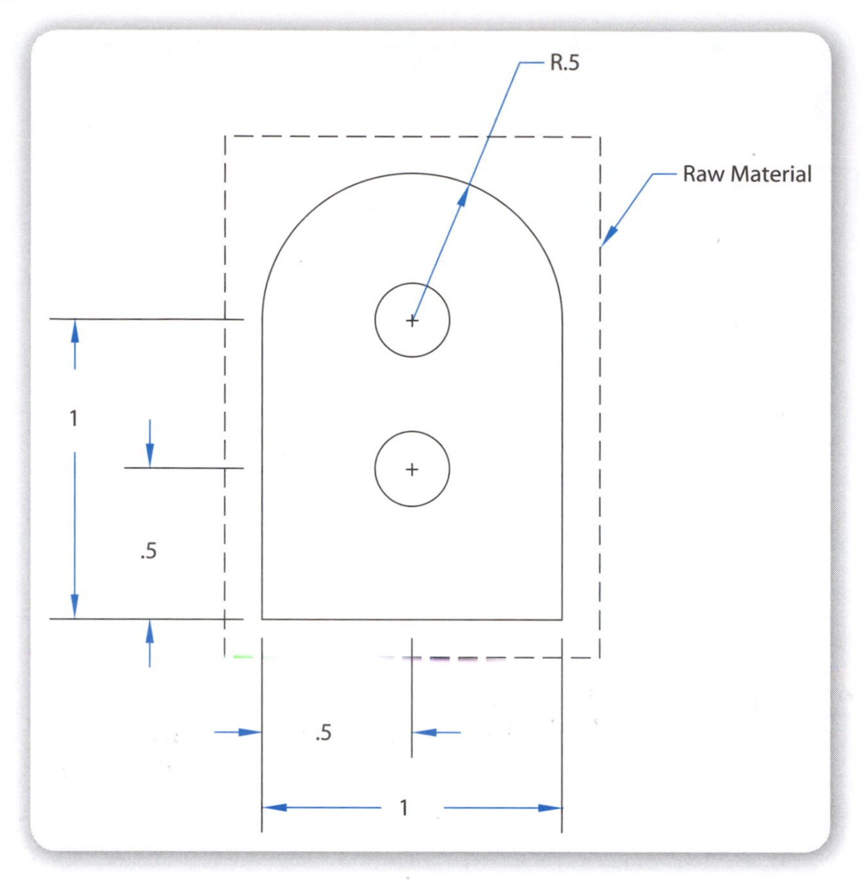

Figure 6.17 The setup and toolpath for the workpiece in Figure 6.16. The toolpath has been offset by .250 to create the correct feature size.

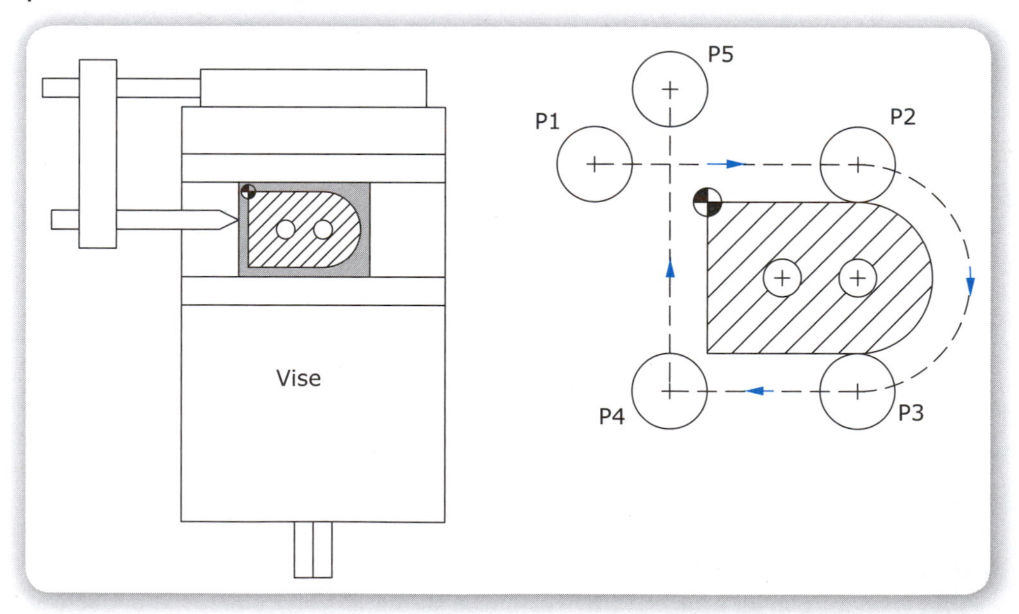

Now we only have to plan the toolpath and plot all of the coordinates from the work zero—absolute coordinates. Just remember that the tool must be moved away from the profile by half the diameter of the cutting tool. If we use a Ø0.500" end mill for the profile and the toolpath has been offset by the radius value of 0.250", then the toolpath might look like that in *Figure 6.17*. We can then make a table of all the center points in preparation for writing the code, as in Table 6.3.

Table 6.3 The Tool Coordinates for the Workpiece Shown in Figure 6.16 and Figure 6.17

Point	X	Y
1	−0.75	0.25
2	1.00	0.25
3	1.00	−1.25
4	−0.25	−1.25
5	−0.25	0.75
Hole 1	0.50	−0.50
Hole 2	1.00	−0.50

Your Turn

Write an NC program using your first name. Plot coordinates first on graph paper. Concentrate on where to start and end and using the least amount of steps possible. Verify your code in your CNC software.

CHAPTER SUMMARY

- Letters/addresses are combined with numbers to form NC words. The words are the codes that the MCU can understand and act upon to produce machined workpieces. One or more complete words form a block.
- An NC part program is made up of blocks of code that are read and executed sequentially by the MCU.
- Each part program will have separate functional areas. Some sections of the program will set up the machine or perform specific functions such as tool changes. Other sections are used to create geometry by giving the machine instructions to mill or drill.
- Most codes are modal and will remain active until replaced by another code from the same group.
- Syntax is the grammar of NC programming. G & M code syntax is somewhat flexible, but violating some rules will cause errors. Programmers are free to write in their own style as long as the lines of communication are preserved.
- The CNC machining job should be completely planned before programmers write any code.
- The machining process, cutting tools, work holding, and the material condition are inseparable components of the planning process.
- All planning should be completed before any writing of the part program begins.

BRING IT HOME

1. What characters indicate the beginning and end of an NC program?
2. How is the program number specified?
3. What is the purpose of a program comment and which characters are used to define the comment?
4. Why might you want to use a safe line at the beginning of a program?
5. Find the formatting mistakes in the following block.

```
N100 G 1 X5.5.0 Y .350 (F10)
(FEED TO FIRST POINT)
```

6. Rewrite the following blocks in order to take advantage of modal commands.

```
N100 G01 X0 Y0
N110 G01 X1 Y0
N120 G01 X1 Y1
```

7. Give an example of a model code.
8. Give an example of a miscellaneous non-modal code.
9. Match each of the following codes to its function.

Code	Function
G01	Peck drill a hole.
G83	Rapid to a position.
M00	Stop the spindle.
M03	Cancel the drilling cycle.
G00	Use incremental positioning.
M09	Change tools.
G02	Cut a clockwise arc.
G03	Pause the program.
G91	Cut a counterclockwise arc.
M08	Turn the coolant off.
M30	Cut a straight line.
M05	Turn the spindle on.
M06	End the program and reset.
G80	Turn the coolant on.

10. What is some of the necessary documentation for a CNC machining job? Why is this documentation important?
11. Create an operation sheet and a setup sheet for the workpiece found in Figure 6.15. Assume that the outside surfaces are already finished and that you need only to drill the holes and mill the pocket.
12. Plan a toolpath for the workpiece in Figure 6.15 using the processes and tools that you came up with for Question 11. Define your workpiece zero and create a list of coordinates to which the tools will travel. You can assume that the Ø0.25" and Ø0.50" holes are at a depth of 0.50" and that the Ø1.0" hole is at a depth of 0.25".

CHAPTER 7
Codes for Positioning and Milling

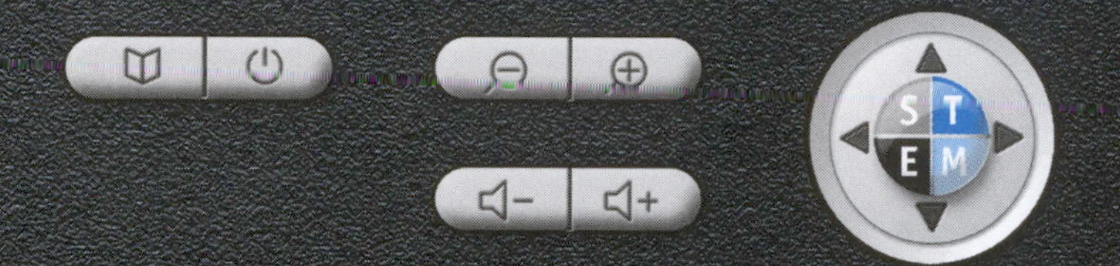

Menu

START LOCATION | DISTANCE | END LOCATION

Before You Begin

Think about these questions as you study the concepts in this chapter:

1. What are some of the limitations of the most common prep codes?
2. Why is rapid traverse used?
3. What types of programming is used to create arcs?
4. What are the limitations of circular interpolations?
5. How do incremental and absolute units affect axis return?
6. Understand the limitations of circular interpolation.

Key Terms

Preparatory Code
Interpolation
Lead In/Out
Linear
Syntax

7.1 CONVENTIONS

In this chapter and the chapters that follow we will look at the codes that are used for NC programming for milling centers. There are a few conventions that we will use to indicate meaning in both text and graphical format. Please take a few minutes to study the text conventions in Table 7.1 and the graphical conventions in *Figure 7.1*

In general, all programming examples will use absolute positioning unless otherwise specified. Also, in the last chapter, we learned that many codes are modal and do not have to be repeated unless they are changed. However, it is easier to keep track of the machine location if we include all of the words and coordinates on every line. Writing out all the information will take a little more time and effort but will save some frustration while we learn to write NC code. So for now, the examples will include modal codes, and we will take shortcuts later on.

preparatory code: Codes that carry out machining operations or establish machine settings; G-codes.

linear: Along a straight line.

syntax: The rules of structure that must be followed when writing in a specific language; grammar.

Table 7.1 *Text Conventions for NC Programming*

Number following a letter—with decimal	Xn.n	n.n is a decimal number. Examples: X2.50, Z5.0001
Number following a letter—no decimal	Ln	n is an integer. Examples: L5, P200
Coordinates	(X, Y, Z)	Coordinates placed between parentheses are assumed to be in alphabetical order. Examples: (1.25, 1.0, −.50), (0, 1)
Syntax	Gn Xn.n Yn.n (Ln)	Syntax is the grammar of the code. The specified words are required, but most are modal. Words in parentheses are optional.

Figure 7.1 *Drawing conventions that are used in the programming examples.*

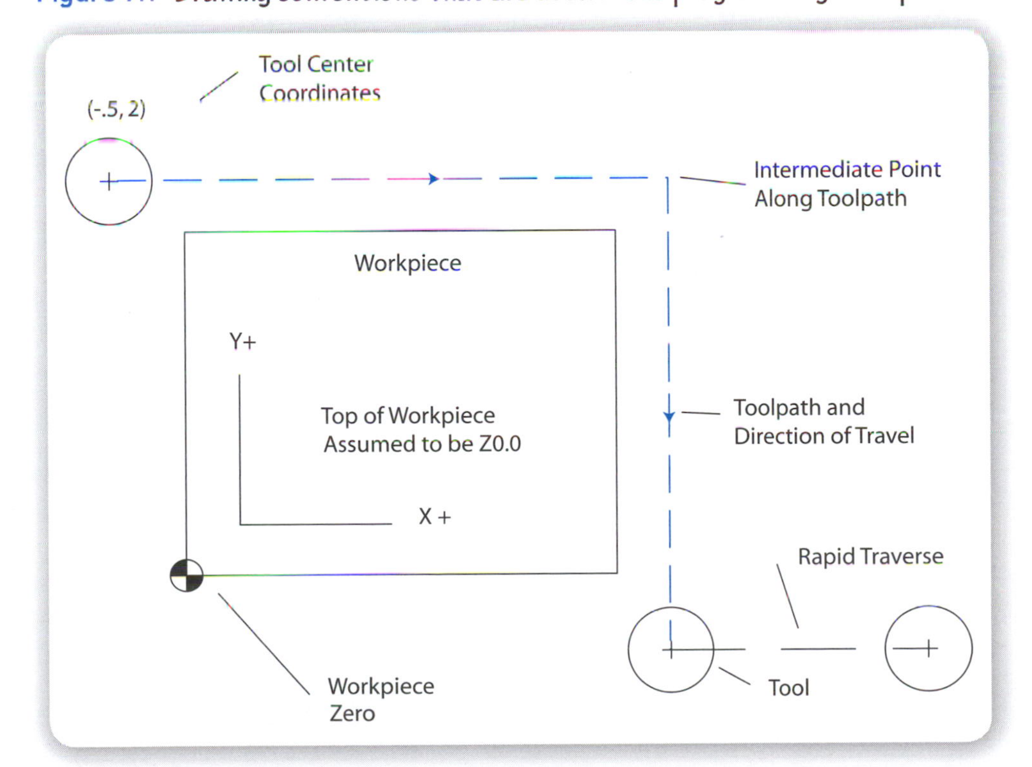

7.2 PREPARATORY CODES

The codes that are used to control and modify axis movements for machining and positioning operations are called preparatory codes. They are specified with the letter "G" and are usually referred to simply as G-codes. There are G-codes for many different machine operations, including positioning, straight-line cutting, circular arc cutting, drilling operations, and many more.

The G-codes that are used to perform machining operations all work in a similar manner. The G-code is called, and then the machine moves to the specified coordinates. Other G-codes are used to set up machining conditions, and they do not directly cause axis movement. These codes will modify the behavior of the codes that actually perform the machining. For example, the G00 code will cause the machine to move to a specified position. However, another code such as G20 or G21 can tell the machine whether that position should be in inches or millimeters.

7.3 G00—RAPID TRAVERSE

G00 is the code used to perform rapid traverse. Rapid traverse is used for quickly positioning the tool in preparation for making a cut or moving the tool to a safe location for tool and part changes. G00 can be used any place where the tool is not directly in contact with the work piece and where you wish to save time. Rapid traverse is never used to perform a cut—the velocity is far too fast and is not constant. However, the G00 code is used extensively for positioning the tool to make milling cuts or to drill holes.

For example, ***Figure 7.2*** shows a tool that starts at the tool change position and must be positioned to machine a pocket into a workpiece. Your experience with conventional milling machines tells you that to machine such geometry, you must

Figure 7.2 Rapid traverse is typically used for positioning.

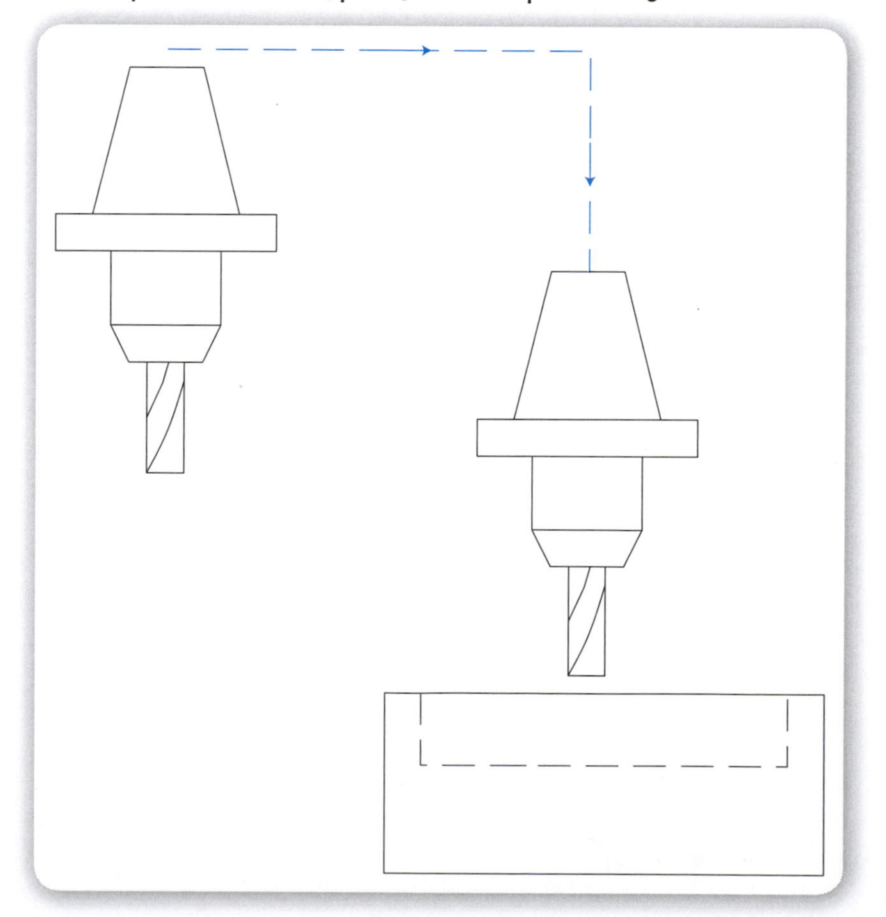

first move the tool into position and then slowly plunge the tool into the part to the required depth. The same is true about NC machining. However, we can instruct the machine to move at a very fast rate to a point near where we want to begin cutting; this is called rapid traverse. The actual plunge cut would use a different code that can be programmed to move more slowly.

Rapid traverse is fairly easy to use; we simply tell the machine how we want it to move and then tell it where that point is located. The cutting tool starts in one position—perhaps where you changed the tool—and is then moved quickly to where you want to start machining. The syntax (grammar) for the rapid positioning is as follows:

```
G00 Xn.n Yn.n Zn.n
```

It is important to note that G-codes always give the end point of a move, not the starting point; the machine already knows the starting point, as in *Figure 7.3*. For example, look at the following code. Assume that we are using absolute positioning and the tool is currently positioned at X0, Y0:

```
N01 G00 X1.0 Y2.0
N02 G00 X4.0 Y2.0
N03 G00 X4.0 Y0.0
```

Figure 7.3 **A toolpath is created by instructing the machine to move to specific point in a Cartesian coordinate system.**

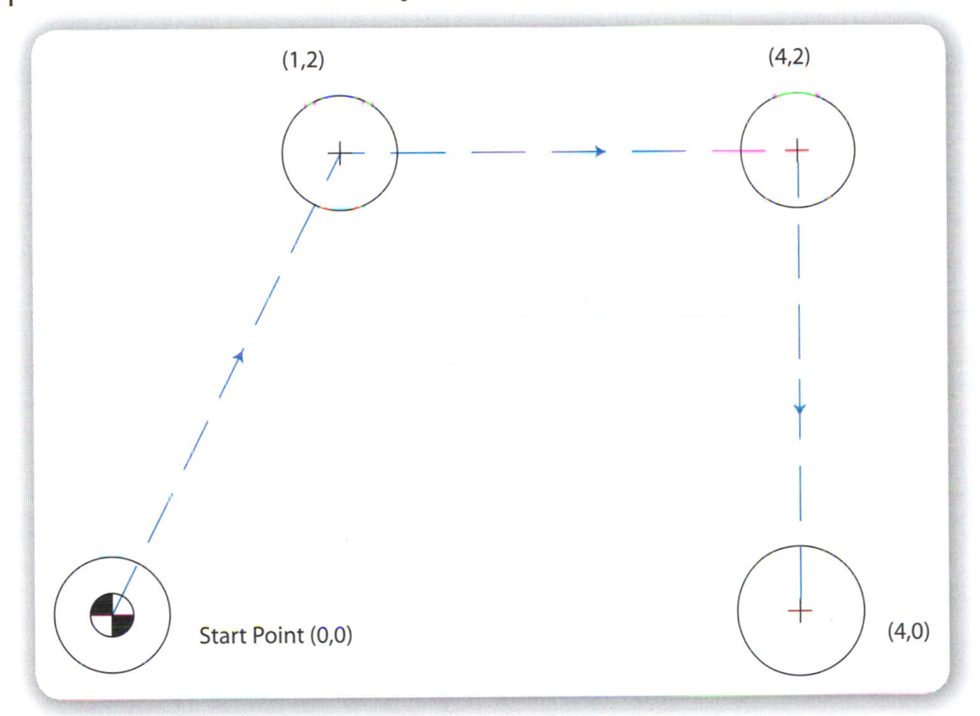

Block N01 instructs the machine to move to X1.0, Y2.0. Block N02 instructs the machine to move from X1.0, Y2.0 to X4.0, Y2.0, which happens to be a point 3.0" to the right. What instruction does block N03 give? It instructs the machine to move from X4.0, Y2.0 to X4.0, Y0.0, a point 3" back.

We should always give a little extra room between the rapid positions and the actual workpiece. The size of the workpiece may vary, or it may not be positioned exactly the same as the previous part. We must also take extra care when machining rough or irregular workpieces such as castings. Even small interference between the tool and workpiece can result in broken tools during rapid positioning. If you intend to rapid traverse to the surface of the workpiece (Z0.0), you should stop short of this coordinate. It is common practice to position to Z.200 and then call a G-code that is designed for cutting (the topic of the next section), as in the following example:

```
N10 G00 X3.0 Y4.0
N20 G00 Z.2                    (Rapid to 0.2 above the surface.)
N30 G01 Z-.5 F4.0              (Use linear interpolation to plunge.)
```

Notice that the tool is positioned to the X- and Y-coordinates first and then brought down in the Z-axis. All three coordinates could have been positioned at the same time on a three-axis machine tool (*see* ***Figure 7.4***), as shown in the following example:

```
N10 G00 X3.0 Y4.0 Z.2  (Rapid)
N20 G01 Z-.5 F4.0
```

This code is a perfectly legitimate method of programming, but beware that the resulting toolpath might not be a straight line to the specified coordinates, and that a crash could result if there are any obstructions. Therefore, *it is good practice to move the X- and Y-axes before moving the Z-axis* while you are learning to program.

We must take care when using rapid positioning for several reasons. First, many machine tools have very fast rapid rates that are in the range of 200" to 1,000" per minute. This velocity is faster than an operator can react to stop the machine traverse. The programmer and setup person should take care to ensure the proper tool path has been coded, or else a serious crash can occur. A good safety rule to

Figure 7.4 Diagonal movement by programming simultaneous X and Z-axis moves.

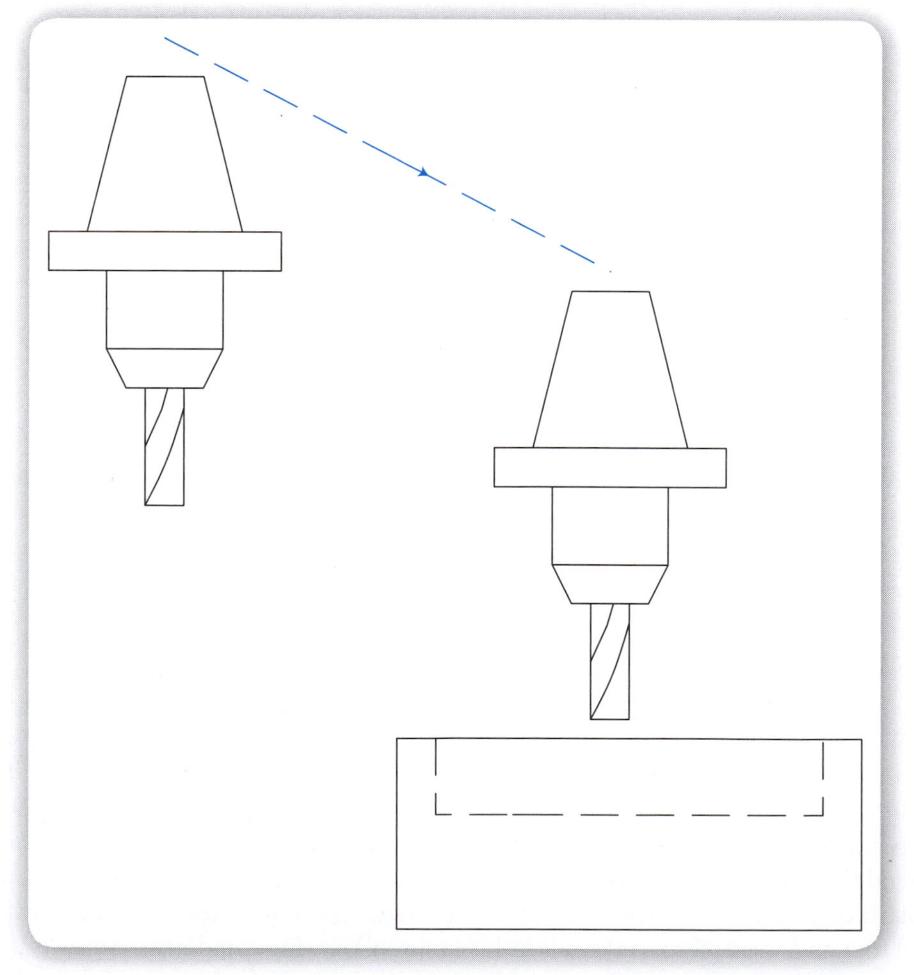

follow while you are a beginning programmer is to never use rapid positioning when the tool is below the surface of the workpiece. This makes it easy to check your part program by reading through and simply checking that the Z-axis position is always above zero when G00 is called. For example, the code below might crash the tool into the workpiece in block N110 because the Z-axis position is below the surface of the workpiece. However, it will be easy to find this mistake by tracking the Z-axis position when G00 is invoked.

```
N100 G01 Z-.5 F5.
N110 G00 X5.0 Y2.0
```

Second, rapid positioning will not always move in a straight line from the current position to the programmed position. The machine control unit (MCU) will instruct the servomotors to move at their maximum velocity when G00 is encountered. This does not pose any problem if movement is in only one axis. Yet if there are two or more axes of movement specified, the result is not always in a straight line. The servomotors will move at their maximum velocity until the coordinate is reached in each axis and then stop. Take the following code as an example; assuming that each axis will move at the same rate, the resulting movement will be a straight line at 45° to X3, Y3:

```
G01 X0.0 Y0.0
G00 X3.0 Y3.0
```

These results can be misleading because the distance traveled by each axis is the same; therefore, each axis will arrive at its destination at the same time. If the distance moved is different, the shortest distance will be reached first, and then that axis will stop while the other axis continues to move in a straight line to its destination. The tool may crash into the work piece or tooling (such as strap clamps) if this concept is ignored. The following code and *Figure 7.5* demonstrate the idea.

```
G01 X0.0 Y0.0
G00 X6.0 Y3.0
```

Figure 7.5 **Rapid traverse does not always result in a straight path.**

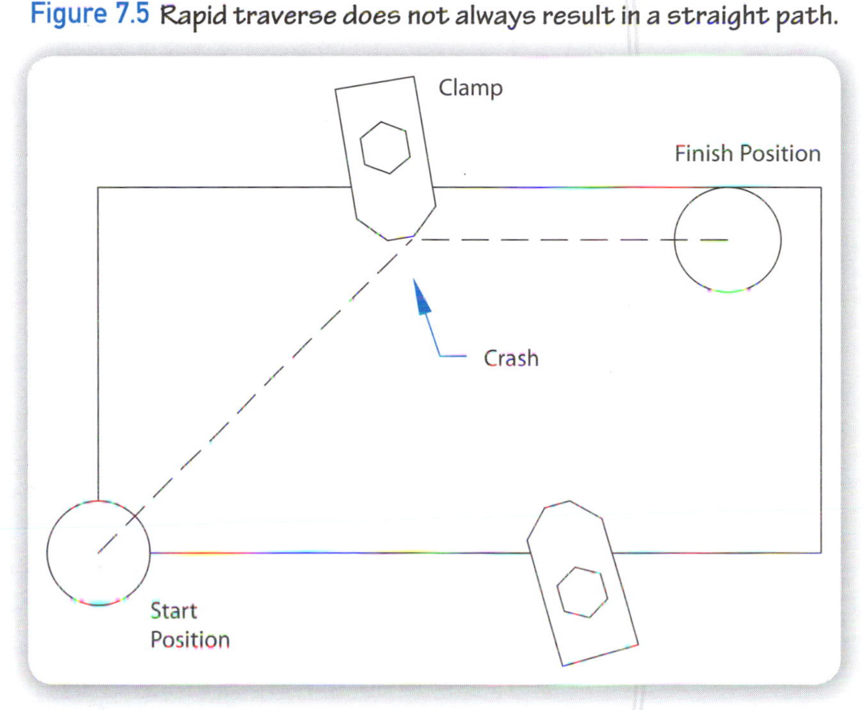

7.4 G01—CUTTING STRAIGHT LINES

The next G-code to learn is G01—linear interpolation. **Interpolation** is a term used to describe the process of *passing through all points*. In the case of linear interpolation, the tool will travel through all points that make up a straight line to the Cartesian coordinates specified in the block. The G01 code moves the tool from one point to another at a specified velocity. The syntax for linear interpolation is as follows:

```
G01 Xn.n Yn.n Zn.n Fn.n
```

interpolation: Tool movements that travel through all theoretical points of a programmed path. Linear and circular interpolation are common functions of CNC machine tools.

You are probably asking, "Why not just use the G00 code?" The G00 code cannot be programmed to move at a constant velocity—it moves as fast as it can to the destination. Furthermore, G00 will not always follow a straight line as was discussed earlier. Conversely, linear interpolation maintains a very accurate velocity, which leads to a constant chip load and a good surface finish.

The G01 code works much like G00, but with one added word. The feed word controls the feed rate and is specified in inches per minute (IPM). As an example of G01, the following code is used to produce the toolpath in ***Figure 7.6.*** (Assume the tool has been positioned to X0, Y0.)

```
N020 G01 X1.0 Y0.0 F3.0
N030 G01 X1.0 Y1.0
N040 G01 X2.0 Y2.0
```

Figure 7.6 **Linear interpolation in use.**

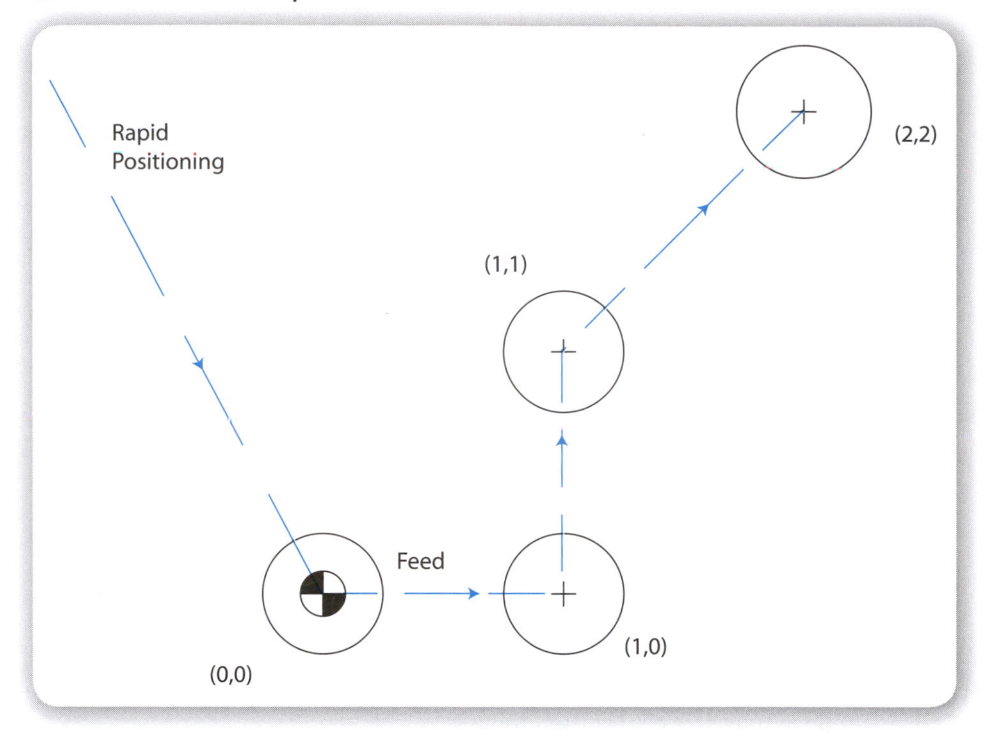

The machine will now move at 3 IPM to X1, Y0 and later to X1, Y1 and X2, Y2. The feed code is also modal, so it may be specified at the first G01 and not specified again until a change is needed.

There may be occasions when incremental programming is necessary. In this case, the program could be rewritten to use incremental coordinates to produce the same tool path:

```
N010 G91
N020 G01 X1.0 Y0.0 F3.0
N030 G01 X0.0 Y1.0
N040 G01 X1.0 Y1.0
```

Linear interpolation is also used for plunge moves into the workpiece. However, we must be sure to use either a center-cutting end mill or to drill a pilot hole first, and reduce the plunge feed to only ½ the normal feed. For example, in ***Figure 7.7*** the cutting tool will plunge into the workpiece and make a cut to the right. It will then feed back out of the pocket and rapid back to the start point.

```
N100 G00 X1.0 Z.2
N110 G01 Z-.25 F2.5 (PLUNGE)
N120 G01 X3.0 F5.0
N130 G01 Z.2
N140 G00 X-.5 Z6.
```

It is generally a good idea to get in the habit of using linear interpolation for any positioning move that takes place below zero in the Z-axis (the top of the workpiece normally being Z0). This is not written in stone, but it will help you

Figure 7.7 Rapid traverse and linear interpolation combined to mill a slot.

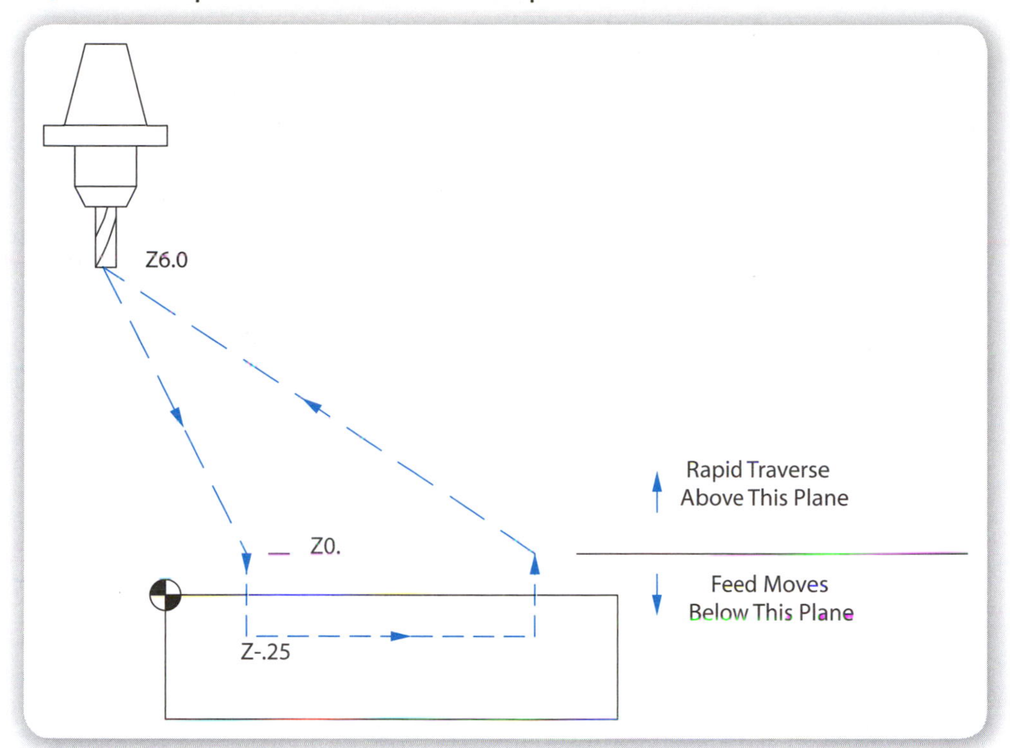

Figure 7.8 It is a good habit to only use feed move when positioned below Z-zero.

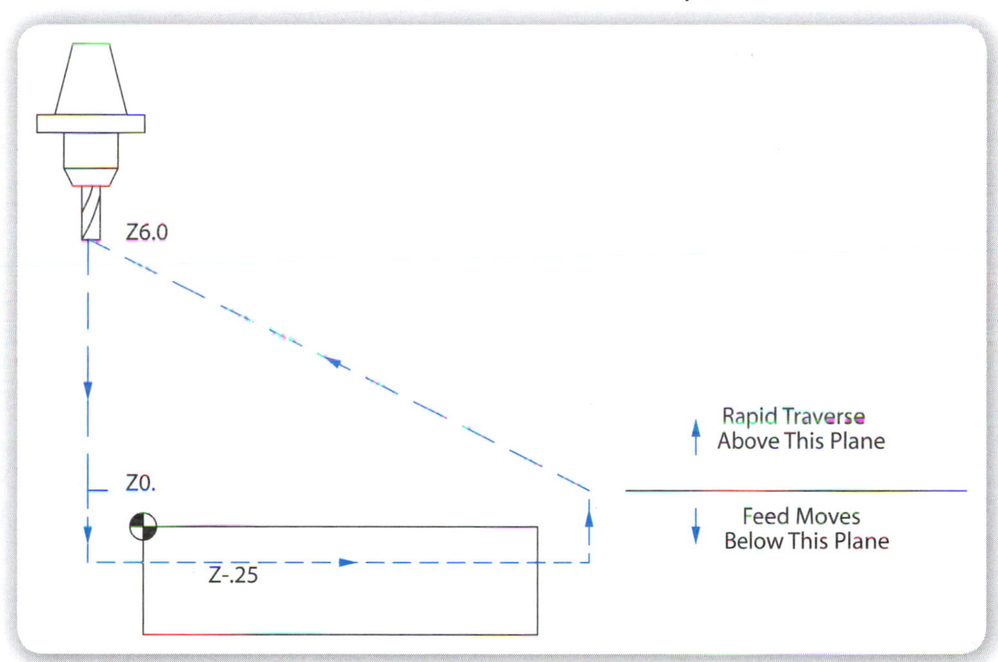

avoid some serious mistakes while you are learning to program. For example, in *Figure 7.8* a shoulder is to be milled into the workpiece. We could immediately position to the minus Z-level, but instead we will rapid traverse to Z.2 and then use linear interpolation to move to the final Z-level:

```
N100 G00 Y-.5 Z.2
N110 G01 Z-.25 F2.5 (PLUNGE)
N120 G01 X5.0 F5.0
N130 G01 Z.2 (FEED UP TO Z.2)
N140 G00 X-.5 Z6.0
```

7.5 G02 & G03—CUTTING CIRCULAR ARCS

Circular arcs are another common feature in machining. To cut an arc we use the codes G02 and G03—circular interpolation. These two codes are identical in function, except that G02 will cut in a clockwise direction (CW) and G03 in a counterclockwise (CCW) direction. Like the G01 code, a feed must also be specified. There are also two variations in the syntax for circular interpolation that are shown in the code example below. The only difference between these two relates to the method used to specify the tool radius. Each method is common and some controls will support both variations.

```
G02 Xn.n Yn.n Zn.n In.n Jn.n Fn.n
G02 Xn.n Yn.n Zn.n R.n. Fn.n
```

As we did similarly for the other two other preparatory codes that we learned, we simply give an X- and Y-axis coordinates to specify where the arc should end. Under normal circumstances, the Z-axis coordinate can be ignored altogether. However, some machines will support helical interpolations. If helical interpolation is supported, any difference between the starting and ending Z position will create a spiral that follows the arc—a helix. Helical interpolation is a useful feature for thread milling and for plunging into a pocket—more on this later.

Specifying the Arc Center with I and J

The X and Y coordinates are not enough to completely describe an arc. We have to somehow describe the radius of the arc and the coordinates of the arc center. For this we need two new words. The I- and J-words specify the incremental distance from the *start* of the arc to the *center* of the arc. The I-word corresponds to the X direction, and the J-word corresponds to the Y direction, as in ***Figure 7.9***.

Figure 7.9 The I and J-words are used with circular interpolation to define the center of the arc.

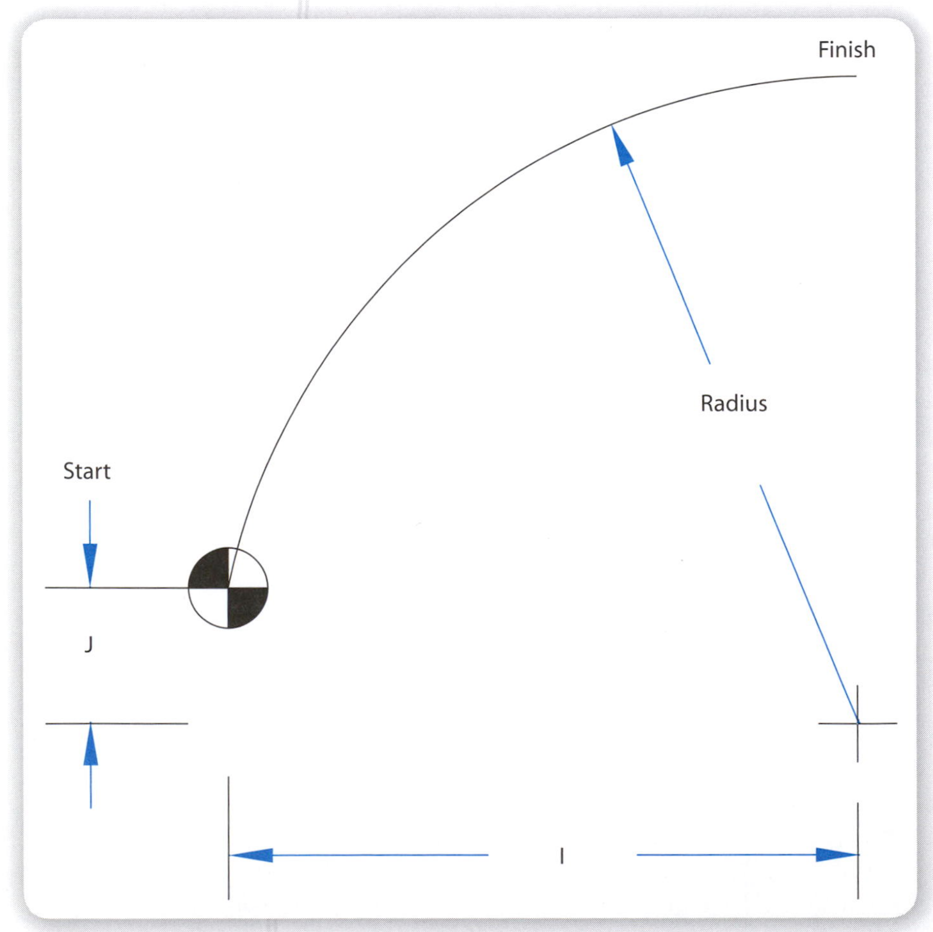

We can think of the I and J values as if a new coordinate system were created so that the origin coincides with the start of the arc. We can then locate the coordinates of the arc center as you would any other point in a Cartesian coordinate system—except that the axes are labeled I and J rather than X and Y. Of course, this means that I and J can have positive or negative signs, depending on the orientation. ***Figure 7.10*** shows several 45° arcs where a new coordinate system has been placed at the start to illustrate how I and J are defined. The values in Table 7.2 indicate the sign or direction of the I- and J-words for each arc.

One confusing situation occurs with I and J when the start point and center of the arc are equal on one of the axes. This is a very common situation that will occur anytime that an

Figure 7.10 The I- and J-directions from the start of the arc, to the center.

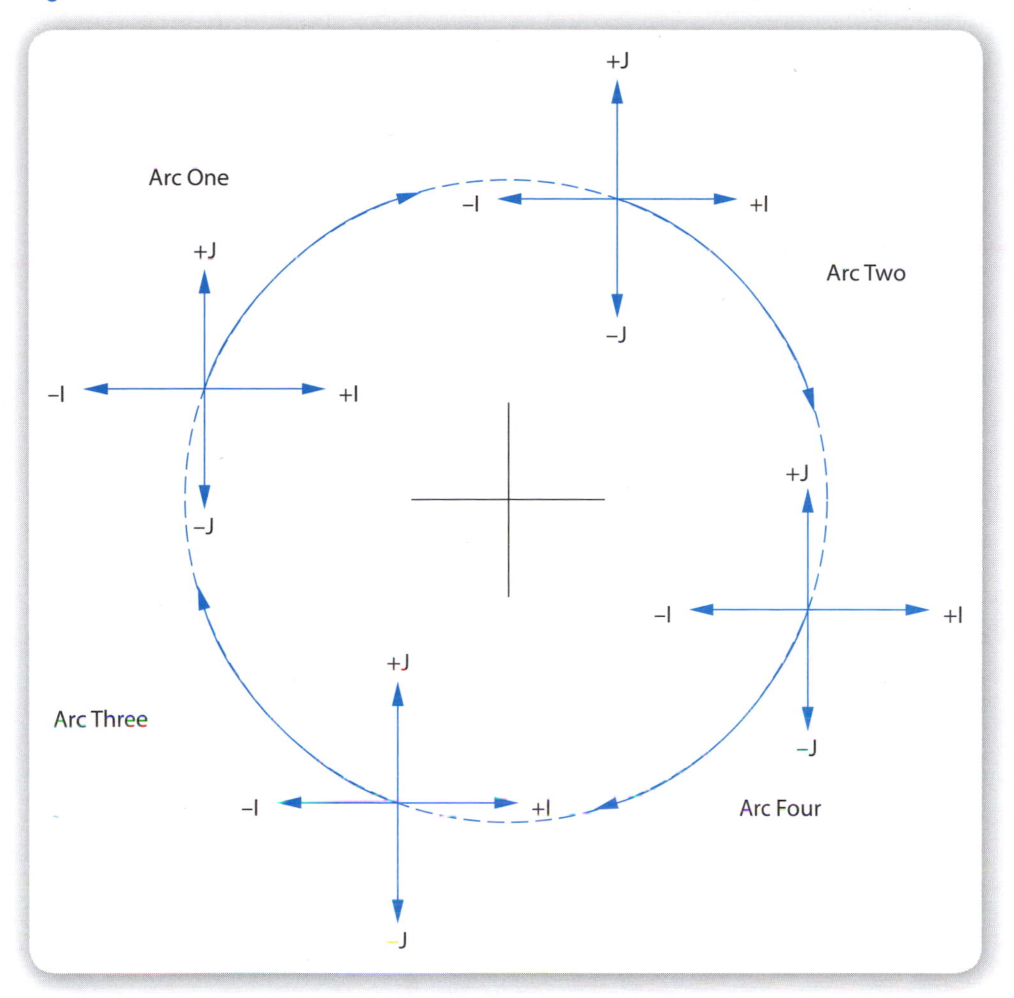

Table 7.2 The Signs of the I- and J-words in Figure 7.10

Arc	I	J
Arc One	Positive	Negative
Arc Two	Negative	Negative
Arc Three	Positive	Positive
Arc Four	Negative	Positive

arc starts at 0°, 90°, 180°, or 270°. In each of these situations, either the I or the J will be equal to zero. The reason for this situation can be found in the very definition of I or J—the incremental distance from the start to the center. If the arc happens to start on any of the previously mentioned angles, then one of the values will have to be zero regardless of the direction to be traveled. *Figure 7.11* demonstrates this idea.

Figure 7.11 The I and J values when the arc starts at the beginning of any of the four quadrants. Note that the values are independent of the arc direction.

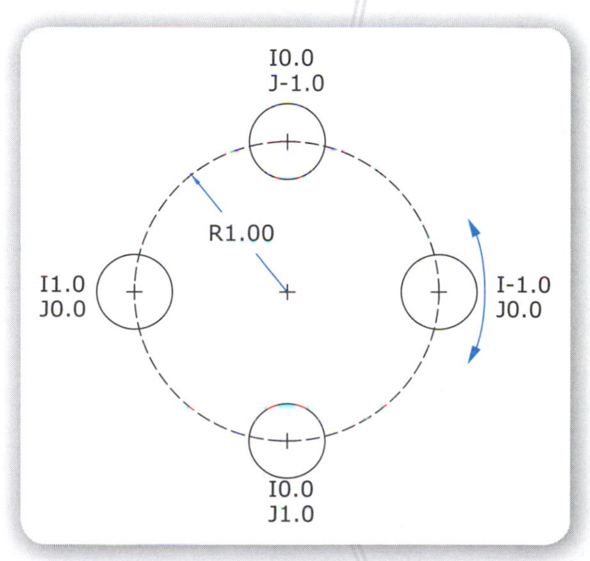

Arc Centers in Absolute Coordinates

On occasion, you will run into a CNC machine tool that is setup to take the I- & J-words as absolute coordinates. That is, rather than using an incremental distance relative to the start of the arc as is usually done, we have to give the absolute coordinates in relation to the program origin. This is not any more difficult that the method previously described. We will see and example of this absolute arc centers in the next section.

Specifying the Arc Center with R

Perhaps an easier method to use to define the center of an arc is to use the R-word. This is very convenient if it is supported by the control. When R is supported, the control will perform the necessary calculation in order to define the center. This does not really save much effort, however, because the programmer will have to make numerous calculations for difficult arcs. Furthermore, a full 360° circle is usually not possible to complete when using R and, therefore, the arc must be broken into at least two segments. Nonetheless, you can save some programming time by using R.

Figure 7.12 **The R-value can be positive or negative depending on the shape of the arc. These two arcs have exactly the same start and end points, and radius. The only parameter that has changed is the sign of R.**

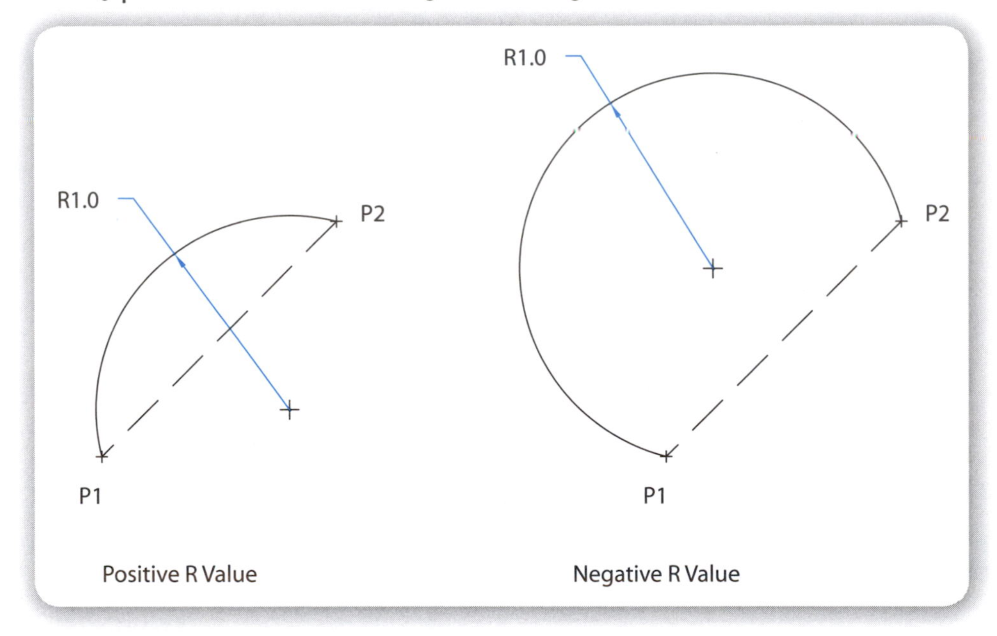

Using R with circular interpolation is straightforward—you simply use the radius of the arc instead of I and J. However, there is one variation to be considered, based upon the shape of the arc. The R-word can be either positive or negative to indicate the shape. This is necessary because there are two solutions to any arc of a given radius between two points. ***Figure 7.12*** illustrates this point.

The two arcs have identical end points and radius, but the arc center is in a different location on the two arcs. To determine the sign of R, draw a chord between the end points of the arc. If the center lies exactly on or outside of the chord and arc, then the sign will be positive. If the chord and the arc bound the center, then R will be negative.

7.6 USING CIRCULAR INTERPOLATION

We can see how circular interpolation works by looking at an example. The following code is used to produce the toolpath in ***Figure 7.13***. Assume that point 1 has the location of X0 and Y0.

```
N01 G01 X0.0 Y0.0 F3.0
N02 G02 X1.5 Y1.5 I1.5 J0
```

Figure 7.13 *G02 used to produce a clockwise toolpath.*

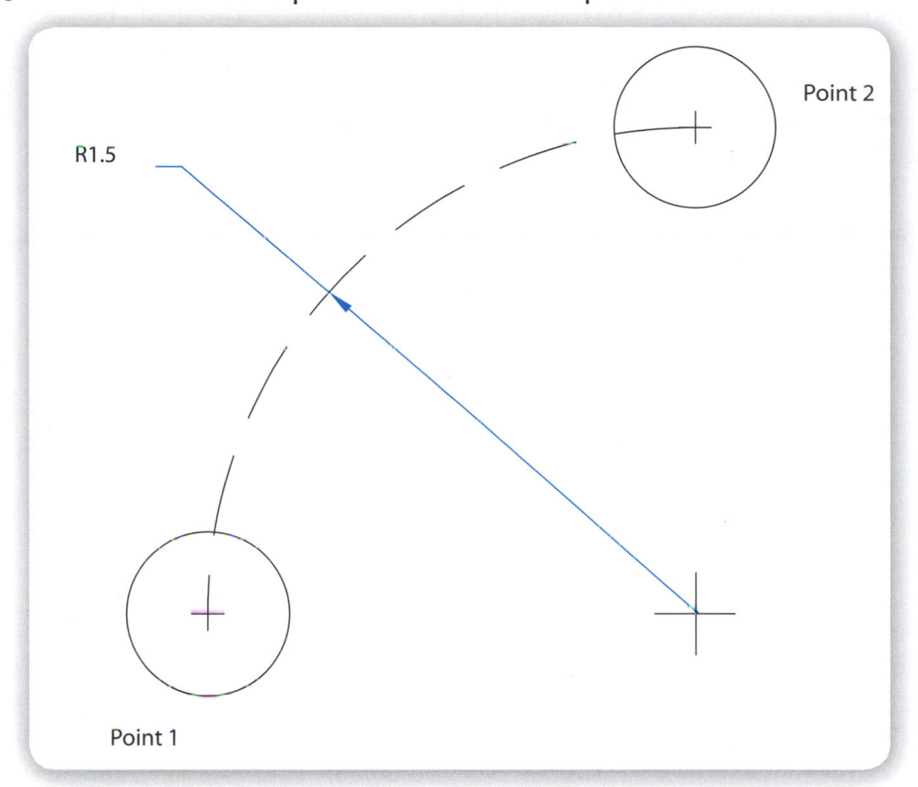

The tool moves to point 1 from wherever it was located and then cuts a CW arc to point 2. The center of the arc is 1.5" from the start in the I direction and 0" from the start in the J direction, which leaves us with I1.5 and J0.

Let's continue with another full-quadrant arc by looking at the following code example, which will produce the toolpath in ***Figure 7.14***. This code will produce two 90° arcs in a row. Again, point 1 has the coordinates of X0 and Y0.

```
N01 G01 X0.0 Y0.0 F3.0
N02 G02 X1.5 Y1.5 I1.5 J0.0
N03 G02 X3.0 Y0.0 I0.0 J-1.5
```

Figure 7.14 Two full-quadrant arc. Most controls will allow this arc to be programmed in on block.

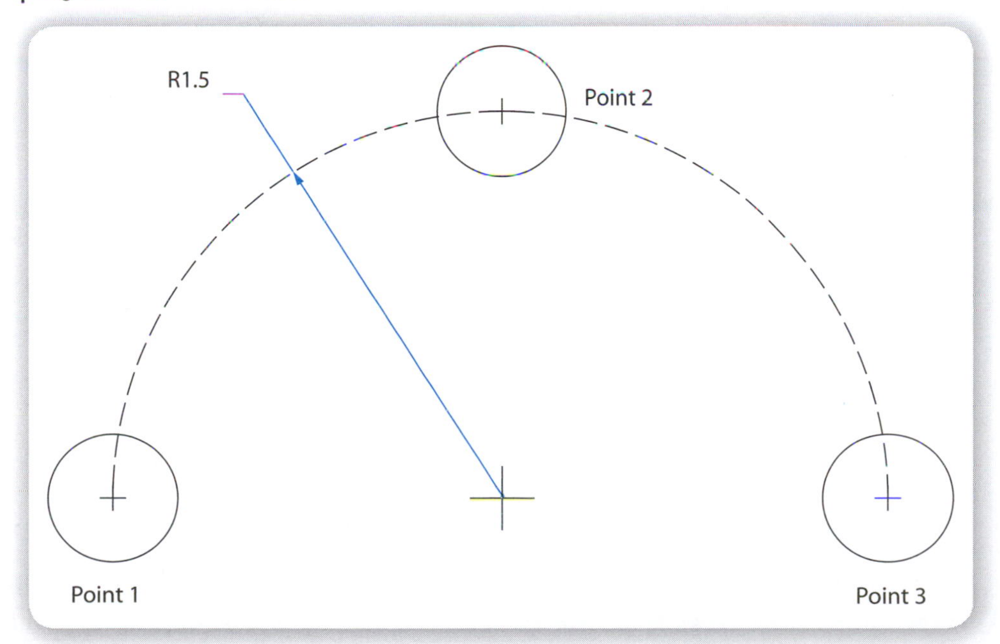

The first arc move in block N02 is identical to that in the previous example. The second arc move in block N03 moves the tool from point 2 to point 3. This time I is equal to zero and the J is –1.5. Remember the definition of I? It is the incremental distance from the start of the arc to the center in X direction. The I-value is equal to zero because the X-coordinate of point 2 is the same as the X-coordinate of the center of the arc. There is no difference in the X direction, so I will equal zero.

Some older machine tools do not allow arcs to be continued to a different quadrant. Therefore, the arcs must be programmed one quadrant at a time as we did in the last example. Most modern machine tools can create this arc in one move from point 1 to point 3, as in the following example:

```
N01 G01 X0.0 Y0.0 F3.0
N02 G02 X3.0 Y0.0 I1.5 J0.0
```

Full-circle arcs are also common. Full-circle arcs always start and end in the same place, and the center point is found in the same manner as any other arc. For example, the circle in ***Figure 7.15*** would be programmed as follows:

```
N01 G02 X-1.0 Y0.0 I 1.0 J0.0
```

Full-quadrant arcs are relatively easy to program without too many calculations. However, arcs that start or end at an odd angle may require the use of trigonometry to find the end points or the I- and J-values. We must be careful to make these calculations with a high degree of accuracy and not let rounding errors *stack up*. The calculations will need to be accurate to at least 0.001" for the X-, Y-, I-, and J-coordinates. Otherwise the machine will produce an error message telling us that the arc cannot be made with these values. ***Figure 7.16*** shows an arc that traverses a 135° angle. The code to create this is as follows:

```
G02 X.7071 Y.7071 I1.0 J0.0
```

We might also be confronted with circular arcs that do not start or end at a quadrant, as illustrated in ***Figure 7.17***. In this case, we will have to carefully calculate the coordinates of both points. The following code will produce the arc.

Figure 7.15 A full arc will have the same start and finish location.

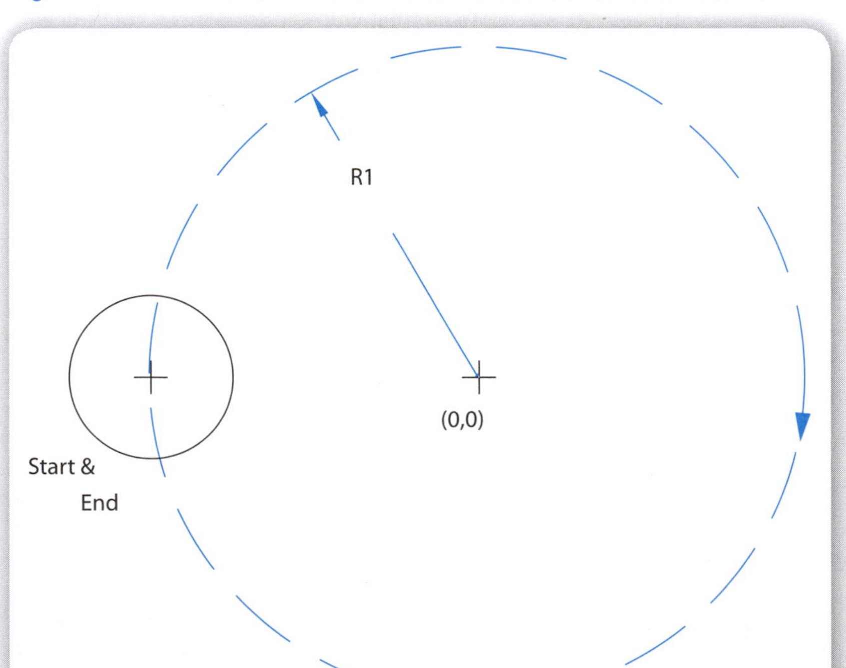

Figure 7.16 An arc that does not end at a quadrant may require the precise use of trigonometry.

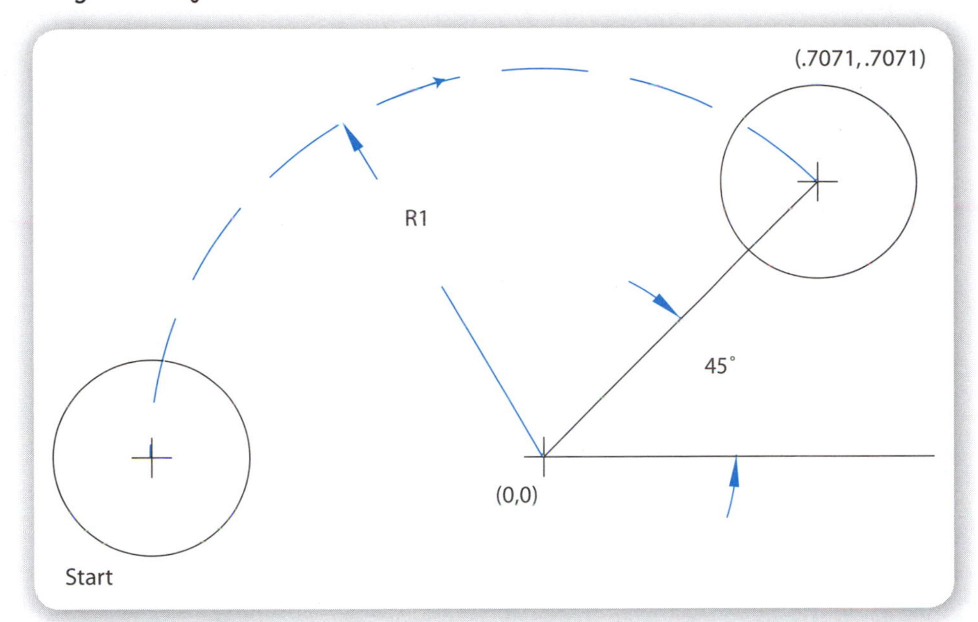

Note that the end point (P2) is identical to that in the previous example, but the I- and J-values have changed.

```
G02 X.7071 Y.7071 I.7071 J-.7071
```

Figure 7.17 A circular arc that does not start or end at a quadrant.

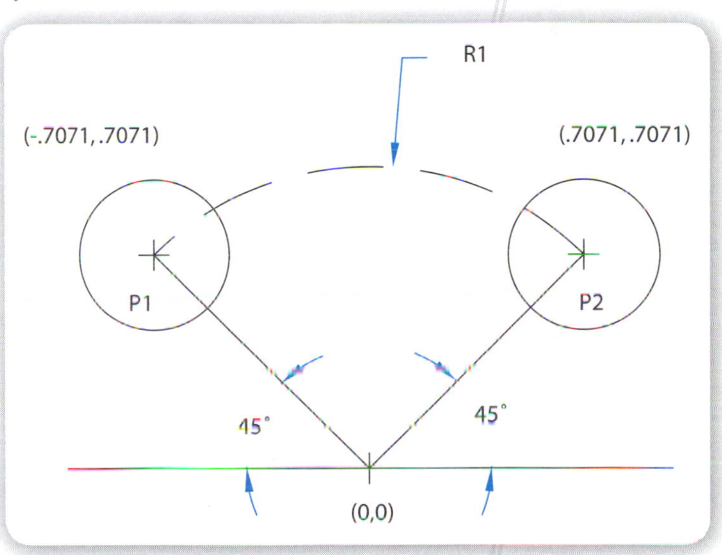

Using Circular Interpolation with a Radius Value

So far the examples we have seen used I and J to specify the radius. I and J tend to be a more universal format, but it is also more difficult to learn. In this section, we will look at a couple of examples that use the R format for arcs—this should seem simple now that you have mastered I and J.

In *Figure 7.18*, there are two arcs to be machined with circular interpolation. Arc one is CW and will use G02, and arc two is CCW and will use G03. The center of each arc is the part zero, and you can assume that the tool is positioned at the start point. The examples below are the single blocks of code used to produce the tool movement.

Left Arc

```
G02 X0.0 Y-1.0 R1.0
```

Right Arc

```
G03 X1.0 Y0.0 R1.0
```

Programming Arcs in Incremental Coordinates

There are situations in which you may need to use incremental positioning with circular interpolation. In this case, the end point of the arc will simply be the incremental distance from the start of the arc to the end point of the arc. No changes will need to be made for the center position; it will be specified exactly the same as

Figure 7.18 An example of arc programming with R.

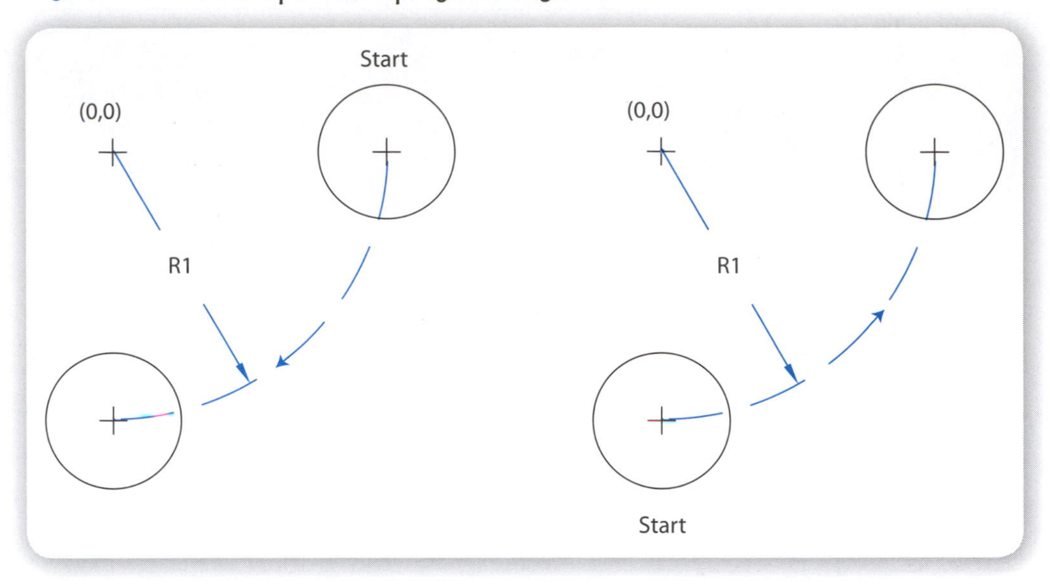

Figure 7.19 Arc example for absolute and incremental coordinates.

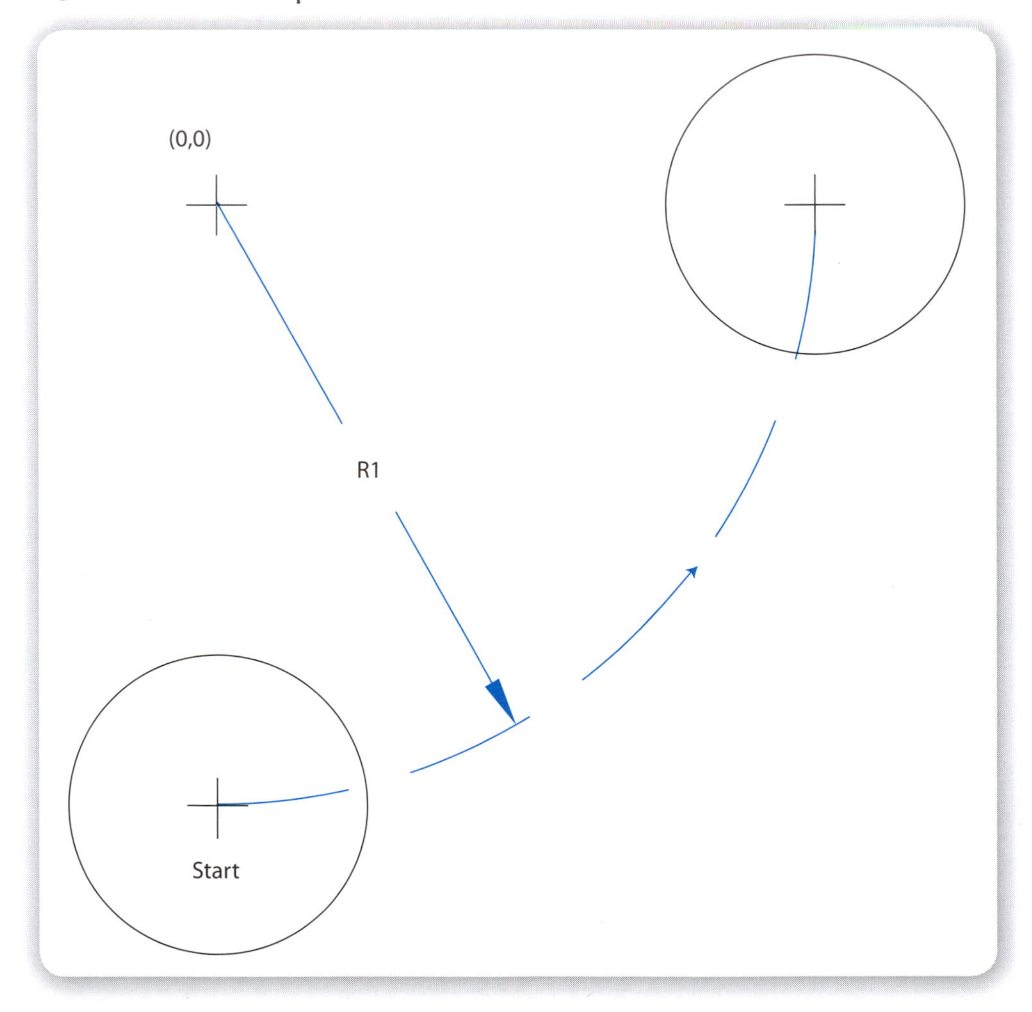

with absolute positioning. The examples below are the blocks used to perform the circular interpolation for the arc in ***Figure 7.19***. The first example uses absolute positioning and the second uses incremental positioning.

```
G90 G03 X1.0 Y0.0 I0.0 J1.0 (ARC IN ABSOLUTE)
G91 G03 X1.0 Y1.0 I0.0 J1.0 (ARC IN INCREMENTAL)
```

The following examples show a complete toolpath (*Figure 7.20*) in which the arcs have been programmed with three different methods: I & J Incremental, Radius, and I & J Absolute. The results are identical.

Figure 7.20 **Arcs can be programmed in several different manners including incremental or absolute I- and J-words or with direct radius values with the R-word. See the associated examples for an illustration of each technique.**

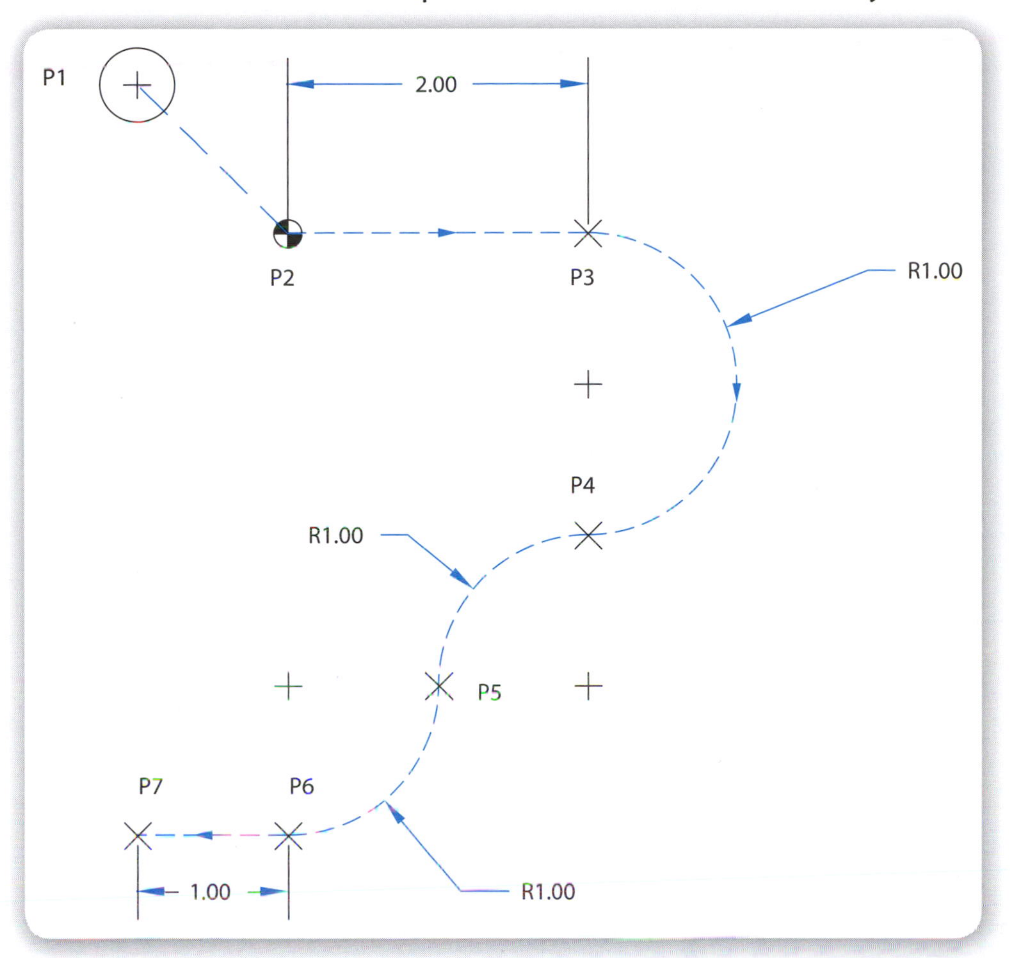

Program Listing 7-1a | **Toolpath from Figure 7.20 Produced with Incremental I & J Values**

```
%
O0700 (I J ARCS INCREMENTAL CENTERS)
N10 G20 G40 G49 G54 G80 G90 G98
N20 M06 T05
N30 G43 H05
N40 M03 S1200
N50 G00 X-1.0 Y1.0 (P1)
N60 G00 Z.2
N70 G01 X0.0 Y0.0 F10.0 (P2)
N80 G01 X2.0 Y0.0 (P3)
```

(continued)

Program Listing 7-1a | *continued*

```
N90 (I AND J IN INCREMENTAL COORDINATES)
N100 G02 X2.0 Y-2.0 I0.0 J-1.0 (P4)
N110 G03 X1.0 Y-3.0 I0.0 J-1.0 (P5)
N120 G02 X0.0 Y-4.0 I-1.0 J0.0 (P6)
N130 G01 X-1.0 Y-4.0 (P7)
N140 G91 G28 Z2.0
N150 M05
N160 M30
%
```

Program Listing 7-1b | **Toolpath from Figure 7.20 Produced with Radius (R-word) Values**

```
%
O0701 (R-WORD ARCS)
N10 G20 G40 G49 G54 G80 G90 G98
N20 M06 T05
N30 G43 H05
N40 M03 S1200
N50 G00 X-1.0 Y1.0 (P1)
N60 G00 Z.2
N70 G01 X0.0 Y0.0 F10.0 (P2)
N80 G01 X2.0 Y0.0 (P3)
N90 (ARCS IN EXPRESSED IN R VALUES)
N100 G02 X2.0 Y-2.0 R1.0 (P4)
N110 G03 X1.0 Y-3.0 R1.0 (P5)
N120 G02 X0.0 Y-4.0 R1.0 (P6)
N130 G01 X-1.0 Y-4.0 (P7)
N140 G91 G28 Z2.0
N150 M05
N160 M30
%
```

Program Listing 7-1c | **Toolpath from Figure 7.20 Produced with Absolute I & J Values**

```
%
O0702 (I J ARCS ABSOLUTE CENTERS)
N10 G20 G40 G49 G54 G80 G90 G98
```

Program Listing 7-1c | *continued*

```
N20 M06 T05
N30 G43 H05
N40 M03 S1200
N50 G00 X-1.0 Y1.0 (P1)
N60 G00 Z.2
N70 G01 X0.0 Y0.0 F10.0 (P2)
N80 G01 X2.0 Y0.0 (P3)
N90 (I AND J IN ABSOLUTE COORDINATES)
N100 G02 X2.0 Y-2.0 I2.0 J-1.0 (P4)
N110 G03 X1.0 Y-3.0 I2.0 J-3.0 (P5)
N120 G02 X0.0 Y-4.0 I0.0 J-3.0 (P6)
N130 G01 X-1.0 Y-4.0 (P7)
N140 G91 G28 Z2.0
N150 M05
N160 M30
%
```

Limitations of Circular Interpolation

The ability to interpolate arcs and circles has been a great advance for machining. This was previously a difficult and expensive process to produce arcs (other than round holes) by conventional machining processes. However, we must proceed with caution. Circular interpolation is produced by a motion control system that is moving at least two axes of the machine tool at the same time. To complicate matters, not only is each axis moving at a different rate, but the rate is not steady; each is either accelerating or decelerating. The ability of a motion control system to produce high-quality circular interpolation is a difficult problem.

Every motion control system will have error. The question we must ask is "Is the error within our tolerance?" To answer this question is difficult without specialized equipment. One method is to cut an arc on a piece of material and then measure it. A coordinate measuring machine can give a good indication of the deviation from a true circular arc-form. We can also check the actual motion of the machine tool while it is moving. There are commercial systems available that use a linear transducer to measure the actual machine position while in motion (***Figure 7.21***). The process is to simply attach one end of the transducer to the spindle and the other to the table. Then a programmed arc movement is executed while the measurements are processed by a computer to show how accurately the machine is moving in the circular path. This report can give the manufacturing engineer accurate information about the process capability of the machine.

The sources of error with circular interpolation (or any other interpolation for that matter) are numerous. First the motion control system can be inaccurate as we previously discussed. We can also have errors that come from the machining operation itself. For example, if we are cutting uneven amounts of material from

Figure 7.21 *Commercial systems using either linear transducers (as shown on left) or laser interferometers can accurately measure the dynamic accuracy of a CNC machine tool. A computer reads the measurements and reports the error to the user (right).*

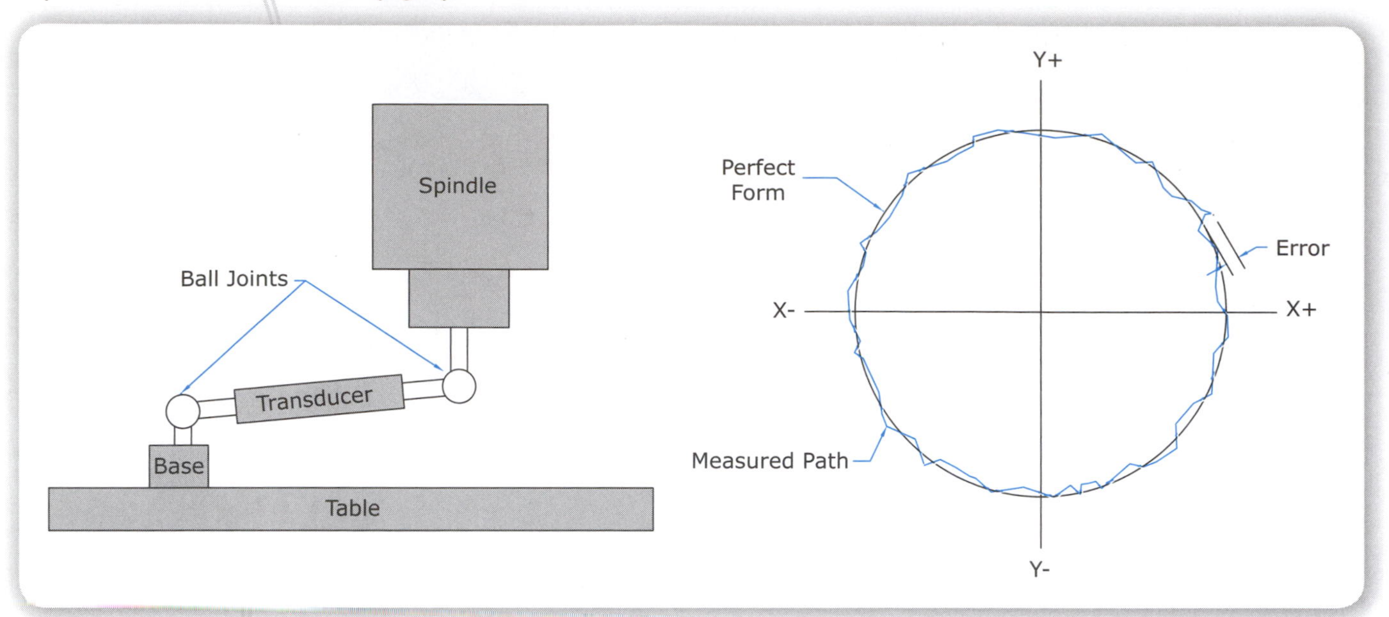

the edge of the workpiece, then we will get uneven tool deflection resulting in inaccurate profiles. This can also occur with uneven tool engagement as when entering or leaving a corner. Changes in programmed toolpath geometry can also reduce accuracy—particularly at high speeds. If the machine tool is rapidly approaching a change of geometry, then it might need to decelerate in order to process the code and to physically be able to make the change of direction without over-shooting. We all know that you have to slow down when driving around a curve. This can mean more cutting time at a particular point, resulting in more material removal (similar to taking a "spring pass").

Generally speaking, boring operations (boring head or bar) will produce an internal arc or circle that is more true to form than interpolation. High-precision holes for bearings and close-fitting components are usually produced by boring. The simple physical apparatus that is required for boring leaves little few places for errors in circular form to creep into the system when compared to a complex motion control system. Boring also tends to produce a superior surface finish. However, interpolation is extremely flexible for creating inside, outside, full, or partial arcs as part of a contoured toolpath. With a highly accurate machine and careful technique, circular interpolation is adequate for most machined features.

S T E M

If you ever had an Etch-a-Sketch toy as a child, then you were introduced to the problems of two-axis motion control at an early age (*see Figure 7.22*). Drawing a straight line was easy, but making an inclined line was another story. You could either create little X and Y movement, which results in a stair-step effect, or turn both knobs at simultaneously to make a smooth (but seldom straight) line. You might even have tried to create a circle on the gray screen by turning the knobs at the same time—a difficult task. CNC motion control systems deal with the same problems.

Figure 7.22 A child's toy presents the basic concepts of two-axis motion control.

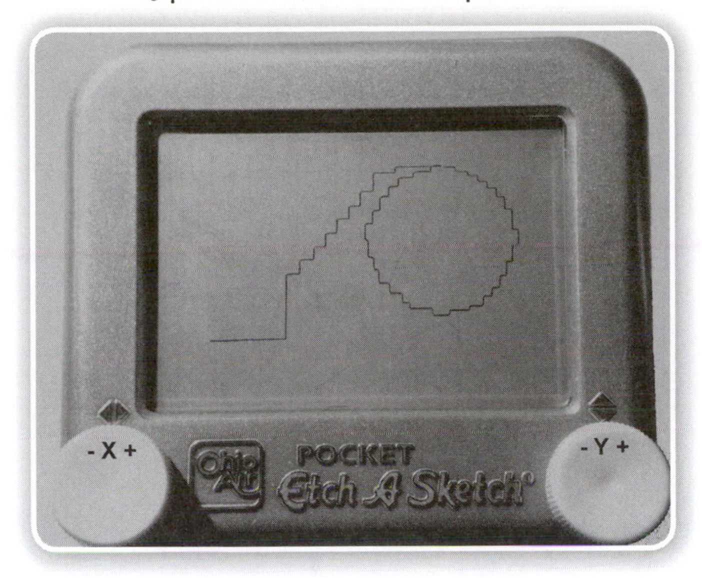

Lead-in and Lead-out

Any time you perform a finishing cut on a surface you should approach the surface at a shallow angle or tangent. The same holds true for leaving the finished surface at the conclusion of the finish pass. This technique is sometimes referred to as lead-in and lead-out. An uneven material load and a slight servo-lag will often leave a small divot in the surface when entering or exiting at a right angle. A **lead-in** and **lead-out** move will give a more evenly distributed tool pressure and better finish at the point of entry and exit, as illustrated in *Figure 7.23*.

lead-in/lead-out: The act of entering or exiting a finished toolpath at a shallow angle or arc to ensure a consistent finish and to prevent gouging that can be caused by tool deflection.

Linear interpolation is often used for lead-in and lead-out simply because it is a little easier to program than an arc. The technique is to approach the surface at an angle of less than 45° and then continue along the regular tool path. *Figure 7.24* shows the path for the finishing cut of an inside pocket. The tool is positioned to the proper depth at point 1 and then moved to point 2 to initiate the cut. The

Figure 7.23 Several possible lead in methods.

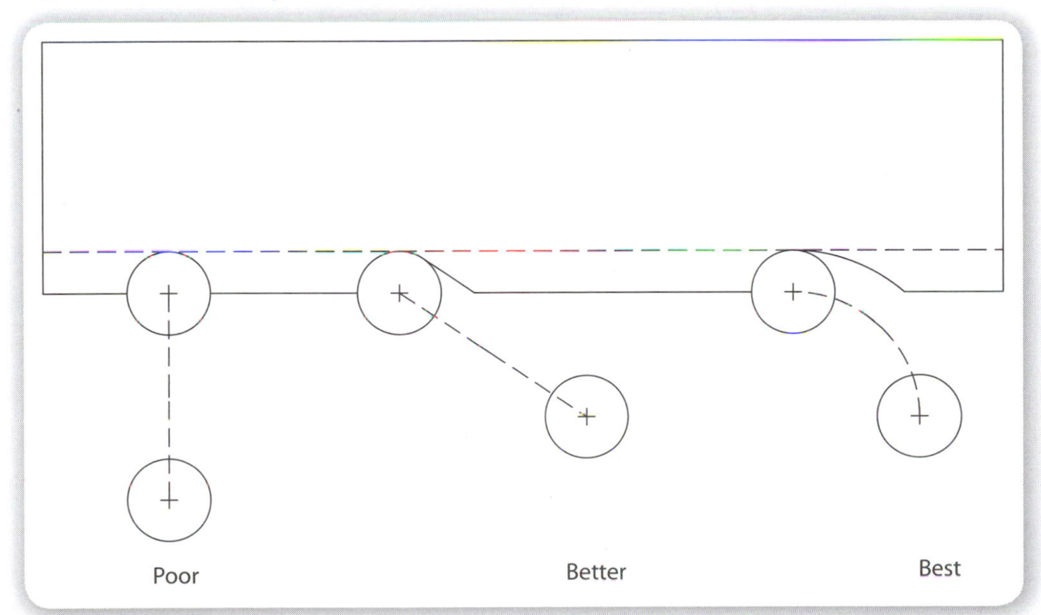

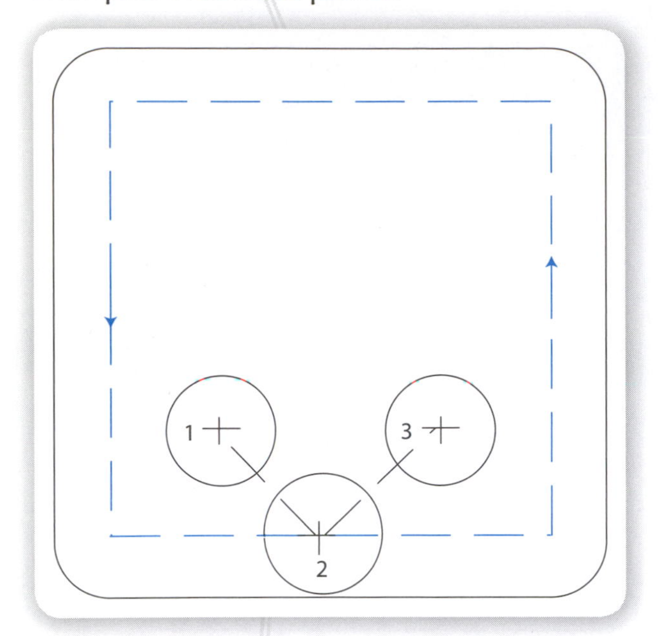

Figure 7.24 An angled lead in and out for the finish pass of an inside pocket.

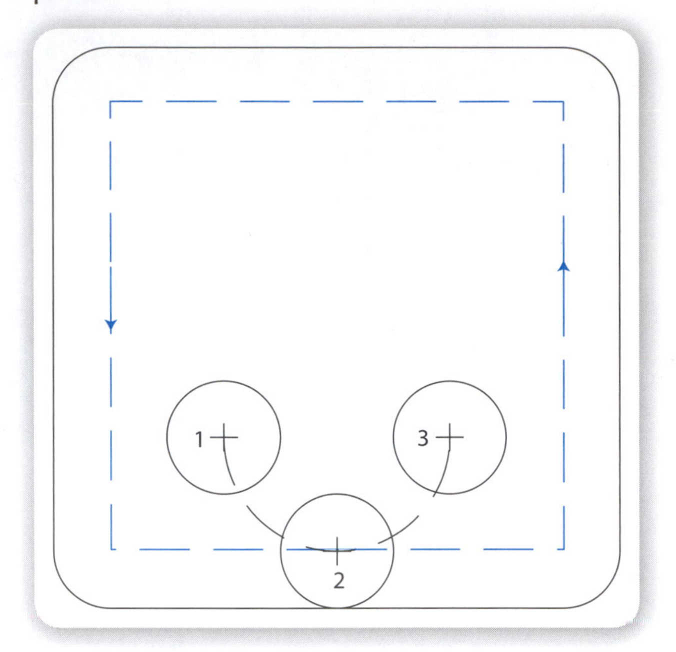

Figure 7.25 Arc moves used to create a tangential lead in and out for the finish pass of an inside pocket.

cutter then continues around counterclockwise back to point 2 again. At this point the exit move is made to point 3. The following blocks describe the tool movement.

```
N50 G00 X.75 Y.75 (START POSITION P1)
N60 G00 Z.2
N70 G01 Z-.25 F5.0
N80 G01 X1.0 Y.5 (LEAD IN TO P2)
N90 G01 X1.5 Y.5
N100 G01 X1.5 Y1.5
N110 G01 X.5 Y1.5
N120 G01 X.5 Y.5
N130 G01 X1.0 Y.5
N140 G01 X1.25 Y.75 (LEAD OUT TO P3)
```

The linear approach works well and is easy to program, but arc moves will give a smoother transition. You can use any segment of an arc that you like, but it is easiest to program if you use full-quadrant arcs. Just be sure that the finish point of the arc is tangent to the surface for best results. The blocks below describe the tool movement to create the inside profile with arcs for a lead, as shown in ***Figure 7.25***.

```
N50 G00 X.75 Y.75 M08 (START POSITION TO P1)
N60 G00 Z.2
N70 G01 Z-.25 F5.0 (PLUNGE)
N80 G03 X1.0 Y.5 I.25 J0.0 (ARC LEAD IN TO P2)
N90 G01 X1.5 Y.5
N100 G01 X1.5 Y1.5
N110 G01 X.5 Y1.5
N120 G01 X.5 Y.5
N130 G01 X1.0 Y.5
N140 G03 X1.25 Y.75 I0.0 J.25 (ARC LEAD OUT TO P3)
```

7.7 G28—RETURN TO MACHINE HOME POSITION

There are some situations in which you may want or need to return to the machine zero on one or more axes to change tools, to inspect the workpiece, or to set a preprogrammed work offset. In these situations, we can use the G28 code to instruct the machine to automatically return to its home position.

When G28 is called, the machine will immediately move at rapid traverse to the home position on the specified axes. If no axis is specified, the machine will return to the home position on every axis; usually the Z-axis is homed first for safety. A typical example of a G28 call is as follows:

```
G91 G28 X0.0 Y0.0 Z0.0
```

Of course, it may not be necessary to zero every axis. You may just want to move the tool out of the way and position the table to change the workpiece. In this case you might only zero the Z and Y-axes:

```
G91 G28 Y0.0 Z0.0
```

You are probably wondering why the G91 code to indicate incremental positioning was conspicuously placed in the block. The reason for this is that G28 also allows you to specify an intermediate point for the machine to travel to before returning to the home position. The intermediate point is a location in the coordinate system that is specified by the X-, Y-, and Z-words. These too can be either absolute or incremental coordinates. In fact, the two previous examples used an intermediate point of zero on each axis, but the incremental code caused them to move nowhere and then return to the home position. The code below will cause the tool to be positioned to an intermediate point that is up and to the right before returning to the home position, as in ***Figure 7.26***.

```
G91 G28 X2.0 Z2.0
```

A more common use of an intermediate point is to move the tool up some distance in the Z-axis before returning to the home position to avoid a collision with any obstructions. You must remember that G28 behaves the same way as rapid traverse in

Figure 7.26 *G28 can include an intermediate point. The machine will pass through this point on its way to zero.*

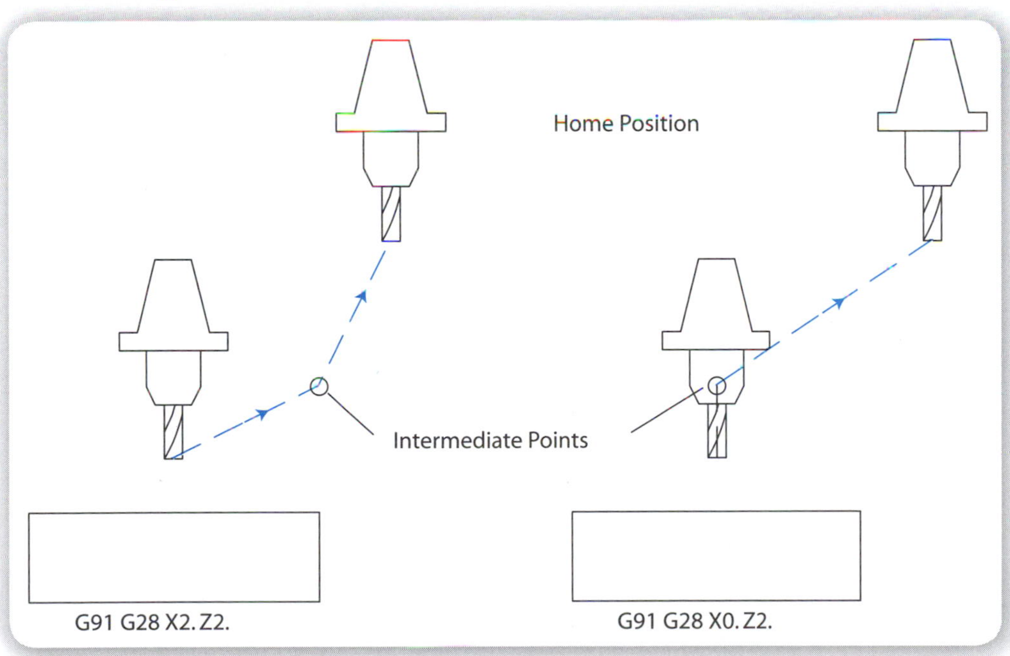

the respect that it does not always travel in a straight line. You should get in the habit of using incremental positioning and moving the tool straight up from the workpiece before returning to the X and Y home position. In the following code example, the tool will be moved up two inches before it returns to the home position:

```
G91 G28 X0.0 Y0.0 Z2.0
```

Care should be taken when absolute positioning is used with G28. It is very easy to make the mistake of setting the intermediate coordinates to zero and expecting upward movement as with incremental positioning. The result may be a crash. The two blocks below look very similar, but they will behave quite differently. Block A will return directly to the home position, but block B will take an excursion to the work zero before going to the home position. This could result in a serious crash if there are any obstructions. Blocks A and B will produce the toolpaths on the left and right sides of ***Figure** 7.27*, respectively.

Block A:

```
G91 G28 X0.0 Z0.0
```

Block B:

```
G90 G28 X0.0 Z0.0
```

Figure 7.27 **Care should be taken to initiating a zero return. An absolute positioning code can cause some unexpected results.**

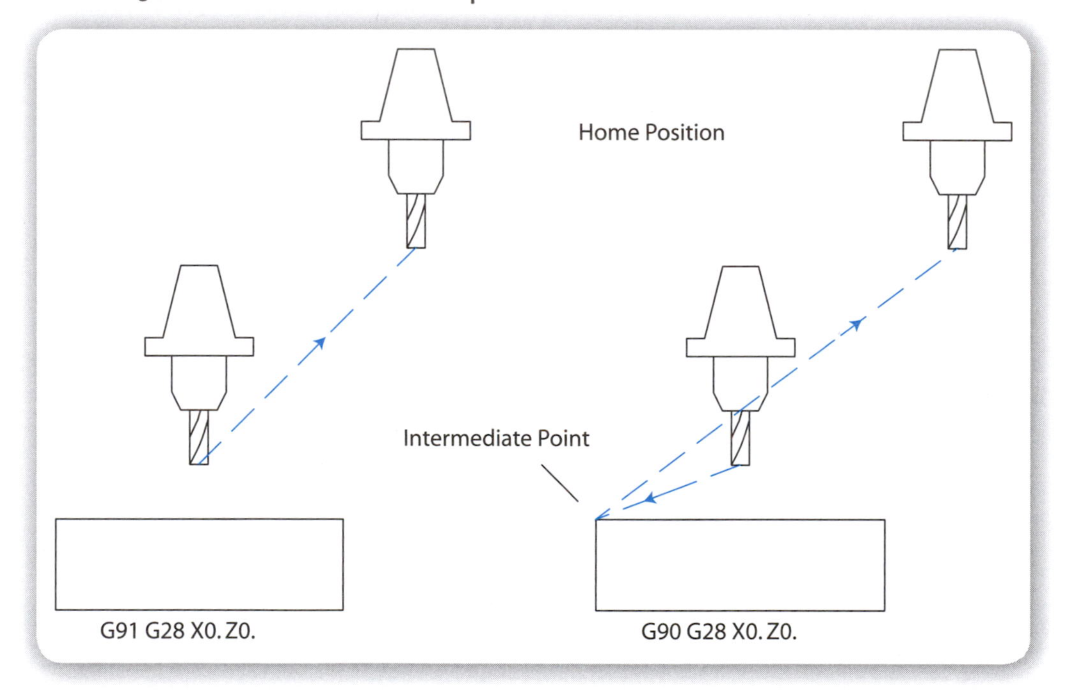

7.8 PROGRAM EXAMPLES

The following are programming examples for the toolpaths of the accompanying sketches. The part programs presented here are complete and represent actual working programs that may be run on most CNC machining centers. You may not recognize all of the components in these programs, but that is to be expected at this point. We will concentrate only on the codes that have already been discussed; the emphasis will be on the section that creates the toolpath.

Please recognize that there may be more elegant and stylistic solutions to these programming problems. For example, many modal codes have been included to make the code easier for the beginning programmer to follow. We will work on creating more elegant programs after mastering the basics.

Unless otherwise specified, the cutting tool will be Ø0.500, the work zero will be in the lower left corner, coordinates are in absolute units, and the Z-depth will be −0.25.

Programming Example 7-1 (*see Figure 7.28*)

Position	X	Y
Start	.50	−1.75
P1	0.	−1.25
End	−.50	−1.75

Note: Work zero is located at the center.

Figure 7.28 *Programming example one.*

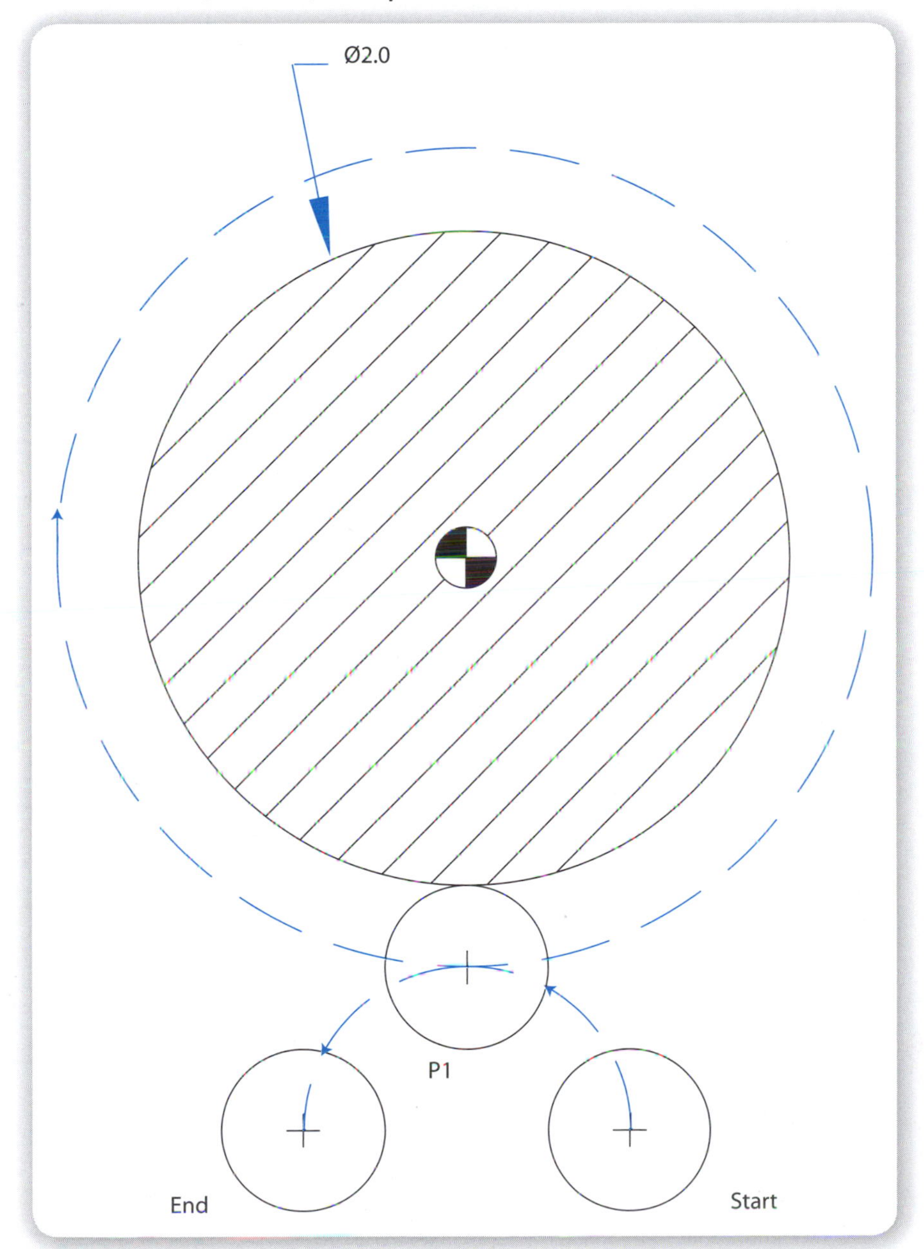

Listing 7-1 Program Code	Explanation
%	
O0701(CHAPTER 7, EXAMPLE 1)	
N10 G20 G40 G49 G54 G80 G90 G98	Safe line
N20 M06 T05 (.500 EM)	Tool change sequence
N30 G43 H05	Tool length offset
N40 M03 S1200	Turn spindle on
N50 G00 X.5 Y-1.75	Position to the start point
N60 G00 Z.2	
N70 G01 Z-.25 F5.	Plunge to the Z-depth
N80 G03 X0. Y-1.25 I-.5 J0.	Lead in
N90 G02 X0. Y-1.25 I0. J1.25	Cut full arc
N100 G03 X-.5 Y-1.75 I0. J-.5	Lead out
N110 G01 Z.2	
N120 M05	
N130 G91 G28 Z1.0 Y0.	Return to home position
N140 M30	End program and reset
%	

Programming Example 7-2 (*see Figure 7.29*)

Figure 7.29 Programming example two.

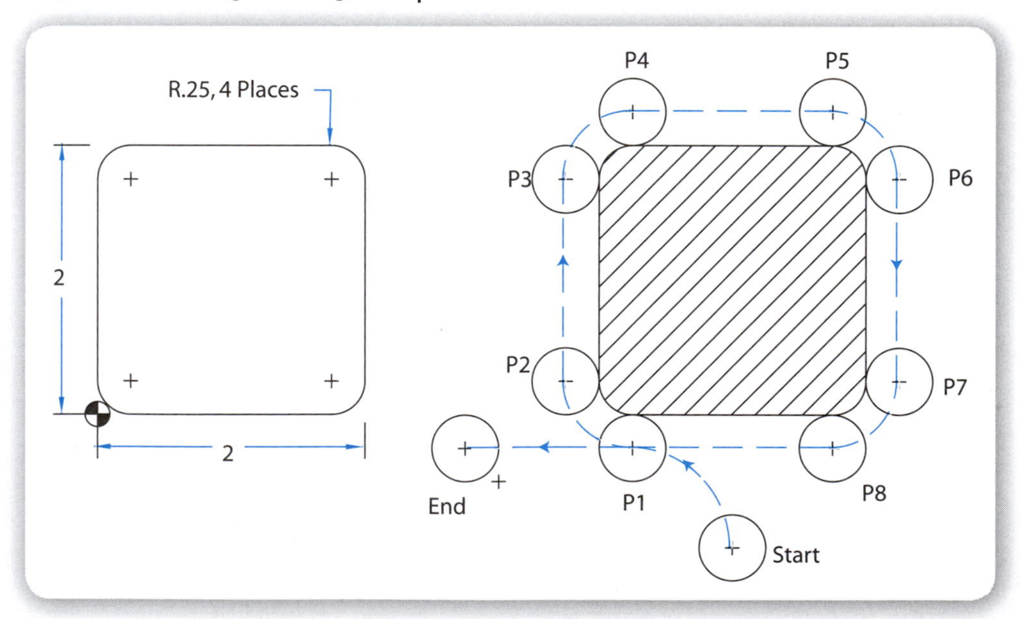

Position	X	Y
Start	1.0	−1.0
P1	.25	−.25
P2	−.25	.25
P3	−.25	1.75
P4	.25	2.25
P5	1.75	2.25
P6	2.25	1.75
P7	2.25	.25
P8	1.75	−.25
End	−1.0	−.25

Listing 7-2 Program Code	Explanation
%	
00702(CHAPTER 7, EXAMPLE 2)	
N10 G20 G40 G49 G54 G80 G90 G98	Safe line
N20 M06 T05 (.500 EM)	Tool change sequence
N30 G43 H05	Tool length offset
N40 M03 S1200	Turn spindle on
N50 G00 X1.0 Y-1.0	Position to the start point
N60 G00 Z.2	
N70 G01 Z-.25 F5.0	Plunge to the Z-depth
N80 G03 X.25 Y-.25 I-.75 J0.	Lead in to P1
N90 G02 X-.25 Y.25 I0. J.5	P2
N100 G01 X-.25 Y1.75	P3
N110 G02 X.25 Y2.25 I.5 J0.	P4
N120 G01 X1.75 Y2.25	P5
N130 G02 X2.25 Y1.75 I0. J-.5	P6
N140 G01 X2.25 Y.25	P7
N150 G02 X1.75 Y-.25 I-.5 J0.	P8
N160 G01 X-1.0 Y-.25	P9
N170 G01 Z.2	
N180 M05	
N190 G91 G28 Z1.0 Y0.	Return to home position
N200 M30	End program and reset
%	

Programming Example 7-3 (*see Figure 7.30*)

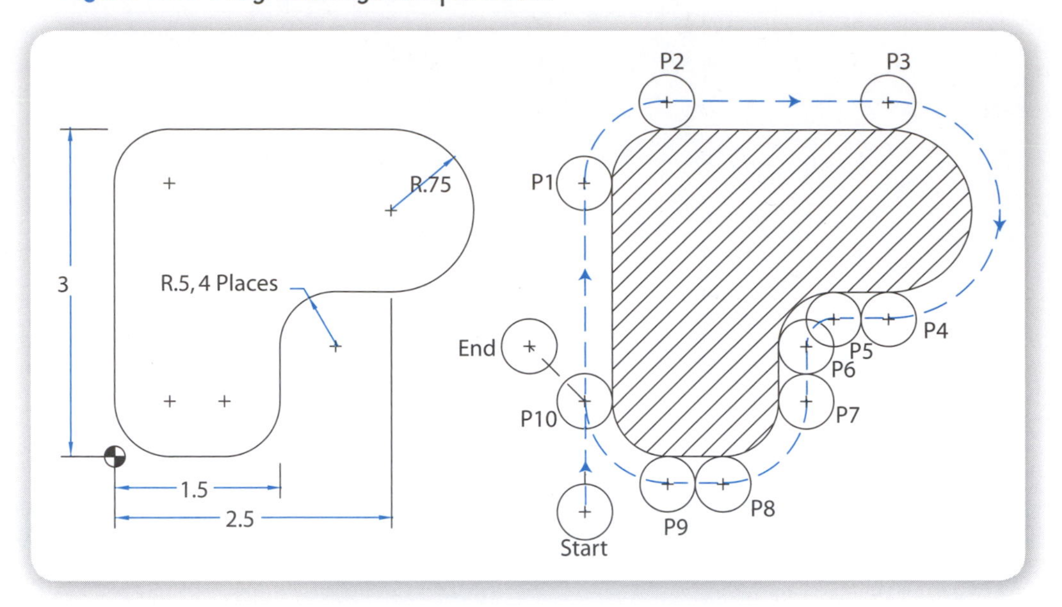

Figure 7.30 Programming example three.

Position	X	Y
Start	−.25	−.5
P1	−.25	2.5
P2	.5	3.25
P3	2.5	3.25
P4	2.5	1.25
P5	2.0	1.25
P6	1.75	1.0
P7	1.75	.5
P8	1.0	−.25
P9	.5	−.25
P10	−.25	.5
End	−.75	1.0

Listing 7-3 Program Code	Explanation
%	
O0703 (CHAPTER 7, EXAMPLE 3)	
N10 G20 G40 G49 G54 G80 G90 G98	Safe line
N20 M06 T05 (.500 EM)	Tool change sequence
N30 G43 H05	Tool length offset
N40 M03 S1200	Turn spindle on
N50 G00 X-.25 Y-.5	Position to the start point
N60 G00 Z.2	

Listing 7-3 *continued*

```
N70 G01 Z-.25 F5.0                    Plunge to the Z-depth
N80 G01 X-.25 Y2.5                    P1
N90 G02 X.5 Y3.25 I.75 J0.            P2
N100 G01 X2.5 Y3.25                   P3
N110 G02 X2.5 Y1.25 I0. J-1.          P4
N120 G01 X2.0 Y1.25                   P5
N130 G03 X1.75 Y1.0 I0. J-.25         P6
N140 G01 X1.75 Y.5                    P7
N150 G02 X1.0 Y-.25 I-.75 J0.         P8
N160 G01 X.5 Y-.25                    P9
N170 G02 X-.25 Y.5 I0. J.75           P10
N180 G01 X-.75 Y1.0                   Lead out to end point
N190 G01 Z.2
N200 M05
N210 G91 G28 Z1.0 Y0.                 Return to home position
N220 M30                              End program and reset
%
```

Programming Example 7-4

This is an example of one technique for roughing the material out of a rectangular pocket, as illustrated in ***Figure 7.31***. The tool is plunged at the start point and then moved in a counterclockwise spiral that overlaps the previous path by half the diameter of the end mill. The very last move will reposition the end mill at the start point in case another pass is needed.

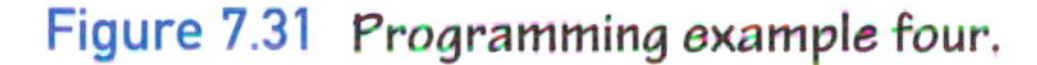

Figure 7.31 **Programming example four.**

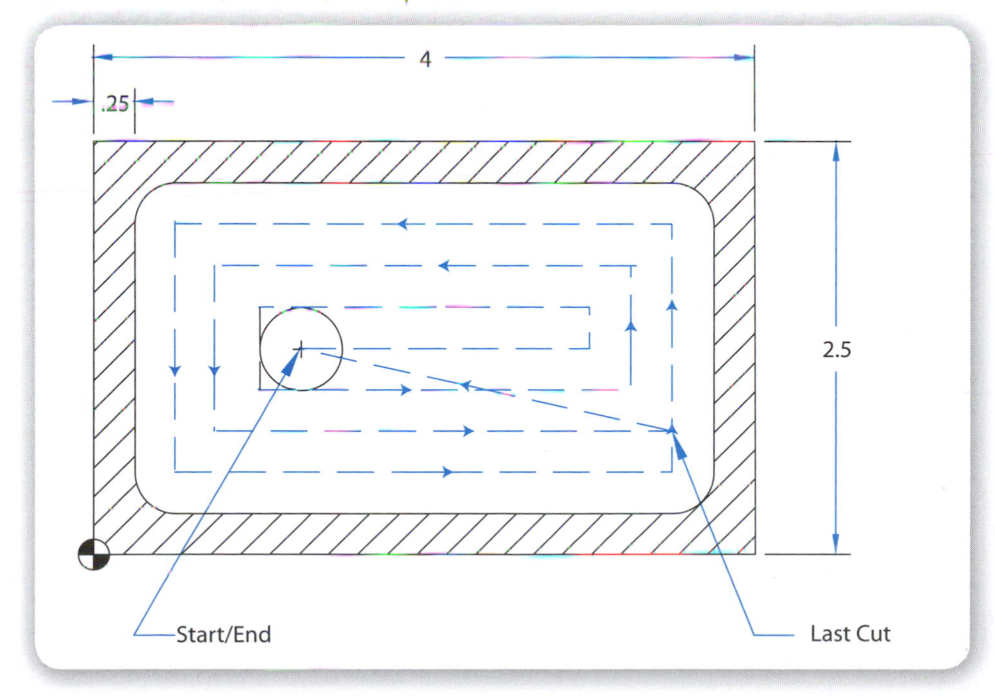

Listing 7-4	Program Code	Explanation
	%	
	O0704(CHAPTER 7, EXAMPLE 4)	
	N10 G20 G40 G49 G54 G80 G90 G98	Safe line
	N20 M06 T05 (.500 EM)	Tool change sequence
	N30 G43 H05	Tool length offset
	N40 M03 S1200	Turn spindle on
	N50 G00 X1.25 Y1.25	Position to the start point
	N60 G00 Z.2	
	N70 G01 Z-.25 F2.5	Plunge to the Z-depth
	N80 G01 X3.0 Y1.25 F5.0	Start of linear roughing passes
	N90 G01 X3.0 Y1.5	
	N100 G01 X1.0 Y1.5	
	N110 G01 X1.0 Y1.0	
	N120 G01 X3.25 Y1.0	
	N130 G01 X3.25 Y1.75	
	N140 G01 X.75 Y1.75	
	N150 G01 X.75 Y.75	
	N160 G01 X3.5 Y.75	
	N170 G01 X3.5 Y2.0	
	N180 G01 X.5 Y2.0	
	N190 G01 X.5 Y.5	
	N200 G01 X3.5 Y.5	
	N210 G01 X3.5 Y.75	End of linear roughing passes
	N220 G01 X1.25 Y1.25	Return to start point
	N230 G01 Z.2	
	N240 M05	
	N250 G91 G28 Z1.0 Y0.	Return to home position
	N260 M30	End program and reset
	%	

Programming Example 7-5

This example, as illustrated in *Figure 7.32*, shows one possible method to use to rough machine a circular pocket. The tool will plunge at the center of the arc and then step out half the diameter of the end mill to cut a counterclockwise arc.

Figure 7.32 **Pocket roughing can be accomplished with multiple circular arcs.**

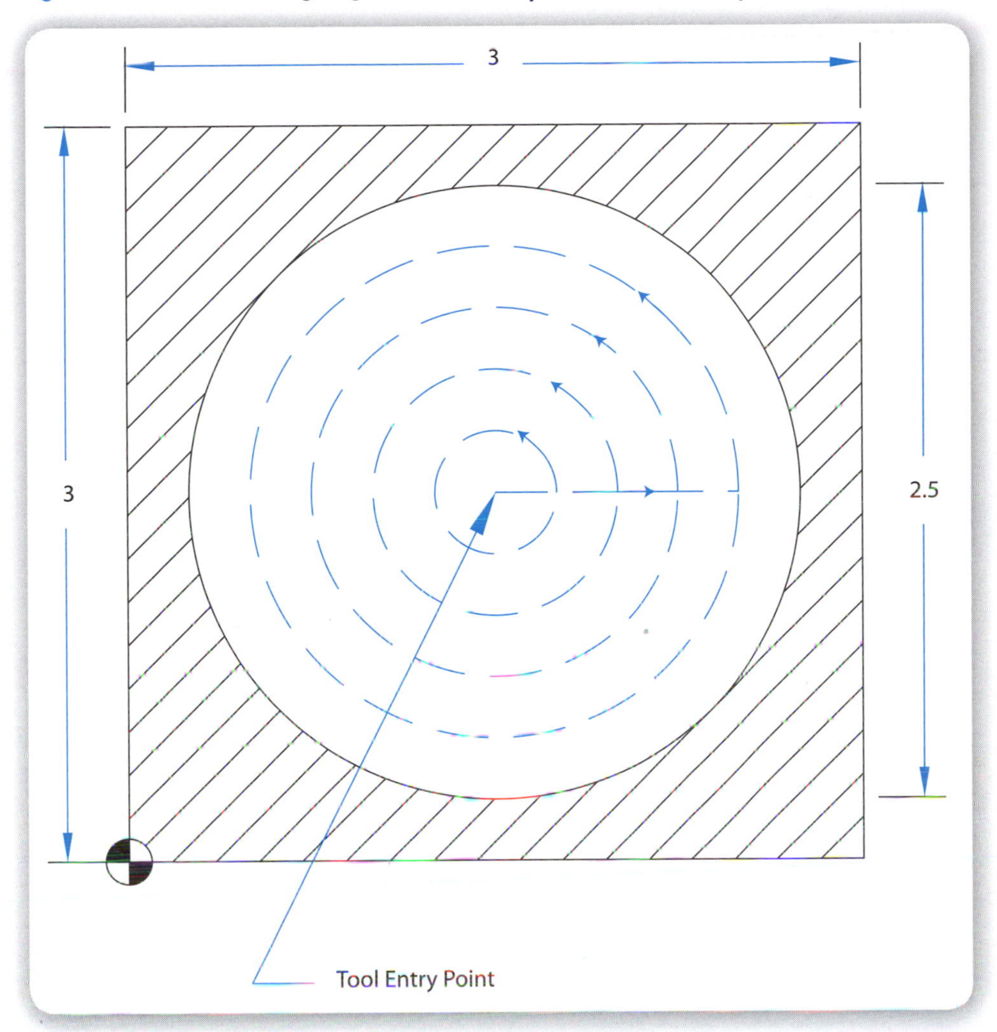

Listing 7-5	Program Code	Explanation
	%	
	O0707(CHAPTER 7, EXAMPLE 5)	
	N10 G20 G40 G49 G54 G80 G90 G98	Safe line
	N20 M06 T05 (.500 EM)	Tool change sequence
	N30 G43 H05	Tool length offset
	N40 M03 S1200	Turn spindle on
	N50 G00 X1.5 Y1.5	Position to the start point

(continued)

Listing 7-5 *continued*

Program Code	Explanation
`N60 G00 Z.2`	
`N80 G01 Z-.25 F2.5`	Plunge to the Z-depth
`N90 G01 X1.75 Y1.5 F5.0`	Position for the first pass
`N100 G03 X1.75 Y1.5 I-.25 J0.`	Cut the arc
`N110 G01 X2.0 Y1.5`	Position for the second pass
`N120 G03 X2.0 Y1.5 I-.5 J0.`	Cut the arc
`N130 G01 X2.25 Y1.5`	Position for the third pass
`N140 G03 X2.25 Y1.5 I-.75 J0.`	Cut the arc
`N150 G01 X2.5 Y1.5`	Position for the final pass
`N160 G03 X2.5 Y1.5 I-1.0 J0.`	Cut the arc
`N170 G01 X1.5 Y1.5`	Return to the center
`N180 G01 Z.2`	
`N190 M05`	
`N200 G91 G28 Z1.0 Y0.`	Return to home position
`N210 M30`	End program and reset
`%`	

Programming Example 7-6

In this example, we will machine the workpiece in ***Figure 7.33***. The first section of the program will face the top of the raw material with a Ø1.0" end mill. Then, the second section will mill the square and round profiles with a Ø0.500" end mill. ***Figure 7.34*** shows a detailed illustration of the toolpath in several steps.

Listing 7-6 Program Code	Explanation
`%`	
`O0707 (CHAPTER 7, EXAMPLE 6)`	
`N10 G20 G40 G49 G54 G80 G90 G98`	Safe line
`N20 (FACE TOP)`	
`N30 M06 T08 (1.0 EM)`	Tool change sequence
`N40 G43 H08`	Tool length offset

Listing 7-6 *continued*

```
N50 M03 S800                              Turn spindle on
N60 G00 X-.75 Y2.0                        Position for facing cuts
N70 G00 Z.2
N80 G01 Z0.0 F5.0                         Begin facing passes
N90 G01 X2.75 Y2.0
N100 G01 X2.75 Y1.5
N110 G01 X-.75 Y1.5
N120 G01 X-.75 Y1.0
N130 G01 X2.75 Y1.0
N140 G01 X2.75 Y.5
N150 G01 X-.75 Y.5
N160 G01 X-.75 Y0.0
N170 G01 X2.75 Y0.0                       Begin facing passes
N180 G01 Z.2
N190 (MILL SQUARE AND CIRCLE)
N200 M06 T05 (.50 EM)                     Tool change sequence
N210 G43 H05                              Tool length offset
N220 M03 S1200                            Turn spindle on
N230 G00 X0.0 Y-.5                        Position for profile cuts
N240 G00 Z.2
N250 G01 Z-.5 F5.0                        Plunge
N260 G01 X0.0 Y2.0 (SQUARE)               Mill square profile
N270 G01 X2.0 Y2.0                        . . .
N280 G01 X2.0 Y0.0                        . . .
N290 G01 X-.5 Y0.0                        . . .
N300 G01 Z-.25                            Change Z-depth for circle
N310 G01 X0.0 Y1.0 (CIRCLE)
N320 G02 X0.0 Y1.0 I1.0 J0.0              First circular pass
N330 G01 X.125 Y1.0
N340 G02 X.125 Y1.0 I.875 J0.0            Second circular pass
```

(continued)

Listing 7-6 *continued*

```
N350 G01 X-.5 Y1.5
N360 G91 G28 Y0.0 Z1.          Return to home position
N370 M30                       End program and reset
%
```

Figure 7.33 The workpiece to be produced in programming example six.

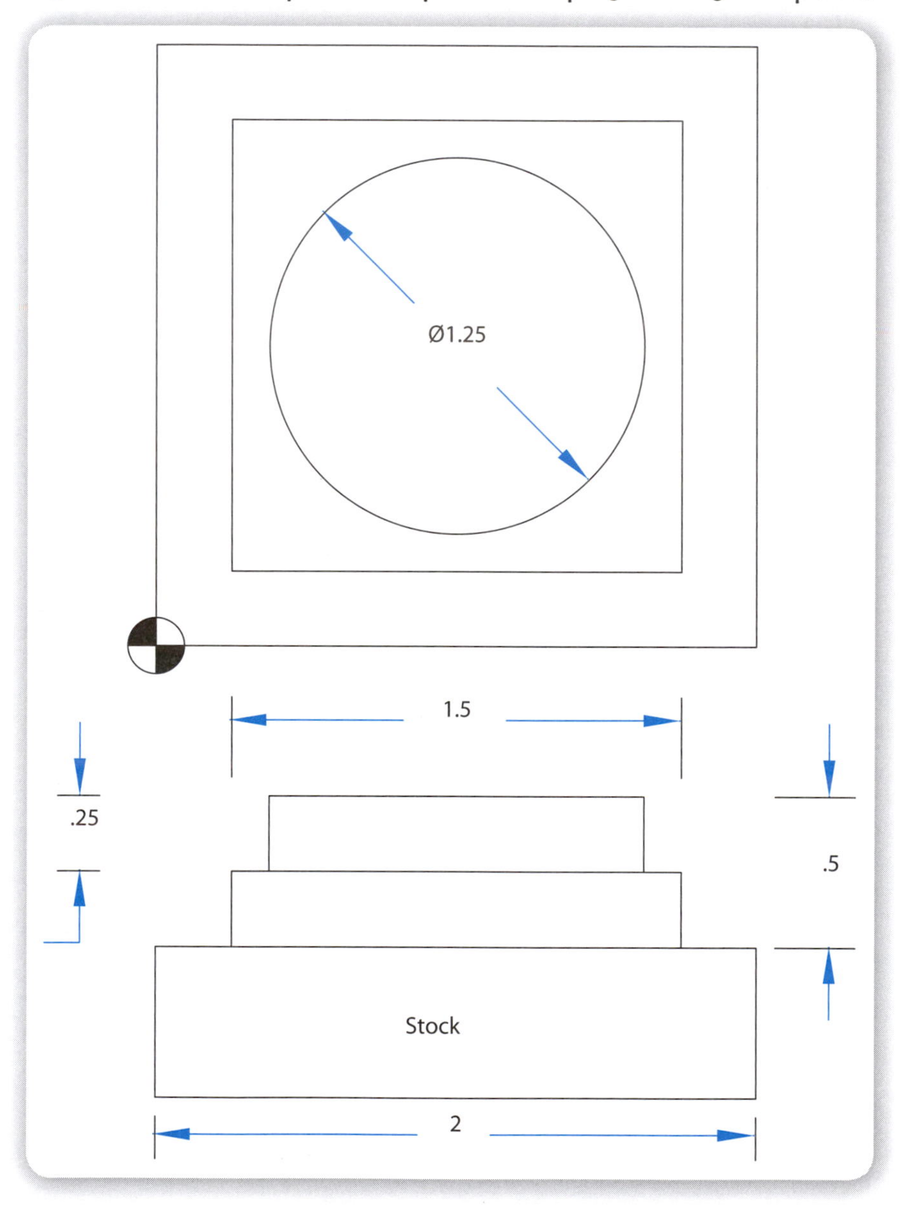

Figure 7.34 The toolpaths that will be required for this workpiece: Facing, square profile, and round profile.

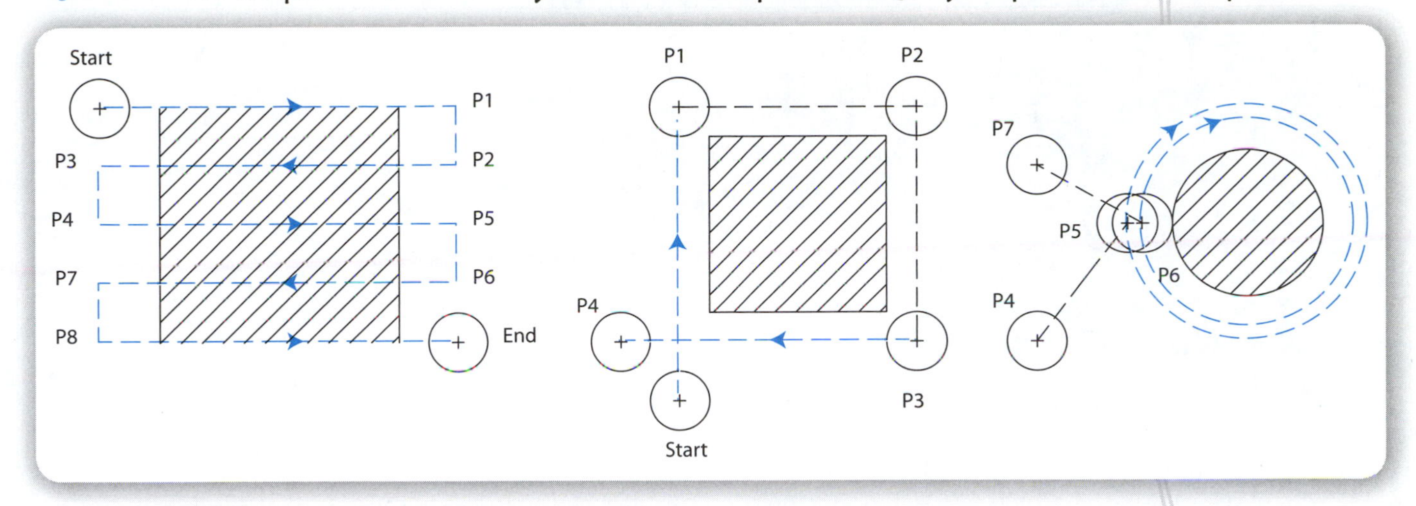

Your Turn

Create a name placard for one of your teachers. The name must have both linear and circular interpolations. Each student should choose a different font.

CHAPTER SUMMARY

- Some preparatory codes are used to position the tool and to make cuts in the material. G00 is used for positioning, G01 is used to cut straight lines, and G02 and G03 are used to cut circular arcs.
- Rapid traverse (G00) is never used to perform a machining operation. In addition, rapid traverse will not necessarily travel in a straight line.
- The X-, Y-, and Z-words are used in conjunction with the G-codes to indicate the axis coordinate—the coordinate is always the end point of the move.
- All machining operations must have a linear feed rate, which is specified with an F address. Plunge cuts should use only ½ the normal feed rate.
- I and J are the most common methods for indicating the center point of an arc, but R is also used. R can be defined as positive or negative to indicate the shape of the arc. R is also limited to arcs of less than a full circle on most controls.
- CNC machining will always be imperfect. Errors are induced by the motion control system, machining processes, and the deflection of materials. All factors must be controlled to produce in-tolerance products. The answer on the computer screen is not always right!
- G28 is used to return an axis to its home positions. G28 can also be programmed to pass through an intermediate point, but we must be careful to use incremental positioning or a crash may result.

BRING IT HOME

1. What is the difference between G00 and G01? Where would you use each of these codes?
2. Explain some of the dangers that might be encountered when using G00. How could these dangers be avoided?
3. In your own words, explain how the I- and J-words are used with circular interpolation.
4. What factors will affect the accuracy of features that are machined with a CNC machine tool?
5. What is the difference between a positive R-value and a negative R-value?
6. Why is incremental positioning usually used with an axis return?
7. Write a block of code to return the Z- and X-axes to their home positions by way of an intermediate point.

EXTRA MILE

1. Write four blocks of code to create the toolpaths for the 90° arcs shown on the left side of *Figure 7.35*. Use I and J and assume the tool is already positioned at the start point.
2. Write four blocks to create the toolpaths for the 90° arcs shown on the right side of Figure 7.35. Use I and J and assume the tool is already positioned at the start point.
3. Write only the code to produce the toolpath shown in *Figure 7.36*. Start by planning the toolpath and charting the tool locations for a Ø 0.500" end mill. You can assume that the tool is already located at the start point that you have selected (no positioning required). The toolpath should also climb cut at 5 IPM and have lead-in and lead-out.
4. Now that you are sure that the toolpath from the previous question is correct, create a complete NC program to cut the profile to a depth of 0.250". Be sure to include all program elements, including the safe line, tool change, and the initial and return positioning moves.

Figure 7.35 Create the toolpaths for these arcs.

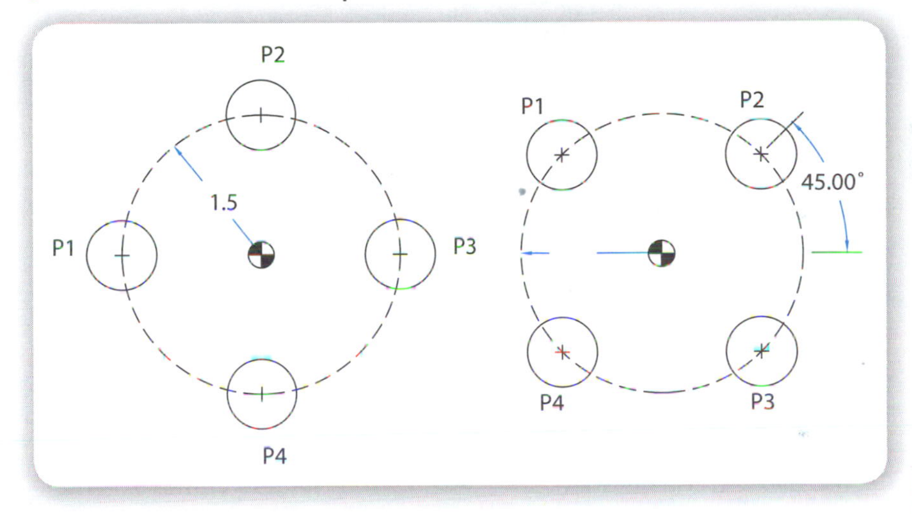

Figure 7.36 Create the toolpath for this profile.

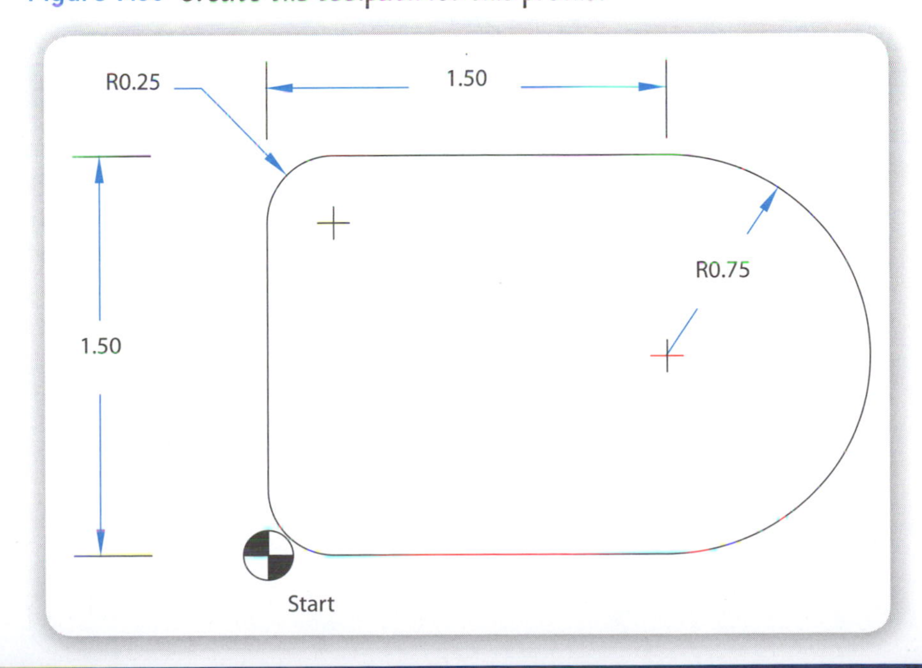

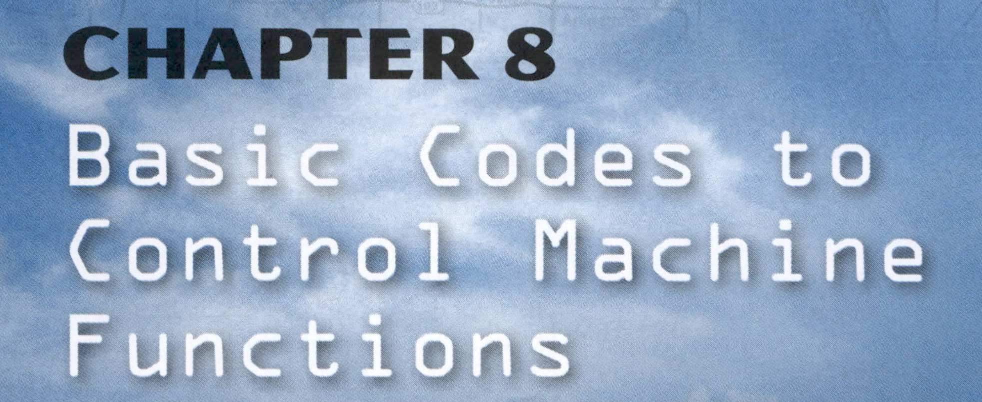

CHAPTER 8
Basic Codes to Control Machine Functions

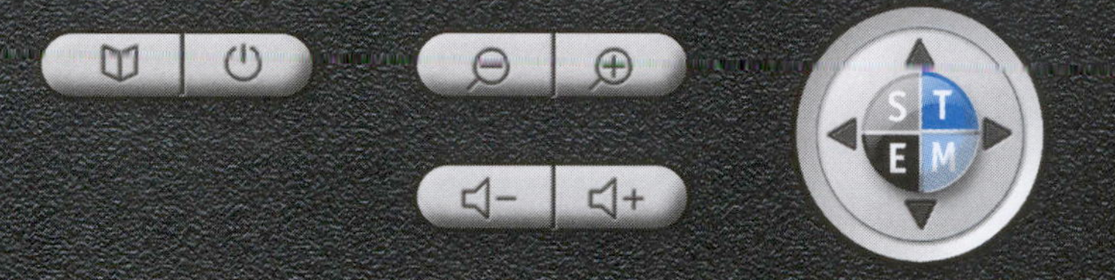

Menu

START LOCATION | DISTANCE | END LOCATION

Before You Begin

Think about these questions as you study the concepts in this chapter:

1. What are the more common M-codes used today?
2. How are automatic tool changers limited in usage?

Key Terms

Automatic Tool Changer (ATC)
Block Delete
Miscellaneous Code (M-Code)
Program Stop

8.1 UNDERSTANDING M-CODES

In addition to G-codes, there are codes used to control various machine functions ranging from tool change to coolant control. These codes are called miscellaneous codes, or simply M-codes. An M-code comprises the M address followed by a number. The address and the number are combined to form the M-word or simply M-code:

M Address + Number = M-Word

Miscellaneous code (M-code):

Codes that control machine functions other than tool movements. For example, M-codes are used to control coolant and spindle rotation.

Table 8.1 lists some of the more common M-codes used in NC programming. (See Appendix A for a comprehensive listing.) There are many other M-codes, but we will concentrate on these for the time being.

Table 8.1 *Common M-code Functions*

M-Code	Function
M00	Program Stop
M01	Optional Program Stop
M03	Spindle On—Forward
M04	Spindle On—Reverse
M05	Spindle Stop
M06	Tool Change
M07	Coolant On—Mist
M08	Coolant On—Flood (default)
M09	Coolant Off
M30	Program Stop and Reset

M-codes are called in a similar manner to G-codes. They can be placed alone in a block or they can be included at the end of a block to save time. Many M-codes are standalone instructions to carry out some function such as turning the spindle off. A few M-codes will also work in conjunction with other addresses to perform some function. For example, the code to turn on the spindle does not make much sense unless a spindle speed has been entered with an S address.

There are a few details that we have to be aware of to be able to properly use M-codes. First, a survey of several control manufacturers reveals that there is little consistency pertaining to when an M-code will be executed. When an M-code is included in the same block as a G-code, the M-code may be executed at the beginning or after the completion of the operation—depending on the manufacturer. For example, take the following block:

```
N100 G01 X2.0 Y2.0 F10.0 M08
```

This block instructs the machine to feed to an X and Y position and to turn the coolant on (M08). However, with some controls the coolant will not turn on until the block is completed, while on others it will turn on immediately. If the coolant does not turn on immediately, the tool might be destroyed.

The safe way around this inconsistency is to assume that the M-code will be executed after the other instructions are finished. We can then cover our bases by calling the M-code in a previous block to ensure that critical functions are initiated:

```
N90 M08 (COOLANT ON)
N100 G01 X2.0 Y2.0 F10.0
```

One other important aspect of M-codes is that most controls will allow only one M-code to be included in a single block. If we place multiple M-codes in a block, it will cause either the control to have an error or only one of the M-codes to initiate. Neither situation is acceptable.

8.2 M06—TOOL CHANGES

Tool changes are accomplished by calling the M06 code. However, because each machine tool behaves a little bit differently, it is extremely important to understand the particular control you are working with or damage may result.

There are several possibilities for the type of tool change that a particular machine will require. The possibilities are based on the physical construction of the machine and the internal logic of the control. Several common configurations are listed below.

- Manual tool change
- Automatic tool changer with absolute locations
- Automatic tool changer with incremental locations

Automatic Tool Changer (ATC)

A device used to automatically insert the required cutting tools into the spindle or cross slide position as the part program executes on a CNC machine tool.

Some low-end and older CNC machine tools do not have an automatic tool changer. This machine will simply stop and wait for the operator to manually change the tool. Program execution will not resume until the cycle start button is pushed. This is obviously not a very efficient design, and consequently it is not very common today.

Most modern CNC machine tools are equipped with an automatic tool changer (ATC) to eliminate manual operator intervention (***Figure 8.1***). The tools are stored in a turret, carousel, magazine, or similar device until needed for machining. The control will keep track of the tools in one of two ways. The most common method is to assign each storage location an identification number. A feedback device such as a rotary encoder or proximity sensor is usually attached to the storage device to ensure that the correct tool is loaded when called. This is by far the safest and most efficient configuration.

The control will use the information from the feedback device to keep track of which tool is in the spindle in one of several ways, based mostly on the sophistication of the system. Many machines use an absolute indexing system that allows the programmer to directly specify the position of the tool

Figure 8.1 Automatic tool changers (left to right) Side Mount, Belt, Turret (lathe).

changer. Other, less-modern machine tools will use an incremental indexing system for tool locations. With an incremental system, each tool position is specified from the current tool position. Incremental systems are more difficult to program and more prone to operator error—fortunately this system is mostly obsolete.

In either of the above designs it is usually important to move the ATC to a reference point at machine startup. The storage device is manually indexed to the first position, and then the machine assigns this position as position number one. From there, it can keep track of which tool is in the spindle.

Another pitfall that you must be aware of is how the machine will behave when the M06 code is encountered—again this is largely dependent on the machine tool manufacturer. Two common configurations are as follows:

- The storage device travels with the spindle (or tool post on a lathe).
- The storage device is stationary.

This distinction is critical because a serious crash can occur if the tool change is called with the machine in the wrong position.

The traveling storage device configuration is sometimes found on machining centers, but it is more common on turning centers. These designs lead to faster tool changes, but more care must be taken. A lathe turret is a common example of the traveling design. All the tools are moved along with the one tool that is being used for the current operation. You may also find machining centers that have been designed with the storage device attached to the headstock. In either case, the tool holder must be backed away from the workpiece before you call a tool change. Otherwise you risk crashing the tool into the workpiece or fixture, as shown in *Figure 8.2.* Most controls have a safety mechanism to allow the setup person to disable tool changes while in a dangerous position, but this can be subject to human error.

Other machines are designed with a stationary storage device. In this case, the machine will have to return to a predefined position in order to complete the change. This means that tool changes will take longer, but there is a smaller

Figure 8.2 Care must be taken to fully understand the movements of the headstock and Automatic Tool Changer (ATC) before use. Some configurations will allow a crash if positioned too close to the table or fixture when a tool change is called.

incidence of errors. The programmer must be aware of the path the machine will take to reach the tool change location. Machining centers with stationary storage devices will typically only need to move the Z-axis straight up in the positive direction to make a tool change. However, you must not make any assumption about the machine without consulting the operator's manual or observing the machine.

The general form of a tool change is to place the M-code in a block followed by the T-word to specify the tool you wish to install. The number associated with the tool can either be an absolute or incremental position in the storage device. For all practical purposes, most modern machines use absolute positions, so you can simply specify the absolute tool location:

```
M06 T01 (CHANGE TO TOOL NUMBER 1)
```

The codes can usually be in any order, but logically it seems to make more sense to call the tool change and then specify the tool number. This is more often a question of style, and many programmers have been trained to put the T-word before the M-word. Most controls will allow any order, but you should consult the operator's manual for the exact specifications.

Immediately after the tool change, you will need to specify the length offset for the tool. The method varies, but it is common to use the G43 code followed by an H-word to indicate the number of the offset. To avoid confusion, it is common practice to use the same number for the tool and its length offset:

```
G43 H01 (SELECT HEIGHT OFFSET NUMBER ONE)
```

Some older machines with less sophisticated controls require the programmer to specify a direction of the length offset. This is accomplished with the G43 code for the positive direction and G44 for the negative direction. You might imagine that this has led to some confusion; now the standard method is to use only G43 for all height offsets.

Let's look at a tool change in the context of a larger program:

```
%
O0001 (Program 8.1)
N10 G20 G40 G49 G54 G80 G90 G98
N20 M06 T01 (TOOL CHANGE)
N30 G43 H01 (SELECT LENGTH OFFSET)
N40 M03 S1200
N50 G00 X0.00 Y0.0 M08
N60 G00 Z.2
N70 G01 Z-.5 F5.0
N90 G01 X1.0 Y1.0
N100 G01 Z.2
N100 M05
N110 M30
%
```

Changing tools within a program is slightly different. It is a good habit to always move the tool to a safe position whether it is required or not. This technique will get you into the proper mindset and keep you from making mistakes when using machines with traveling storage devices. In the following example the tool is programmed to move at a feed rate to a position 0.200" above the top of the workpiece. Next, it is moved at rapid traverse to an out-of-the-way position in blocks N100 and N110.

```
N90 G01 Z.2
N100 G00 Z10.0 (MOVE TOOL UP)
N110 G00 X-5.0 Y5.0 (MOVE TOOL TO SAFE LOCATION)
N120 M05
N130 G49 (CANCEL TOOL LENGTH OFFSET)
N140 M06 T02 (SECOND TOOL CHANGE)
N150 G43 H02
N160 M03 S1500
```

You might also have noticed the G49 code immediately before the tool call. G49 cancels the tool length offset, and it is used as a safety precaution just in case you forget to call a new height offset. Some controls will automatically cancel the length offset when M06 is called, but do not make this assumption unless you are absolutely positive about the behavior of a particular control. Failure to call the new length offset for the new tool will leave the old offset in the memory. *This will cause a crash if the new tool is longer than the previous tool.*

Perhaps a better way to accomplish this move to a safe location is to automatically send the machine to its home position with the G28 code, as shown below. Some machines only need to return home in the Z-axis to change tools, while others will also require X and Y home.

```
N90 G01 Z.2
N100 M05
N110 G49
N120 G91 G28 Z.0 (RETURN TO Z-AXIS HOME POSITION)
N130 M06 T02 (SECOND TOOL CHANGE)
N140 G43 H02
N150 M03 S1500
```

Alternatively, you can use the following code:

```
N90 G01 Z.2
N100 M05
N110 G49
N120 G91 G28 X.0 Y.0 Z.0 (RETURN TO HOME POSITION)
N130 M06 T02 (SECOND TOOL CHANGE)
N140 G43 H02
N150 M03 S1500
```

If there are any obstructions such as clamps or fixtures, you may want to select an intermediate point to ensure that no contact will occur:

```
N120 G90 G28 X2.0 Z5.0 (RETURN THROUGH AN INTERMEDIATE
  POINT)
```

Some manufacturers have elected to simplify the tool change process by forcing a home return in the Z-axis and turning off the spindle whenever a tool change is called. Just one word of caution: The machine will move in rapid traverse as soon as the tool change code is encountered. For the sake of safety, the tool should first be positioned above the workpiece and be clear from any obstructions before you call the tool change.

8.3 M03, M04, AND M05—TURNING THE SPINDLE ON OR OFF

After the tool change, it is a good idea to turn the spindle on before attempting to perform any metal cutting. This is accomplished with the codes M03 and M04. The only difference between M03 and M04 is the direction of rotation. M03 is for forward rotation and M04 is for reverse. So, which direction is which? If you were to face the end of the spindle, the forward direction (M03) would be defined as counterclockwise rotation. Conversely, the reverse direction (M04) would be defined as clockwise rotation. ***Figure 8.3*** illustrates these directions.

Figure 8.3 M03 is the most common direction of rotation for end mills and drills. Occasionally, left-handed cutting tools are encountered and they will require M04.

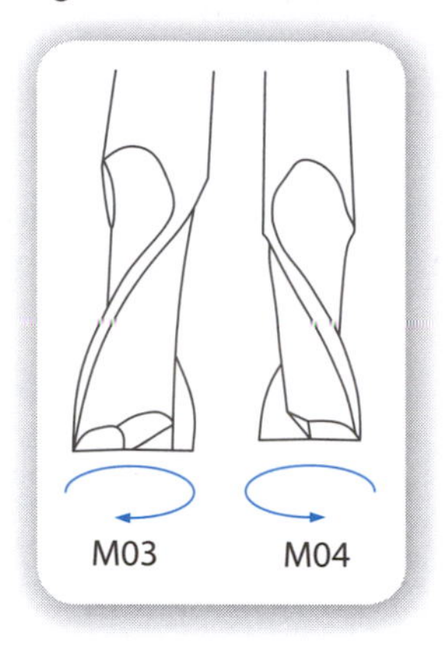

Virtually all end mills, drills, and similar tools you will encounter are right-handed and consequently use M03. However, there are situations in which left-handed cutting tools are used and you will need to use M04. It is also common to use M04 in turning operations when the turret is located on the back side, but you must decide which code to use on a tool-by-tool basis.

The spindle must also be given a speed in revolutions per minute before it will turn on. The spindle speed is specified with the S letter followed by the speed. There is no need to include a decimal point, because most controls only recognize an integer. Some programmers prefer to reverse the order of the M and S words, but it is of little consequence on a modern control.

```
N20 M06 T01
N30 G43 H01
N40 M03 S1200 (SPINDLE ON FORWARD AT 1200 RPM)
N50 G00 X.0 Y.0 M08
N60 G00 Z.2
```

Likewise, the spindle might need to be run in the reverse direction:

```
N20 M06 T02
N30 G43 H02
N40 M04 S1200 (SPINDLE ON REVERSE AT 1200 RPM)
N50 G00 X.0 Y.0 M08
N60 G00 Z.2
```

One last detail to think about when turning the spindle on is gear selection. Some machines are equipped with a multiple-speed gearbox that must be selected prior to machining. However, there is a great variation in the design and construction of the various machine tools, and you should thoroughly investigate the specification of the machine tool before attempting to program it.

Some machines will automatically select a gear range based on the programmed spindle speed. Others will require that the gear range be programmed before the spindle can be turned on. Consult the operator's manual before making any assumptions about gear ranges.

M41 and M42 are the usual codes for gear selection. M41 is used for low gear and M42 is used to select the high range. It is important to use the proper gear for the job at hand or you might put excessive strain on the spindle motor. The situation is analogous to driving a car. If you try to go up a hill in high gear, the motor will have to push too hard and it may stall. Similarly, on a downhill slope, you will probably use a higher gear or the motor will have to spin dangerously fast.

The following code example shows gear selection prior to turning the spindle on. However, you should consult the operator's manual before making decisions about gear selection. Additional gear ranges, default values, M-codes, and electronic switching may be available.

```
N20 M06 T10 (3.0 FACE MILL)
N30 G43 H10
N40 M41 (SELECT LOW GEAR RANGE)
N50 M03 S200
```

8.4 M07, M08, AND M09—COOLANT CONTROL

Coolant can also be controlled with one of several M-codes. Coolants are often used in machining operations to extend the tool life, flush chips away, and improve the surface finish. Most machine tools are equipped with a flood coolant system that sprays a continuous stream of liquid coolant into the cutting zone through flexible hoses. Other machines may also use a mist coolant system that can spray a mist of compressed air and coolant onto the cutting tool.

M08 is the general code for turning on the coolant system. However, M07 may be used to differentiate between flood and mist systems on machine tools equipped with both systems.

M09 is used to turn off the coolant. It is good practice to turn the coolant off during tool changes to avoid spraying coolant into any critical areas. It is also a good idea to turn the coolant off when a pause is needed in the middle of the program to inspect the tool or workpiece. (It can be annoying and dangerous to be splashed with coolant.) Many newer controls will automatically suspend coolant when a tool change is called or when the program is reset with the M30 code; however, many programmers include M09 out of habit, as in the following code example.

```
N120 G49
N130 M09 (COOLANT OFF)
N140 M06 T02
N150 G43 H02
```

If you do decide to stop the coolant for any reason, be sure to turn it on again before resuming machining or damage may result.

It is often difficult to ensure that the coolant is continuously flowing at the right target. As the tool lengths change, the coolant stream must be readjusted to cool the tool and flush the chips. To remedy this problem, some manufacturers have developed programmable coolant nozzles (*see* ***Figure 8.4***) that can be automatically repositioned at any point in the program.

Figure 8.4 **A programmable coolant nozzle being used to direct flood coolant at the proper target.** *Courtesy of Haas Automation.*

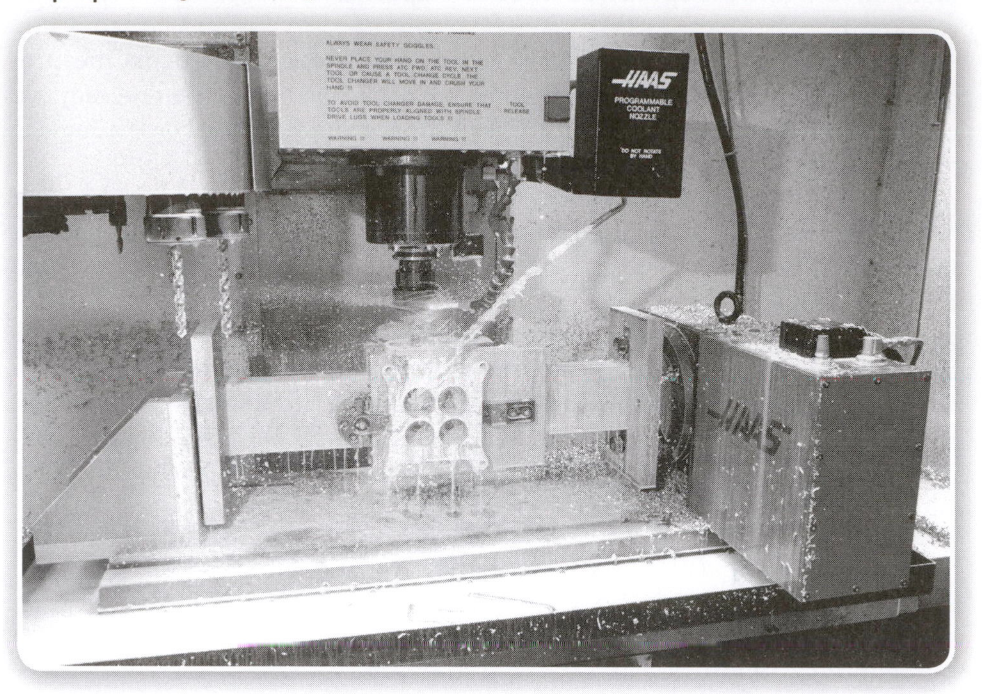

8.5 M30 AND M02—ENDING A PROGRAM

Preparation for ending a program requires a couple of different actions. First, you must position the tool and workpiece so that it is easy to load and unload the workpiece. Second, the program must be reset to the beginning in preparation for the next cycle.

M30 is the most common code used to signify to the control that the program has ended. M30 will stop the machine wherever it is currently positioned and reset the memory to the top of the program. The machine will now be ready for the next cycle:

```
N210 M05
N220 G49
N240 M09
N240 G91 G28 X.0 Y.0 Z.0 (RETURN TO HOME POSITION)
N250 M30 (END PROGRAM AND RESET)
%
```

Returning completely to the home position may be wasteful. An alternative might be to move the machine to some intermediate position that will give the operator easy access to the fixture without moving the machine excessively. It is common practice to move the Z-axis to home regardless of X and Y positioning. This is not wasteful, because the tool change at the beginning of the program will probably initiate a Z-axis home on its own.

```
N210 M05
N220 G49
N240 M09
N240 G91 G28 Z0.0 (RETURN TO Z HOME POSITION)
N250 G90 G00 X4.0 Y8.0 (POSITION TABLE)
N260 M30
%
```

It is important to note that any modal codes that are active when the program terminates will still be active when the next cycle is started. This is the reason the safe line is used at the beginning of the program. However, it is good to get into the habit of canceling codes that might cause a problem as soon as you do not need them anymore.

M30 can be used anywhere in the program. For example, we may have a need to only run the first half of an existing program. Rather than deleting part of the program, we could simply insert the M30 code wherever we wish to terminate the execution.

M02 is another code used to signify the end of a program. M02 has mostly been superseded by M30, but it is still sometimes encountered. It is less flexible than M30 and can only be used at the end of the program. M02 does not reset the memory; the operator must manually press the reset button.

8.6 M00 AND M01—INTERRUPTING THE PROGRAM EXECUTION

There are many circumstances in which you will need to pause the program during execution in order to perform some task. For example, you may need to stop the program to clean chips out of the cutting zone or off a drill. Other times, several operations can be combined into one program; this may involve turning the part over or moving clamps to a new position.

Two codes will automatically cause program execution to pause: M00 and M01. The first, M00, is an unconditional program stop. When the control reads the M00 code, it will stop all axis motion until the operator presses the cycle start button again. The spindle will continue to revolve, and any other functions that are currently in operation will continue. Upon restarting, the program will continue at the line immediately after the program stop.

program stop:
An M-code that causes the program execution to pause until started again by the operator. This stop can be conditional (M01) or unconditional (M00).

```
N240 G00 Z8.0 M09
N250 M05 (STOP SPINDLE)
N260 M00 (STOP TO CLEAN OUT CHIPS)
N270 M03 S1500 (RESTART SPINDLE)
N280 G00 Z.2 M08
```

It is an important safety precaution to turn off the spindle and coolant before reaching into the machining area to clear the chips or to make adjustments. Just be sure to start them again before resuming machining.

There are some circumstances in which you may want to only occasionally stop program execution. The M01 code is called an optional program stop, and it allows the operator to over-ride a stop until needed. The M01 functions identically to M00 except that it functions only when the operator has enabled the *optional stop* button on the control. This is very useful for known problem spots or to check tools for damage before continuing. The following program segment will behave differently based on the position of the optional stop button, as detailed in Table 8.2.

```
N150 M06 T02
N160 G43 H02
N170 M01 (CHECK THE TOOL FOR DAMAGE)
N180 M03 S1500
```

Table 8.2 Behavior of Optional Stop

Optional Stop Switch on Control	Machine Behavior
On	Program execution will stop, and the program will wait for the operator to press the cycle start button.
Off	Program execution is not interrupted.

8.7 BLOCK DELETE (/)

Block delete is a useful function that allows the operator to completely skip certain blocks within the program. This is accomplished by placing a forward slash (/) at the beginning of a line. Block delete does not really delete the lines within the program; the block will still be there. However, it does cause them to be ignored by the control.

block delete:

A function used to cause the control to ignore a block of code when the block-delete switch is active. The block is proceeded with the "/" character.

Block delete functions in the same manner as optional stop in the respect that there is a switch on the control that must be activated in order for the line to be ignored. If the block delete switch is on and the control reads a block delete, it skips ahead to the next line. If the block delete switch is off and the control reads a block delete, the block will be executed normally.

In the example below, the same program segment is used for two similar parts, except that part B has an additional row of drilled holes. The blocks will be ignored if the block delete switch is set to the on position, as detailed in Table 8.3.

```
N80 G81 X0.0 Y1.0 Z-1.0 R.2 F5.0 (DRILLED HOLES, ROW ONE)
N90 X.0 Y2.0
N100 X.0 Y3.0
/N110 X1. Y1.0 (ROW TWO, USED ONLY ON PART B)
/N120 X1. Y2.0
/N130 X1. Y3.0
N130 M05
```

Table 8.3 Behavior of Block Delete

Block Delete Switch on Control	Machine Behavior
On	Blocks starting with "/" will not be executed
Off	Blocks starting with "/" will be executed normally

8.8 MISCELLANEOUS FUNCTIONS

Numerous other functions are accessible using M-codes. Some of these functions are standardized and others are not. A few of the notable M-codes are listed below; however, many others may be available on a particular control. See Appendix A for a listing of other common M-codes.

M19—Oriented Spindle Stop

M19 causes the spindle to rotate to a reference position and lock. There is a small notch or divot on the tool holder that is used to align the tool with the oriented position. This function can be useful when the tool must be aligned axially for use or storage.

M97, M98, and M99—Subprogram Control

Subroutines and subprograms are a method of modularizing a NC program to make it more portable and easier to modify. These methods will be covered in detail in Chapter 11, which concerns advanced programming techniques.

Machine tool and control manufacturers may also include a plethora of other M-codes to automate operations and enhance the functions of the machine tool. Some common—but less standardized—functions include clamping and unclamping of power chucks and vises, safety over-rides, opening and closing the doors, parts catchers, coolant nozzles, etc.

Most controls will also include user M-codes that close a relay or generate a simple electrical signal. These M-codes can be used externally to activate some device. For example, an external M-code might signal a robot to load the next workpiece. The beauty of external M-codes is that the manufacturing operations can be customized and automated independently of the control.

Your Turn

Create a program that utilizes common M codes in machining. You could even leave some missing. Have students be able to describe what action is taking place in each block containing an M code.

CHAPTER SUMMARY

- M-codes can be used to initiate miscellaneous functions that are generally not related to tool movement.
- M-codes act like switches that stay on until they are turned off by another M-code (e.g., the coolant—M08—will stay on until turned off by M09).
- Most often, we are allowed only one M-code per block of code. The convention is to place the M-code at the end of the block.
- Some M-codes are used in conjunction with other addresses. For example, an M06 tool change is usually called in conjunction with a T-word.
- Program operation can be modified with the optional stop and block delete codes. These codes will only be activated when the appropriate switch on the control is active.
- The machine tool manufacturer often provides many additional M-codes. However, many of these codes are non-standard; therefore, we should consult the user manual before making any assumptions.

BRING IT HOME

1. How is an M-code different from a G-code? How are they similar?
2. What is the difference between M00 and M01? Describe a situation in which you might use each of these codes.
3. What is *block delete* and how is it used?
4. Which M-code is most often used to start spindle rotation?
5. Why is it important to select the proper gear range for the spindle?
6. What advantage is there in ending a program with M30 rather than M02?
7. Why is it important to move the spindle to a clear location before changing tools? Describe two methods of accomplishing this task.

EXTRA MILE

Describe the common use of the following functions:

Code	Definition/Use
M00	
M01	
M02	
M03	
M04	
S	
M05	
M06	
T	
G43	

Code	Definition/Use
H	
M07	
M08	
M09	
M19	
M30	
M41	
M42	
/	

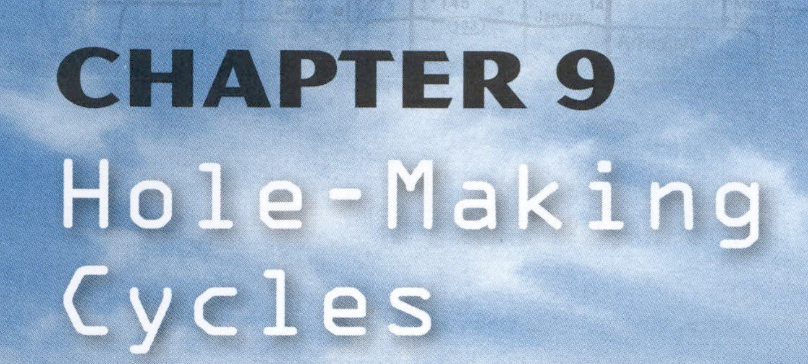

CHAPTER 9
Hole-Making Cycles

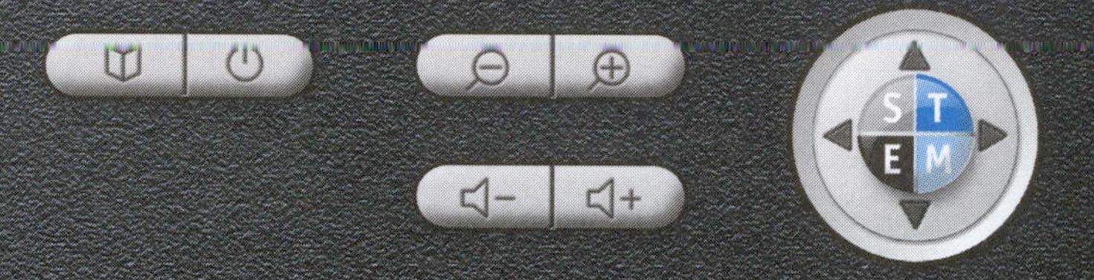

Menu | START LOCATION | DISTANCE | END LOCATION

Before You Begin

Think about these questions as you study the concepts in this chapter:

1. How are canned cycles used in hole-making operations?
2. What are some of the more common G codes used to program canned cycles?
3. How do you calculate time and velocity when using canned cycles?
4. Why are loops used in canned cycles?
5. How do you used initial and retract planes to safely avoid obstacles?

Key Terms

Canned Cycle
Chip Breaker
Dwell
Orient
Peck Drilling

9.1 INTRODUCING THE CANNED CYCLE

Canned cycles (or fixed cycles) were developed to perform common tasks that would require many lines of standard code to accomplish. A canned cycle is a specialized function that is contained within the control program and machine control unit (MCU). It is designed to perform some specific operation such as drilling a hole or performing a tapping operation. Canned cycles save time because they can be accessed with a single block of code rather than with many individual tool movements.

canned cycle: Preprogrammed subroutine that helps to automate machining tasks. Examples include drilling and roughing cycles.

Take for example the hard way to drill a hole in the program below. The code needed to drill this hole is programmed with a series of rapid traverse and linear interpolation codes to drill a hole 1.0" deep. It also clears the chips every 0.2" and rapids back to the bottom to start the next peck.

```
%
O0101 (Program 9.A)
N10 G20 G40 G49 G54 G80 G90 G98
N20 M06 T05
N30 G43 H05
N40 M03 S2000
N50 G00 X1.0 Y1.0 M08
N60 G00 Z.2 (RAPID TO START POINT)
N70 G01 Z-.2 F5. (PLUNGE TO FIRST DEPTH)
N80 G00 Z.2 (CLEAR CHIPS)
N90 G00 Z-.18 (RAPID TO BOTTOM OF HOLE)
N100 G01 Z-.4 F5. (REPEAT UNTIL FINISHED)
N120 G00 Z.2
N120 G00 Z-.38
N130 G01 Z-.6 F5.
N140 G00 Z.2
N150 G00 Z-.58
N160 G01 Z-.8 F5.
N170 G00 Z.2
N180 G00 Z-.78
N190 G01 Z-1.0 F5. (FINAL DEPTH)
N200 G00 Z.2
N210 M05
N220 M30
%
```

As you can see, there are quite a few blocks in this program to drill one simple hole. The program also performs rapid movement while below the surface of the part; this can be dangerous if any mistakes are made. A simpler and safer way to perform this operation would be to use a canned cycle. The previous program could be rewritten using a canned cycle, as shown in the program below. The single block of code (N60) replaces 15 other blocks to perform the same drilling operation.

```
%
O0102 (Program 9.B)
N10 G20 G40 G49 G54 G80 G90 G98
N20 M06 T05
N30 G43 H05
N40 M03 S2000
N50 M08
N60 G99 G83 X1.0 Y1.0 Z-1.0 Q.2 R.200 F5.0 (DRILL HOLE)
N80 G80
```

```
N90 M05
N100 M30
%
```

There are canned cycles to perform a number of different operations, including drilling, tapping, boring, bolt-hole circles, patterns, pockets, and many others. Some of these have been standardized, but many others are designated by the machine tool manufacturer. In the remainder of this section, we will concentrate on the most common canned cycles used in the manufacture of holes. Table 9.1 gives a short list of the G-codes that we will take a closer look at in the remainder of this chapter.

Table 9.1 *Common G-Codes for Hole Making*

G-Code	Definition
G80	Cancel the canned cycle
G81	Standard drilling cycle
G82	Drill with timed dwell
G83	Peck-drilling cycle
G73	Drill cycle with chip breaker
G84	Tapping cycle (RH)
G74	Tapping cycle (LH)
G85	Standard boring cycle
G98	Return tool to initial plane
G99	Return tool to retract plane

9.2 DRILLING AND REAMING CYCLES

G81—Standard Drilling Cycle

All of the canned cycles for hole making follow roughly the same format. The machine will behave as follows when the cycle is called:

1. Rapid to the start location (usually X and Y and then Z).
2. Perform the operation.
3. Return to a predetermined location and wait for another instruction.

G81 is the code used for standard drilling, and it is a good place to start our discussion. The example below shows the typical format for the G81 drilling cycle.

```
G81 X1.0 Y1.0 Z-1.0 R.2 F5.0 (DRILL HOLE)
```

The X- and Y-words are the coordinates where the hole will be drilled. The final depth of the hole is specified with the Z-word. The only other word needed is R to set the retract plane. The retract plane is the Z-axis height at which drilling will begin. The retract plane should always be set slightly above the material surface in order to give the tool a chance to come to a stop from rapid and to give room for chips and coolant—0.200" clearance is often used. We must be careful not to set the retract plane too high or a lot of time will be spent drilling air.

A couple words of caution are needed with drilling cycles. First, the tool will move at rapid traverse to the start location and between holes, so beware of any obstruction. It may be wise to pre-position the tool to an intermediate position if there is any doubt about clearance. Second, if an X and Y location is not specified, the hole may be drilled wherever the tool is currently located. This can cause some unexpected results, so we must be sure to include an X and Y location in the initial drill call.

Canned cycles and their associated parameters are modal, so it is very easy to drill multiple holes with the minimal amount of code. Once a cycle is called, subsequent holes are specified by simply giving the new locations in the blocks that follow:

```
N60 G81 X1.0 Y1.0 Z-1.0 R.2 F5.0 (DRILL HOLE)
N70 X2.0 Y1.0 (HOLE TWO)
N80 X3.0 Y1.0 (HOLE THREE)
N90 X4.0 Y1.0 (HOLE FOUR)
N100 G80
```

You may also have noticed that a G80 code has been included. G80 is used to cancel a canned cycle and it should be called as soon as you are finished with the cycle.

Initial Plane (G98) Versus Retract Plane (G99)

There are two important planes that we must consider when using hole-making cycles: the initial plane and the retract plane. These two planes are used to control vertical tool movement within the drilling cycle and between multiple holes. The machine can be instructed to return to either plane with a G-code: G98 for the initial plane, and G99 for the retract plane. ***Figure 9.1*** illustrates this.

Figure 9.1 **Comparison of G98 and G99. The tool will return to the initial plane when G98 is active but only to the retract plane when G99 is invoked.**

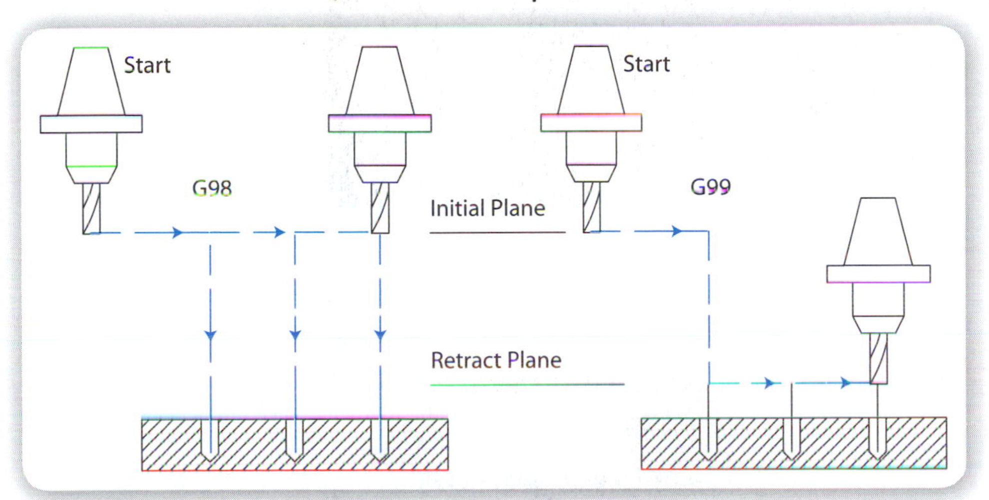

The initial plane is the Z-level where the tool is located when the canned cycle is called. Imagine that the tool is located at Z5.0 when the drilling cycle is called; then the initial plane will become equal to Z5.0. The retract plane is simply specified in the block with the R-word, as described previously. The retract plane in the previous example was set to Z.2.

Upon encountering the drill call, the machine will immediately rapid the retract plane. When the hole is completed, the machine has the possibility of returning to either the retract plane or the initial plane. The format is as follows:

```
N60 G99 (RETURN TO RETRACT PLANE)
N70 G81 X1.0 Y1.0 Z-1.0 R.2 F5.0 (DRILL HOLE)
N60 G98 (RETURN TO INITIAL PLANE)
N70 G81 X1.0 Y1.0 Z-1.0 R.2 F5.0 (DRILL HOLE)
```

A more common format is to call the G98 or G99 mode in the same block as the drill call:

```
N70 G99 G81 X1.0 Y1.0 Z-1.0 R.2 F5.0 (DRILL HOLE)
```

When G98 is active, the tool will return to the initial plane at the end of each cycle and between multiple holes. G98 is used the when extra clearance is needed to avoid collisions between the tool and workpiece. For this reason, G98 is set as the default in the safe line.

When using G98, a safe level should be selected, but it should not be so high as to waste time with excessive tool movement. For example, after a tool change you might want to position the tool close to the workpiece before calling the drill cycle. Otherwise, the initial plane will be at the same height as the tool change and the tool will have to travel excessively far, as in the example below:

```
N20 M06 T05
N30 G43 H05
N40 M03 S2000
N50 M08
N60 G00 Z1.0 (POSITION FOR INITIAL PLANE)
N70 G98 G81 X1.0 Y1.0 Z-1.0 R.2 F5.0
```

G99 is the preferred method to use when there are no obstructions or clearance problems (*see Figure 9.2*). G99 will cause the tool to return only to the retract plane at the end of each cycle or between holes. This method can save a lot of time, especially when numerous holes are to be drilled.

Figure 9.2 ***Care must be taken to set the retract plane above any obstructions when using G99.***

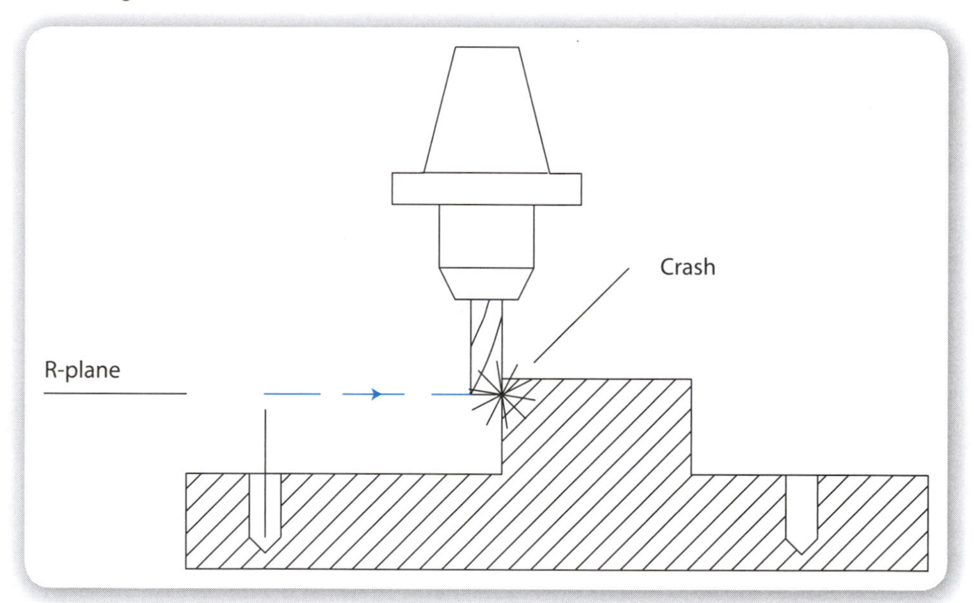

We can also switch between G98 and G99 between the holes. For example, the code below shows a series of holes being drilled with a canned cycle. If there happens to be some obstacle between holes three and four, we could call G98 in the same block as hole number three to move the tool up to the initial plane before proceeding to the last hole.

```
N50 G99 G81 X1.0 Y.5 Z-.5 R.200 F10.0 (HOLE 1)
N60 X2.0 Y.5 (HOLE 2)
N70 G98 X3.0 Y.5 (HOLE 3 RETURN TO INITIAL PLANE TO
  CLEAR CLAMPS)
N80 X4.0 Y.5 (HOLE 4)
```

Canned Cycles and Incremental Programming

The need may occasionally arise to program canned cycles in incremental coordinates. In this situation, a couple of changes will need to be made for the canned cycle to perform properly:

1. The retract plane will be set as an incremental distance from the initial plane.
2. The Z-depth will be set as the incremental distance from the retract plane to the bottom of the hole.

Table 9.2 illustrates the difference between incremental and absolute positioning in respect to canned cycles. Assume that the tool is positioned 1.00" above the workpiece in each case.

Table 9.2 *Absolute Versus Incremental Positioning in Drilling Cycles*

Absolute	Incremental
G90	G91
G81 Z-1. R.2 F5.0	G81 Z-1.2 R-.8 F5.0

As illustrated in ***Figure 9.3***, the tool is positioned 1.00" over the top of the workpiece, and we wish to drill a hole that is 1.00" deep. In absolute programming, we simple invoke the code:

```
G90
G81 Z-1.0 R.2 F5.0
```

The result of this code is that the tool will rapid traverse to 0.2" above the part and then proceed to drill until the hole is 1.00" deep in the workpiece. The actual movement at a cutting feed rate is 1.2"—the "R" height plus the "Z" depth.

The same outcome could be had with incremental programming:

```
G91
G81 Z-1.2 R-.8 F5.0
```

Again, the tool is positioned 1.00" over the top of the workpiece and we wish to drill a hole that is 1.00" deep. With incremental programming the "R" plane is established as a distance from the initial point. Therefore, we must use R-.8 to cause the retract plane to be established 0.2" above the workpiece. Then Z-1.2 is used to drill to a depth of 1.00" below the surface. Again, the actual movement at a cutting feed rate is 1.2".

Point Depth Versus Full Diameter Depth

Drilled holes are typically specified as the full diameter depth rather than the point depth, which will be slightly deeper. However, the drill is usually touched off at the tip, and the depth is programmed to the tip. This means that we will have to program the tool to a depth that is slightly deeper to achieve the correct full diameter depth.

This extra amount will be determined by the diameter of the drill and by the included angle of the point, as illustrated in ***Figure 9.4***. The two most common point angles are 118° and 135°. The extra tip depth can then be determined with trigonometry, as presented in the following formula:

$$\text{Tip Height} = \frac{\text{Radius}}{\tan\left(\frac{\alpha}{2}\right)}$$

where:

α equals the included angle of the tip.

Figure 9.3 Canned cycles for drilling can be programmed using absolute or incremental coordinates. The R and Z addresses will look quite different, but the result will be identical. See the code differences in Table 9.2.

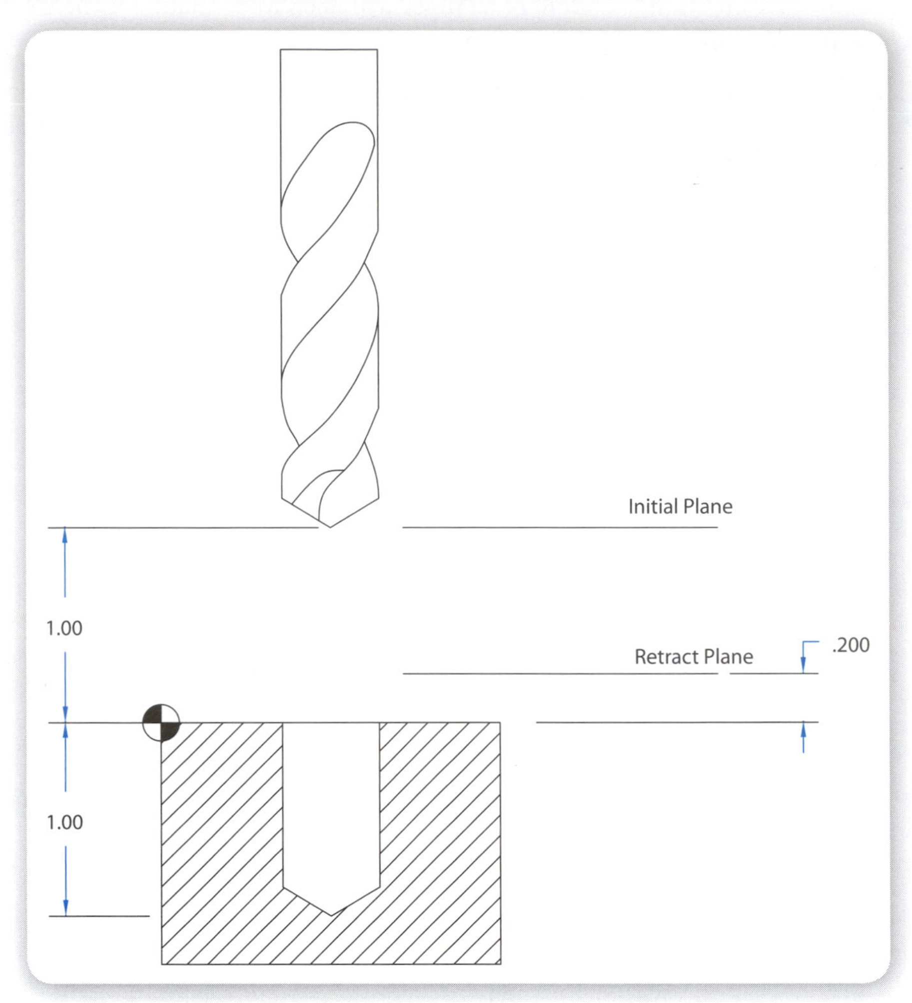

Figure 9.4 The drill will have to be programmed to drill slightly deeper to achieve the correct full diameter depth. The height (H) of the drill tip can be considered one leg of the triangle and the radius (R) the other leg, as is shown on the right. Therefore, any tip angle and drill diameter can be used in the tip height formula.

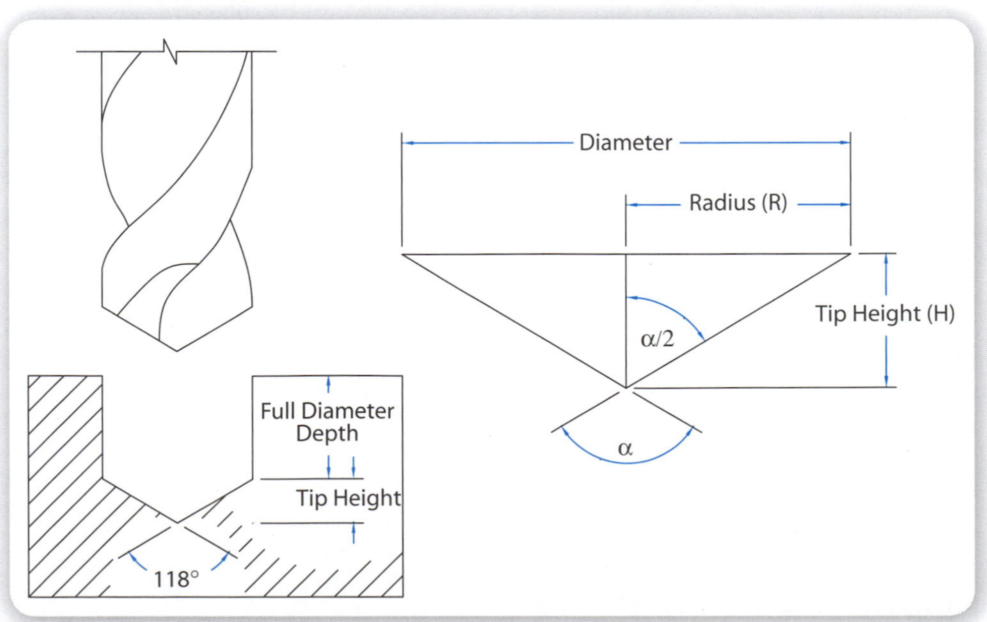

Take for example a standard drill with a tip angle of 118° and a diameter of 0.75":

$$\text{Tip Height} = \frac{.75/2}{\tan(118/2)}$$

$$\text{Tip Height} = \frac{.375}{\tan(59)}$$

$$\text{Tip Height} = \frac{.375}{1.664}$$

$$\text{Tip Height} = .255''$$

For drills with a 118° and 135° included angle, this value approximates as shown in Table 9.3.

Table 9.3 Drill Tip Heights

118° Point	135° Point
Tip Height = 0.300 × Diameter	Tip Height = 0.207 × Diameter

Note: See Appendix B for a table of drill tip heights for standard drills.

How Deep Should You Drill?

A very common problem for the machinist is the ambiguous hole callout on a blueprint. A contentious engineer will specify if the hole depth is to be a "point depth" or a "full diameter depth." For example:

Drill Ø .500, Full diameter, .75 deep

This will of course mean that the 0.500" diameter will extend to a depth of 0.75", and the tip of the drill will go a little deeper. This is fine in most cases. However, there are times when the extra depth can ruin a workpiece—such as when drilling holes in hydraulic manifolds. Other callouts can be unclear:

Drill Ø .500, .75 deep

Or for the minimalist:

Ø .500, × .75

In these situations, we don't really know what the designer intended. We don't even know if the hole is to be drilled in the second example. Should this be tip depth or full diameter depth? I don't know either. Common shop practice is to use full diameter depth, but unless the drawing is clearly dimensioned, it is best to error on the safe side.

Callouts for threaded holes are also problematic. Often only the thread depth is specified, without any information to regarding the allowable depth of the tap drill. For example:

Drill and tap ¼-20 UNC, .50 deep

This is great for the actual threaded portion of the hole, but what about the tap drill depth? We know that the tap drill must extend somewhat below the very bottom of the full-thread depth, but how much is acceptable? In general, it is easier to tap a hole if it is drilled deep enough to provide room for the tapered portion of the tap and to accommodate any chips. A general rule is that the full diameter of the tap drill should be deeper than the thread by 1X its diameter. For example, in our ¼-20 UNC, .50 deep callout the tap drill size is Ø.201". So we could drill 0.50" + 0.201" for a total of 0.701" deep. But again, it is best if the callout is clearer to start with:

Drill Ø.201 Full diameter .75 deep,

Tap ¼-20 UNC, .50 deep

dwell:
A short pause in machine movement usually to allow a cut to be completed.

G82—Drilling with Dwell

G82 is the next canned cycle we will look at. G82 is similar to a standard drilling cycle except that it will pause briefly (dwell) at the final Z-depth. The reason for the dwell is to allow the tool to rotate an additional amount before being retracted out of the hole.

A normal drilling cycle will retract the tool as soon as the Z-level is reached. This will leave a slightly sloping surface that is unacceptable in some circumstances. The G82 cycle was developed specifically for features that must have a uniform surface, such as counterbored and countersunk holes. In fact, G82 is sometimes referred to as a counterboring cycle.

A general example of a drilling cycle with dwell is as follows:

```
G82 X1.0 Y1.0 Z-.5 R.2 P2.0 F5.0
```

The P-word is used to specify the dwell time. The normal unit used to specify the dwell time is seconds; however, there are some variations among the different manufacturers. Some controls will use milliseconds or even hundredths of a second for the dwell time. Others will assume that an integer number is in milliseconds and any decimal number is in seconds. For example, P2.5 would be interpreted as 2.5 seconds, but P500 would be interpreted as 500 milliseconds or ½ of a second. Be sure to consult your operator's manual for exact specifications.

How long should the dwell last? The dwell should last as long as it takes the tool to revolve enough to create a flat bottom, plus a little extra time for a spring pass. This does not generally exceed more than one or two revolutions. We do not want the tool to be rotating against the material unless there is enough stock left to form a chip. Otherwise, the tool will be dulled and excess heat may build up.

The minimum amount of rotation is simply the number of degrees between the flutes. For example, a two-flute drill would require 180° of revolution. Of course, you probably noticed that the dwell is in seconds, so we have to convert the degrees to seconds. Fortunately the following formula can be used for this purpose:

$$\text{Time} = \frac{N \times 60}{\text{RPM}}$$

where

- Time = dwell in seconds
- N = number of revolutions
- RPM = spindle speed

For example, a hole is being counterbored with a three-flute tool at 1000 RPM. What is the minimum dwell time?

Solution:

The three-flute tool will have to revolve at least 120° or 0.333 revolutions.

$$\text{Time} = \frac{0.333 \times 60}{1000}$$

$$\text{Time} = 0.02$$

peck drilling:
Drilling operations that reciprocate in and out of the drilled hole to clear and break chips while machining.

G83—Peck Drilling

G83 is the code used to call a peck-drilling cycle. **Peck drilling** is similar to a standard drilling cycle except that the tool is backed out of the hole at certain intervals to clear the chips and cool the tool. Almost every deep hole will need to be peck drilled in order to maintain a reasonable tool life. This is particularly true when using HSS drills that cannot handle as much heat as carbide. In fact, the G83 cycle is used much more often than the standard G81 drilling cycle for the average machining application.

A typical peck-drilling call will have the following format:

```
G83 X1.0 Y1.0 Z-1.0 Q.25 R.2 F5.0 (PECK DRILL)
```

The Q-word is the only new code. Q is an incremental distance used to specify the peck depth. In other words, if Q is equal to 0.25, then the drill will feed 0.25" deep and then back out to the retract plane. The drill will then rapid back to the bottom of the hole and plunge an additional 0.25". This will continue until the Z-depth is reached.

G73—Chip Breaker

chip breaker: A geometrical feature of a cutting tool that encourages the chip to break rather than remain intact.

The G73 code is a variation of peck drilling cycle called a chip breaking cycle, or sometimes a high-speed drilling cycle. This cycle is similar to peck drilling except that the tool will not pull completely out of the hole to clear the chips. Instead, the tool will back up slightly at every peck interval to cause the chip to be severed. This is a very useful cycle when the work material is somewhat stringy and the long chips become difficult to clear. The only downfall to the G73 cycle is that the tool may overheat on deep holes. However, a significant amount of time can be saved when numerous shallow holes are to be drilled.

The example below shows the typical format for a G73 call. The other parameters are identical to peck drilling.

```
G73 X1.0 Y1.0 Z-1.0 Q.25 K.75 R.2 F5.0 (CHIP BREAKER
  CYCLE)
```

The G73 cycle does include a provision to periodically return to the retract plane. The K-word is optional, but it can be included to force a full retract after the K distance is reached. Take the example above. This cycle will retract fully after every third peck because Q is equal to 0.25 and K is 0.75.

Many controls offer additional words and parameters to more fully control the G73 cycle. Examples include the ability to decrease the peck depth as the depth becomes greater and to adjust the distance moved at each peck. There is some variation between different controls, so the operator's manual should be consulted for details before making assumptions.

9.3 TAPPING CYCLES

G84—Standard Tapping Cycle

The tapping cycle is called with the G84 code. The tapping cycle will feed a tap to the bottom of the hole and then reverse the spindle to remove the tap from the hole. The example below is a typical tapping call for a tap that has 13 threads per inch:

```
M03 S100
G84 X1.0 Y1.0 Z-1.0 R.2 F7.6923 (TAPPING CYCLE)
```

The one critical factor that must be considered when tapping is that the spindle speed and the feed must be carefully synchronized, or the tap may be damaged. The first step in writing a block to perform tapping is to calculate the pitch of the tap. The pitch of a standard (single-lead) tap is simply the distance from the center of one thread to the center of the next thread. Pitch can be calculated with the following formula:

Pitch = 1/Number of threads per inch

The second step is to find the feed by multiplying the pitch by the spindle speed. If you later decide to change the spindle speed, it is important to recalculate the feed.

For example, you need to calculate the feed to perform a tapping operation for a 0.500-13 UNC thread at 100 RPM.

Solution:

Pitch = 1/number of threads per inch
Pitch = 1/13
Pitch = 0.076923 inches
Feed = pitch × spindle speed
Feed = 0.076923 × 100
Feed = 7.6923" per minute

We could also perform a little algebra on the two previous formulas to find the feed:

Feed = spindle speed/threads per inch
Feed = 100 RPM/13 TPI
Feed = 7.6923" per minute

It is important to take this calculation out to as many decimal places as the control will accept. The small rounding error coupled with any machine error can create a large enough discrepancy to cause tool or thread damage.

The actual use of the tapping cycle will also be tempered by the ability of the control. A machine that is equipped with rigid tapping ability may still have limitations in the accuracy of the tapped hole depth. Care should be taken not to run the tap into the bottom of a blind hole in an attempt to reach the correct thread depth; a broken tap will result. Furthermore, if the machine tool does not support rigid tapping, then a floating tap holder must be used. A floating tap holder allows a small amount of up and down movement to accommodate any error in the feed. This may prevent a broken tap or damaged workpiece.

G74—Left-handed Tapping Cycle

G74 is used to call a left-handed tapping cycle. It is identical in every way to the standard tapping cycle except that the spindle direction will be reversed.

9.4 BORING CYCLES

G85—Standard Boring Cycle

G85 is used to call a canned cycle for boring. The typical format is as follows:

```
G85 X1.0 Y1.0 Z-1.0 R.2 F10.0 (STANDARD BORING CYCLE)
```

The G85 boring cycle will feed to the bottom of the hole and then feed back out to the retract plane. This will result in a small spring cut being taken on the return pass. This can cause a poor surface finish in some circumstances and may not be desirable. Not to worry; the manufacturers will also provide a number of additional boring cycles that allow the programmer to control exactly how the tool is removed from the hole.

Careers in the Designed World

CAREERS IN COMPUTER INTEGRATED MANUFACTURING

Robot World

Paul Santi is a process engineering manager for FANUC Robotics, overseeing a group of mechanical engineers who develop robotic automation processes for industry.

The motion control systems used in FANUC robots are directly related to the company's CNC controller platform. Many of the robotic systems incorporate equipment that uses CNC programming for machine tool controls, and Santi has to coordinate his processes with the CNC work.

"The robotic systems my group develops need to interface with the CNC machine tools," Santi says. "We need to ensure that the electrical communications are correct and that our mechanical tooling mates properly with the fixtures."

On the Job

As a process engineer, Santi must develop the proper automation for various operations. "We may get a request from a customer for a part that must be produced a certain way," he says. "We have to create a robotic system to produce that part to meet the customer's needs."

In his managerial role, Santi decides which engineers will work on which projects. He resolves issues related to a project's timing and technical issues. He also helps train his engineers on new technologies.

Recently, FANUC has received many requests for alternative energy projects, such as assembling solar panels. "We're dealing with new customers and never-before-attempted applications," Santi says. "It's exciting to develop new processes for products that historically have been very low-production items that now need to be produced in mass quantities at high speed. Keeping them cost-effective is the key."

Santi enjoys the challenge of creating new automated processes. "We have the ability to come up with creative ideas to meet customers' needs," he says. "It's very fulfilling to devise a solution that has never been done before."

Inspirations

Santi always had a feel for mechanics. His father worked in mechanical engineering, and Santi was exposed to the field at an early age. In high school, he even worked summer jobs at companies where his father was employed. In one such job, he was asked to improve the design of existing components. The hands-on experience of reading blueprints and making the necessary design modifications to production equipment proved a rare opportunity for a high school student.

Education

Santi received his bachelor of science degree in mechanical engineering from Michigan Technological University. His focus was manufacturing, and he took lab courses on CNC machine tool operation. He also had a chance to work in the robotics lab, which helped convince him to work in the robotics industry.

Advice for Students

Santi advises students not only to gain a strong foundation in technical knowledge, but also to learn to work well with others.

"Understanding how to work constructively with different groups of people is really important," he says. "In the workplace, there are many contacts you make and many different types of people you have to interact with. During your education, take advantage of as many opportunities as you can to work in teams with different groups of people."

Boring operations use an eccentric boring head that can be adjusted to the proper radius with a screw. Bored holes also tend to have very tight specifications that may require the operator to stop the machine and check the diameter before proceeding to the next operation. If the hole is undersized, the boring head will have to be adjusted and the cycle will need to be run again. One technique to accomplish this is as follows:

```
N90 G98 (RETURN TO THE INITIAL PLANE)
N100 G85 X1.0 Y1.0 Z-1.0 R.2 F10.0 (STANDARD BORING
  CYCLE)
N110 M00 (STOP TO CHECK THE HOLE DIAMETER)
/N120 X1.0 Y1.0
/N130 M00
N140 G80 (CANCEL CANNED CYCLE)
```

In the previous program segment, the hole is bored and then the program is stopped to check the diameter. If the diameter is within specifications, the block delete switch is activated and the program continues to the next operation (skipping the repeated boring cycle). If adjustments need to be made, the operator can adjust the boring head and then deactivate the block delete switch to run the boring cycle again.

Boring Cycle Variations

There are many variations of boring cycles that differ only slightly from the standard boring cycle. Table 9.4 describes a few of the more common variations. There may also be control-specific parameters that affect the behavior of these cycles.

G76—Boring Cycle with Spindle Orientation

By far the most quality-oriented boring cycle is G76. This cycle feeds the boring bar to the bottom as is programmed but then moves away from the edge of the hole before retracting as in ***Figure 9.5***. This ensures that the tool will not cause a spiral scrape or vertical scratch as will occur in other cycles.

The spindle will need to stop and then orient itself into a known angular position before moving away and retracting. It is critical that the tool is installed in the oriented position during setup. Failure to do so will result it a crash. It is always a good idea to dry-run a boring cycle before attempting to cut material just to make sure it is proper.

orient:
The act of moving the spindle to a known angular position. This is required for functions such as threading and boring, where the angular position must be accounted for.

Table 9.4 Standard Boring Cycles

Boring Cycle	Descriptions
G85	Standard boring cycle. Feeds to the bottom and then feeds back out
G86	The spindle will stop at the bottom of the hole and then the tool will be removed at rapid speed rather than feed. Tends to leave a vertical tool mark along the wall
G87	The spindle will stop at the bottom of the hole, and then the tool must be retracted manually.
G88	Same as G87, but with a dwell at the bottom of the hole
G89	Same as G85, but with a dwell at the bottom of the hole
G76	Orients the tool at the bottom of the hole and then shifts slightly away from the wall before being retracted at rapid speed. Care must be taken to install the tool in the oriented position, or else a crash will result. This cycle gives the best performance but is more difficult to set up.

Each machine tool will have its own procedure for spindle orientation. It usually involves installing the tool in the tool holder so that the cutting point is aligned with a divot. Then the tool will be installed in the spindle or magazine in the correct orientation according to the manufacturer's instructions. Another detail to pay attention to is the distance that the tool will backup at the end of the cycle. This is usually set in the machine parameters. It becomes critical when a small hole is being bored with a relatively large boring bar. Even if properly oriented, the back side of the bar might come into contact with the workpiece when the tool is extracted.

9.5 LOOPING

Another feature that is available with canned cycles is the ability to repeat the cycle numerous times with just one additional code. The L-word is used to indicate the number of times a cycle should be repeated executed, and it can be used with any

Figure 9.5 The G76 canned cycle for boring will orient the tool to a known spindle position before backing away from the edge and retracting. Care must be taken to install the tool in the proper position or a crash will result when the tool "backs away" (pushes into the opposite side).

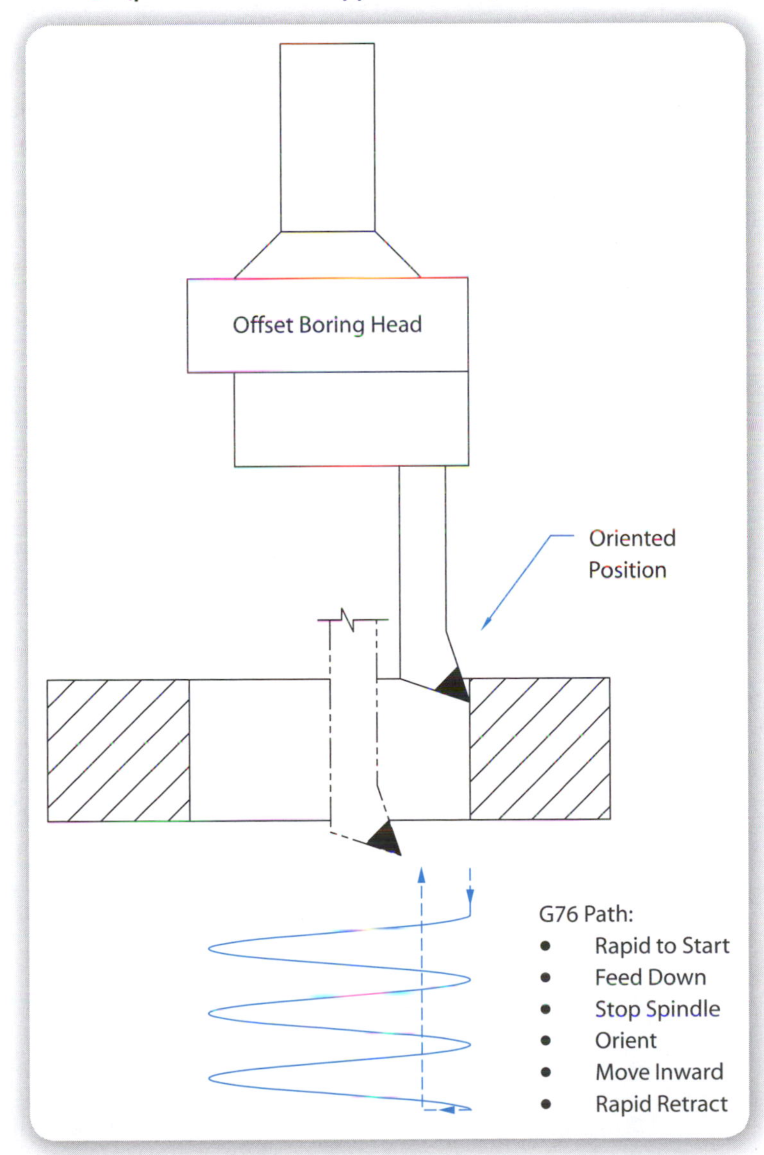

of the hole-making cycles we have discussed in the chapter. The basic format is as follows:

```
N90 G91 G81 X1.0 Y0.0 Z-1.2 R-.8 F5.0 L5
```

The previous block instructed the machine to move one inch to the right and then drill a hole. This will repeat until there are a total of five holes, as is indicated by the looping instruction "L5." If the tool is positioned at X0.0, Y1.0, and Z1.0, then the tool will move over 1" and then start drilling the holes. However, it can be cumbersome to initiate canned cycles in incremental mode. An easier solution is to call the canned cycle in absolute mode, and then switch to incremental for the subsequent locations:

```
N90 G90 G81 X1.0 Y1.0 Z-1.0 R.2 F5.0
N100 G91 X1.0 L4 (REPEAT HOLE 4 TIMES AT 1.0" INCREMENTS)
N110 G80
```

Loops are a good time saver, and they can lead to very elegant programs. However, programming in incremental coordinates requires us to use extra care. It is hard to justify the extra time it takes to program loops when only a few holes are to be programmed. The real power of loops comes when dozens or hundreds of holes must be programmed in a pattern. Loops can then save many keystrokes and make the program easier to understand and alter.

9.6 PROGRAMMING EXAMPLES USING CANNED CYCLES

Programming Example 9-1

In this programming example we will use canned cycles to produce the workpiece shown in ***Figure 9.6***.

Listing 9-1	Program Code	Explanation
	%	
	O0 901 (CHAPTER 9, EXAMPLE 1)	
	N10 G20 G40 G49 G54 G80 G90 G98	
	N20 M06 T02	
	N30 G43 H02	
	N40 M03 S1200	
	N50 G99 G81 X1.0 Y.5 Z-.5 R.2 F10.0	Drilling cycle to drill holes 0.50" deep
	N60 X2.0 Y.5	Hole 2
	N70 X3.0 Y.5	Hole 3
	N80 X3.0 Y1.0	Hole 4
	N90 X2.0 Y1.0	Hole 5
	N100 X1.0 Y1.0	Hole 6
	N110 G80	Cancel drilling cycle

Listing 9-1 *continued*

```
N120 G91 G28 Z2.0 Y0.0
N130 M05
N140 M30
%
```

Figure 9.6 Workpiece produced by programming example one and two.

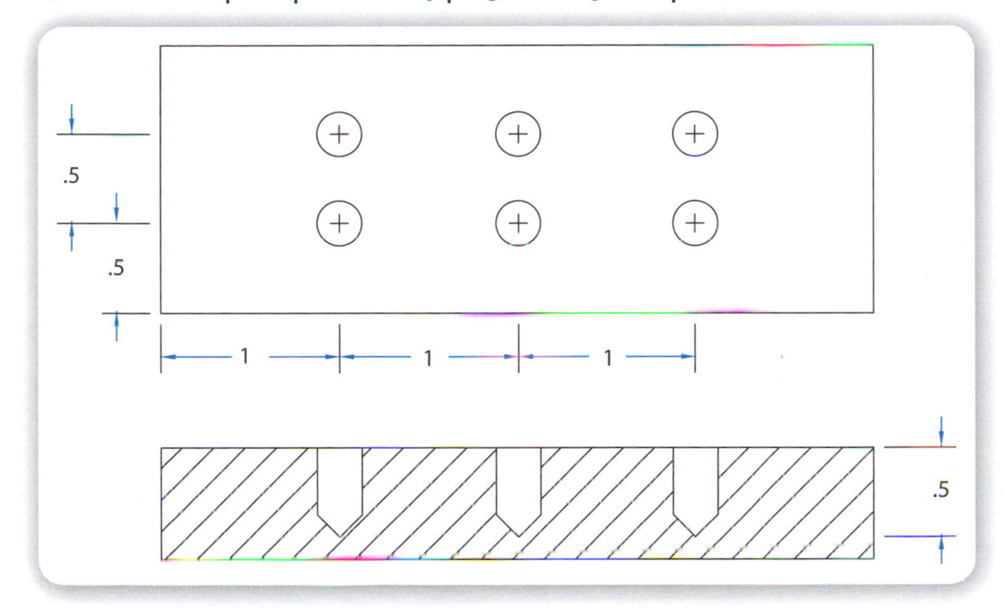

Programming Example 9-2

The program in Listing 9-2 creates the same hole pattern as Programming Example 9-1 (Listing 9-1), but this time a loop is used.

Listing 9-2	Program Code	Explanation
	`%`	
	`O0 902 (CHAPTER 9, EXAMPLE 2)`	
	`N10 G20 G40 G49 G54 G80 G90 G98`	
	`N20 M06 T02`	
	`N30 G43 H02`	
	`N40 M03 S1200`	
	`N50 G99 G81 X1.0 Y.5 Z-.5 R.2 F10.0`	Call drilling cycle while in absolute mode.
	`N60 G91 X1.0 L2`	Switch to incremental and loop twice.
	`N70 Y.5 (MOVE TO SECOND ROW)`	
	`N80 X-1.0 L2`	Loop to drill the last two holes.

(continued)

Listing 9-2 *continued*

```
N110 G80
N120 G91 G28 Z2.0 Y0.0
N130 M05
N140 M30
%
```

Programming Example 9-3

This programming example will use canned cycles to produce the workpiece shown in ***Figure 9.7***. We will start by first making a table of all the hole locations (starting at zero and continuing counterclockwise). Table 9.5 lists these hole positions. We will also use the center of the workpiece as work zero.

Figure 9.7 Workpiece produced by programming example three.

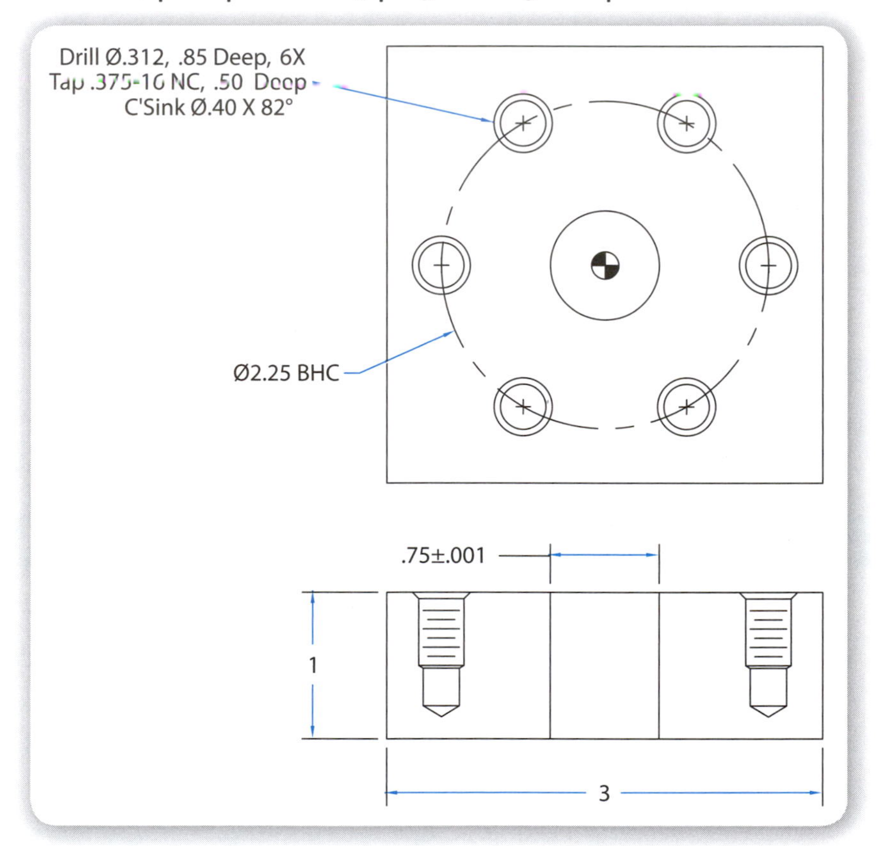

Table 9.5 Hole Positions for Bolt Hole Circle in Figure 9.7

Position	X	Y
# 1	1.25	0.
# 2	.563	.974
# 3	−.563	.974
# 4	−1.25	0.
# 5	−.563	−.974
# 6	.563	−.974

Listing 9-3	Program Code	Explanation
	%	
	O0903 (CHAPTER 9, EXAMPLE 3)	
	N10 G20 G40 G49 G54 G80 G90 G98	
	N20 M06 T01 (SPOT DRILL)	
	N30 G43 H01 M08	
	N40 M03 S1200	
	N50 G99 G81 X0.0 Y0.0 Z-.2 R.200 F3.0	Spot drill all of the holes.
	N60 X1.25 Y0.0	
	N70 X.563 Y.974	
	N80 X-.563 Y.974	
	N90 X-1.25 Y0.0	
	N100 X-.563 Y-.974	
	N110 X.563 Y-.974	
	N120 G80 M09	
	N130 G00 Z10.0 Y5.0	Spindle moved in preparation for tool change.
	N140 M06 T02 (TAP DRILL)	
	N150 G43 H02 M08	
	N160 M03 S1200	
	N170 G99 G83 X0.0 Y0.0 Z-.85 Q.25 R.200 F3.7	Peck drill all seven hole locations with a 0.312″ drill.
	N180 X1.25 Y0.0 Z-.594 (DRILL TAPPED HOLES)	Z-level is set to give a full diameter depth of 0.5″.
	N190 X.563 Y.974	
	N200 X-.563 Y.974	
	N210 X-1.25 Y0.0	
	N220 X-.563 Y-.974	
	N230 X.563 Y-.974	
	N240 G80 M09	
	N250 G00 Z10.0 Y5.0	
	N260 M06 T04 (82 DEG. COUNTER SINK)	
	N270 G43 H04 M08	
	N280 M03 S370	

(continued)

Listing 9-3 *continued*

```
N290 G99 G82 X1.25 Y0.0 Z-.230 P.17 R.200 F1.1      Countersink the holes prior
                                                     to tapping.
N300 X.563 Y.974
N310 X-.563 Y.974
N320 X-1.25 Y0.0
N330 X-.563 Y-.974
N340 X.563 Y-.974
N350 G80 M09
N360 G00 Z10.0 Y5.0
N370 M06 T03 (3/8-16 TAP)
N380 G43 H03 (USE CUTTING OIL)
N390 M03 S150
N400 G99 G84 X1.25 Y0.0 Z-.5 R.200 F9.375           Tap the six holes. Notice
                                                     that the speed and feed are
                                                     synchronized to the pitch.
N410 X.563 Y.974
N420 X-.563 Y.974
N430 X-1.25 Y0.0
N440 X-.563 Y-.974
N450 X.563 Y-.974
N460 G80
N470 G00 Z10.0 Y5.0
N480 M06 T06 (.719 DRILL)
N490 G43 H06 M08
N500 M03 S611
N510 G99 G83 X0.0 Y0.0 Z-.975 Q.187 R.200 F2.44     Rough drill the center hole.
N520 G80 M09
N530 G00 Z10.0 Y5.0
N540 M06 T07 (CARBIDE BORING BAR)
N550 G43 H07
N560 M03 S916
N570 G99 G85 X0.0 Y0.0 Z-.975 R.200 F1.37          Bore the center hole.
```

Listing 9-3 *continued*

```
N580 G80
N590 G00 Z10.0 Y5.0
N600 G91 G28 Z2.0 Y0.0
N610 M05
N620 M30
%
```

Your Turn

Create an NC program that will create a holder that will accommodate several size pens, pencils, and markers using hole-making cycles. Verify results to your instructor.

CHAPTER SUMMARY

- Canned cycles are functions that perform specialized operations with just a few blocks of code.
- There are a number of standard canned cycles for hole-making operations, including drilling, tapping, and boring.
- Canned cycles are modal. This property makes it easy to drill multiple holes by specifying only the new hole locations.
- We can control the Z-level of the tool as it moves between holes by calling G98 (initial plane) or G99 (retract plane). Returning to the initial plane is safer but takes more time—it is only used when there is some obstacle to avoid.
- When performing a tapping cycle, it is important to carefully synchronize the feed with the lead of the tap; otherwise damage may occur. If the machine tool does not support rigid tapping, then a floating tap holder must be used to accommodate any error in the feed.
- A variety of boring canned cycles are available, and each has its own advantage. The particular cycle is selected based on the machining conditions and tolerance requirements.
- Spindle orientation is required for some canned cycles. Be sure to follow the manufacturer's instructions when setting up cutting tools for these processes.
- Loops can be used with the hole-making canned cycles to create hole patterns. Loops can significantly reduce the number of blocks in a program.

BRING IT HOME

1. Why is it important to give an X and Y address in the initial call of a canned cycle? What could be the consequence if they were not included?
2. Which code is used to cancel a canned cycle?
3. Where is the initial plane?
4. How is the retract plane set?
5. If we wanted to return the tool to the initial plane each time, which code would we use?
6. What advantage does the G76 boring cycle have over G85?
7. Why is it important to properly align the cutting tool when using spindle orientation?

EXTRA MILE

1. Calculate the tip height for a Ø1.25" drill with a 135° point.
2. Write one block of code to perform the indicated operations. Assume the location is X1.0, Y1.0, and the depth is Z-.5 for each operation.
   - Standard drilling cycle
   - Drill with dwell of 0.150 seconds
   - Peck drilling cycle with a 0.187" peck depth
   - Chip breaker with a 0.100" peck and full retract every 0.200"
   - RH tapping cycle for ¼-20 NC thread; define a spindle speed and feed
   - Bore a hole and feed out
   - Bore a hole and rapid out
3. Modify the following block to use incremental coordinates. (Assume that the top of the workpiece is Z0.0 and the tool is positioned at Z2.0.)

```
G90 G81 X1.0 Y1.0 Z-1.0 R.2
F10.0
```

4. Take the block you created in the previous question and create a loop to drill five holes, 1" apart along the X-axis.
5. Write an NC program to drill, countersink, and tap the workpiece in *Figure 9.8*. Be sure to use the upper left corner as the work zero. Assume that the tool material is aluminum.

Figure 9.8 Write a complete program to create the features of this workpiece.

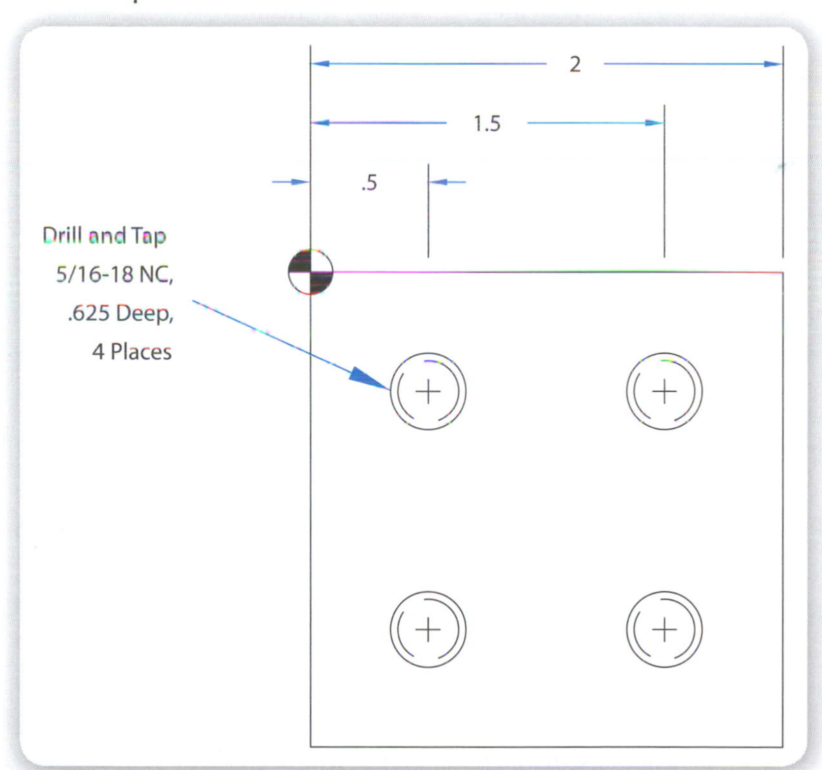

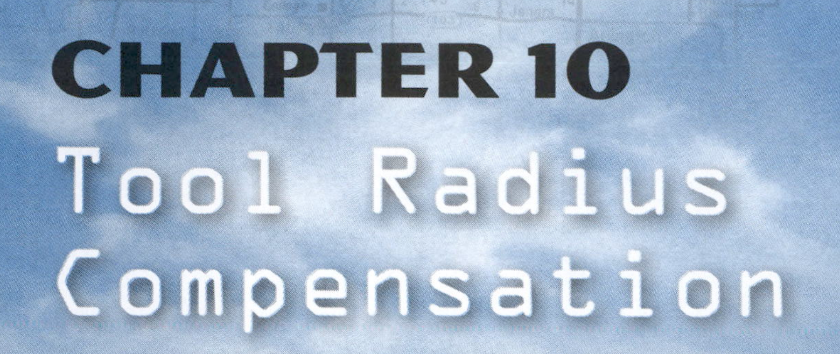

CHAPTER 10
Tool Radius Compensation

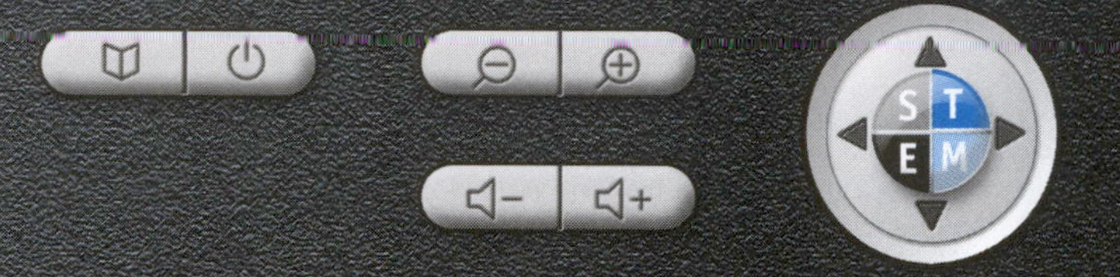

Menu | START LOCATION | DISTANCE | END LOCATION

Before You Begin

Think about these questions as you study the concepts in this chapter:

1. What is tool radius compensation and when is it appropriate to apply?

2. How is the tool diameter register related to the tool radius compensation?

3. How are the toolpaths written for manual and automatic tool radius compensation?

4. What are some of the limitations of tool radius compensation?

Key Terms

Diameter Offset
Tangent
Tool Radius Compensation
Vertex

10.1 WHAT IS TOOL RADIUS COMPENSATION?

Tool radius compensation is the act of accommodating for the cutting tool radius in order to produce the workpiece of the correct geometry. Tool radius compensation is also commonly called cutter compensation or tool nose radius compensation, depending on the source and context.

tool radius compensation: Cutter compensation.

We have already seen numerous examples of tool radius compensation in the previous programming examples. We took it for granted that the toolpath would have to be programmed to compensate for the radius of the cutting tool. For example, the centerline of the toolpath in *Figure 10.1* is to the left of the finished surface. This example contains only parallel and perpendicular lines, and full-quadrant arcs. Therefore, it is quite easy to manually calculate the tool center locations. Other programming situations are not so straightforward—we will have to use trigonometry to calculate the toolpath. Most controls will also allow automatic tool radius compensation, which, under the proper circumstances, can save us some programming time and make it easy to adjust the toolpath.

Figure 10.1 **Simple tool radius compensation on a workpiece containing parallel and perpendicular lines and full-quadrant arcs.**

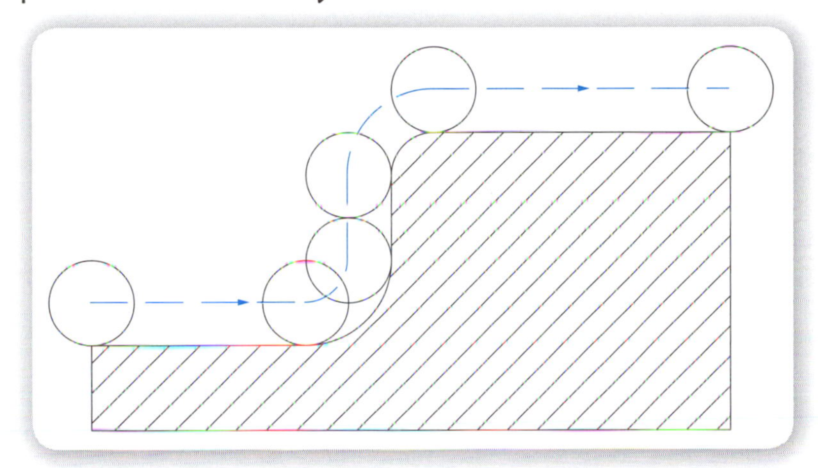

10.2 AUTOMATIC TOOL RADIUS COMPENSATION

Automatic tool radius compensation—sometimes called cutter comp—is a feature that is provided with most controls that will automatically plan the toolpath without any additional manual calculations. Up to this point, we have only seen tool center programming (i.e., the programmed coordinates were the centerline of the tool). With automatic tool radius compensation, we will program to the point where the tool and workpiece boundary are **tangent**, as in *Figure 10.2*. This makes for fewer calculations and easier adjustments.

tangent: The condition created when an arc touches another arc, curve, or line at only one point.

Programming with automatic tool radius compensation differs from tool center programming in two ways. First, the programmed coordinates of the toolpath are the actual corners or edges on the workpiece. The tool is assumed to be tangent at any of these points where one geographical element meets another. Tangent means that just one point of the circle (cutter) touches the line or arc from the workpiece. Take, for example, *Figure 10.3* where the left edge of the workpiece meets with an arc. We will transition from cutting a straight line to cutting an arc, so the end mill must be tangent to the end of the line and also the arc. Automatic tool radius compensation differs in a second manner: we have to transition from tool center coordinate to tool edge (or tangent) coordinates. This will be referred to as lead-in and lead-out.

Figure 10.2 Cutter compensation employs tool edge programming. The programmed coordinates are given as tangent points between the cutting tool and workpiece. The MCU automatically position the tool properly. Manual compensation requires the programmer to calculate the tool center points so that a tangency is maintained.

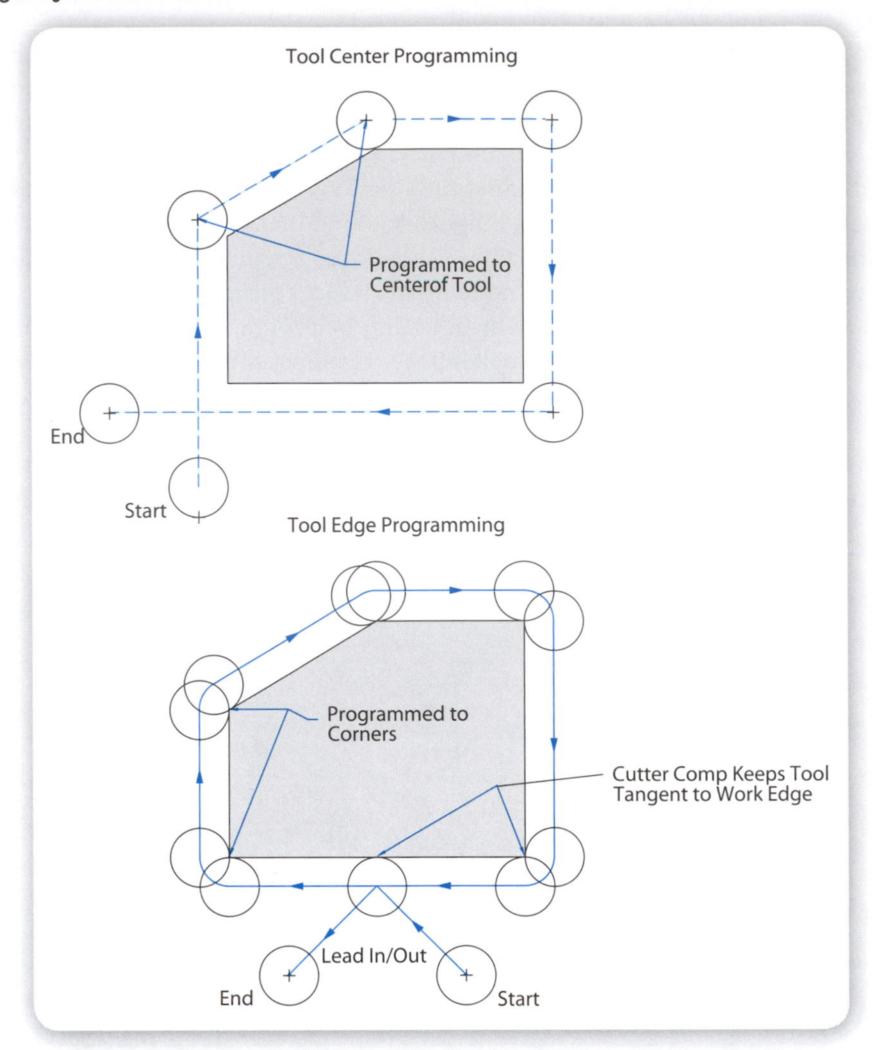

Figure 10.3 A circle, such as the periphery of an end mill, can be tangent to line and arcs. Tangent means that the arc and line will touch the circle only at one point. This is an important concept for tool radius compensation.

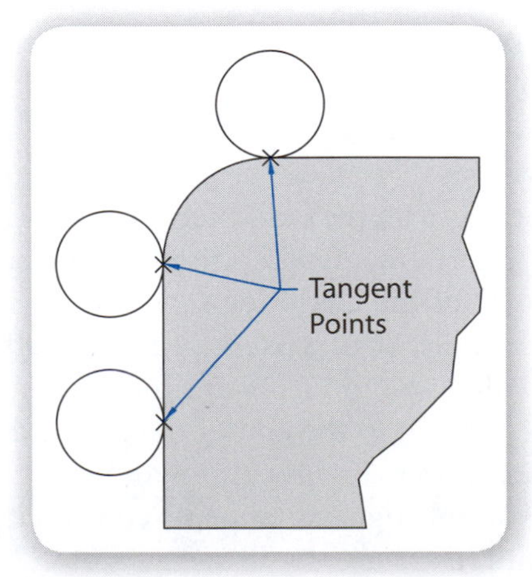

One major problem with manual compensation is that if the diameter of the tool changes, then we have to recalculate the tool coordinates and update the program to accommodate the new diameter. Automatic compensation will take care of this problem with a simple adjustment in the offsets register. This is very useful for maintaining tolerances as the tool wears or as machining conditions change. Automatic compensation is typically used only for finish passes where small adjustments are necessary.

Automatic tool radius compensation is activated with a G-code that will cause the tool to stay on either the left side or right side of the programmed toolpath, as illustrated in ***Figure 10.4***. G41 indicates compensation to the left of the toolpath, and G42 indicates compensation to the right. G40 cancels automatic compensation. Table 10.1 details these codes.

Figure 10.4 *The decision to use G41 or G42 will be determined by the direction of cut.*

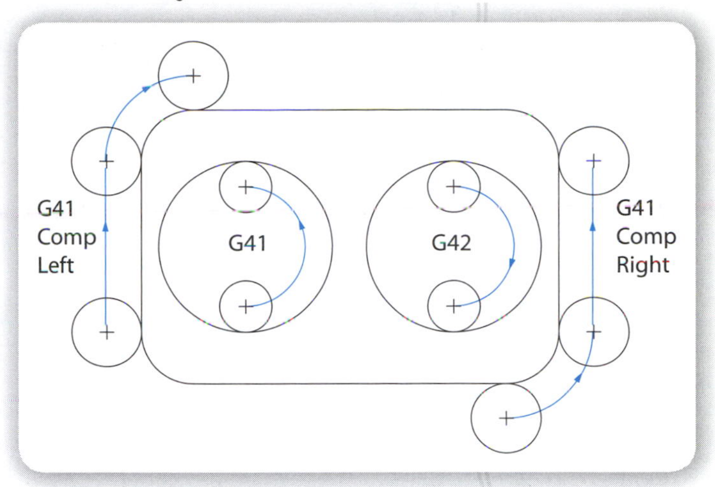

The basic syntax of automatic compensation is perhaps best understood by looking at a larger example. Let us assume that the workpiece shown in ***Figure 10.5*** is 3.00" wide and 1.00" tall and then look at how automatic compensation is used to produce the toolpath:

```
N100 G00 X-1.0 Y2.0
N110 G41 D05 G01 X0.0 Y1.00 F10.0
N120 G01 X3.0 Y1.0
N130 G40 G01 X4.0 Y2.0
```

Table 10.1 *Codes for Automatic Cutter Compensation*

Code	Explanation
G41	Compensate left
G42	Compensate right
G40	Cancel compensation
D	Diameter offset number (stored in register)

Figure 10.5 *An example of cutter compensation. The tool is automatically offset to the left side of the toolpath between P1 and P2.*

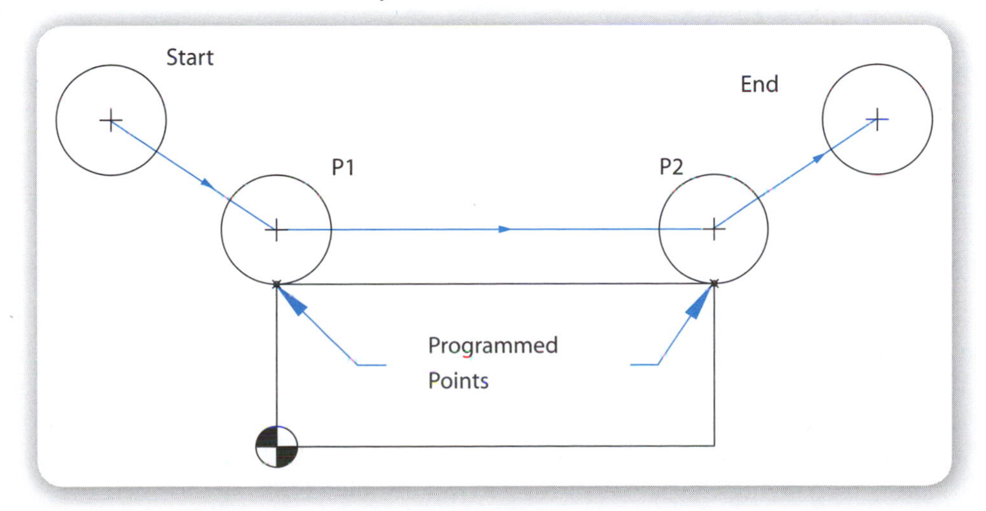

Figure 10.6 **Most controls will automatically roll around the corners when using cutter compensation.**

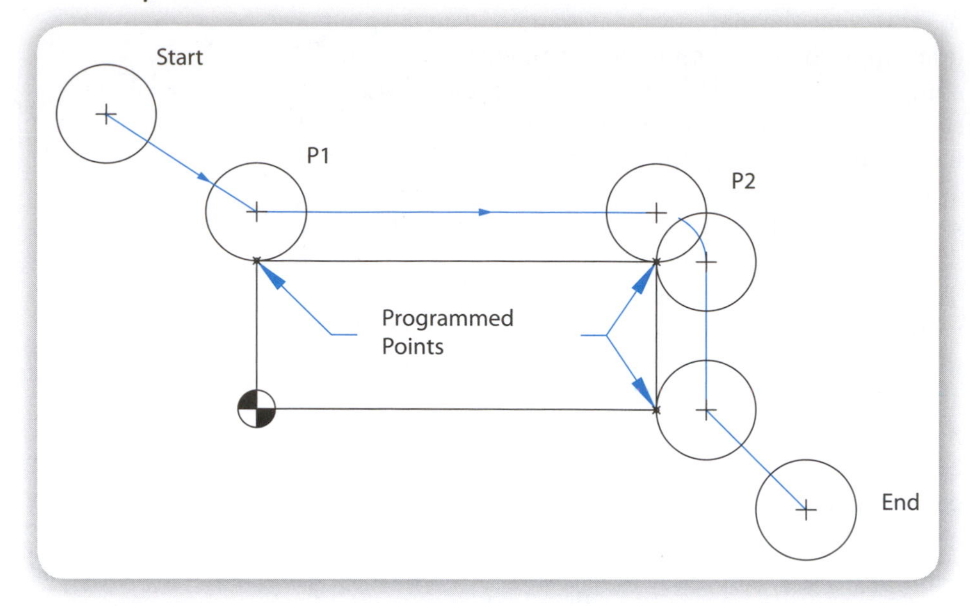

- Line N100 is a positioning move that gets the tool to the start point. This move is still using tool center coordinates.
- The next line, N110, is where the compensation is activated as the tool moves the programmed coordinates. The coordinates X0.0 and Y1.0 represent the actual corner of the workpiece—not the center of the tool. The D address is an offset value that is set by the operator and stored in the register.
- The tool then travels to the workpiece coordinates X3.0, Y1.0 in line N120.
- Finally, the tool is moved to the endpoint with center coordinates, and the compensation is cancelled with G40.

Let's look at another example using a workpiece of the same size. This time we will continue the toolpath around the edge of the workpiece as in ***Figure 10.6***:

```
N100 G00 X-1.0 Y2.0
N110 G41 D05 G01 X0.0 Y1.00 F10.0
N120 G01 X3.0 Y1.0
N130 G01 X3.0 Y0.0
N140 G40 G01 X4.0 Y-1.0
```

- Line N130 directs the tool to the lower corner of the workpiece. Many controls will automatically roll the tool around the corner in an arc. This prevents excessive travel beyond the workpiece boundary.
- Line N140 terminates the automatic compensation.

diameter offset:
A value stored in the offsets register in the control that will be used for automatic cutter compensation. Each tool will have its own diameter offset value.

Adjusting the Diameter Offset

The real power of automatic tool radius compensation is that the size of the workpiece can be controlled by simply changing the tool's diameter offset value in the register. Automatic compensation also makes it easy to use cutting tools that have been reground and are no longer a nominal size. In fact, CNC machinists constantly adjust the diameter offsets to control the finished size of the workpiece because of changing conditions or for a newly installed tool.

Figure 10.7 The actual offset amount is dictated by the value that has been entered in the tool diameter registry. This value is often manipulated to control the feature size.

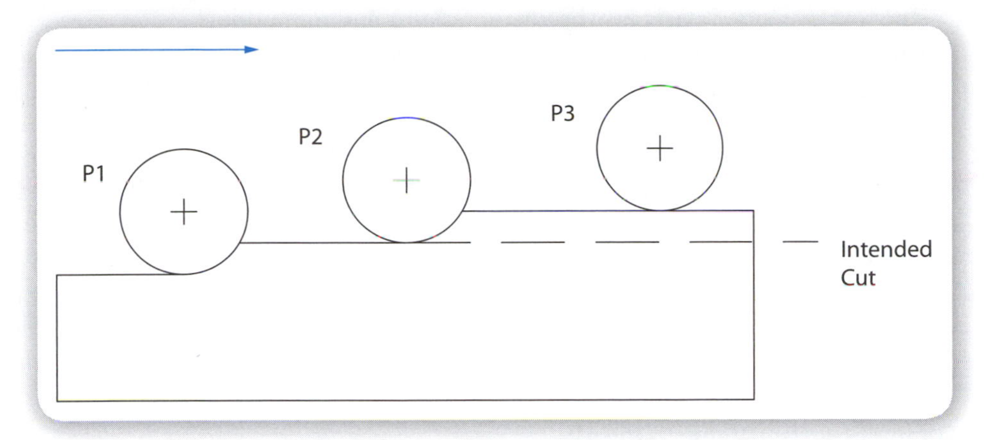

So, how is the size of the workpiece controlled with automatic tool radius compensation? Take *Figure 10.7*, for example. The dashed line indicates the nominal surface that is programmed to be cut. If the cutting tool is Ø0.500" in actual diameter, then the registry entry would be 0.500" and the nominal cut would be taken.

However, we can control the size of the feature by manipulating the offset value. If we enter a value of 0.400" into the register, then the control thinks that the cutting tool is smaller than it actually is and it moves the tool closer to the workpiece. (Remember that the tool is actually Ø0.500".) The tool at P1 is shown cutting more material than it was programmed to cut. We can also cause the control to stay farther away from the line by entering a larger value into the register. For example, if the diameter offset were set to 0.600", then the control would move the tool further away from the edge, as is shown at P3.

Figure 10.8 Use automatic tool radius compensation to maintain the dimensions.

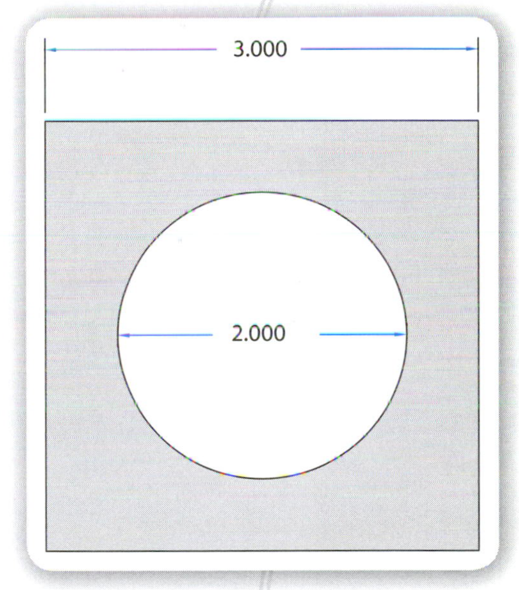

Example of Adjusting Feature Size with Tool Radius Compensation

Imagine that you are machining the workpiece shown in *Figure 10.8*. The parts have been running fine for several hours, but then you notice that the outside dimension is getting larger, and the inside pocket is getting smaller. You are using a Ø.500 end mill, but it seem to be getting dull. What can you do?

First, let's look at the actual measurements. The pocket is now 1.990", and the outside square is now 3.010. We want to get it closer to the stated dimensions. First, if the outside is too big by 0.010", then the left edge and the right edge are each oversized by 0.005". We will need the cutter to move in closer by 0.005". We accomplish this by telling the control that the cutter is now smaller:

Diameter Offset Value	Result
.500"	Outside Feature Oversize by 0.010"
.490"	The outside feature cut to correct size. The corrective action causes the control to move the cutter .005" closer to the edge. An additional amount of 0.005" is taken off the left and 0.005" is removed from the right side, causing the feature to be 0.010" smaller that the previously oversized measurement

So what about the inside pocket? It is undersized, meaning that there is too much material left. Lowering the value in the diameter register to 0.490 will cause the tool to move closer to the edge and create a larger hole. Congratulations, we are now making good parts for the customer and making money!

Later the tool becomes so dull that the surface finish is suffering, so you decide to change to a new cutter. Of course, you measure the cutter diameter with a micrometer and find that it is Ø.500. You change the value of the registry to 0.500 and expect it to cut perfectly. However, the first pocket comes out over-sized by 0.005", and the outside is under-sized by 0.005". Your experience tells you that sometimes a new end mill can cut a little deeper than expected, but this was more than usual—maybe you mismeasured. What is the solution?

The cutter is taking too much material. We have to instruct the MCU to keep the cutter 0.0025" farther away from each edge. Therefore, we will change the value in the diameter in the registry.

Diameter Offset Value	Result
.500"	Pocket in oversized by 0.005"
.505"	The pocket is cut to correct size. The corrective action causes the control to move the cutter .0025" farther away from the edge, resulting in a smaller inside dimension.

Lead-In and Lead-Out with Tool Radius Compensation

We saw in a previous example that a lead-in move must be given to initiate automatic tool radius compensation. This lead-in and lead-out is needed to give the tool space to move from the center coordinates to the tangent coordinates. For this reason, there must be ample space between the initial point and the first tangent point. The control must have a distance that is equal to at least the radius of the tool to complete this move. Additionally, the approach move must not be at an acute angle. The standard practice is to approach at an obtuse angle to provide easy entry into the material. Furthermore, the tool should not be in contact with the material during the lead-in or lead-out move.

Another limitation of automatic compensation is that it cannot be initiated during an arc move. In *Figure 10.9*, the toolpath on the left is an illegal move and will cause the control to error. The legal way to use an arc for lead-in with automatic compensation is to initiate compensation in a straight line to the start point of an

Figure 10.9 **Cutter compensation cannot be initiated with circular interpolation (left). Use a linear move to the start of the arc to initiate compensations and then program the path to the tangent points (right).**

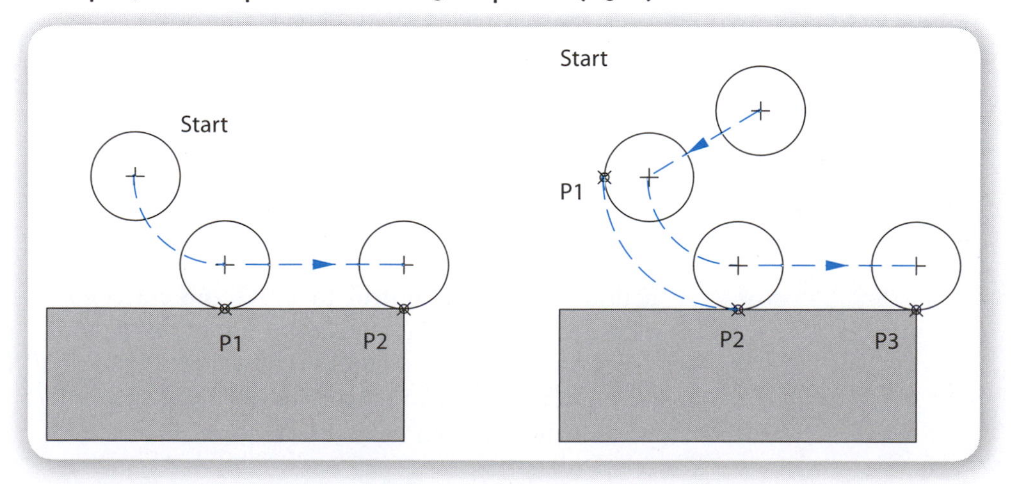

Figure 10.10 Examples of conditions that may cause the control to generate an error when using cutter compensation.

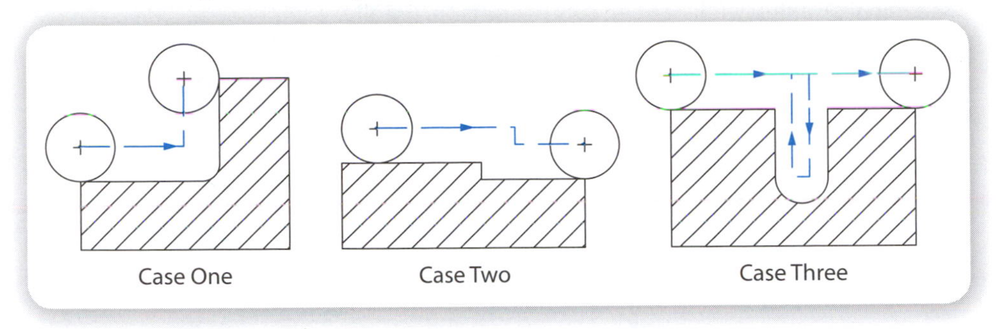

arc. The toolpath on the right illustrates this point. The tool is positioned at the start point and then moved to tangent point P1 while simultaneously initiating automatic compensation. An arc is then called to move to point P2, and the toolpath is continued as normal to the next points. See Programming Example 10-4 for an illustration of this technique.

Error Conditions and Limitations of Automatic Compensation

There are a few limitations that we might encounter when using cutter compensation. ***Figure 10.10*** illustrates the first four of these conditions.

- An error will occur if the programmed radius is smaller than the radius (diameter) in the offset register.
- Some machines will gouge the workpiece if a step smaller than the radius is programmed.
- A compensated toolpath cannot reverse itself if it is programmed into a space that is smaller than the programmed diameter.
- Only one Z-axis move can be programmed between the initial call for compensation and the first X-Y move. Additionally, controls deal with any Z-axis moves in different ways. Some may not allow any other Z-axis moves once compensation in started.
- Many controls will allow cutter compensation on the X-Y plane. Any compensation in other planes will have to be done manually.
- Entry or exit problem:
  - Lead-in or lead-out is not long enough. The control requires a travel distance of at least the radius value of the tool that is stored in the offset registry.
  - Attempting to start or stop cutter compensation with an arc.
  - Entry or exit at an acute angle. Many controls will not allow an angle of less than 90°.

Programming with the Real Diameter or Zero Diameter

One variation of our tool radius compensation discussion involves programming a toolpath as though the tool has a diameter of 0.000". You are probably wondering first why and then how this would be accomplished. Well first, we might program with a zero diameter to avoid some of the limitations that we just discussed concerning error conditions. We can essentially trick the machine into doing something that would normally cause an error. The second reason is to convert a centerline part program to support cutter compensation without having to recalculate all of the points.

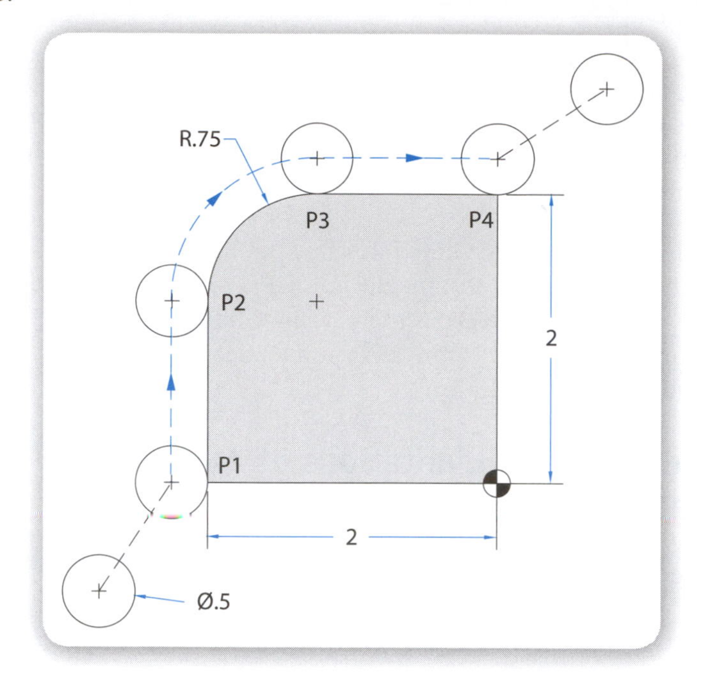

Figure 10.11 There are several different ways to program this toolpath and get the same results. First, is tool centerline programming. We could also use tool edge programming with cutter compensation. Finally, a tool centerline path can be easily converted to use cutter compensation without changing coordinates by setting the diameter offset to zero and adding the appropriate G41 and G40 codes.

Take for example the toolpath shown in ***Figure 10.11***. A typical centerline toolpath for this might look like the following:

```
G00 X-2.75 Y-0.75
G01 X-2.25 Y0.00 F20.0
G01 X-2.25 Y1.25
G02 X-1.25 Y2.25 I1.0 J0.0
G01 X0.0 Y2.25
G01 X0.75 Y2.75
```

The disadvantage is of course that we don't have any control over the size of the cut. If the tool starts to wear or the part is out of tolerance, there is nothing we can do besides edit the program or install a new tool. We could then convert it to use tool radius compensation very easily by adding a few instructions:

```
G00 X-2.75 Y-0.75
G01 G41 D05 X-2.25 Y0.00 F20.0    Turn on cutter compensation left
G01 X-2.25 Y1.25
G02 X-1.25 Y2.25 I1.0 J0.0
G01 X0.0 Y2.25
G01 G40 X0.75 Y2.75               Turn off cutter compensation
```

The offset registry would be adjusted so that the tool has a diameter value of zero. If a minor change is needed to adjust the dimension of the workpiece, then the diameter can be changed slightly. For example, the values −0.005 or 0.005 would cause the tool to move closer to or further from the edge. This is often done in a separate column called "wear offset":

Offset Registry	Diameter Offset	Wear Offset
D05	0.000	−0.005

What Drives Cutter Compensation?

We have seen that tool radius compensation can be used to carefully control the size of a feature such as a pocket or boss. As a new cutting tool wears, its diameter will become smaller the cutting edge duller. The tool will have to be moved closer to the edge and exert more pressure in order to cut the proper dimension. So why not just replace the old tool with a brand new one when it starts to get a little dull?

New cutting tools are very expensive. A small, Ø.500, carbide end mill can easily cost $50. Even at that it might not cut the proper size even when new. Cutter compensation gives the machinist the flexibility to adjust the diameter offset in order to machine a part to the proper dimension. Furthermore, it saves money. Rather than disposing of the end mills as soon as they get a little dull, they can be re-sharpened for about 1/6th of their original cost. The resharpened cutters will have an odd diameter, which the machinist can easily measure and enter into the registry. If cutter compensation was not used, then the program would need to be re-written for any new cutter diameter. Wow, is that expensive!

Let's compare this method to programming with the normal method of tool edge programming. The code and offset would look like the following:

```
G00 X-2.75 Y-0.75
G01 G41 D05 X-2.00 Y0.00 F20.0   Turn on cutter compensation left
G01 X-2.00 Y1.25
G02 X-1.25 Y2.00 I0.75 J0.0
G01 X0.0 Y2.00
G01 G40 X0.75 Y2.75              Turn off cutter compensation
```

Offset Registry	Diameter Offset	Wear Offset
D05	0.500	−0.005

Notice that the programmed coordinates are no longer on the centerline, but on at the tangent point of the work edge. The other change is that the actual diameter of the tool would be entered in the registry. Small changes could again be made with the wear offset.

10.3 REFERENCE LOCATIONS

A fair amount of skill in plane geometry and right-angle trigonometry is necessary to program toolpaths that contain any elements that are not parallel to the axes. The following sections show you how to use these skills to locate points for programming a toolpath with inclined lines and tangent arcs. If you have not yet studied these subjects, then please refer to Chapter 14 for a brief lesson.

The primary objective of NC programming is to find the locations on a Cartesian coordinate system that can be used to program a toolpath and therefore produce a finished workpiece. The first step in this process is to find a reference point from which we can measure intersections or geometry transitions that define the shape of the workpiece. The second step will be to find the relative coordinates of any intersections from this reference point. Finally, we can determine the center point of the cutting tool in relation to the intersections—these will be the coordinates that we use to write the part program.

For example, let us look at the geometry in ***Figure 10.12***. The final goal is to find the coordinates of the cutting tool (P3). However, we must first find a reference location (P1) and then the geometry intersection (P2) before we can determine the coordinates of the cutting tool.

Figure 10.12 **Several calculations may be needed to find the cutter location when a line intersects an arc.**

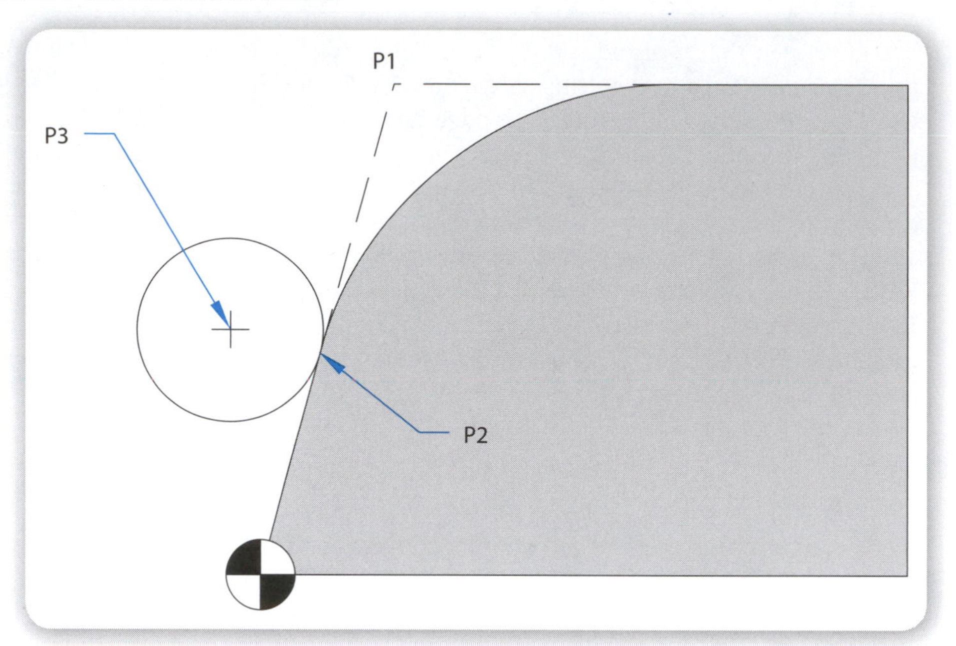

To reiterate, the steps to finding the tool center location are as follows:

- Find a reference point on the workpiece, such as an apparent intersection.
- Find the distance from the reference point to the geometry intersections (this may be zero in some cases).
- Find the tool locations from the geometry intersections when the tool is set tangent to the geometry intersection.

We will start the process by first finding a known point in the workpiece geometry. For example, we can find a point where two lines intersect or find the center of an arc. This information is either given in the blueprint or easily determined with a few calculations. This coordinate will be used later with additional calculations to plan the toolpath.

Take, for example, the workpiece shown in ***Figure 10.13***. The blueprint gives enough information to allow us to find the coordinates of the intersection at P1. We start by drawing a right triangle with sides that are parallel to the X- and Y-axes. The triangle can then be solved to find the X-component (side adjacent) of the triangle:

$$\tan 35^\circ = \frac{\text{Opposite}}{\text{Adajacent}}$$

$$0.7002 = \frac{1.00}{\text{Adjacent}}$$

$$\text{Adjacent} = 1.4281$$

Next, we add the sides of the triangle to any relevant values given on the blueprint to find the absolute coordinates of P1:

X-coordinate = 0" + 1.4281" = X1.4281
Y-coordinate = 0.50" + 1.00" = Y1.500

We use a similar method when we are confronted with a radius that is tangent to two lines as in ***Figure 10.14***. In this situation, we will eventually need to know the coordinate of the arc center. This value is seldom included in a blueprint, so

Figure 10.13 Reference points can be found by creating and then solving a triangle on the workpiece.

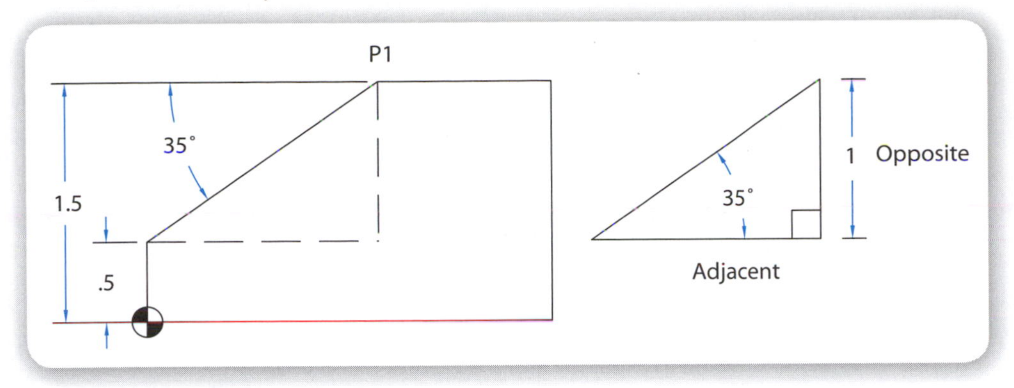

Figure 10.14 Lines can be extended to form an apparent intersection. This intersection can serve as a reference point for other calculations.

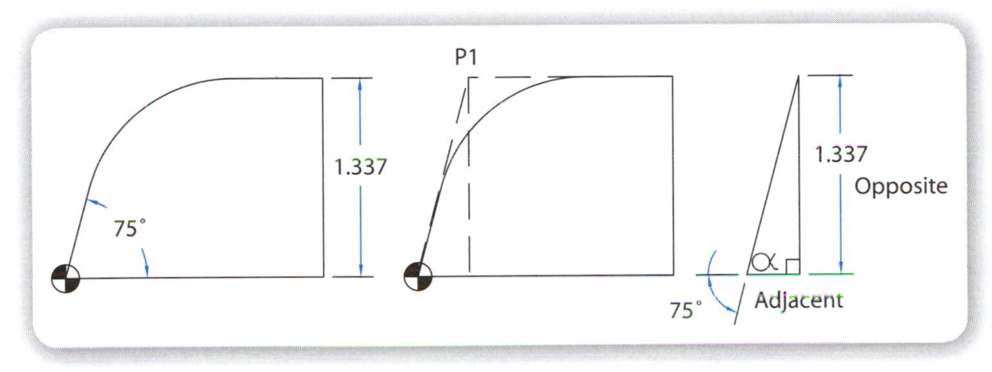

the NC programmer will have to calculate these coordinates. The process starts by extending the lines to form an intersection. Next, we draw a triangle with an endpoint at the intersection and then solve the triangle:

$$\tan 75^\circ = \frac{\text{Opposite}}{\text{Adjacent}}$$

$$3.732 = \frac{1.337}{\text{Adjacent}}$$

$$\text{Adjacent} = 0.3582$$

Remember that the reference points that we found in the previous examples are only the first step to finding the tool locations. We still do not know any of the actual tool locations.

Lines and arcs are the two basic elements that create the shape of the finished workpiece. This leaves us with three possible types of intersections to contend with: line-to-line, line-to-arc, and arc-to-arc. Each of these situations requires a slightly different approach to find the solution. We will look at solutions to these three possible intersections in the remainder of this section.

10.4 TOOL LOCATION ON ANGULAR TOOLPATHS

Toolpaths along angular lines can be a challenge to solve. However, if we use a visual approach and remember a few rules of geometry, the calculations will become second nature. You may want to do a quick review of the geometry and

Figure 10.15 A workpiece with one angular feature requires the tool be positioned tangent to the feature to be machined.

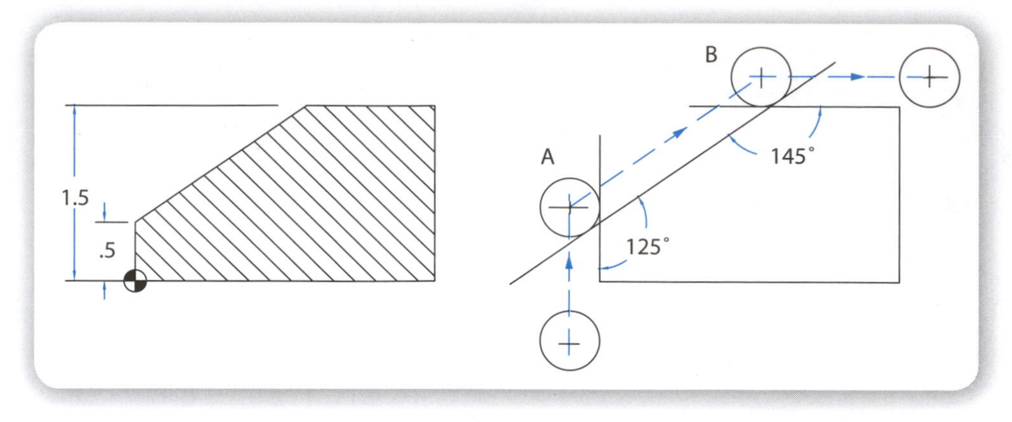

Figure 10.16 A magnified view of corners A and B form. The center of the tool can be found by setting the cutter tangent to the edges and solving the resulting triangle.

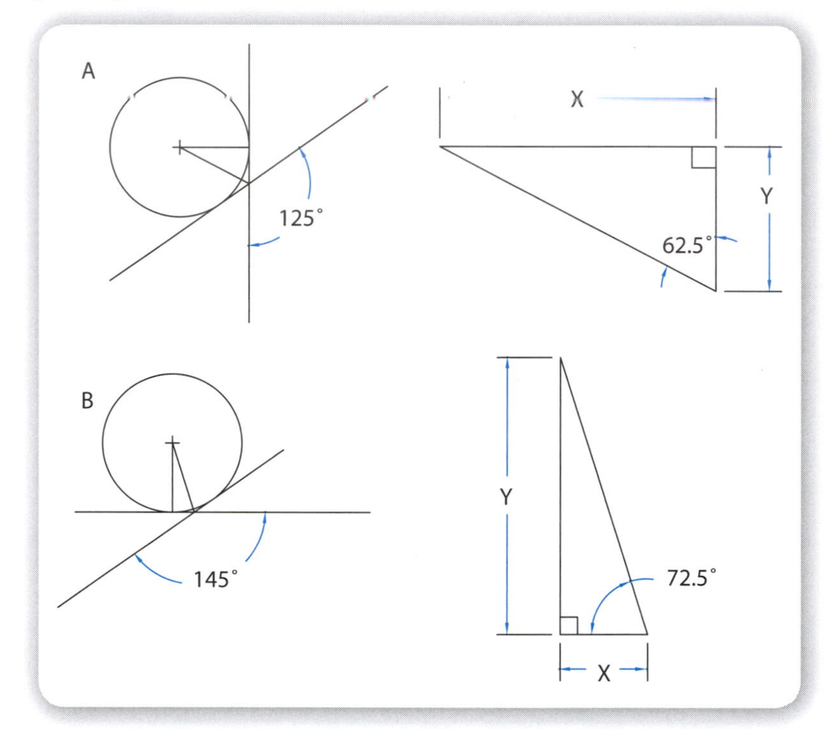

trigonometry in Chapter 14 before diving into this section. Pay particular attention to the relationship between the angles of intersecting lines and the properties of tangent lines and circles.

A workpiece with one angular feature requires the tool to be positioned tangent to the feature to be machined. The basic idea behind manual tool radius compensation is to find the point where the tool will be tangent to the lines we wish to machine. We are seldom machining a single line, so it is important to position the tool so that it is ready to take the next cut. Take the workpiece in ***Figure 10.15*** as an example. The tool will start at a position below the workpiece and then travel to the next position to start the angled line. The question becomes, "Where should the tool be positioned?"

Upon closer inspection we can see that the centerline of the tool must travel slightly past the corner to a point where it will be in position to cut the next line with the proper compensation. We can find this point by extending the lines at the intersection and placing the cutter at a point that is tangent to both lines (much like a dowel pin sitting in a V-block). From here, we can calculate the X- and Y-coordinates of the tool centerline from the corner of the workpiece. The basic process is as follows:

1. Extend the edges of the part to form an X-shaped intersection.
2. Draw a circle (to represent the cutter) that is tangent to the two extended lines.
3. Create a triangle by drawing a line from the corner of the workpiece to the center of the circle, and then draw a perpendicular line from the center to the vertical (or horizontal) line.
4. Solve the triangle, and find the coordinates of the cutter.

Let's look at the two corners, A and B, of our workpiece and find the coordinates of the tool center (relative to the corner). A magnified view of the corners and triangles is shown in ***Figure 10.16***.

We start solving the problem by remembering what we learned about the tangent relationship between arcs and lines:

- A line drawn from the arc center to the tangent point will be perpendicular to the adjacent line.
- The arc will bisect the angle.

These two nuggets of information will allow us to create and solve a triangle to find the relative distance from the **vertex** to the arc center.

vertex:
A fixed point of rotation of an angle.

Let's start with corner A. The blueprint specifies a corner angle of 125°, and we decide to use a Ø0.500" end mill. We then draw a triangle with sides parallel to the X- and Y-axes and fill in the known values. We know the radius of the end mill is 0.250", and the angle will be 62.5° (half of 125°). The X-leg of the triangle is the side opposite of angle $\alpha(62.5°)$, and the Y-leg is the side adjacent to angle α. The Y component can then be found as follows:

$$\tan 62.5^\circ = \frac{X}{Y}$$

$$Y = \frac{0.25}{1.921}$$

$$Y = 0.1301$$

Of course, these dimensions are relative only to the vertex of the intersecting line—they are not the absolute coordinates used in the NC program. The next step is to find the actual absolute coordinate that will be used in the part program.

The coordinates of corner A can be found on the blueprint (see ***Figure 10.17***) as X0.0 and Y0.500. We can find the center point of the end mill by adding or subtracting the values that we calculated:

$$X = 0.0" - 0.250" = X - 0.250$$
$$Y = 0.5" + 0.1301" = Y0.6301$$

Therefore, the absolute coordinates of the end mill at corner A will be X − 0.250 and Y0.6301.

Corner B can be solved in a similar manner, except that this time the Y component is already known. (It is the radius of the cutter.) Also, the Y component is now the side opposite angle α and the X component is the side adjacent to angle α.

$$\tan 72.5^\circ = \frac{Y}{X}$$

$$X = \frac{0.25}{3.1716}$$

$$X = 0.0788$$

Figure 10.17 A method for finding cutter coordinates.

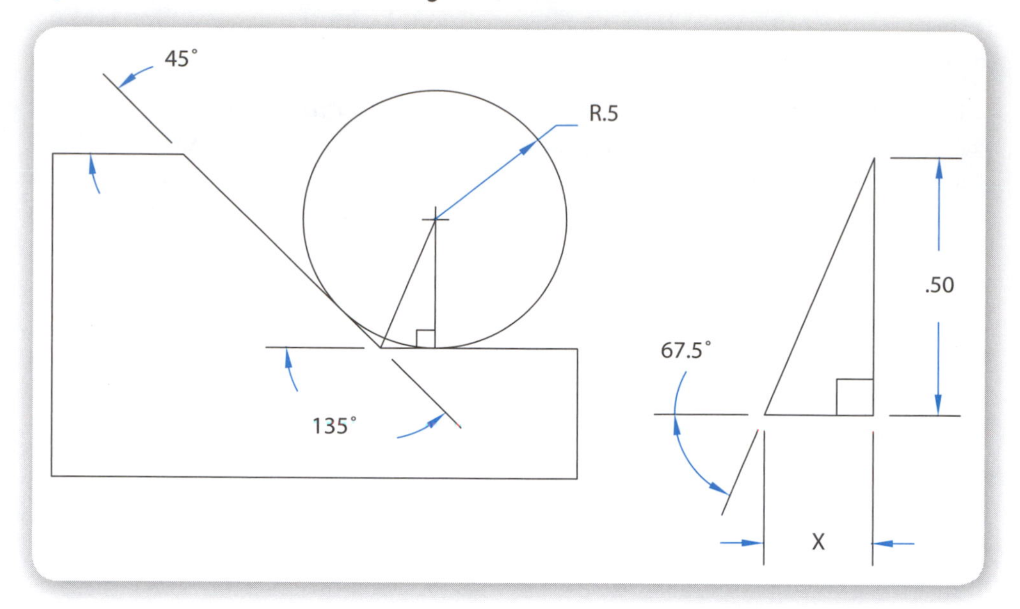

The coordinates of corner B were found in a previous example (***see Figure 10.17***) to be X1.4281 and Y1.500. We can find the center point of the end mill by adding or subtracting the values that we calculated:

Corner B cutter center coordinates:

$$X = 1.4281'' - 0.0788'' = X1.3493$$
$$Y = 1.500'' + 0.250'' = Y1.750$$

Therefore, the absolute coordinates of the end mill at corner B will be X1.3493 and Y1.750.

Let's look at one more example. In ***Figure 10.17***, an angled line intersects with a horizontal line. A Ø1.000" end mill will have to be positioned tangent to these two lines in order for the tool path to be created. A triangle can then be drawn between the vertex, tangent point, and the arc center. Again, the arc will bisect the angle to yield the one known angle of 67.5° in the triangle, and the line will form at a right angle from the tangent point to the arc center.

The X component of the triangle can then be found as follows:

$$\tan 67.5^\circ = \frac{0.500}{X}$$

$$2.4142 = \frac{0.500}{X}$$

$$X = \frac{0.500}{2.4142}$$

$$X = 0.2071$$

The first few examples all included one intersecting line that was parallel to either the X- or Y-axis. However, there are some situations in which neither of the lines will be parallel. These cases require a little more effort to solve. Take for example the workpiece found in ***Figure 10.18***. This toolpath will require us first to find the hypotenuse of the triangle and then to use the hypotenuse to find the legs of the triangle.

The first step is to determine the angle that is formed between the intersecting lines. We can show that this angle is 120° by drawing a horizontal line from the vertex and applying the principles of intersecting lines from Chapter 14 (line intersecting two parallel lines).

Figure 10.18 *Tool coordinates for unaligned, intersecting lines may require several calculations.*

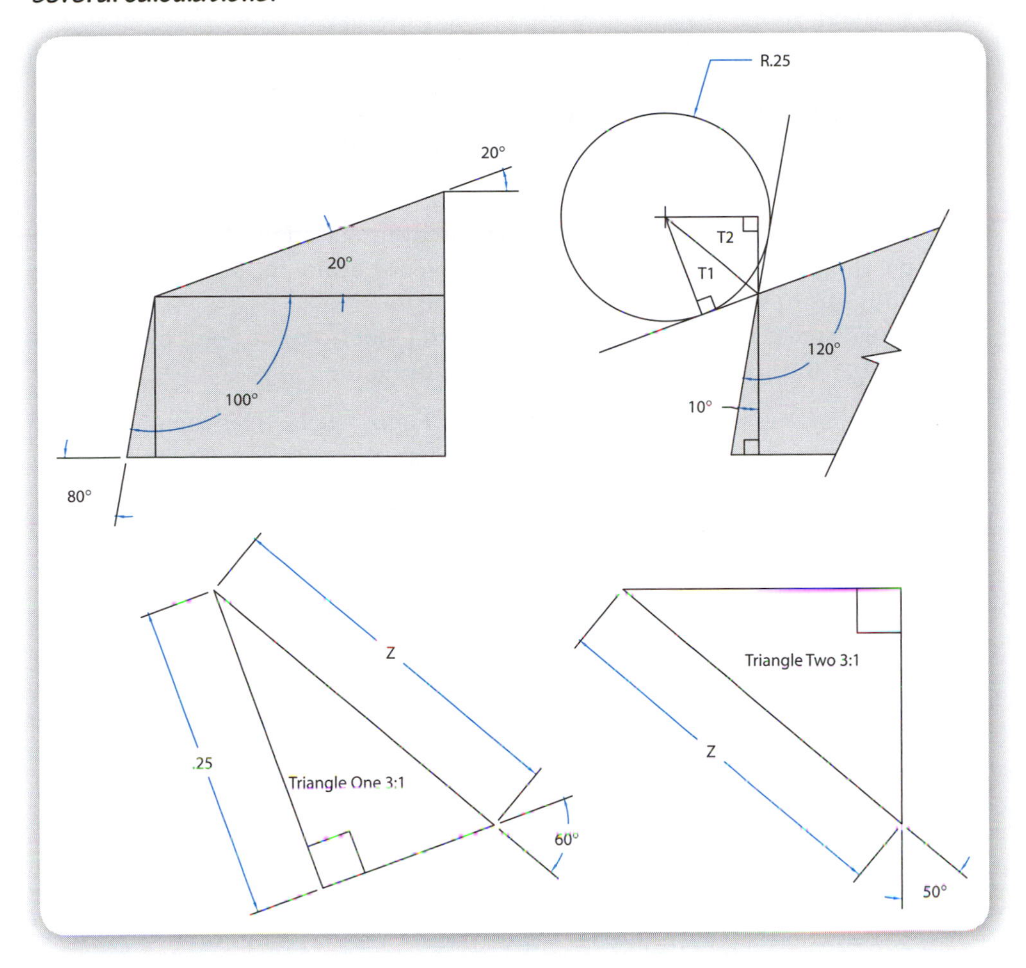

Next, we solve the triangle from the vertex, tangent point, and arc center. The value we are looking for is the length of the hypotenuse (labeled Z in the drawing):

$$\sin 60^{\circ} = \frac{0.25}{Z}$$

$$0.866 = \frac{0.25}{Z}$$

$$Z = 0.2887$$

Then we will create another triangle by extending a vertical line from the vertex. The triangles will have a parallel orientation that will allow us to find the X and Y components from the vertex. The vertical leg will form a 10° angle to the upper line; therefore, the triangle will contain a 50° angle (60° − 10° = 50°). The solution will be as follows:

X Leg:

$$\sin 50^{\circ} = \frac{X}{0.2887}$$

$$X = 0.2212$$

Y Leg:

$$\cos 50^{\circ} = \frac{Y}{0.2887}$$

$$Y = 0.1856$$

Tool Location on Intersecting Arcs and Angles

It is extremely common for a workpiece to be specified with a radius on one or more corners. It is relatively easy to find the tangent point when all of the lines are horizontal and vertical, but angled lines present a more difficult problem.

The procedure for finding the tangent point is similar to the method used previously, except that now we have to do some additional work. The general process is as follows:

1. Extend any straight lines to form an apparent intersection.
2. Find the coordinates of the point of tangency from the intersection. This requires us to solve two additional triangles.
3. Calculate the center coordinates of the cutting tool from the point of tangency.
4. Calculate the absolute coordinates of the cutting tool.

Take, for example, the workpiece shown in ***Figure 10.19***. This toolpath has an angled line segment that intersects with a radius. We start by extending the line segments and then finding the coordinates from the vertex to the point of tangency. Next, we draw Triangle One between the vertex (point A), arc center, and the point of tangency (point B). Then we solve the leg that runs between the vertex and point of tangency (labeled Z in the drawing):

$$\tan 52.5^\circ = \frac{1.00}{Z}$$

$$1.3032 = \frac{1.00}{Z}$$

$$Z = 0.7673$$

This value can then be used to find the X and Y components from the vertex. To do this we will need to solve Triangle Two, which we found to have a hypotenuse of 0.7673" and legs that are parallel to the axes.

Figure 10.19 The programmer may have to solve triangles to find the tool coordinates.

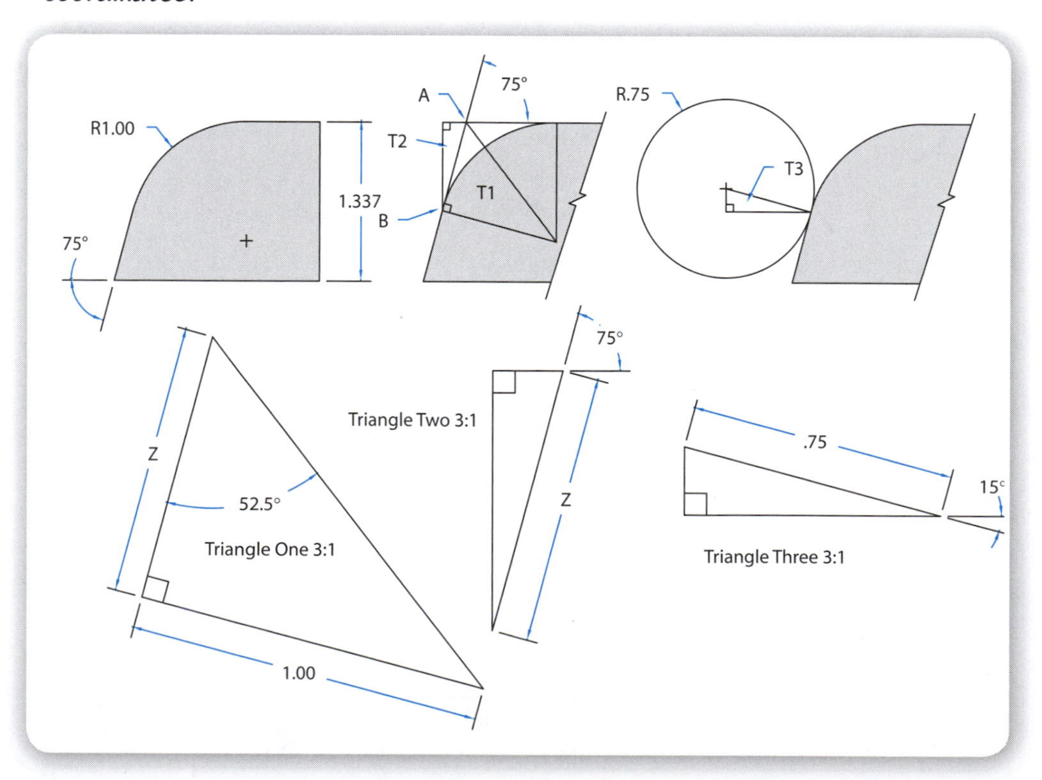

The distances from the vertex to the tangent point (Triangle Two) are as follows:

X-Leg:

$$\cos 75^\circ = \frac{X}{0.7673}$$
$$X = 0.1986$$

Y-Leg:

$$\sin 75^\circ = \frac{Y}{0.7673}$$
$$Y = 0.7412$$

Now we have the location of the point of tangency from point A. Next we will need to find the center of the cutting tool by solving Triangle Three. This triangle has legs parallel to the axes and one angle of 15° that can be found via its relationship to the angled line. The hypotenuse is equal to the radius of the tool—0.75".

The distances from the tangent point to the tool center (Triangle Three) are as follows:

X-Leg:

$$\cos 15^\circ = \frac{X}{0.75}$$
$$X = 0.7244$$

Y-Leg:

$$\sin 15^\circ = \frac{Y}{0.75}$$
$$Y = 0.1941$$

The only step left is to find the absolute coordinate of the cutting tool. For this task will need to know the coordinates of a reference point. Fortunately, we have already made this calculation by finding the coordinates of point A in a previous example (*see Figure 10.20*). Now we can find the tool's absolute coordinates by adding or subtracting the appropriate values that we have calculated.

Absolute coordinates of point A:

X0.3581 Y1.337

Absolute coordinates of the tool:

X = 0.3581" − 0.7244" − 0.1986" = X−0.5649
Y = 1.337" − 0.7412" + 0.1941" = Y0.7899

In order to program an arc, we also need to calculate the I and J values from the start to the center of the arc. This will be accomplished by finding the incremental distances from the tool center to the arc center. The next section will include a discussion on the basic technique that you can apply to the previous example.

10.5 TOOL LOCATION ON RADIAL TOOLPATHS

Tool locations on intersecting radial toolpaths are another common challenge to the NC programmer. The process for finding the tool location is similar to some of the other situations that we have looked at. However, we now have to know the coordinates of at least one of the arcs relative to a reference point or work zero. We can then use the center coordinates of this arc to find any tool locations.

Figure 10.20 The geometric relationships between two tangent arcs. A line drawn between the centers of the arcs will be perpendicular to their shared line of tangency.

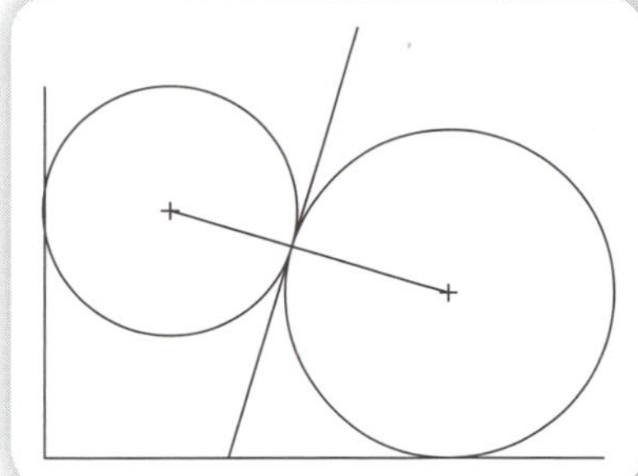

In these situations, the two arcs will form a point of tangency. We can then exploit this tangent relationship to find the coordinates of both the tangent points and the tool. Take, for example, the tangent circles in *Figure 10.20*. Several glaring properties are apparent. First, the length of a line between the centers of the arcs is simply the sum of the radii. Second, the line segment between the centers will be perpendicular to a tangent line—this can be used to find the angle of the line segment.

Let's look at an example of intersecting arcs and find the tool coordinates to produce the workpiece. The toolpath in *Figure 10.21* includes two tangent arcs. To solve problems of this nature, we must first find a reference location such as the center coordinates of one arc. Enough information must be provided on the blueprint, or we will not be able to solve the problem. In this case, the origin is in the upper left corner and the center location of the smaller arc has been defined as X0.500, Y-0.250.

We can see that the larger arc is tangent to the top of the workpiece, thereby giving us the Y-coordinate of the arc center as Y-1.00. We can then draw Triangle One with a hypotenuse between the two centers and legs that are parallel to the axes. We will be able to establish angle α and the length of the X-leg by solving this triangle.

We can find the hypotenuse of the triangle by adding the radii, and calculate the length of the Y-leg by subtracting the Y-coordinates of the two arcs (1.00" − 0.25" = 0.75"). We can then find angle α and the X-leg of the triangle with trigonometry. (The X-leg will be needed later to program the arc.)

Angle α:

$$\sin\alpha = \frac{0.75}{1.50}$$

$$\sin\alpha = 0.500$$

$$\text{a}\sin\alpha = 30^\circ$$

Figure 10.21 A technique for solving intersecting arcs.

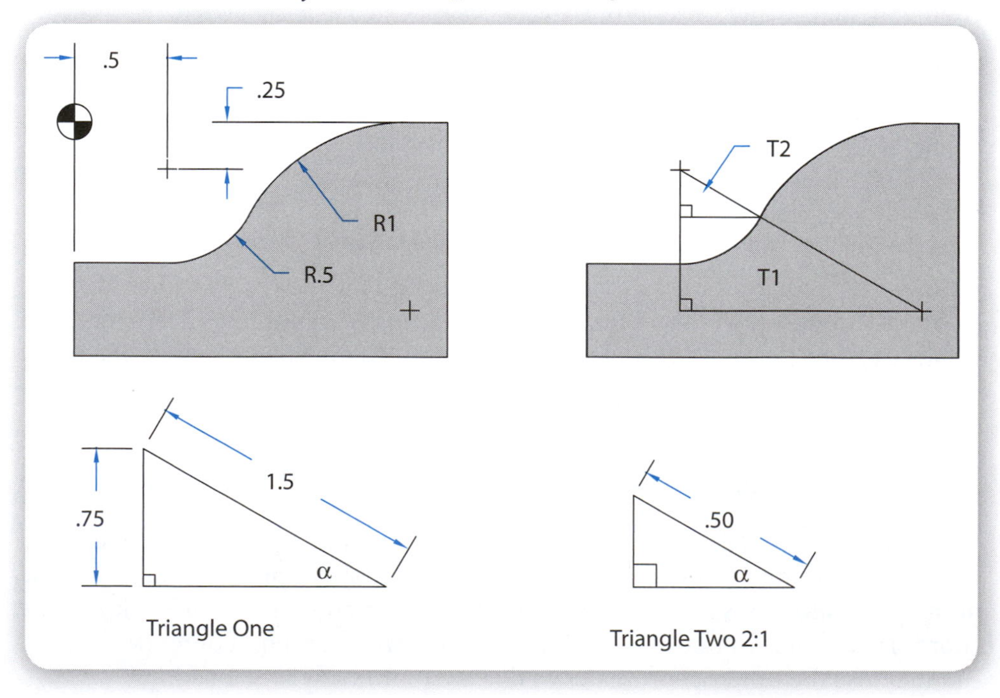

We can now use this angle to find the tangent point by finding the X- and Y-legs of the similar triangle (Triangle Two) and finding the absolute coordinates of the tangent point:

X-leg of Triangle One:

$$\cos 30^\circ = \frac{X}{1.50}$$
$$X = 1.2990$$

X-leg of Triangle Two:

$$\cos 30^\circ = \frac{X}{0.5}$$
$$X = 0.4330$$

Y-leg of Triangle Two:

$$\sin 30^\circ = \frac{Y}{0.5}$$
$$Y = 0.250$$

Coordinates of the tangent point:

X = 0.500" + 0.4330" = X.9330
Y = −0.250" − 0.250" = Y−.500

So far we have made sufficient calculations to find the absolute coordinates of both arcs and the tangent point. This information could then be used to write the NC code to create the toolpath. The tangent point is most useful when using automatic tool radius compensation, as we will see in the next section. If we are to write the code without automatic compensation, then we must find the tool center when the tool is at the tangent point.

Figure 10.22 shows the tool path that would be created if we were to use a Ø0.500" end mill to machine this tool path. This will create a similar triangle with a hypotenuse of 1.25" and an incremental distance to the center of the arc and absolute coordinates as follows:

X-leg:

$$\cos 30^\circ = \frac{X}{1.25}$$
$$X = 1.0825$$

Y-leg:

$$\sin 30^\circ = \frac{Y}{1.25}$$
$$Y = 0.625$$

Absolute coordinates of Ø0.500" tool:

X = 0.500" + 1.2990" – 1.0825" = X.7165
Y = −1.00" + 0.625" = Y−.375

These will then become the absolute coordinates of the start of the arc, and the legs of the triangle would become the I and J values. This toolpath is given in the context of a complete program in Programming Example 10-1 (see page 266).

Figure 10.22 *Tool center location for intersecting arcs.*

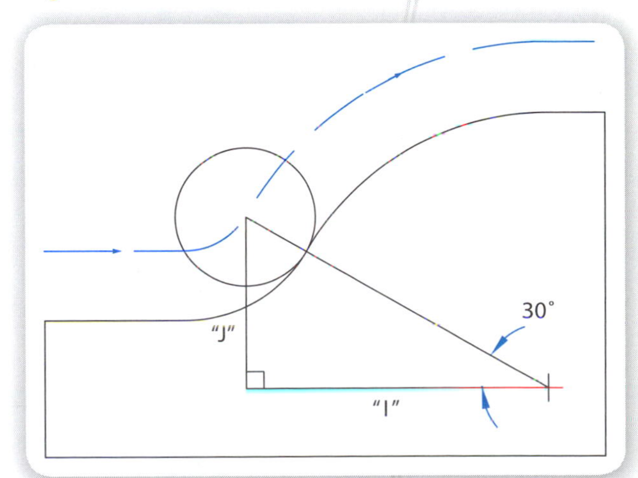

10.6 PROGRAMMING EXAMPLES

Programming Examples 10-1 and 10-2 are complete programs that create the toolpath in *Figure 10.21*. Example 10-1 Toolpath Programmed on Centerline for Profile in Figure is programmed to the tool center, and Example 10-2 Toolpath Programmed with Automatic Tool Radius Compensation for *Figure 10.23* uses automatic tool radius compensation. We have already completed the lengthy calculations for this toolpath; they are summarized in *Figure 10.23*. Example 10.3 Toolpath Programmed with Cutter Compensation for Profile in *Figure 10.24* uses cutter compensation to produce the profile in *Figure 10.24*. The Ø0.500" cutting tool will be positioned to create an angled lead-in move at 45°. Example 10-4 Toolpath Programmed with Finish Pass seen in Figure creates the finish pass for the pocket in *Figure 10.25*. The tool is positioned with center coordinates to the center of the pocket, and then the tool is programmed with cutter compensation to the tangent point of the first arc.

Figure 10.23 Summary of calculations.

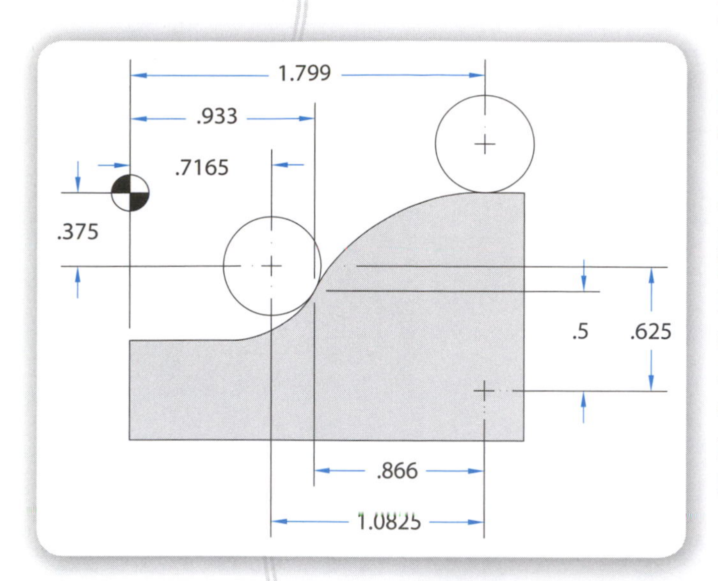

Example 10-1 Toolpath Programmed on Centerline for Profile in Figure 10.22

Program Code	Explanation
%	
O1001 (CHAPTER 10 EXAMPLE 1)	
N10 G20 G40 G49 G54 G80 G90 G98	
N20 M06 T05 (.500 END MILL)	Tool change sequence
N30 G43 H05	
N40 M03 S1200	
N50 G00 X-.5 Y-.5	Positioning move
N60 G00 Z.2	
N70 G01 Z-.25 F10.0	
N80 G01 X.5 Y-.5	
N90 G03 X.7165 Y-.375 I0.0 J.25	First arc
N100 G02 X1.799 Y.25 I1.0825 J-.625	Second arc
N110 G01 X2.5 Y.25	
N130 G01 Z.2	
N130 G91 G28 Z2.0	
N140 M05	
N150 M30	
%	

Example 10-2 Toolpath Programmed with Automatic Tool Radius Compensation for Figure 10.23

Program Code	Explanation
%	
O1002 (CHAPTER 10 EXAMPLE 2)	
N10 G20 G40 G49 G54 G80 G90 G98	
N20 M06 T05 (.500 END MILL)	Tool change sequence
N30 G43 H05	
N40 M03 S1200	
N50 G00 X-.5 Y-.25	
N60 G00 Z.2	
N70 G01 Z-.25 F10.0	
N80 G41 D05 G01 X-0.0 Y-.75	Lead in and invoke compensation.
N90 G01 X.5 Y-.75	
N100 G03 X.933 Y-.5 I0.0 J.5	First arc
N120 G02 X1.799 Y0.0 I.866 J-.500	Second arc
N120 G01 X2.0 Y0.0	
N130 G40 G01 X2.5 Y.5	Lead out and cancel compensation.
N140 G01 Z.2	
N150 G91 G28 Z2.0	
N160 M05	
N170 M30	
%	

Figure 10.24 Workpiece produced by programming Example 3.

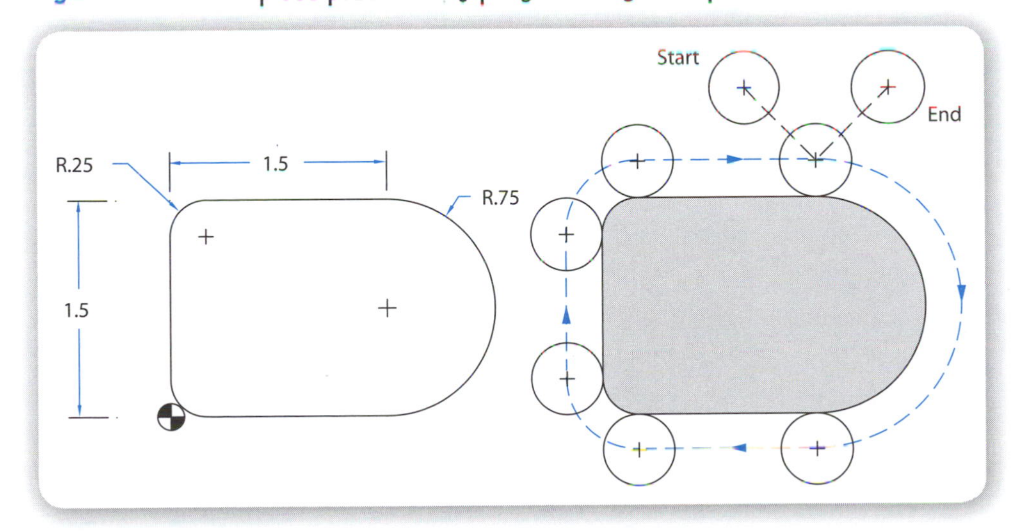

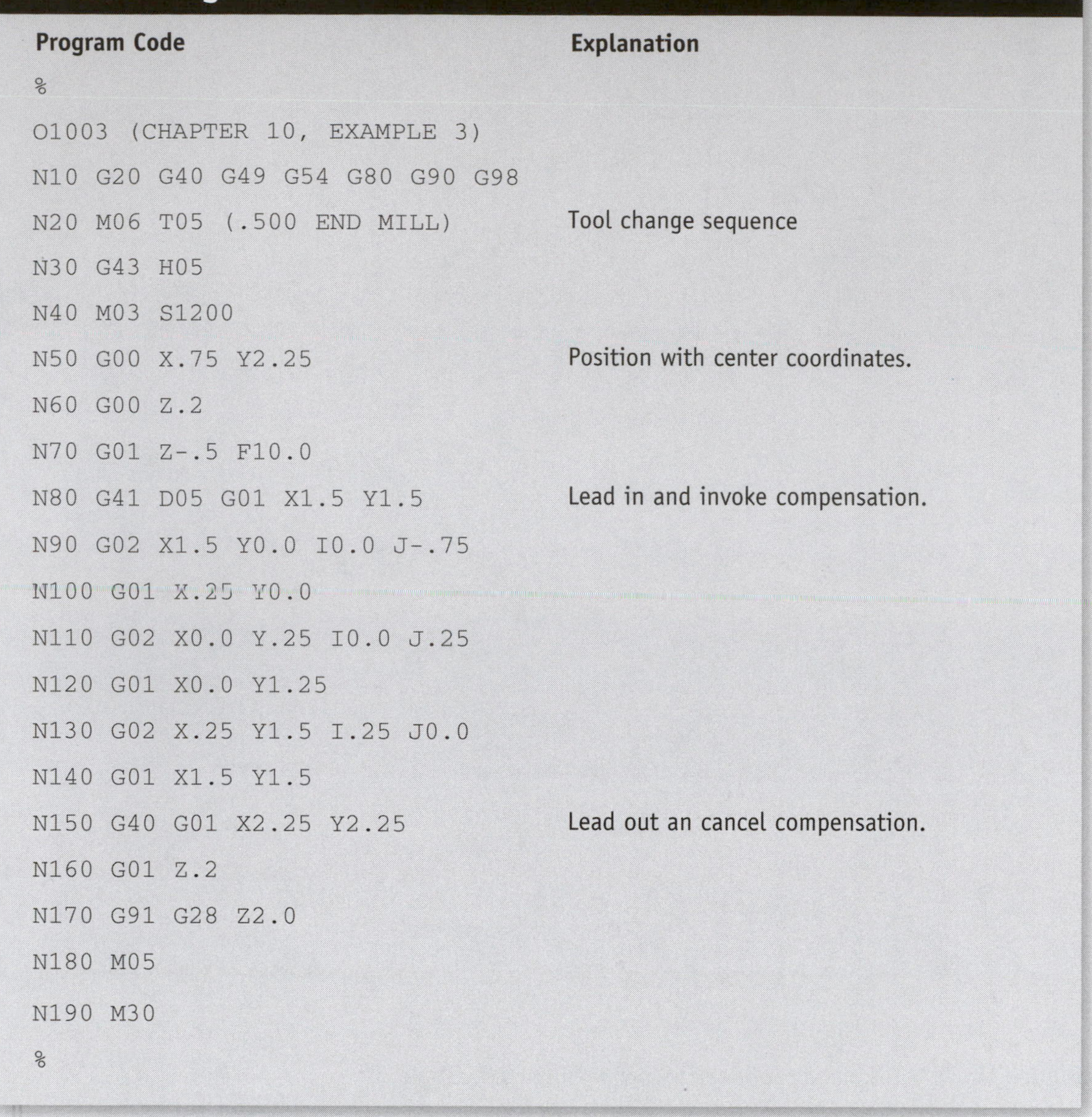

Example 10-3 Toolpath Programmed with Cutter Compensation for Profile in Figure 10.24

Program Code	Explanation
%	
O1003 (CHAPTER 10, EXAMPLE 3)	
N10 G20 G40 G49 G54 G80 G90 G98	
N20 M06 T05 (.500 END MILL)	Tool change sequence
N30 G43 H05	
N40 M03 S1200	
N50 G00 X.75 Y2.25	Position with center coordinates.
N60 G00 Z.2	
N70 G01 Z-.5 F10.0	
N80 G41 D05 G01 X1.5 Y1.5	Lead in and invoke compensation.
N90 G02 X1.5 Y0.0 I0.0 J-.75	
N100 G01 X.25 Y0.0	
N110 G02 X0.0 Y.25 I0.0 J.25	
N120 G01 X0.0 Y1.25	
N130 G02 X.25 Y1.5 I.25 J0.0	
N140 G01 X1.5 Y1.5	
N150 G40 G01 X2.25 Y2.25	Lead out an cancel compensation.
N160 G01 Z.2	
N170 G91 G28 Z2.0	
N180 M05	
N190 M30	
%	

Figure 10.25 Workpiece produced by programming Example 4.

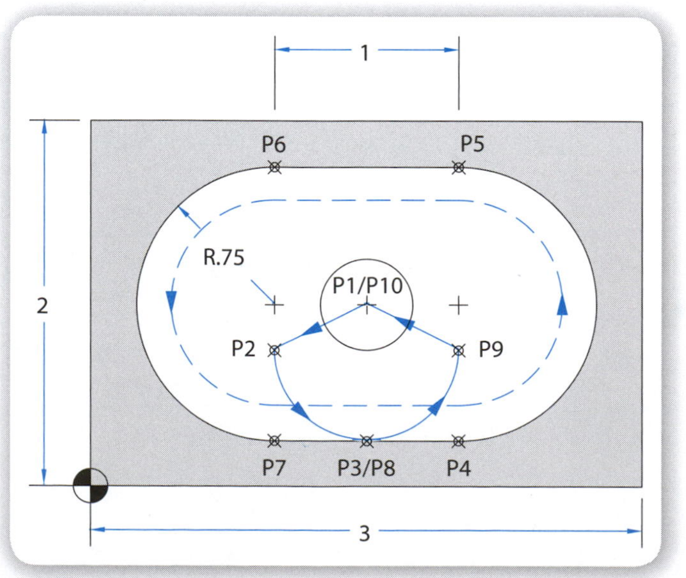

Example 10.4 Toolpath Programmed with Finish Pass Seen in Figure 10.25

Program Code	Explanation
%	
O1004 (CHAPTER 10, EXAMPLE 4)	
N10 G20 G40 G49 G54 G80 G90 G98	
N20 M06 T05 (.500 END MILL)	Tool change sequence
N30 G43 H05	
N40 M03 S1200	
N50 G00 X1.5 Y1.0	Position to center of pocket.
N60 G00 Z.2	
N70 G01 Z-.5 F10.0	
N80 G41 D05 X1.0 Y.75	Move to tangent point of arc and invoke compensation.
N90 G03 X1.5 Y.25 I.5 J0.0	Arc lead in under compensation.
N100 G01 X2.0 Y.25	
N110 G03 X2.0 Y1.75 I0.0 J.75	
N120 G01 X1.0 Y1.75	
N130 G03 X1.0 Y.25 I0.0 J-.75	
N140 G01 X1.5 Y.25	
N150 G03 X2.0 Y.75 I0.0 J.5	Arc lead out
N160 G40 G01 X1.5 Y1.0	Move back to center and cancel compensation.
N170 G01 Z.2	
N180 G91 G28 Y0.0 Z2.0	
N190 M05	
N200 M30	
%	

Your Turn

Calculate the cutter compensation on your initials problem from previous chapters. Compare results with NC program verification cycle.

CHAPTER SUMMARY

- Manual and automatic tool radius compensation (cutter compensation) are used to offset the cutting tool in order to machine the correct features of the workpiece.
- Both compensation methods require that we first determine the coordinates of any intersecting or changing geometry along the toolpath. Manual compensation requires an additional calculation to find the center coordinates of the tool when it is tangent to the profile.
- Automatic compensation allows the operator to control the size of the features by adjusting the diameter offset value for the tool in the register. Each tool will have its own offset number, which is specified with the D-word. A decrease in the offset value will cause the tool to move closer to the programmed path, and an increase in the offset value will cause the tool to move farther away from the programmed path.
- Automatic compensation is specified as either left (G41) or right (G42). The offset side is determined by the direction of travel: left for climb cutting and right for conventional cutting.
- Automatic compensation is invoked with a linear lead-in move that gives the tool enough room to switch from a centerline coordinate to tangent coordinates. The length of this lead-in move must be at least equal to the radius and at an angle of 90° or greater. Compensation cannot be initiated with an arc.
- Compensation is cancelled with the G40 code. This code is usually included with a linear move to a centerline coordinate.

BRING IT HOME

1. What does it mean to manually compensate for the radius of the cutting tool?
2. Why does the cutting tool have to be positioned at the tangent point of every intersection?
3. What is the main disadvantage of using manual compensation? Can you think of a situation in which it might be used?
4. Is automatic compensation usually used for roughing or for finishing toolpaths?
5. What is one disadvantage of automatic tool radius compensation?
6. How is the diameter offset value related to cutter compensation? Where are the values stored?

EXTRA MILE

1. Evaluate the following two situations and determine how you would correct the problem by adjusting the diameter compensation value.
   - The internal pocket we have just milled is measured and found to be 0.010 too small.
   - The external periphery of a feature we have just milled is found to be 0.005 too small.
2. Write two blocks of code to initiate cutter compensation to the left side and move a tool from X0.0, Y0.0 to the point X5.0, Y0.0. Assume the tool is positioned in block N100, and continue the program from this point:

```
N100 G00 X-1.0 Y0.0
```

3. Find the tool center coordinates that will be needed to manually compensate for the profile in *Figure 10.26*. Assume that the tool is Ø0.500" and that the work zero is at the lower left corner.
4. Write a complete program using automatic compensation to create the profile in *Figure 10.27*. Cut the profile in two equal steps to a depth of Z-0.500, and use a Ø0.500" end mill to complete the task.
5. Find the absolute coordinates of every intersection and center point of each radius for the profile shown in *Figure 10.28*. Assume the lower left corner is the work zero. Redraw the profile on a separate sheet of graph paper and show all of your work.
6. Now that you have the necessary coordinates for the profile, create a complete program to mill the profile in *Figure 10.28* with automatic compensation. Mill the profile with climb cutting to a depth of Z-0.5. The profile can be completed in one pass with a Ø0.750" end mill.

Figure 10.26 Find the tool coordinates for manual compensation.

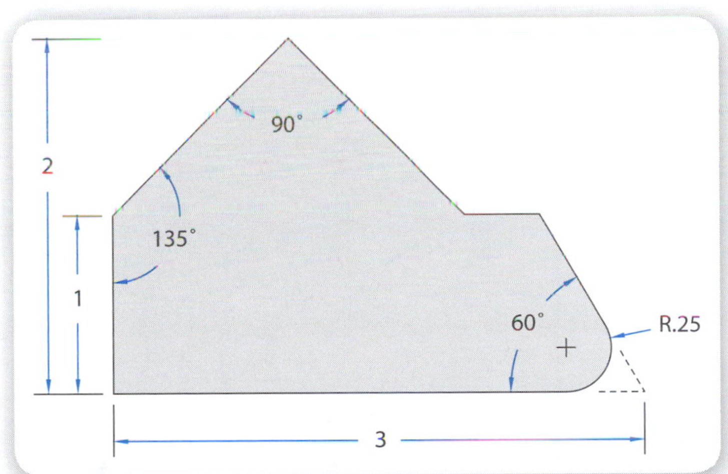

Figure 10.27 Use cutter compensation to create this toolpath.

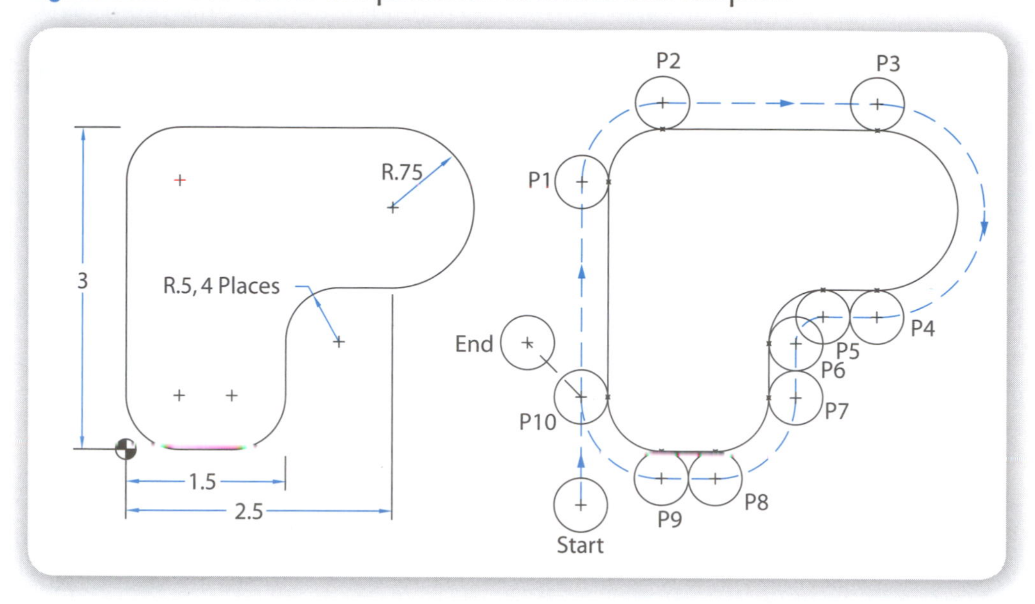

Figure 10.28 Find the coordinates of each intersection and point of tangency. Use these coordinate and cutter compensation to create the toolpath.

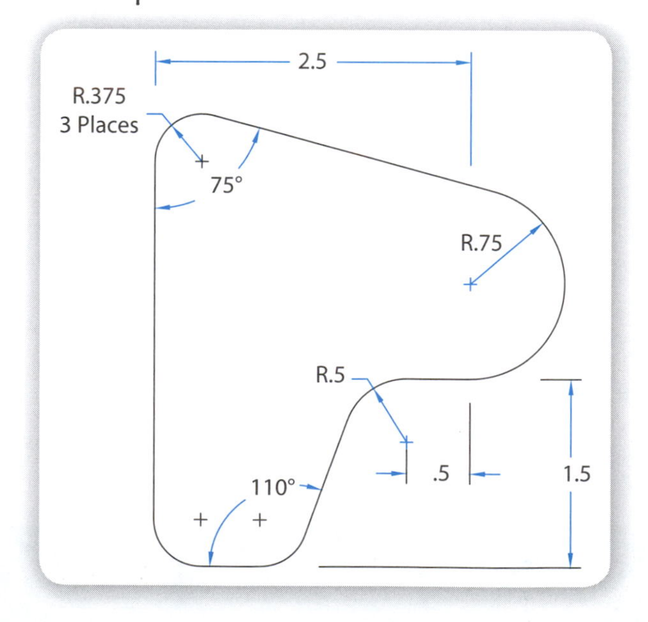

Kemp
ALLEN
Harrod
Maysville
Alger
McGuffey
HARDIN
Grant
Westminster
West Newton
Forsker
Kenton
Hume
Cridersville
Uniopolis
Waynesfield
AUGLAIZE
Neil Armstrong Air & Space Mus.
St. Johns
Hepburn
Meeker
Merseilles
Little Sandusky
Wyandot
Morral
Kirkpatrick
Caledonia
Marta
MORROW
Espyville
Big Island
Bellaire
Harding Home and Mus.
New Bloomington
Denmark
New Winchester
Galion
Iberia
Mount Gilead
Claridon

CHAPTER 11

Advanced Programming Concepts

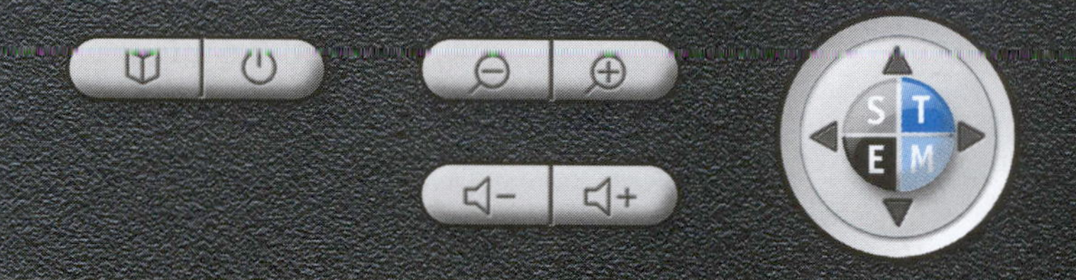

Menu | START LOCATION | DISTANCE | END LOCATION

Before You Begin

Think about these questions as you study the concepts in this chapter:

1. How do you move the work offset during program execution?
2. How are subprograms and subroutines used to organize program functions and automate repetitive tasks?
3. Which prep codes are used to load tool and fixture offsets into the MCU?
4. How are exact stop and dwell used to control tool movements?
5. How does macro programming add flexibility to standard G & M code part programs?

Key Terms

Dwell
Exact Stop
Macro
Rotary Table
Subprogram
Subroutine
Variable

11.1 SPECIFYING A NEW WORK ZERO WITH G92

In an earlier chapter, we discussed the idea of the work zero and work offsets. We saw that the work zero is the origin for all of the programmed coordinates and that the work offsets were the distance from the machine zero to the work zero. These values are most often established by the setup person.

In practice, it is often necessary to move or adjust the value of the work offsets in order to establish the work zero in a new location. This is done for a number of reasons:

- To compensate for accumulated errors in the system
- Multiple workplanes (G54, G55, . . .) are not available
- To automate programming tasks such as creating multiple patterns

There are three commonly available G-codes for resetting: G92, G50 on many lathes, and G10 on some newer machining centers. All of these codes work in a similar manner, but we will only look at G92, which is the most common code for this purpose. G92 was in use long before many controls had the ability to use multiple work offsets such as G54, G55, etc. Therefore, it is supported by most controls. You should consult the operator's manual for specific information on the G-codes mentioned in this section.

Using G92

G92 is used to move a currently established work zero usually by changing the X- and Y-offsets. You may remember that the X- and Y-offsets for a particular coordinate system are stored in the control in the offsets registry. G92 will permanently change the values in the offset registry to establish a new work zero. The syntax for G92 is as follows:

```
G92 Xn.n Yn.n Zn.n
```

When the G92 is called, the control will reset the offsets to reflect the values specified with the X- and Y-words. The most important item to keep straight with G92 is that *the X- and Y-words indicate the current tool position in the new coordinate system.* This means that we will have to position the tool to a known position before calling G92.

Take, for example, the setup in ***Figure 11.1***. In this setup, workpiece A is the original work zero that was established by the setup person, and workpiece B represents the location where we wish to establish a new work zero. We will begin by moving the tool to a known location such as the origin and then we will call G92 to reset the work zero. The code for this might look like the code in Listing 11-1:

Figure 11.1 **G92 being used to move the work zero 6" to the right in order to machine the second workplace.**

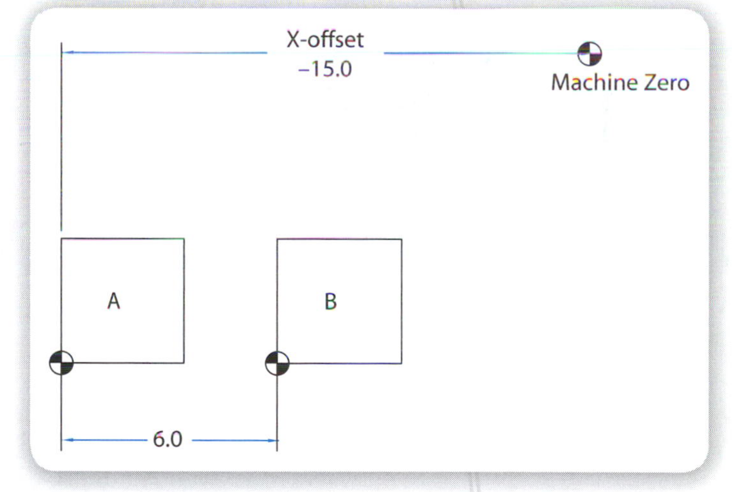

Listing 11-1	Program Code	Explanation
	`N200 G00 X0.0 Y0.0`	Position to a known location in the original coordinate system.
	`N210 G92 X-6.0 Y0.0`	Set the new offsets.

At this point, the control adds −6.0 to the X-offset and creates the new work offset. If we were to watch the offsets registry in the control, then we would see the X-offset change from −15.0 to −9.0. This would indicate that the work zero has shifted to the right by 6". You might notice that we entered a value of X−6.0 and not X6.0. This was done because X-6.0, Y0.0 represents the current location in the new coordinate system. The work zero has been moved, and yet the tool has stayed in the same place. Thus, the coordinates represent the *current* position in the *new* coordinate system.

There are other ways that we could have programmed this move to obtain the exact same results. In the previous example, we moved the tool to the original X0.0, Y0.0 and then called G92 X-6.0, Y0.0. However, we could have done this differently by first moving the tool to where we wanted to establish the new work zero. For example, Listing 11-2 produces the same results:

Listing 11-2	Program Code	Explanation
	`N200 G00 X6.0 Y0.0`	Position the tool to the location of the new coordinate system relative to the original.
	`N210 G92 X0.0 Y0.0`	Set the new offsets.

Let's look at one more example. A similar setup is shown in *Figure 11.2*, but this time we will write the code to move both the X- and Y-offsets to establish a new work zero in Listing 11-3. Again, we will position the tool to the physical location where we want to locate the new work zero and then we will use G92 to reset the zero.

Figure 11.2 A work shift can be performed in both the X- and Y-axes.

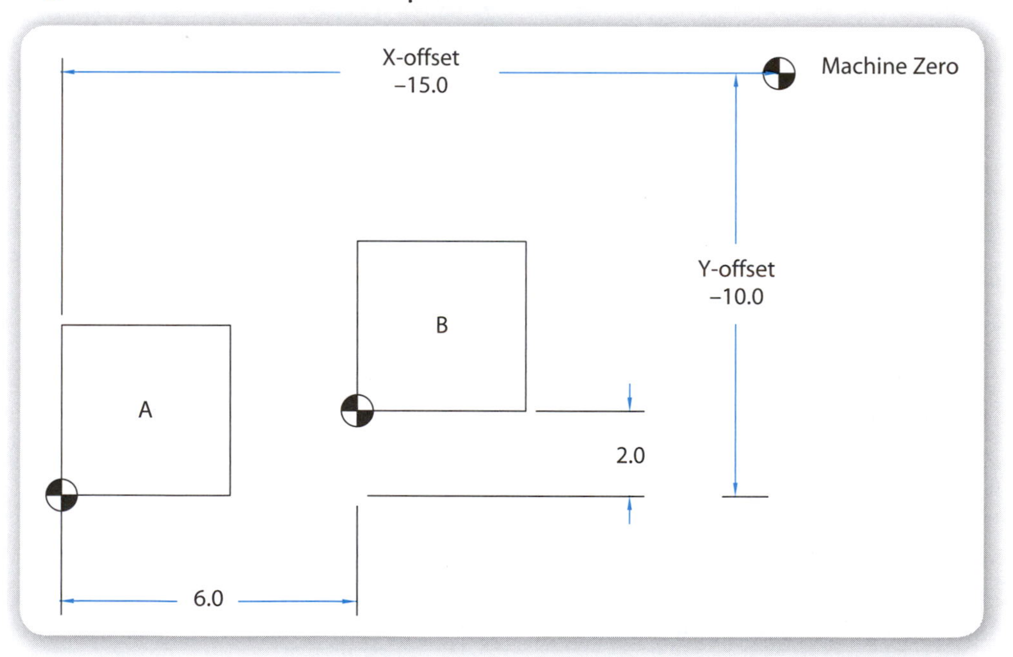

Listing 11-3	Program Code	Explanation
	`N200 G00 X6.0 Y2.0`	Position the tool at the location of the new coordinate system relative to the original.
	`N210 G92 X0.0 Y0.0`	Set the new offsets.

The X- and Y-offsets in the register will now be changed to reflect the programmed coordinates. Table 11.1 Offset Values Changed by the G92 Code compares the original values to the new values after G92 has been executed.

Table 11.1 *Offset Values Changed by the G92 Code*

Offset Values	X-offset	Y-offset
Original Value	−15.000	−10.000
New Value	−9.000	−8.000

It should be noted that these offset values are permanently changed and must be reset before running the program from the start again. This can be accomplished by placing a block of code at the beginning of the program that takes the machine to its home position and then uses a G92 to set the proper offsets. For example, the code in Listing 11-4 might be placed at the beginning of a program to reset the work zero every time the program starts again.

Listing 11-4	Program Code	Explanation
	`%`	
	`O1000 (Establish Work Shift)`	
	`N10 G20 G40 G49 G54 G80 G90 G98`	
	`G91 G28 X0.0 Y0.0`	Move to machine home position.
	`G92 X15.00 Y10.0`	Reset work zero.

The previous method is often used by programmers on every program regardless of whether or not any other G92 codes are used. It helps ensure that if any unlikely positioning errors are made, they will be negated every time the program starts again. However, this can cause a lot of extra travel and may be unnecessary on machines with absolute positioning systems. We should also note that in our example the X- and Y-offsets were nice round numbers (15.0 and 10.0). In reality, these numbers are established by edge finding the workpiece, and they are likely to be messy numbers out to four decimal places. It is then up to the setup person to edit the program with the actual offset values.

Another practical reason for resetting the work zero at the beginning is that if we have to stop the program for any reason during execution, then we will not know where the proper work zero is located. The machine will be *lost*, and we may have to manually reset the work zero with an edge finder or indicator—very time consuming.

Example 11-1 G92 Work Shift

Now we will look at a complete programming example using G92. The program in Listing 11-5 will be used to produce the workpiece in *Figure 11.3*. This example will drill a bolt-hole circle at two different locations. Rather than recalculating the absolute values of each hole location, we will simply treat the center of that pattern as our zero. However, this will require us to reset the work zero at the proper location with G92.

Figure 11.3 *Programming example using G92 to move the work zero.*

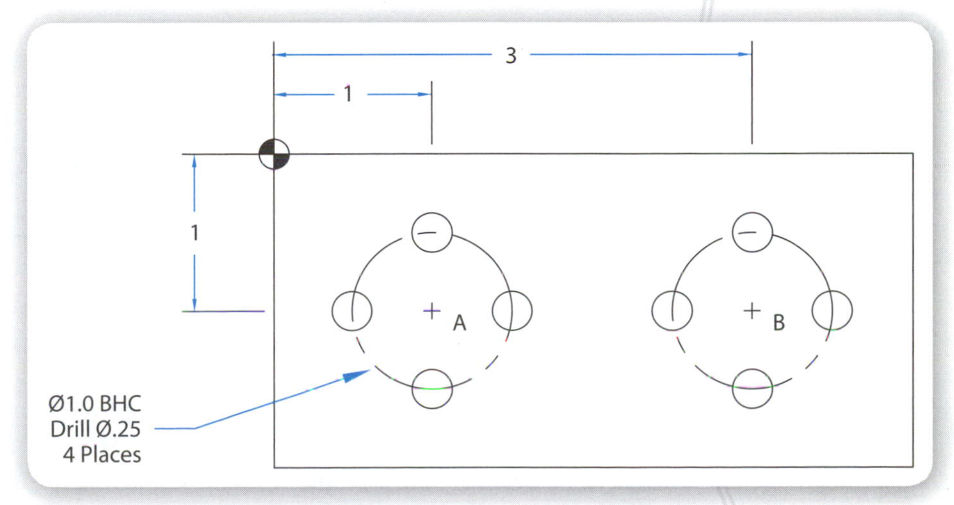

Listing 11-5 Program Code	Explanation
`%`	
`O1001 (Chapter 11, Example One Work Shift)`	
`N10 G20 G40 G49 G54 G80 G90 G98`	
`N20 G91 G28 X0.0 Y0.0 Z0.0`	Move to home position.
`N30 G92 X12.645 Y8.411`	Reset work zero (values from setup).
`N40 M06 T03 (.25 Drill)`	
`N50 G43 H03`	
`N60 M03 S1200`	
`N70 G90 G00 X1.0 Y-1.0`	Move to center of first pattern.
`N80 G00 Z.2 M08`	
`N90 G92 X0.0 Y0.0 (Set Work Zero at Point A)`	Reset work zero.
`N100 G99 G83 X0.5 Y0.0 Z-.5 F5.0 Q.1 R.2`	Drill first BHC.
`N110 X0.0 Y0.5`	
`N120 X-0.5 Y0.0`	
`N130 X0.0 Y-0.5`	
`N140 G80`	
`N150 G00 X1.0 Y0.0`	Move to center of second pattern.
`N160 G00 Z.2 M08`	
`N170 G92 X0.0 Y0.0 (Set Work Zero at Point B)`	Reset work zero.
`N180 G99 G83 X0.5 Y0.0 Z-.5 Q.1 F5.0`	Drill second BHC.
`N190 X0.0 Y0.5`	
`N200 X-0.5 Y0.0`	
`N210 X0.0 Y-0.5`	
`N220 G80`	
`N230 G91 G28 Z2.0`	
`N240 M05`	
`N250 M30`	
`%`	

11.2 G10 COMMAND TO LOAD WORK AND TOOL OFFSETS

It doesn't take long to realize that many repetitive tasks can be automated to save time and money. Two such situations are with work offset values and tool geometry values. As we discussed earlier in Chapter 5, each cutting tool will have a unique length and diameter. As we discussed, these values called "offsets" are established by the setup person during the setup routine. The tools lengths are typically measured by manually touching the tool to the workpiece top of the workpiece. Of course, the expensive machine tool is sitting around not any making money while this is happening.

Many companies have gone to tool presetting to cut downtime. A tool presetter is a device that precisely measures the length and diameter of the cutting tools before they are installed in the machine. The established values can be entered into the control manually, but that is time consuming and prone to mistakes. A better way would be to do it automatically from within the NC program, which can be edited at a computer and not interfere with the operation of the machine tool. This is accomplished by using the G10 code.

To load the values into the CNC control, we can write a few lines of code to perform the task. This can be done in a stand-alone program used only for loading offset values or in the main program after the program stop (M30 or M02).

General Usage:
G10 Ln Pn Rn.n (Used for loading tool length and diameter offsets)

Or
G10 Ln Pn Xn.n Yn.n Zn.n (Used for loading fixture offsets)

G10—Preparatory Code for Loading Offsets
L—Offset Type
P—Offset Number
R—Value of the Tool Diameter or Tool Length
X, Y or Z—Axis Values for Work Offsets

L2 Selects a Work Offset (G54, G55 . . . G59)
P1 through P6 indicates Work Offset Number

For example: G10 L2 P1 X-10.120 Y-5.560 Z0.000 will set work offset number G54 to a position of −10.120" in the X-axis and −5.560" in the Y-axis, and 0.000" in the Z-axis relative to machine's home position. The P-value starts with the G54 work offset and is incremented by +1 for each successive position. P2 → G55, P3 → G56. . . . .P6 → G59.

L10 Loads Tool Lengths
P1, P2, P3. . . . Length Offset Number. Note that the range of offset numbers will be set by the machine tool manufacturer.

For example: G10 L10 P5 R-5.234 will set the length offset for tool number T5 to a value of −5.234".

L12 Loads Tool Diameter
P1, P2, P3. . . . .Diameter Offset Number. Note that the range of offset numbers will be set by the machine tool manufacturer.

For example: G10 L12 P5 R.750 will set the diameter offset for tool number T5 to a value of Ø.750".

To illustrate the use of the G10 code, let's imagine that three tools have been installed in their holders and then each length and diameter was established in a presetter. We also have planned to use a fixture with a known location that will be setup on the G56 workplane.

Tool #	**Length Offset**	**Diameter Offset**
T1	−10.438	.250
T2	−11.625	.4998
T3	−8.890	1.500
Work Offset No.	X	Y
G56	−15.500	−10.375

We would implement the code as follows:

```
(Load the X and Y Values for the G56 Workplane)
G10 L2 P3 X-15.500 Y-10.375
(Load Tool Length Values)
G10 L10 P1 R-10.438
G10 L10 P2 R-11.625
G10 L10 P3 R-8.890
(Load Tool Diameter Values)
G10 L12 P1 R.250
G10 L12 P2 R.4998
G10 L12 P3 R1.500
```

11.3 CONTROLLING TOOLPATH TRANSITIONS WITH DWELL AND EXACT STOP

There are some occasions in machining in which it is useful to slow or stop the cutting tool at the end of a toolpath. For example, imagine that we are cutting an unusually tough or springy material that is likely to deflect the tool or material. In either case the workpiece geometry could suffer from an out of tolerance condition as a result of the deflection. In machining we often take a "spring pass" to make up for this deflect, but this can take extra time and have unintended consequences.

There are two methods that allow us to take short pauses at the end of a toolpath. The first is a **dwell** command with G04, the other is an **exact stop** with G09 and G61. First of all, the dwell command is used to cause the tool to briefly stop advancing while the spindle is still running. This can be useful for situations in which it is necessary pause to allow a few spring passes with the cutting tool or to wait for some other operation condition.

dwell:
A short pause in machine movement usually to allow a cut to be completed.

exact stop:
A method of CNC tool control that forces the machine to decelerate and stop without over-traveling before proceeding to the next tool move.

```
G04 Pn or Pn.n (milliseconds or seconds)
```

Example:

```
G04 P1.50 (Pause for 1.50 seconds to finish cut)
G04 P150 (Pause for 150 milliseconds to finish cut)
```

In a similar concept, the exact stop command is used to cause the tool to decelerate at the end of a toolpath and come to a brief but complete stop before continuing on to the next portion. This can be useful when a large machining table is moving at a relatively high speed to keep the tool from over-traveling. It is also useful when milling pockets to give the tool a decreasing chip load when entering a corner. Exact stop is also used on outside corners to prevent the tool from "lagging" around the corner and causing an uncontrolled corner break. There are two different codes for exact stop. G09 is a non-modal command affecting only the block where it is specified and G61 which is modal and will remain in effect until it is cancelled with the G64 code.

Example:

```
G09 G01 X1.0 Y1.0 F40.0 (Perform exact stop on this block
  only)
G01 X1.0 Y2.0

G61 G01 X1.0 Y1.0 F40.0 (Perform exact stop on all
  blocks)
G01 X1.0 Y2.0
G01 X0.0 Y2.0
G64 (Cancel exact stop)
```

Some caution should be exercised when using exact stop. Toolpaths with many minute lines or arc segments (see surface milling examples in Chapter 13) will suffer greatly if the modal form of exact stop in invoked. The tool will have to start and stop frequently in such a condition, and this will cause excessive vibration, use extra power, and result is a much slower machining process. It would be a bit like pushing the gas and then hitting the brake every few feet when driving a car—not very efficient!

11.4 AUTOMATION WITH SUBPROGRAMS (M98) AND SUBROUTINES (M97)

The primary purpose of any structured computer programming is to automate repetitive tasks in the most elegant manner possible. NC programming is not any different—we want to minimize the number of steps needed to get the job done and make the program logic easy to understand.

The basic rule in computer programming is that any time a process must be repeated, we should use a **subroutine** to perform the task. Generically speaking, a subroutine is a small part of a computer program that performs a specific task. Most computer programming languages—including G & M codes—have this built-in ability to create and use subroutines.

G & M code programming actually contains two closely related methods for utilizing subroutines. The first is called an internal subroutine, which is called with the M97 code. The second is an external subroutine, which is known as a **subprogram** in G & M code programming. Subprograms are called with the M98 code. A program can contain multiple subroutines and subprograms. These two functions are very similar, but differ as follows:

- M97 subroutines are contained within one main part program. This part program is contained in one file.
- M98 subprograms use one main part program and separate subprograms for each task. The main program and subprograms are each a discrete file.

subroutine: Blocks of code within the main program used to repeat a specific programming task.

subprogram: A external program referenced by the main program in order to perform a specific programming task.

The basic syntax and use of the M97 subroutine is as follows:

```
N100 M97 P1234 (CALL SUB ON LINE 1234)
...
...
N1234 (START OF SUBROUTINE)
...
...
M99 (END OF SUBROUTINE)
```

The subroutine is called when the program reads the line containing the M97 code. The control will then jump to the line specified with the P-word and execute the code one line at a time until it is cancelled. When the control encounters the M99 code, the subroutine is cancelled and the program is reset to the block immediately following the initial M97 call. *Figure 11.4* illustrates this flow control.

Subprograms work in a similar manner except that there is a main program and at least one subprogram. **Both programs must be loaded into the control prior to execution.** Table 11.2 Structure and Syntax of a Subprogram shows the basic syntax and use of the M98 subprogram.

Figure 11.4 The M97 code will cause the program to jump to the line specified with the P-word and then execute the subroutine. Execution will continue until M99 is encountered, and then the program will jump back to the block immediately following the initial M97.

```
O1097 (Sub-Routine)
N10 G20 G40 G49 G54 G80 G90 G98
N20 M06 T03
N30 G43 H03
N40 M03 S1200
N50 G00 X0.0 Y0.0 (M08)
N60 G00 Z.2
N80 M97 P500 (Call Sub-Routine)
N80 G91 G28 Z2.0
N90 M05
N100 M30
N500 G01 X-.25 F5.0 (Start of Sub)
N510 G01 X1.0 Y0.0
N520 G01 Z.2
N530 M99 (End of Sub)
%
```

Table 11.2 **Structure and Syntax of a Subprogram**

Main Program	Subprogram
N100 M98 P1234 L1	O1234
. . . Program Code	. . . Program Code
. . .	. . .
. . .	. . .
	M99

When the control encounters the M98 code, the subprogram is called and the control immediately jumps to the beginning of the separate subprogram that is specified by the P-word. Upon reaching the M99 in the subprogram, the control will jump back to the main program and begin execution at the block immediately following the initial M98 code. ***Figure 11.5*** illustrates this.

Subprograms and subroutines add a level of complexity to a basic part program, but they have many benefits when used appropriately. First, they can greatly reduce the size of the code and the number of subsequent typographical errors that are bound to occur in long programs. Second, they make it easy to make changes to a program. For example, say we wrote a program to mill a dozen pockets into a workpiece, and we did not use any subs. If we want to change the pockets at a later time, then we will have to make the change a dozen times. However, with a subprogram, we would only have to make the change once. Finally, subprograms and subroutines can help organize complex programs. It is common practice in many shops to use a subprogram for every tool in the program. This is an excellent device for organizing and it makes it easy to rerun an operation in case of a broken tool or other malfunction.

There is also one additional option to M97 and M98: the loop, which is indicated by the L-word. This loop works the same way as loops with canned cycles. The subprogram will be executed as many times as is indicated by Lnn. If only one operation is required, then L may be omitted. Part of the power of subprograms is their ability to help us program patterns or arrays with the minimum amount of code. Loops are

Figure 11.5 **The M98 code will call a separate subprogram. When the control encounters the M98, it will jump to the program number in the control that is specified by the P-word. Upon reaching the M99 in the subprogram, the control will jump back to the main program and begin execution at the block immediately following the initial M98.**

Main Program

```
%
O1098 (Sub-Programs)
N10 G20 G40 G49 G54 G80 G90 G98
N20 M06 T03
N30 G43 H03
N40 M03 S1200
N50 G00 X0.0 Y0.0 (M08)
N60 G00 Z.2
N80 M98 P500 (Call Sub-Program)
N80 G91 G28 Z2.0
N90 M05
N100 M30
%
```

Sub-Program

```
%
O0500 (Sub)
N10 G01 X-.25 F5.0
N20 G01 X1.0 Y0.0
N30 G01 Z.2
N40 M99 (End of Sub)
%
```

the primary method we use for these operations. Of course, if we expect to repeat a pattern in a different location, then we must write the proper code to cause this to happen. This can be accomplished by any of the following methods:

- Write the subprogram in G91 incremental coordinates and move to the absolute position of each pattern from within the main program.
- Write the subprogram in G90 absolute coordinates and then use G92 work shift to move the work zero from the main program.
- Write the subprogram in G90 absolute coordinates and then use multiple workplanes (G54, G55, . . .) from the main program to locate each pattern.

It is common practice to use the first method of incremental coordinates, but we have seen that incremental coordinates require a great amount of care to execute properly. The second method is excellent for patterns contained on the same workpiece, but if multiple fixtures are used, then the use of multiple workplanes is desirable.

Figure 11.6 *We will use a subprogram to drill the four bolt-hole circles in this workpiece.*

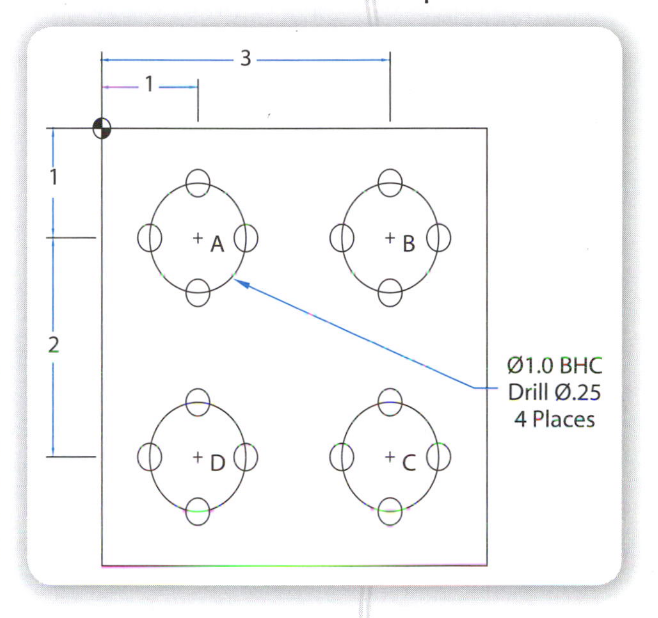

Example 11-2 Subprogram Programming

In this first example using subprograms, we will examine the code in Listing 11-6 and Listing 11-7 that is needed to produce the workpiece in *Figure 11.6*. First, we will create a main program that will position the tool at the absolute location of the first hole of each pattern. Then we will call a subprogram to machine each of the bolt-hole circles. Notice that the subprogram uses incremental coordinates and that all tool changes and spindle commands are called in the main program.

Listing 11-6 Main Program Code	Explanation
%	
O101102 (Chapter 11, Example 2 Main Program)	
N10 G20 G40 G49 G54 G80 G90 G98	
N20 M06 T03 (.25 Drill)	
N30 G43 H03 M08	
N40 M03 S1200	
N50 G00 X1.5 Y-1.0	Position to first pattern location.
N60 G00 Z.2	
N70 M98 P9999 (Set A)	Call subprogram.
N80 G00 X2.5 Y-1.0	Position to second pattern location.
N90 M98 P9999 (Set B)	Call subprogram.
N100 G00 X2.5 Y-3.0	Position to third pattern location.
N110 M98 P9999 (Set C)	Call subprogram.
N120 G00 X1.5 Y-3.0	Position to fourth pattern location.
N130 M98 P9999 (Set D)	Call subprogram.
N140 G91 G28 Z2.0	
N150 M05	
N160 M30	
%	

Listing 11-7	Subprogram Code	Explanation
	%	
	O9999 (Chapter 11, Example 2 Sub Program)	
	N10 G91	Switch to incremental coordinates.
	N20 G99 G83 X0.0 Y0.0 Z-.7 F5.0 Q.1 R0.0	Drilling operations (Hole 1)
	N30 X-0.5 Y0.5	
	N40 X-0.5 Y-0.5 (Hole 3)	
	N50 X0.5 Y-0.5 (Hole 4)	
	N60 G80	
	N70 G90	Change to absolute coordinates.
	N80 M99	Return to main program.
	%	

Example 11-3 Subprogram Programming

In this second example, we will see how loops are used to create arrays with very few blocks of code in Listing 11-8 and Listing 11-9 to produce the workpiece in *Figure 11.7*. The main program will call the subprogram and cause it to loop four times for each row. We should note that the first hole of each pattern is at the 3-o'clock position (zero degrees) and that we are moving around the pattern in a counterclockwise direction. The last hole in the pattern is at 6 o'clock (270°); therefore, we must move incrementally to X2.5, Y0.5 to get to the beginning of the next pattern in the loop.

Figure 11.7 Multiple patterns can be easily handled by looping a subprogram.

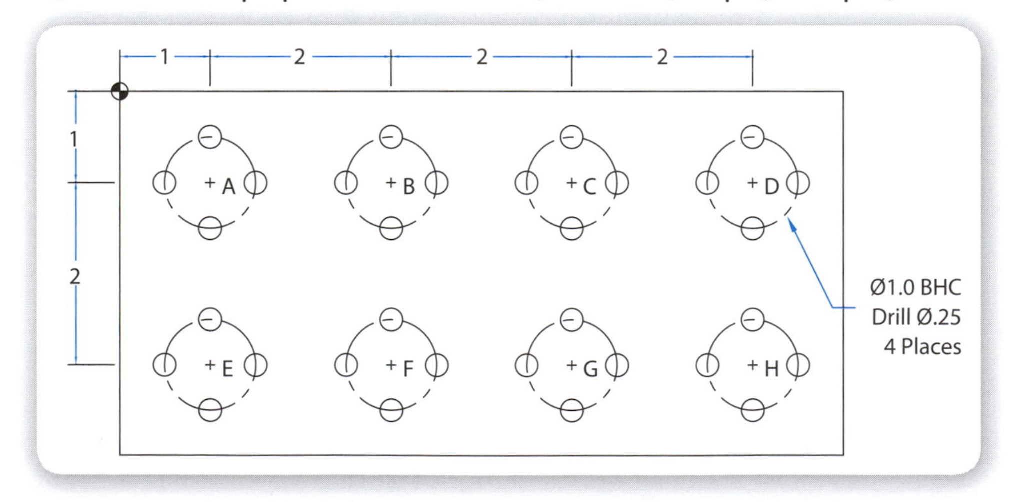

Listing 11-8	Main Program Code	Explanation
	%	
	O1103 (Chapter 11, Example 3 Main Program)	
	N10 G20 G40 G49 G54 G80 G90 G98	

Listing 11-8 *continued*

Code	Explanation
N20 M06 T03 (.25 Drill)	
N30 G43 H03 M08	
N40 M03 S1200	
N50 G00 X1.5 Y-1.0	Position for first row.
N60 G00 Z.2	
N70 M98 P9998 L4 (Top Row)	Call subprogram and loop four times.
N80 G90 G00 X1.5 Y-3.0	Position for second row.
N90 M98 P9998 L4 (Bottom Row)	Call subprogram and loop four times.
N100 G91 G28 Z2.0	
N110 M05	
N120 M30	
%	

Listing 11-9 Subprogram Code	Explanation
%	
O9998 (Chapter 11, Example 3 Sub Program)	
N10 G91	
N20 G99 G83 X0.0 Y0.0 Z-.7 F5.0 Q.1 R0.0	Start of BHC
N30 X-0.5 Y0.5	
N40 X-0.5 Y-0.5	
N50 X0.5 Y-0.5	
N60 G80	
N70 G00 X2.5 Y0.5	Position to the start of the next pattern.
N80 M99	
%	

11.5 MACROS WITH FUNCTIONS AND VARIABLES

Eventually, you will find that it can be useful to have some flexibility in your programming. **Macros** are blocks of code that use any of the logical features such as variables, decision structures, or functions. You might find that there are some products that are very similar to others. For example, a shop might manufacture rails for decks that vary only in length. The hole spaces are identical, but each customer has a different length requirement.

A savvy programmer will recognize that a program can be written with variables that can change values. For example, the rails in *Figure 11.8* are very similar. Rail A has 5 post holes and rail B has 11. We could write one program to drill both sets of holes. The operator would simply change the **variables** to match the customer's order.

macro:
A computer language tool in NC programming that provides the user with some flexibility and automation through the use of variables, loops and decision structures.

Figure 11.8 Parts with similar geometries are great candidates for macro programs. Variables in the macro can be modified easily to update the program. In this case the holes are similar, just the spacing and quantity change.

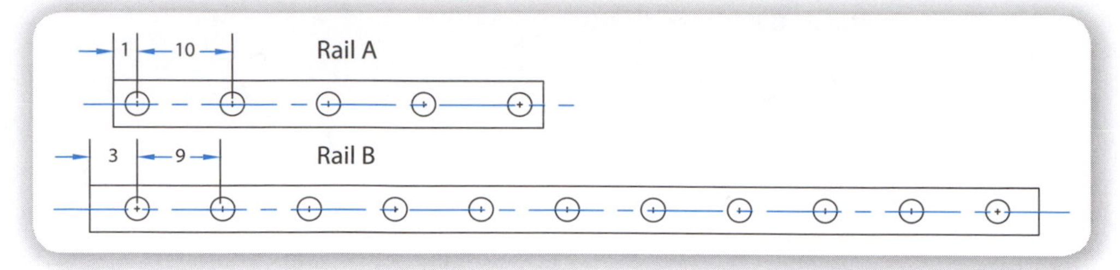

variable:
The symbolic representation of an element in an equation without regard to an actual numeric value of the element.

For example, the code to produce part A might be written as follows:

`#1=1.0 (First Hole Location)`	Sets Variable #1 to the Value 1.0
`#2=5 (Number of Holes)`	Set Variable #2 to the Value 5
`#3=10.0 (Hole Spacing)`	Set Variable #3 to the Value 10.0
`#4=#2-1 (Number of Holes Minus 1)`	Math Operation for the "L" Value
`G00 X#1 Y0.0 Z.200`	Position of the First Hole is at X1.0
`G90 G81 X0.0 Y0.0 Z-1.00 F5.0`	Drill the First Hole
`G91 G81 X#3 L#4`	Drill Four Holes at 10.0" Intervals
`G80`	

Now if the next order is for style "B", then the operator can change only variables #1, #2 and #3 to reflect the new dimensions as follows:

`#1=3.0 (First Hole Location)`	Sets Variable #1 to the Value 3.0
`#2=11 (Number of Holes)`	Set Variable #2 to the Value 11
`#3=9.0 (Hole Spacing)`	Set Variable #3 to the Value 9.0
`#4=#2-1 (Number of Holes Minus 1)`	Math Operation for the "L" Value
`G00 X#1 Y0.0 Z.200`	Position of First Hole is at X3.0
`G90 G81 X0.0 Y0.0 Z-1.00 F5.0`	Drill the First Hole
`G91 G81 X#3 L#4`	Drill Ten Holes at 9.0" Intervals
`G80`	

There are also numerous functions available within the macro library that can be used for performing mathematical operations. For example, the sine and cosine functions can be used to calculate the sides of a triangle for drilling holes patterns at an angle.

```
#1=2.5 (Hole Spacing)
#2=4 (Total Number of Holes)
G00 X0.0 Y0.0 Z.200
G90 G81 Z-1.2 F5.0
G91 G81 X[SIN[45 * #1]] Y[COS[45 * #1]] L[#2-1]
G80
```

Logical operators are used to make decisions based upon some set of conditions. Let's say that we are using a robot to load parts into a milling center. We might be waiting for an electrical signal from the robot that it has completed the loading and that the machining operations are now ready to commence. In this case we are waiting for some internal machine variable (#1180) to receive a digital "on" or 1 from the robot. If the machine has received the "on" signal then the program jumps to line N200.

```
If[#1180=1] GOTO200
M00
N200 M03 S1000
G00 X5.00 Y-2.50
...
```

Our discussion here has just barely scratched the surface of macro programming. There are entire books written on the subject as well as much useful information in the machine tools user manual.

11.6 PROGRAMMING WITH A ROTARY AXIS

The final topic in this chapter is the use of a rotary axis. A rotary axis is most often found in the form of a **rotary table** or an indexing head. The rotary table is a device that can be fastened to the milling table and then rotated to the desired angular position with a G-code. The positioning designation for a single-axis rotary table is the letter A. Some more sophisticated rotary tables can rotate in two axes, which would be designated A and B.

rotary table: A device that is mounted on the milling table to provide programmable rotation of a workpiece.

Machines that support a rotary axis typically allow the table to be positioned with both the G00 code for rapid traverse and G01 for linear interpolation, and in canned cycles. When we simply want to rapid traverse the rotary table to an angular position, we are *indexing* the axis. However, when we wish to perform a milling operation while we rotate the rotary table and move a linear axis, we would then call this simultaneous fourth-axis interpolation. We make this distinction because many of the CAD/CAM systems are designated as either indexing or simultaneous.

The syntax for programming a rotary table is the same as for rapid traverse and linear interpolation, and a canned cycle—we only need to add the A-word to the block. The rotary axis is simply an addition to the standard code that we are already familiar with, as in the following example:

Rapid Traverse

```
G00 Xn.n Yn.n Zn.n An.n
```

Linear Interpolation

```
G01 Xn.n Yn.n Zn.n An.n Fn.n
```

Typical Canned Cycle with Z-Axis Indexing

```
G83 X0.0 Y0.0 Z-1.25 A0.0 F5.0 Q.1 R.2
A30.0
A60.0
G80
```

The rotary axis is easy to use, but there are a few technical points that we must understand when programming. First, the A-axis will usually have a resolution of 0.001 decimal degrees. We must always remember that as we move farther away from the center of the axis, the linear error greatly increases. Second, the A-axis can be programmed to move in the positive or negative direction and can use absolute or incremental coordinates. This follows the standard right-hand rule for rotation, but you may have to consult the operator's manual to determine the proper designation. Lastly, the A-axis usually cannot be rotated indefinitely in the same direction. This is typically a software issue that will account for only so many degrees of rotation. Once the limit is reached, we must unwind the axis back to the start. Other than these few items, a rotary table is straightforward.

Programming examples 11-4 and 11-5 demonstrate the programming of a rotary table. The first example only indexes the workpiece into position so that a series of holes can be drilled on the periphery. The second example performs simultaneous X- and A-axis interpolation to mill a helix into a cylindrical workpiece.

Example 11-4 Fourth-Axis Programming

In this example, we will drill holes on the periphery of a disk. The disk will be fixed upright on a rotary table, and the hole-drilling cycle will be programmed to index to the absolute positions to drill each hole 45° apart, as in ***Figure 11.9***. Listing 11-10 details the code for this operation.

Figure 11.9 **A typical workpiece that might be produced with the indexing operation on a rotary table.**

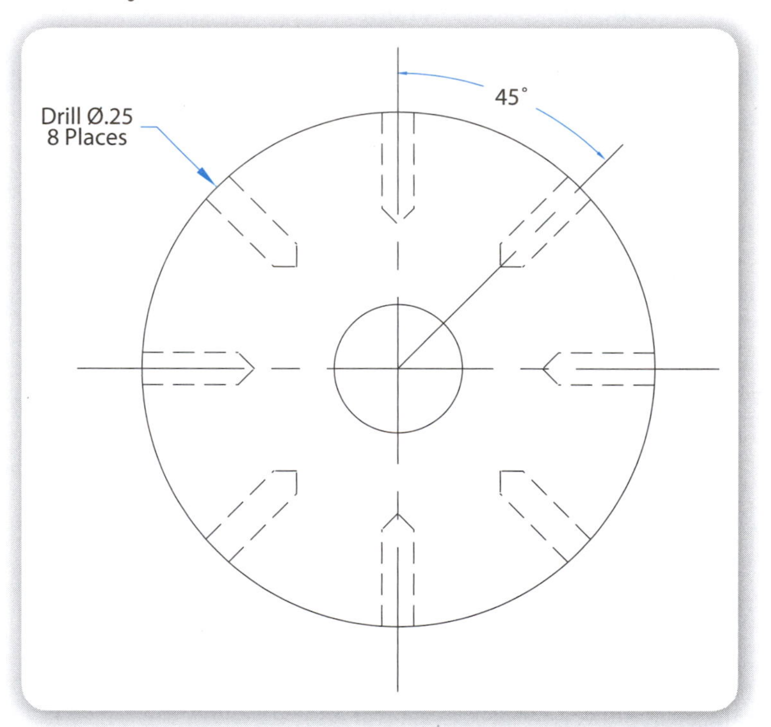

Listing 11-10 Program Code	Explanation

```
%
O1104 (Chapter 11, A-Axis Example)
N10 G20 G40 G49 G54 G80 G90 G98
N20 M06 T03 (.25 Drill)
N30 G43 H03
N40 M03 S1200
```

Listing 11-10 *continued*

`N50 G00 X0.0 Y0.0`	Position for first hole
`N60 G00 Z.2 M08`	
`N70 G99 G83 X0.0 Y0.0 Z-1.25 A0.0 F5.0 Q.1 R.2`	Call drilling cycle and drill hole at A0.0.
`N80 A45.0`	Subsequent hole locations
`N90 A90.0`	...
`N100 A135.0`	...
`N110 A180.0`	...
`N120 A225.0`	...
`N130 A270.0`	...
`N140 A315.0`	...
`N150 G80`	
`N160 G91 G00 A-315.0`	Unwind the A-axis to the start.
`N170 G91 G28 Y0.0 Z2.0`	
`N180 M05`	
`N190 M30`	
`%`	

Example 11-5 Fourth-Axis Programming

In this fourth-axis example, we will look at the program in Listing 11-11, which uses a ball end mill to cut a helix into the workpiece that is illustrated in *Figures 11.10* and *11.11*. This cut is accomplished by rotating the A-axis while simultaneously moving the X-axis. The helix will have a left-hand twist of 90° for a distance of 2.0 inches along the X-axis.

Figure 11.10 Helical grooves can be cut by moving along the linear axis while rotating the rotary axis at the same time.

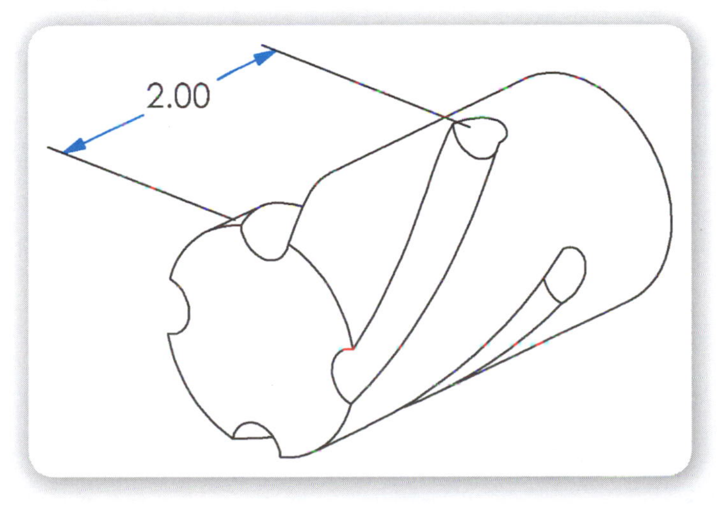

Figure 11.11 The workpiece mounted and machined on a rotary table.

Listing 11-11 Program Code	Explanation
%	
O1105 (Chapter 11, Example Five Helix)	
N10 G20 G40 G49 G54 G80 G90 G98	
N20 M06 T04 (.375 Ball EM)	
N30 G43 H03	
N40 M03 S1200	
N50 G00 X-.5 Y0.0	Position tool in front of the workpiece.
N60 G00 Z-.187 M08	
N70 G01 X0.0 Y0.0 A0.0 F5.0	Initial feed move to the face.
N80 G01 X2.0 A90.0	Cut the first helix.
N90 G01 Z.2	
N100 G00 X-.5 Y0.0	Position for the next cut.
N110 G00 Z-.187	
N120 G01 X0.0 Y0.0 A90.0 F5.0	

Listing 11-11 *continued*

N130 G01 X2.0 A180.0	Cut the second helix.
N140 G01 Z.2	
N150 G00 X-.5 Y0.0	Position for the next cut.
N160 G00 Z-.187	
N170 G01 X0.0 Y0.0 A180.0 F5.0	
N180 G01 X2.0 A270.0	Cut the third helix.
N190 G01 Z.2	
N200 G00 X-.5 Y0.0	Position for the next cut.
N210 G00 Z-.187	
N220 G01 X0.0 Y0.0 A270.0 F5.0	
N230 G01 X2.0 A360.0	Cut the fourth helix.
N240 G01 Z.2	
N250 G91 G00 A-360.0	Unwind the A-axis.
N260 G91 G28 Y0.0 Z2.0	
N270 M05	
N280 M30	
%	

Your Turn

Using the teacher name placard problem from previous chapters, try to recreate the NC program using a subroutine or subprogram. For example, use a repeated letter or pattern in the name to shorten your program.

Arrived at Destination

CHAPTER SUMMARY

- The work zero can be shifted with the G92 code during program execution. This is a useful function for machining patterns without having to recalculate coordinates.
- G92 permanently changes the values in the offset registry, and care must be taken to reset the values to the original offset that were determined during the setup process.
- Subroutines and subprograms are used for automating repetitive tasks.
- Subroutines are called with the M97 code and are all contained within the same part program.
- Subprograms are called with the M98 code from a main program and each subprogram is independent of the main program.
- Both subroutines and subprogram are cancelled with the M99 code. The program will return to the block immediately following the subroutine or subprogram call and continue normal program execution.
- The P-word is used to specify the location of the subroutine and subprogram. If a subroutine is used, then Pnnnn represents the starting line number of the subroutine code. If a subprogram is called, then Pnnnn represents the program number of the separate subprogram.
- Tool and work offset information, such as length, diameter, and location can be loaded into the MCU from within a program by using the G10 command.
- Dwell and exact stop are techniques to allow for smoother transitions at the end of tool movements. The resulting workpiece can have a better surface finish and more precise dimensions when used dwell and exact stop are used properly.
- Macros are a feature of many CNC controls that allows the programmer to build flexibility into a part program. Macros include all of the elements that you would find in a sophisticated language, such as variables, math functions, and decision statements to control program flow.
- A rotary axis is often used with CNC machining. The movement of the rotary axis is indicated with the A-word and can be specified in rapid traverse, linear interpolation, and in canned cycles.
- A rotary axis can be used for simple indexing operations, or it can be used in conjunction with the linear axes for applications including helical milling.

BRING IT HOME

1. What does it mean to move the work zero? How will this affect the operation of the program?
2. When the G92 code is called, what do the X-, Y-, and Z-coordinates represent?
3. Explain what will happen when the following blocks are executed in sequence:

```
G00 X5.0 Y4.0
G92 X0.0 Y0.0
```

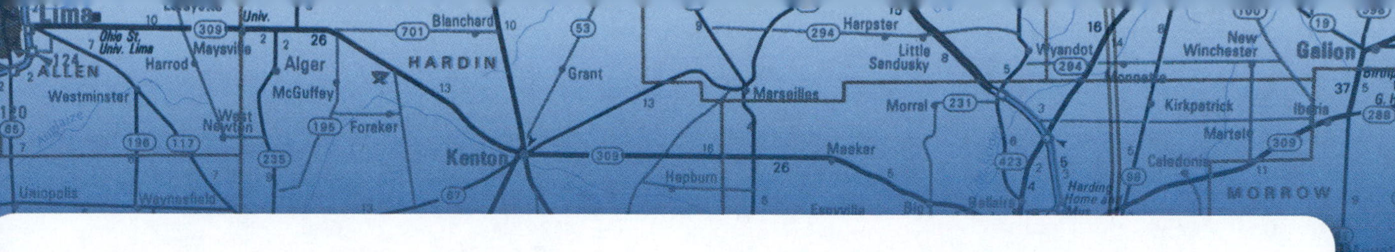

4. Assume from the previous question that the offset values in the first row below represents the original values in the registry. Write the new values in Table 11.3 Write in the Values after the work shift is executed.

Table 11.3 Write in the Values

Offset Values	X-offset	Y-offset
Original Value	−12.875	−7.450
New Value		

5. Are there any advantages in using subroutines and subprograms? Why not just cut-and-paste the code that you want to reuse?
6. Explain the difference between the M97 subroutine and the M98 subprogram.
7. What happens when the M99 code is executed during a subroutine or subprogram?
8. What does the L-word represent in a subprogram?

EXTRA MILE

1. Use the G10 command to write a block of code to set the G58 work offset to X-10.525, Y-5.125.
2. Use the G10 command to write two blocks of code to set the length and diameter of tool #11 to −9.000 and Ø.625 respectively.
3. Give an example a programming situation in which a macro would be useful.
4. What code would you use to cause the cutting tool to pause for 20 milliseconds at the end of a block?
5. Write a complete program to create the finishing toolpath for the workpiece shown in *Figure 11.12*. Use one main program and then a subprogram to machine the pockets. Write the subprogram in absolute coordinates and use the G92 work shift to position the work zero in the appropriate location. A sample toolpath is shown in the top left pocket. Write the program to use a Ø.250 end mill.

Figure 11.12 Write a complete program using subprograms to create the finishing toolpath for the six pockets. One possible option for the toolpath is illustrated in the top left pocket.

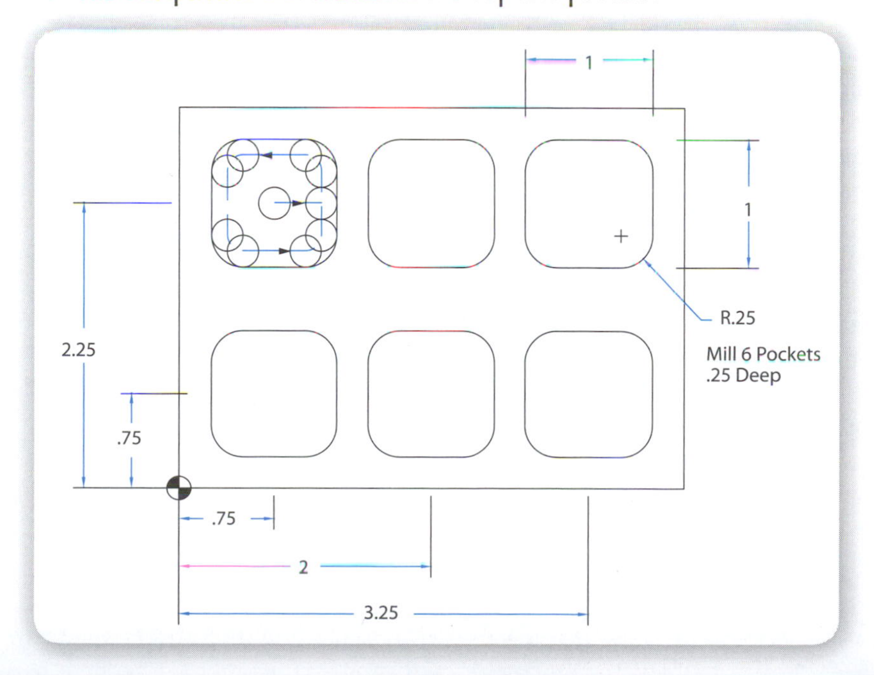

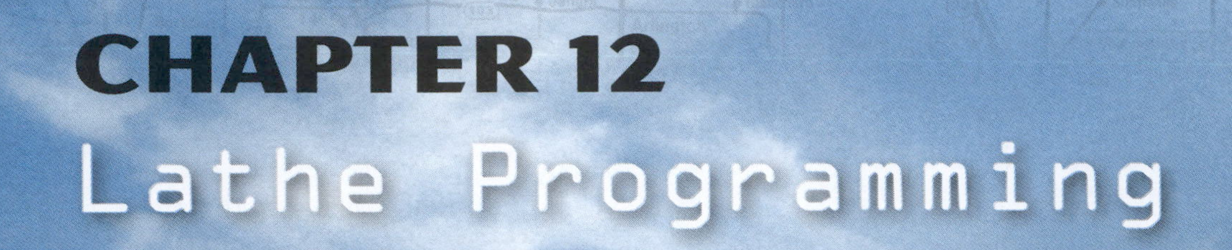

CHAPTER 12
Lathe Programming

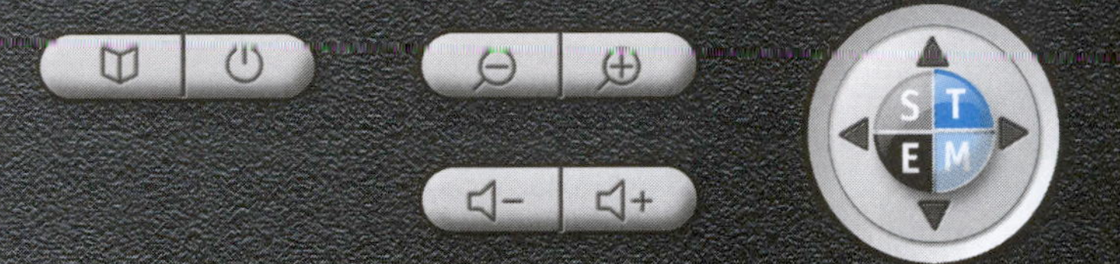

Menu | START LOCATION | DISTANCE | END LOCATION

Before You Begin

Think about these questions as you study the concepts in this chapter:

1. What are some of the differences between mill and lathe programming?
2. What are the limitations of turning without automatic cutter compensation?
3. How do canned cycles help automate turning operations?
4. What is the process typically used to setup the work piece and tool on a CNC lathe?
5. Understand the process used to set up the workpiece and tool on a CNC lathe.

Key Terms

Auto-Turning Cycle
Constant Surface Speed
Imaginary Tool Tip
Reference Tool
Zero Setting Tool

12.1 INTRODUCTION TO THE CNC TURNING CENTER

In this chapter we will look at another common CNC machine tool—the CNC turning center. CNC turning center is a common name for a CNC lathe. Lathes differ from mills in that they are used primarily to produce cylindrical workpieces such as shafts, tubes, rings, and screw threads.

CNC turning centers are very popular in industry. In fact, they are just as important to the manufacturing field as CNC machining centers. However, the overall use of CNC turning centers is slightly different than that of CNC machining centers. Industry surveys reveal that, overall, CNC turning centers are programmed less often than CNC machining centers even though they produce a similar quantity of finished goods. Turning centers tend to be used on the high-production end of the spectrum and therefore require fewer new programs and setups. This is not to say that CNC turning centers are not used for short runs and custom workpieces but only that the average lot sizes for turned workpieces are much higher than for milling centers.

NC programming for the turning center is very similar to NC programming for milling. In fact, it may be easier, because we only need to concern ourselves with two axes rather than three axes of tool movement. Furthermore, the basic G & M codes used for milling are also used for turning. If you can program a CNC mill, then you can easily learn to program a CNC lathe.

Programming Standards

One interesting problem with lathe programming is that CNC controls are much less standardized than with milling. The EIA/ISO standards do specify the codes for lathe programming, but in practice, the machine tool manufacturers have not followed the standards as closely as with milling center controls. This lack of standardization means that the programmer has to spend a great deal of time learning the specific codes for each machine tool or risk making many mistakes.

In industry, there are two dominant controls for turning centers: Fanuc and Okuma. Programming for each of these controls is very similar for basic operations such as linear and circular interpolation, but they diverge as the codes become more sophisticated—as with canned cycles. In this chapter, the examples will be shown in the Okuma style. However, the differences between the two styles will be noted throughout the text.

Lathe Axes Designation

CNC turning centers also use the Cartesian coordinate system for programmed coordinates, but the axes are different from those in milling. First of all, turning centers follow the convention that the axis aligned with the spindle should be designated "Z." Next, the axis that is at a right angle to the spindle is designated "X." ***Figure 12.1*** illustrates these points.

We should also note the direction of each axis. Positive Z points to the right of the spindle, and the positive X direction points toward the back of the machine. This is done because the typical CNC turning center is constructed with the tool mounted on the back side (opposite of the operator). The work zero is generally set to the front and centerline of the finished workpiece.

Figure 12.1 The typical axis designation that is found on CNC turning centers. The Z-axis is aligned with the spindle, and the X-axis is at a right angle to Z.

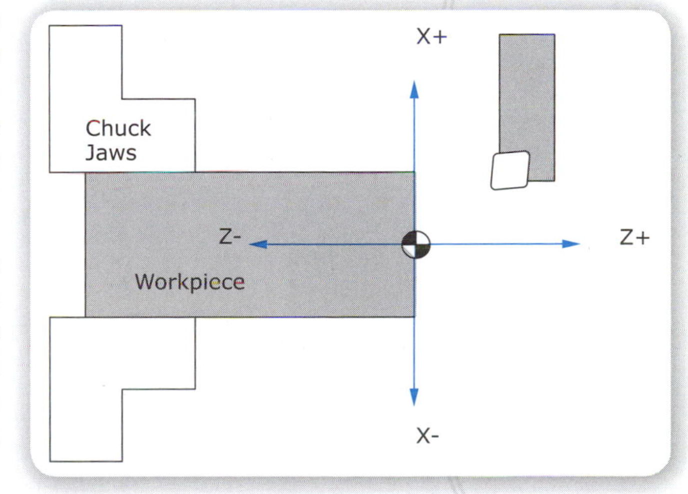

Figure 12.2 The M03 and M04 codes designate the direction of spindle rotation on a CNC turning center.

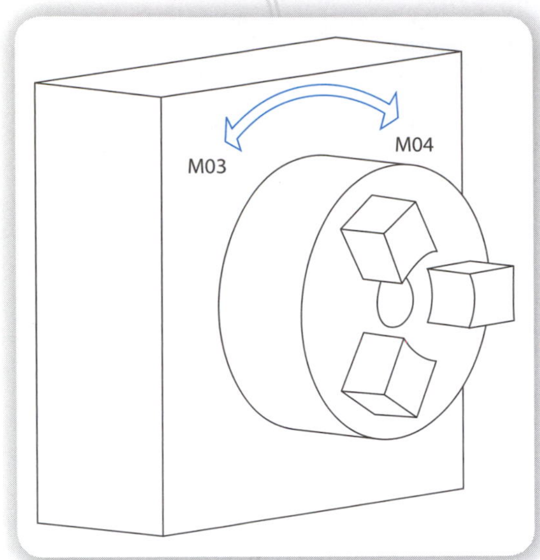

Programming on the Diameter or Radius?

It is common for CNC turning centers to allow X-axis coordinates to be programmed as diameter values rather than radius values. *This only affects X-axis coordinates.* Diameter coordinates are more convenient than the radius values; this can help to avoid the simple math errors that can occur when we have to divide the diameter in half to obtain the radius. For example, if we need to turn a workpiece to a diameter that is specified on the print (Ø1.75, for example), then we simply specify the X-coordinate as that diameter. Example 12-1 shows the difference between these two methods.

Example 12-1 Turn a Workpiece to Ø1.75

Diameter Programming	Radius Programming
`G01 X1.75 Z-1.0`	`G01 X.875 Z-1.0`

Spindle Rotation

Before we can turn a workpiece, we must start the spindle rotation. This is accomplished with the M03 and M04 codes as with machining centers; the direction is illustrated in ***Figure 12.2***. The M03 code is perhaps the most common direction. It is used for most OD turning with the tools mounted upside down in the rear turret, and also for drilling operations with standard right-hand-twist tools.

Spindle rotation is actually related to the axis designations. After all, rotation around the Z-axis is actually rotation in the C-axis. In fact, many CNC turning centers allow interpolation and indexing in the C-axis.

12.2 BASIC TURNING OPERATIONS

In this section, we will look at programming examples to produce workpieces on a CNC turning center in the Okuma format. Most of the codes will already seem familiar, but we will take some time to look at codes and formats that diverge from the already familiar milling codes.

Figure 12.3 In OD turning the tool will be positioned to the starting point and then linear interpolation will be used to move the tool along the toolpath.

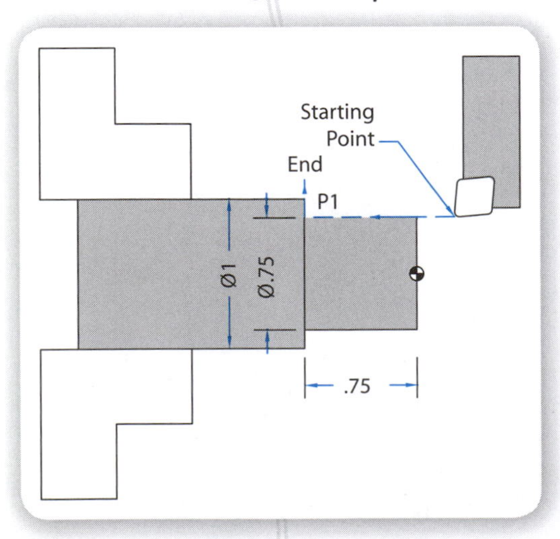

OD Turning and Facing

In this first example, we will perform a simple OD turning operation, as in Listing 12-1, to produce the workpiece in ***Figure 12.3***. The methods for facing operations are identical except that the feed will be along the X-axis.

The tool will first be positioned to the starting point, and then linear interpolation will be used to move the tool along the toolpath to P1. Next, the tool will be moved up from the turned diameter to cut the rear shoulder.

You may notice a few codes that are unfamiliar, as in block N10:

```
N10 G50 S1000
```

The G50 code followed by S1000 sets the maximum spindle RPM to 1,000 RPM. This is important in turning because the centrifugal forces created by the rotating chuck can be very high. If we exceed the rated speed of the chuck, then an accident might occur. We should also note that G50 is also used to perform a temporary work shift (similar to G92 in milling, but non-modal).

Listing 12-1	Program Code	Explanation
	%	
	O1201 (Chapter 12, Example 1)	
	N10 G50 S1100	Set maximum RPM to 1,000.
	N20 G90 G95 G96 M41 S100	Set inches per revolution and constant surface speed modes, change to low gear range, and set cutting speed to 100 SFPM.
	N30 G00 X10. Z10.	
	N40 T0808 M03 (Turning Tool)	Change to tool 08, and turn on the spindle in the forward direction.
	N60 G00 X.75 Z.25	Rapid traverse to the start point.
	N70 G01 X.75 Z-.75 F.01	Turn the Ø0.75 diameter to a length of 0.75.
	N80 G01 X1.25 Z-.75	Cut the shoulder, and move off the workpiece.
	N90 G00 X10. Z10.	
	N100 M30	
	%	

The second set of unfamiliar codes can be found in block N20:

```
N20 G90 G95 G96 M41 S100
```

The G95 code is used to specify the feed rate format in inches per revolution. It can also be set to G94 to format the feed rate in inches per minute, but this is not very convenient for lathe programming.

G96 is another important code. It is used to designate the format of the spindle speed. In the case of G96, **constant surface speed** will be invoked. We could also have used G97 to specify the actual spindle RPM, which is useful when threading and drilling.

constant surface speed: A mode of operation that allows a lathe spindle to continuously adjust the spindle RPM as the tool moves through different diameters. This mode is used to provide an improved surface finish and tool life.

Constant surface speed will cause the machine to continuously adjust the spindle RPM to give the programmed cutting speed in surface feet per minute. What this means is that as the tool moves to different diameters, the spindle RPM will be increased or decreased to satisfy the familiar equation:

$$RPM = \frac{12 \times Velocity}{\pi \times Diameter}$$

As the tool moves to smaller diameters, it must increase the RPM to maintain the proper cutting speed (velocity). As the tool moves toward larger diameters, it will reduce the RPM. This method ensures the most consistent surface finish and tool life.

The M41 code is used to specify the low gear range. Some machine tools are equipped with several gear ranges that can be specified with M42, M43, etc. The S100 code designates the spindle speed and is required to appear in the same block as the gear range code. The S-word will reflect the spindle speed mode that was set with either G96 or G97. For example, G96 G41 S100 sets the spindle speed to 100 surface feet per minute in the constant surface speed mode. However, G97 G41 S100 will set the spindle mode to RPM and cause the spindle to rotate at a steady 100 RPM.

One last block that is a little different is N40:

```
N40 T0808 M03
```

In this block, the tool is called with the T-word. This differs from milling in that we do not need to use the M06 code to call a tool change. The tool and the spindle are independent, so there is no reason to stop the spindle and move to a tool change position in order to change tools as with M06. We simply call the proper tool number, and the tool automatically changes. Also missing from the tool change is G43 to call the tool length offsets. On CNC turning centers, the tool offsets are called with the additional offset number added to the tool number (e.g., T0808). This number represents the location of the offset values in the offset registry.

Okuma turning centers have an additional quirk related to tool offsets. The length offsets and tool nose radius are separate from each other. So, if we wish to use cutter compensation, then an additional offset number must be added to the tool call (e.g., T080808).

Unrelated to the tool change is M03, which is used to start the spindle rotation. This was placed in this block for convenience only, but it could have been in any other block previous to any actual machining operation that did not contain another M-code. The speed was already set in a previous block, so there is no reason to use the S-word unless we wish to change speeds.

Figure 12.4 Taper turning: workpiece produced using the program in Listing 12-2.

Taper Turning

This next example is very similar to the previous one except that this time we will cut a tapered diameter in addition to a straight diameter on the workpiece in *Figure 12.4*. Listing 12-2 details the program.

Listing 12-2	Program Code	Explanation
	%	
	O1202 (Chapter 12, Example 2)	
	N10 G50 S1100	
	N20 G90 G95 G96 M41 S100	
	N30 G00 X10. Z10.	
	N40 T0808 M03 (Turning Tool)	
	N50 G00 X.50 Z.25	Position to start point.
	N60 G01 X.50 Z-.25 F.01	Move to P1.
	N70 G01 X1.0 Z-1.0	Move to P2.
	N80 G01 X1.25 Z-1.0	Move to end point.
	N90 G00 X10. Z10.	
	N100 M30	
	%	

ID Boring

The next example shows one possible method of finish boring the internal diameter of ***Figure 12.5***. In Listing 12-3, we position the tool to the start point and then move to the back of the hole. At this point we take the tool to the center of the workpiece to create a flat surface on the back side. We should note that boring is a delicate operation and that this particular operation assumes that only a very small amount of material has been left for the finish pass. Care must also be taken so that we do not crash the boring bar into the opposite side of the hole as it approaches the center.

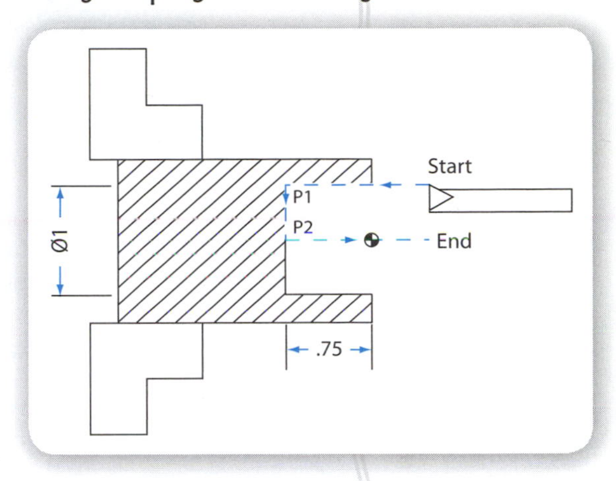

Figure 12.5 **Finish boring: workpiece produce using the program in Listing 12-3.**

Circular Arcs

Circular interpolation is another common operation that is performed on the CNC turning center. Circular interpolation for turning follows all of the same rules as those we are already familiar with, except that we are now working with the X- and Z-axes rather than the X- and Y-axes.

You may remember from milling that there were additional words (I and J) that were used to describe the distance from the start of the arc to the center of the arc, as described in Chapter 7. Each of these words was an incremental distance along each linear axis to the center of the arc segment. On the CNC lathe, we also need to describe the location of the arc center, but we must now use *I and K* rather than I and J. This is because we are using the X- and Z-axes. By definition, the I-word corresponds to the X-axis, and the K-word corresponds to the Z-axis. The definitions that follow further describe the I- and K-words. We can most easily remember the relationships between each axis and its corresponding word if we note that they fall in alphabetical order—XYZ and IJK.

Listing 12-3	Program Code	Explanation
	%	
	O1203 (Chapter 12, Example 3)	
	N10 G50 S1100	
	N20 G90 G95 G96 M41 S100	
	N30 G00 X10. Z10.	
	N40 T0101 M04 (Boring Bar)	Change tools and start spindle in reverse.
	N50 G00 X1.0 Z.5	Position to the start point.
	N60 G01 X1.0 Z-.75 F.005	Move to P1.
	N70 G01 X0.0 Z-.75	Move to P2.
	N80 G01 X0.0 Z.5	Move to end point at feed rate.
	N90 G00 X10. Z10.	
	N100 M30	
	%	

Definitions

Word	Definition
I	X-axis increment distance from the start to the center of the arc segment. This is a radius value and not a diameter value.
K	Z-axis increment distance from the start to the center of the arc segment.

Figure 12.6 (Left) The G03 code invokes CCW circular interpolation. (Right) G02 invokes CW circular interpolation.

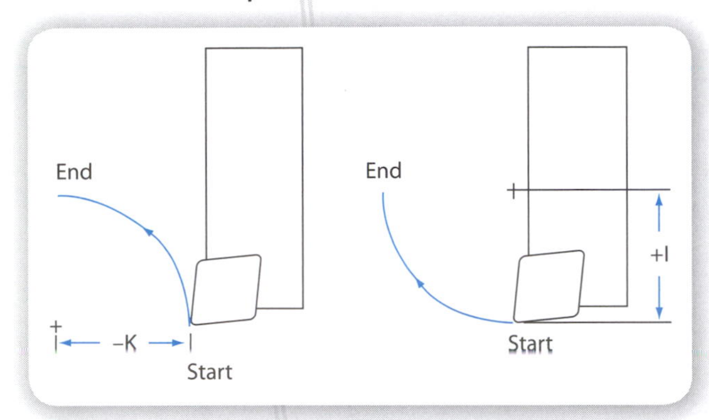

We must also be sure that the signs of the I- and K-words are correct. Again this follows the same rules as with milling, but the directions sometime confuse new programmers. Let's look at the code that would be used to produce the arcs in *Figure 12.6*. We will assume that each arc starts at (X0, Z0), each radius is 1.0" and that each covers a full quadrant.

Left Arc

```
G03 X2.0 Z-1.0 I0.0 K-1.0
```

Right Arc

```
G02 X2.0 Z 1.0 I1.0 K0.0
```

Each of these arcs has the same start point of (X0, Z0) and the same endpoint of (X2.0, Z-1.0). We must remember that the X-value is 2.0 because it is a diameter value. The center of the first arc is in a negative direction from the start and therefore has a K-value of −1.0. There is zero difference in the X-axis, so the I-value is 0.0.

Conversely, the center of the second arc is at an incremental distance of 1.0 from the start and therefore has an I-value of +1.0. This time there is no incremental difference along the Z-axis, so the K-value remains at 0.0.

You may be wondering why we used a value of I1.0 instead of I2.0 for the second arc. This is a common stumbling block that stems from the idea of programming with diameter values rather than radius values. Although the X-coordinates are entered as the diameter, *the I-words are entered as an incremental, radial* distance.

We will now look at a programming example that will turn two arcs and a single straight diameter. The program in Listing 12-4 will produce the finished profile for the workpiece in *Figure 12.7*.

Figure 12.7 Circular Interpolation.

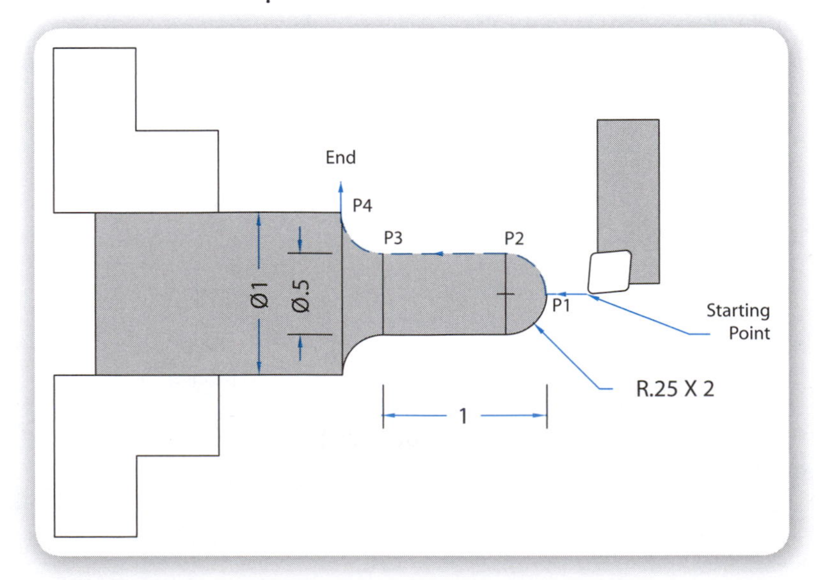

Listing 12-4	Program Code	Explanation
	%	
	O1204 (Chapter 12, Example 4)	
	N10 G50 S1100	
	N20 G90 G95 G96 M41 S100	
	N30 G00 X10. Z10.	
	N40 T0808 M03 (Turning Tool)	
	N50 G00 X0.0 Z.25	Position to the start point.
	N60 G01 X0.0 Z0.0 F.01	Move to P1.
	N70 G03 X.50 Z-.25 I0.0 K-.25	Move in a counterclockwise arc to P2.
	N80 G01 X.50 Z-1.0	Move to P3.
	N90 G02 X1.0 Z-1.25 I.25 K0.0	Move in a counterclockwise arc to P4.
	N100 G01 X1.25 Z-1.25	Move to end point.
	N110 G00 X10. Z10.	
	N120 M30	
	%	

Tool Nose Radius Compensation (Cutter Compensation)

As with milling operations, most finish-turning operations should be performed with the aid of automatic cutter compensation. Cutter compensation allows the control to make minute adjustments to the tool tip position in order to accurately produce the machined features within print specifications. This is particularly important when turning a taper or radius, which cannot be easily and accurately machined with tool tip programming.

Cutter compensation behaves differently depending on whether tool tip programming or tool center programming is used. When using tool center programming, the tool will be automatically offset by the radius value that has been entered into the offset registry. All surfaces will be affected in this mode. Conversely, when tool tip programming is used, only tapers and radii will be affected by the compensation value. Straight diameters and perpendicular faces will remain unchanged unless the tool length (rather than tool radius) offsets are adjusted.

Cutter compensation for turning can be offset to the left or right side of the programmed toolpath with the G41 and G42 codes, respectively. This is illustrated in *Figure 12.8*. Compensation is cancelled with the G40 code.

The same rules apply to cutter compensation for turning as for milling:

- A sufficient distance of at least the radius is needed for the ramp-on and ramp-off moves.
- The ramp-on and ramp-off moves should not be at an acute angle.
- Cutter compensation activates on the first linear move after the compensation code is called.

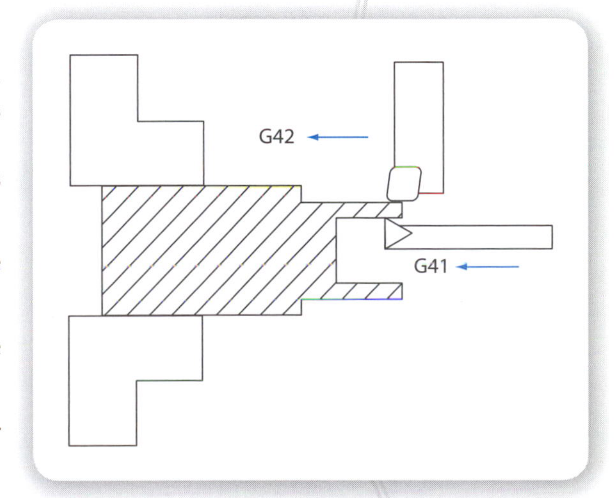

Figure 12.8 *G-codes for cutter compensation while turning. G41 causes the tool to be offset to the left of the programmed toolpath and G42 directs the tool to be offset to the right.*

Figure 12.9 The tool orientation must be designated during setup before using automatic tool nose radius compensation. The tool tip of a typical right-hand, OD turning tool would have a #3 orientation.

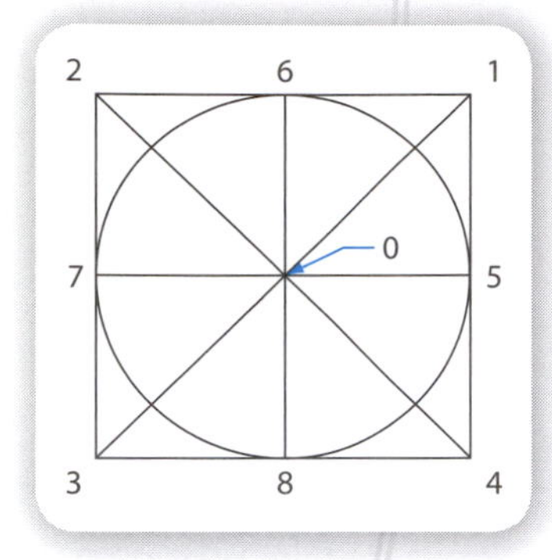

imaginary tool tip: The point of intersection formed by two lines that are tangent to a tool nose radius and aligned with the major axes of a lathe. This is a common programming mode for CNC turning.

- Cutter compensation cannot be invoked during circular interpolation.
- Okuma controls require that an additional offset number be attached to the tool number (e.g., T050505).

Tool Orientation

Unlike milling cutters and drills, which can only have one orientation, lathe tools can have numerous orientations. For example, the right-hand OD turning tool in Figure 12.8 is oriented to cut on the lower left-hand corner of the tool. Likewise, the boring bar in the same figure is oriented to cut on its top left corner. Many other variations are possible, and this creates a problem for the control when attempting to use cutter compensation. After all, the control will need to know which edge it is that must be offset. The only way the control can know this is for the setup person to manually enter a number into the registry that corresponds to the proper orientation.

There are nine possible orientations for turning tools. Each of these orientations has a corresponding numeric value that can be seen in ***Figure 12.9***. The orientation number is located at the point where the **imaginary tool tip** would be located. The top OD turning tool in ***Figure 12.10*** has a #3 orientation, and the boring bar has a #2 orientation. It is important to set up the tool with the proper

Figure 12.10 Examples of tool orientation positions used for common turning operations. From top to bottom: OD right-hand turning, OD left-hand turning, ID boring and OD profile turning.

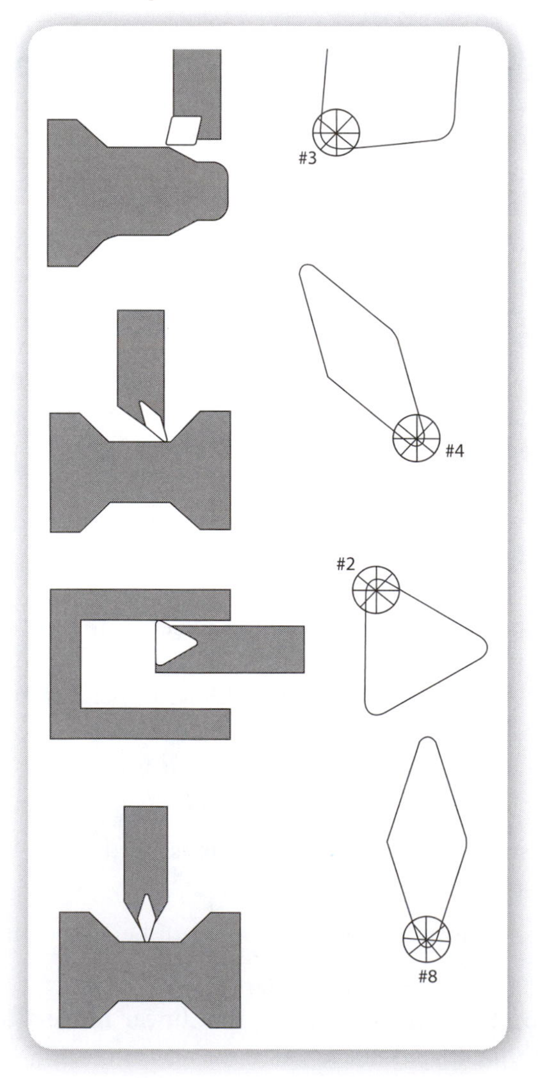

values or else cutter compensation will not work properly. We might also note that the #0 orientation allows the tool to be programmed on the tool center rather than with the tool tip.

Let's look at a programming example that uses cutter compensation. The program in Listing 12-5 will turn a finish profile on the workpiece shown in ***Figure 12.11***.

Figure 12.11 **Finish profile with cutter compensation: workpiece produce using the program in Listing 12-5.**

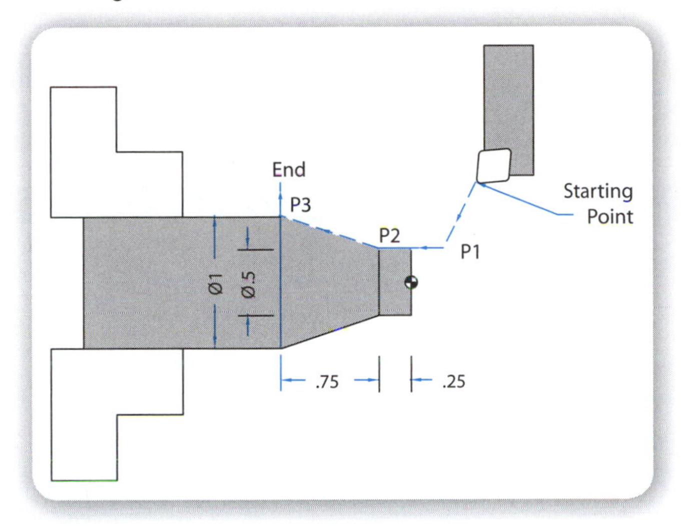

Listing 12-5	Program Code	Explanation
	%	
	O1205 (Chapter 12, Example 5)	
	N10 G50 S1100	
	N20 G90 G95 G96 M41 S100	
	N30 G00 X10. Z10.	
	N40 T080808 M03 (Turning Tool)	Tool change with addition offset for compensation.
	N50 G00 X1.25 Z.5	Rapid to the start point.
	N60 G42 G01 X.50 Z.25 F.01	Move to P1 and invoke compensation.
	N70 G01 X.50 Z-.25	P2
	N80 G01 X1.0 Z-1.0	P3
	N90 G40 G01 X1.25 Z-1.0	Move to the end point and turn off compensation.
	N100 G00 X10. Z10.	
	N110 M30	
	%	

reference tool:
See Zero Setting tool.

zero setting tool:
A cutting tool used to establish the axial work-zeros of a workpiece from which all other cutting tools are referenced through the use of tool offsets.

12.3 TURNING TOOL SETUP

Before we can begin turning any components on a CNC lathe, we will need to install the tooling and properly establish the offsets (*see Table 12.1*). We will pick one tool as a "**reference tool**" or "**zero setting tool.**" The reference tool is usually an OD, right-hand turning tool that is commonly used with most turning jobs;

however, this is completely arbitrary. This tool will be used as a reference point from which all the other tools will be measured. This process is described in the following table.

The reasoning for tool and work offsets in turning is similar to that used in mill programming. The tools used are of different lengths and configurations. The one difference is that in milling the workpiece could float around on the milling table. On a lathe, the workpiece is always aligned with the center of the spindle.

Turning tools are usually mounted in a turret that can then "index" from position to position. The turning and drilling tools are a mounted in separate stations with a series of clamping and holding devices. The work is of course held in a chuck or collet and spun while the rigid tool is moved along the toolpath.

Table 12.1 Lathe Tool Setup Procedure

Establish the X-Axis Work Zero on the Centerline	
Step One Zero Setting Tool	Install the raw material into the lathe chuck and then take a light cut on the diameter with the "Zero-Setting Tool." Leave the tool at this diameter and jog it out past the front of the workpiece and then stop the chuck from rotating.
Step Two 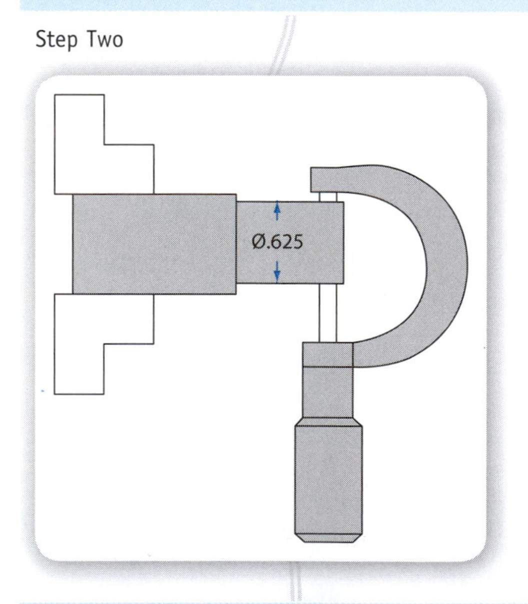	Carefully measure the diameter with a micrometer. Again, leave the Zero Setting Tool at this diameter. Do not move it up or down.
Step Three	Now find the workpiece offsets screen and enter the measured diameter (.625 in this case) into the X-axis measurement as instructed by the manufacturer. This tells the machine that the tool is resting at the specified diameter; therefore, the X-axis origin is implicitly located on the centerline of the spindle.

Table 12.1 *Continued*

Establish the Z-Axis Zero at the Face	
Step Four 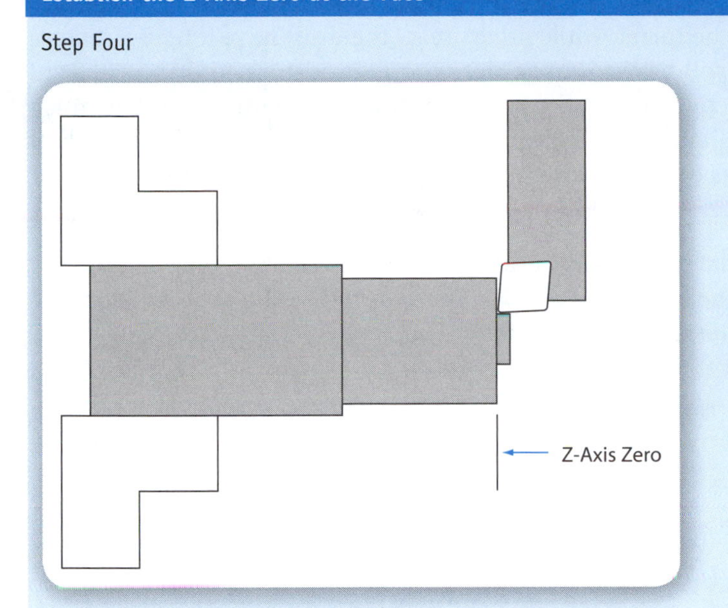	Use the zero setting tool to take several light facing passes along the front of the stock until the face is completely cleanedup. When you are satisfied with the location of the face move the tool up and park it without moving left or right in the Z-axis. Stop the spindle.
Step Five	Now find the workpiece offsets screen and establish the Z-axis work offset of 0.000 on the face of the workpiece as instructed by the manufacturer. This tells the machine that the tool is resting at the face of the part, which is now the Z-axis origin.
Establish the Length and Diameter Offsets for Each of the Remaining Cutting Tools	
Step Six 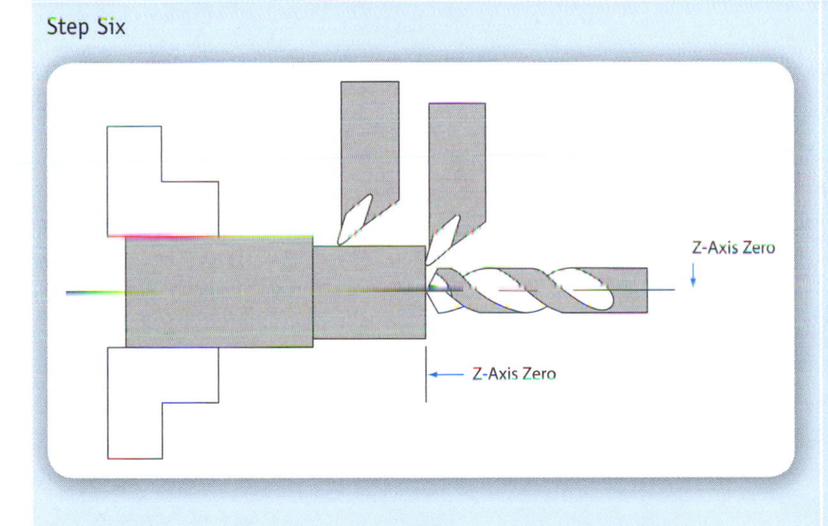	The last step is to measure all of the remaining tools relative to the diameter and face that we have just established in the previous steps. Each cutting tool will have a slightly different location relative to the zero setting tool. That is each tool will stick out a little more or less in each direction. This distance is calculated with the "tool offset." To establish the X-axis offset for each tool, we simply load the tool into the cutting position and then jog the tool manually until it makes contact with the turned diameter. Now we find the Tool Offsets screen and establish the X-axis measurement as instructed by the manufacturer. That is, when we are touching the .625 diameter, the control uses this information to calculate the proper offset value. A similar method is used to set the Z-axis offset values. In this case the tool is jogged to the face and "zeroed." One exception to this process is for tools that cut on the centerline such as drills, center drills, and taps. The X-offset is set to directly 0.0000 to indicate that the there is no difference between the centerline of the spindle and the centerline of the tool.

To understand turning tool setup, let's look at the example in *Figure 12.12*. The three tools are mounted in a turret that revolves into position around a fixed centerline. As you can see, the tools are all different sizes. Some are very close to the face and center of the turret, while others (e.g., the drill) have long overhangs. Essentially, the turret will have to move close to the headstock to bring either of the turning tools into contact with the face. The drill, on the other hand, is much longer. So the turret will have to stay further away to reach the face of the work. If we used the first tool as or zero setting tool, then let's look at the offset values for the three tools.

Figure 12.12 **Additional tool offsets are required because of the varying sizes of the cutting tools. Each offset is the difference between the cutting edge locations relative to the zero-setting tool.**

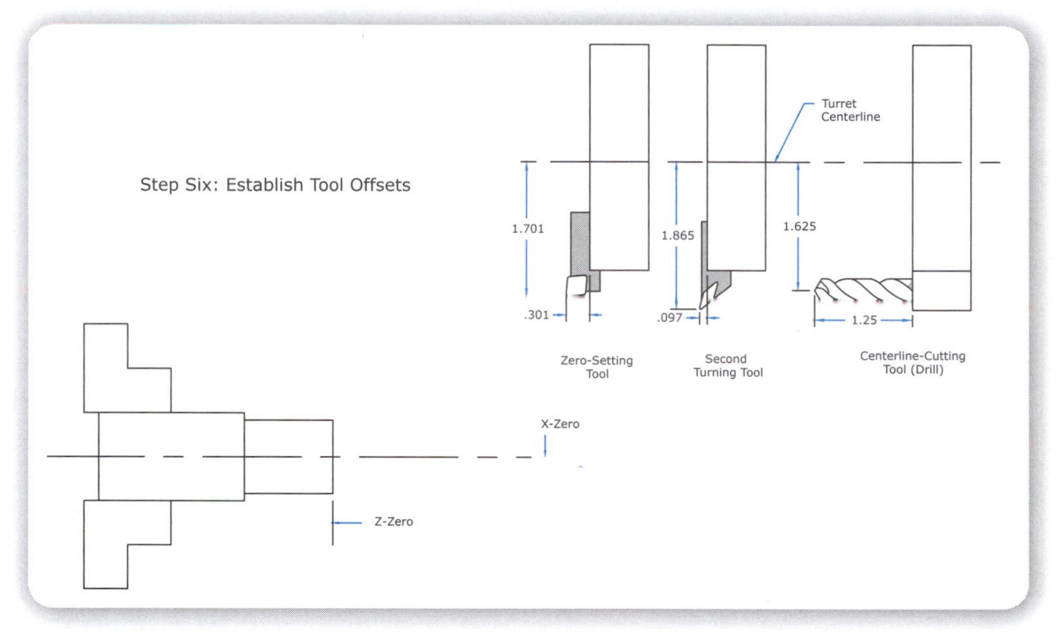

Zero Setting Tool

This tool was used to establish the face of the part, therefore:

Z-offset equals 0.000"

Second Turning Tool

We can see that the Zero Setting Tool protrudes .301" from the face and the second tool protrudes only .097". Therefore, the second tool will have to move in the minus Z direction by the difference ($.301 - .097 = .204$) to reach the face. The Z-offset for this tool will be -0.204.

The Center Tool (Drill)

Long tools like drills will tend to stick out significantly past the edge of the zero setting tool. In fact this drill is .941" longer, so the turret will be offset to the right to compensate for the extra length. The Z-offset for this tool will be set to 0.941.

Alternate Methods for Establishing X-Offsets

There are two schools of thought concerning the establishment of the X-axis offset on a lathe. Some shops will always use the centerline of the center cutting tool holders to establish the X-zero. Essentially, a test indicator is mounted in the chuck and then the tool holder is manually aligned with the center, as in *Figure 12.13*.

Figure 12.13 The centerline of the tool holder is aligned with the axis of the spindle by mounting a test indicator to the chuck and manually adjusting the tool holder to the center.

This method is perhaps the most accurate and efficient method on many machine tools. The center tool holders are usually precision ground and once the center is established, all of the drilling tool X- axis offsets can be set as 0.000. It is very convenient.

The second method is to use the zero setting tool, as described earlier. The tool is positioned to touch a known diameter, and then the location is captured by the control. The problem with this method comes when we want to use center cutting tools. Now the drill or tap will need to be "touched off" on a known diameter, and then we will record the position and manually adjust the value to make up for the radius. This will work, but it may be more complicated. The values of the methods are displayed as follows. Also notice that the Z-offsets are unaffected.

	Center Method	Traditional Method
Zero Setting Tool Offsets	X 0.076	X 0.000
	Z 0.000	Z 0.000
Second Turing Tool Offsets	X 0.240	X 0.164
	Z -0.204	Z -0.204
Drill Offsets	X 0.000	X -.076
	Z 0.941	Z 0.941

Many of today's CNC turning centers include an electronic touch-off for establishing tool offsets (***Figure 12.14***). This method is faster and more accurate than physically making contact with an established surface. Furthermore, it is less likely that the cutting tool or the surface of the workpiece will be damaged.

Figure 12.14 A tool probe arm is a common option on many CNC turning centers. The tool probe establishes the tool offset electronically by touching a high-precision contact.

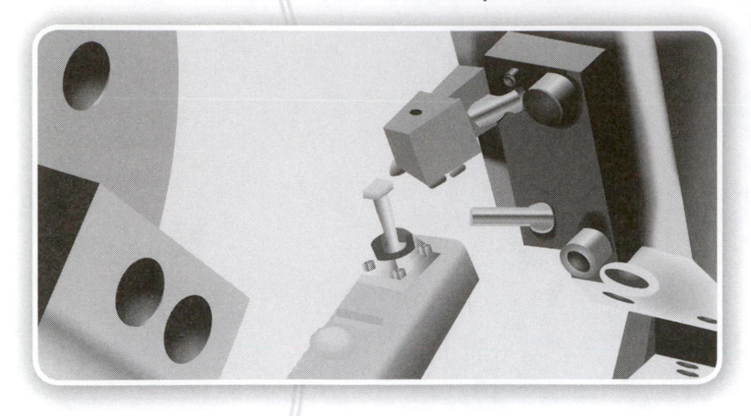

Figure 12.15 Single-point turning tools are often programmed at an imaginary tool tip. This method eliminates the need to manually compensate for the tool radius on straight cuts.

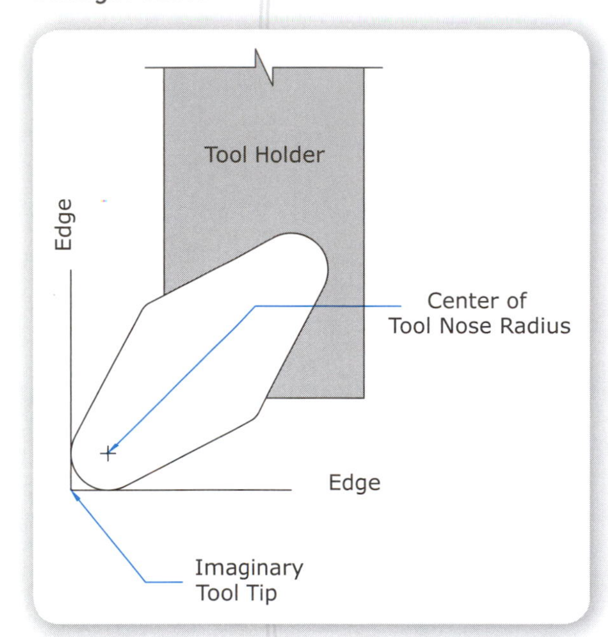

Tool Edge or Tool Center

We saw in the section on milling that cutting tools can be programmed either on the centerline or with automatic cutter compensation. Lathe programming is similar but has one additional option. Many lathes are configured to use an *imaginary tool tip* as the programmed point of contact. We can visualize this imaginary tool tip by setting the tool nose radius tangent to the X- and Z-axes. This is illustrated in ***Figure 12.15***.

Turning tools usually have a small radius ground into the tip of the tool, typically ranging from 1/64" to 3/32". This radius is analogous to the radius of a milling tool and will have to be accounted for when machining. Tool tip programming eliminates the need to manually compensate for this radius when turning straight diameters or facing perpendicular surfaces.

Most turning centers can be set up to use either the imaginary tool tip or the center of the tool nose radius. However, if we use the center of the tool nose radius, then we must manually compensate for the radius with each move (similar to milling). For example, we might need to turn a straight OD of Ø1.75 with a turning tool that has a tool nose radius of 1/32". To program this move with tool tip programming, we would simply use the X-coordinate of X1.75. Yet, if we are using tool centerline programming, then we have to add two times the radius to the specified diameter to reach the correct point of X1.8125. (Remember that these are diameter values, so to actually move 1/32" we have to program an additional 1/16".) These two methods are shown in the code that follows in Example 12-2.

Example 12-2 Tool Tip versus Tool Center Programming to Turn a Workpiece to Ø.1.75

Tool Tip Programming	Tool Center Programming
`G01 X1.75 Z-1.0`	`G01 X1.8125 Z-1.0`

Limitations of Tool Tip Programming

Tool tip programming is convenient, but it does have some serious limitations. The correct geometry is created only when turning and facing parallel to either the X- or Z-axis. If we attempt to turn a taper or radius, then a geometry error will occur, as illustrated in ***Figure 12.16***. This error occurs because the actual tangential edge of the tool falls short of the theoretical toolpath.

When cutting tapers, this error can be compensated for with trigonometry. On the other hand, the error with circular interpolation cannot be easily controlled. This is because the error is constantly changing as the tool tip follows the programmed toolpath. One solution to this problem is to program to the centerline of the tool nose radius and manually compensate for the radius. The other solution is to use automatic cutter compensation and let the control figure out the correct path. This is actually the preferred solution because we will probably already be using cutter compensation for the finished toolpaths, and it allows us to continue programming with the tool tip.

Figure 12.16 Tool tip programming has limitations when cutting tapers and radii. (Left) The actual cutting edge will fall short of the expected taper. (Right) A similar situation will occur when cutting a radius with the tool tip programming. The remedy for this situation is to use either tool center programming or automatic cutter compensation.

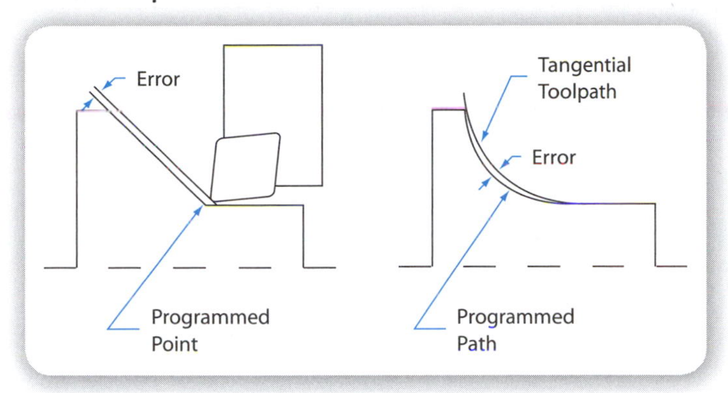

12.4 COMMON CANNED CYCLES

CNC controls also have various canned cycles available for common tasks such as threading, drilling, and grooving. However, this is where the major control manufacturers have diverged and a lack of standardization has occurred. This is not as serious a problem as it sounds. Once we understand one style of canned cycle, then it is relatively easy to adapt to another.

In the remainder of this section, we will look at both the Okuma and Fanuc styles of canned cycles. However, the programming examples will use the Okuma canned cycles. We will follow the examples with a reference table that will describe differences in these two systems.

Threading

Threading is one of the primary operations that is performed on a lathe, so it is no surprise that CNC turning centers can perform a variety of different threading cycles. Perhaps the most useful of these cycles is the multi-pass threading cycle that is invoked with the G71 code (G76 on Fanuc controls). The multi-pass threading cycle allows us to cut a complete screw thread with a single block of code. This threading cycle will cause the tool to cut multiple passes without any operator involvement or any additional code. The screw thread will be automatically roughed and finished based upon the code supplied in the part program and parameters that have been set in the CNC control.

All we have to do to use the threading cycle is move the tool to the correct starting position and then call the canned cycle. The canned cycle will be executed and the tool will be returned to the original starting point when the cycle is finished.

The syntax for both the Okuma and Fanuc threading cycles is given in Example 12-3 and Example 12-4. It is accompanied by a short code example to show the code in context. Both code examples will produce the same thread.

Now let's take a look at the complete programming example in Listing 12-6, which uses a threading canned cycle to produce the screw thread in ***Figure 12.17***. We can see from the drawing that the thread is a 1.0 − 8 UNC. With this information we can go to a handbook and look up the minor diameter, which turns out to be Ø0.844. This information is required in order to specify the height of the thread, which is given in the code with the H-word (K for Fanuc). The height of the thread is a diameter value, so we can calculate it by simply subtracting the minor diameter from the major diameter:

$$1.0'' - 0.844'' = 0.156''$$

Example 12-3

Canned Cycle Syntax: Okuma	Code Example
G71 X Z C B D H F	G00 X1.5 Z.2 (pre-position)
X = Finish Minor Diameter	G71 X.900 Z-1.0 C0.0 B30.0 H.1
Z = Finish Z-axis Coordinate	D.02 F.0833
C = C-axis Start Position (optional)	
B = Feed Angle (1/2 angle of thread)	
H = Diametrical Thread Height (major Ø - minor Ø)	
D = Depth of First Pass (in diameter)	
F = Feed (pitch of the thread)	

Example 12-4

Canned Cycle Syntax: Fanuc	Code Example
G76 X Z A I K D F	G00 X1.5 Z.2 (pre-position)
X = Finish Minor Diameter	G76 X.900 Z-1.0 A60.0 I0.0 K.05
Z = Finish Z-axis Coordinate	D100 F.0833
A = Thread Angle	
I = Thread Taper (usually zero)	
K = Radial Thread Height (major Ø - minor Ø) / 2	
D = Depth of First Pass (radial 1 /10,000")	
F = Feed (pitch of the thread)	

Figure 12.17 Multi-pass Threading Canned Cycle.

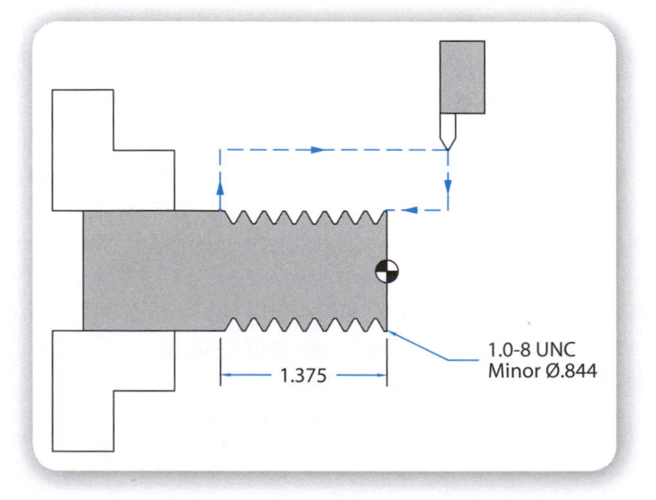

Listing 12-6	Program Code	Explanation
	%	
	O1206 (Chapter 12, Example 6)	
	N10 G50 S1100	
	N20 G90 G95 G97 M41 S350	Constant RPM mode
	N30 G00 X10. Z10.	
	N40 T0404 M03 (OD Threading)	
	N50 G00 X2.0 Z.5	Position to start
	N60 G71 X.844 Z-1.375 B30.0 D.04 H.156 F.125	Execute threading cycle.
	N90 G00 X10. Z10.	
	N100 M30	
	%	

Next we have to calculate the feed, which on a single-lead thread is equivalent to the pitch. The thread is specified as having eight threads per inch. Therefore the pitch (and feed) can be calculated as

$$\frac{1}{8\text{TPI}} = 0.125"$$

Drilling

The G74 code is used for peck drilling on both Okuma and Fanuc controls; however, some of the words are different. This code works in a very similar manner to the G83 code for milling. We simply position the drill to a safe starting point and then call the drilling cycle. The drill then drills to each incremental peck depth and then retracts to clear the chips. This continues until the programmed Z-depth is reached, at which point the drill returns to the starting position. The Okuma drilling cycle actually goes much farther than just peck drilling. It has several words that allow it to perform a chip-breaking cycle and also a dwell.

The syntax for both the Okuma and Fanuc drilling cycles is given in Example 12-5 and Example 12-6. The code examples will perform roughly the same operation except that the Okuma cycle will stop to break the chip every 0.125", but only retract to clear the chip every 0.250" as is specified with the D- and L-words.

Example 12-5

Canned Cycle Syntax: Fanuc

```
G74 X Z K F
X = Finish Diameter
Z = Finish Z-depth
K = Peck Depth
F = Feed
```

Code Example

Example

```
G00 X0.0 Z.200 (pre-position)
G74 X.0 Z-.750 K.125 F.007
```

Example 12-6

Canned Cycle Syntax: Okuma	Code Example
G74 X Z D L F (E)	Example
X = Finish Diameter	G00 X0.0 Z.200 (pre-position)
Z = Finish Z-depth	G74 X.0 Z-.750 D.125 L.250 F.007
D = Peck Depth	
L= Full Retract Depth	
F = Feed	
E = Dwell Time (optional)	

We may be able to get a better picture of the drilling cycle in use by looking at a complete example. The code in Listing 12-7 will produce the drilled hole that is illustrated in ***Figure 12.18***. This cycle will cause the drill to drill to a depth of Z-1.0 and then return to the starting point. Note that we are again using the G97 code to indicate constant RPM. The X-coordinate is set to zero, so the spindle will turn at the maximum RPM if we try to use the constant surface speed mode.

Figure 12.18 *Canned Drilling Cycle.*

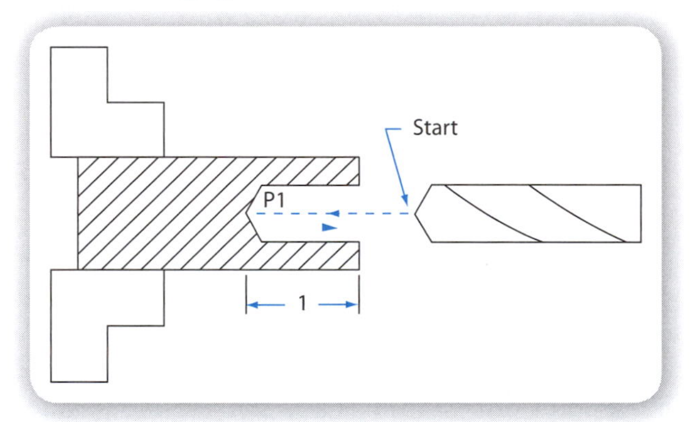

Listing 12-7	Program Code	Explanation
	%	
	O1207 (Chapter 12, Example 7)	
	N10 G50 S1100	
	N20 G90 G95 G97 M41 S800	Constant RPM mode
	N30 G00 X10. Z10.	
	N40 T0202 M03 (.50 Drill)	
	N50 G00 X0.0 Z.5	Position tool.
	N60 G74 X0.0 Z-1.0 D.125 L.25 F.005	Call the drilling cycle. Break the chip every 0.125" and perform a full retract every 0.25".
	N90 G00 X10. Z10.	
	N100 M30	
	%	

Grooving

Grooving is an operation in which the tool is plunged into the surface of the workpiece until the proper depth is reached. The grooving canned cycle automates this operation by including provisions for multiple passes, pecks, and chip breakers and a dwell at the finished depth. This operation is performed on the X-axis by calling the G73 code (G75 on Fanuc controls). The details of this cycle can be found in Example 12-7.

The grooving canned cycle allows for wide grooves to be cut with multiple passes by specifying a shift value with the K-word. We simply position the grooving tool to the start point and then call the canned cycle. The tool will first take an initial cut to the finished diameter. Then it will retract to the starting X position and move over in the Z-axis by the amount specified with the K-word. The tool will make several passes until it reaches the programmed Z-coordinate. This process is illustrated in ***Figure 12.19***.

Example 12-7

Canned Cycle Syntax: Fanuc

```
G75 X Z I K D F
X = Finish Diameter
Z = Finish Z-axis Coordinate
I = Peck Depth (radial)
K = Z-shift Amount
F = Feed
```

Code Example

Example

```
G00 X1.25 Z-.200 (pre-position)
G75 X.5 Z-.500 I.063 K.100 F.005
```

Warning

This cycle varies greatly between controls. Consult the operator's manual before programming.

Figure 12.19 The grooving canned cycle allow for wide grooves to be cut with multiple passes by specifying a shift value with the K-word.

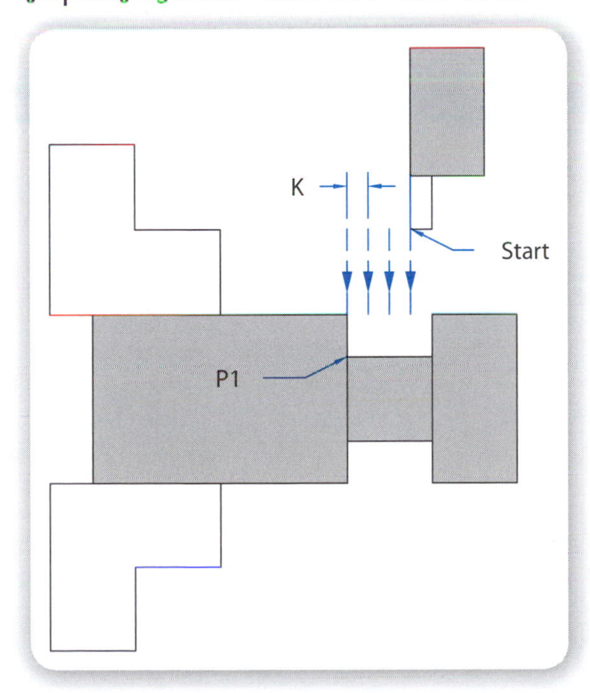

Example 12-8

Canned Cycle Syntax: Okuma	Code Example
`G73 X Z D L E K F`	Example
`X = Finish Diameter`	`G00 X1.25 Z-.200 (pre-position)`
`Z = Finish Z-axis`	`G73 X.5 Z-.500 D.125 L.25 E.120 K.100 F.005`
`Coordinate`	
`D = Peck Depth`	
`L = Full Retract Depth`	
`E = Dwell Time`	
`K = Z-shift Amount`	
`F = Feed`	

Next we will look at the complete program in Listing 12-8, which uses the G73 cycle to produce a groove as shown in ***Figure 12.20***. This groove has a finish diameter of Ø 0.500 and is only as wide as the cutting tool. We will not have to perform a shift for this groove, so the K-value will be set to zero. This cycle will pause to break the chip at 0.05 intervals and perform a full retract every 0.100. Furthermore, when the finish diameter is reached, the tool will dwell for 0.08 seconds to allow a full rotation (see Example 12-8).

Auto-Turning Canned Cycles

Roughing operations are a very time-consuming issue with lathe programming. A typical workpiece starts off as a bar of raw stock that must be whittled-down to the finished dimensions. Unfortunately, single-point turning tools are only capable of removing relatively small amounts of material with each pass (Ø0.050–0.180"). This means that many roughing passes will be needed to remove the excess material as in ***Figure 12.21***.

Listing 12-8	Program Code	Explanation
	`%`	
	`O1208 (Chapter 12, Example 8)`	
	`N10 G50 S1100`	
	`N20 G90 G95 G96 M41 S100`	
	`N30 G00 X10. Z10.`	
	`N40 T0303 M03 (.125 Grooving)`	
	`N50 G00 X1.25 Z-.625`	Move tool to the start position.
	`N60 G73 X0.5 Z-.625 D.05 L.10 K0.0 E.08 F.005`	Call the grooving cycle.
	`N90 G00 X10. Z10.`	
	`N100 M30`	
	`%`	

Figure 12.20 *Grooving.*

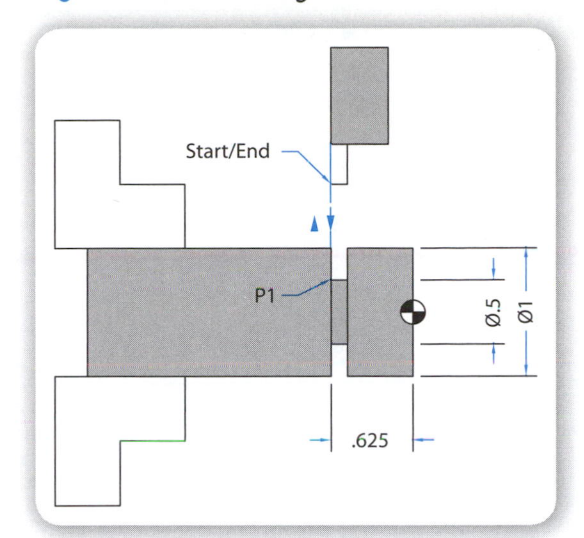

Figure 12.21 *Auto-turning cycles for roughing and finishing.*

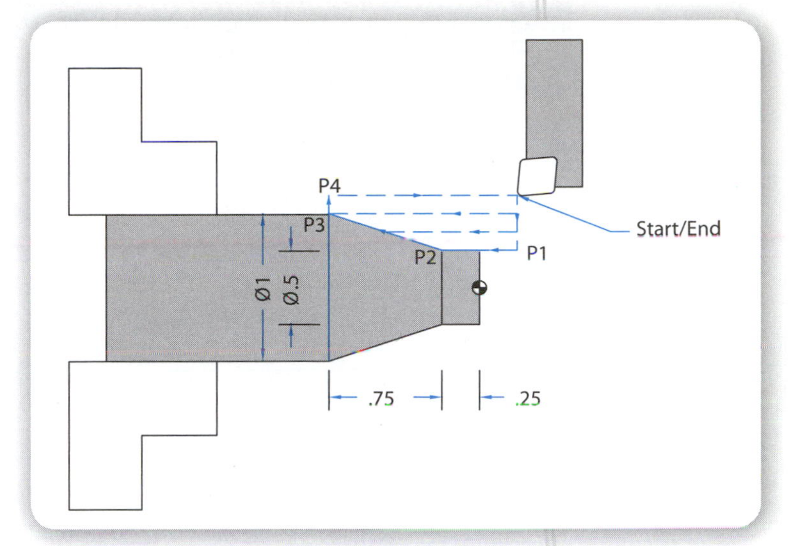

Each of these roughing passes could be programmed individually, but this is very time-consuming and leads to lengthy part programs. Fortunately, the controls on most CNC turning centers support several canned cycles for automatic roughing and finishing. Both the Okuma and Fanuc controls have a number of auto-turning cycles that automate the roughing and finishing processes. In essence, the control automatically creates the roughing passes based upon a user-supplied definition of the finished workpiece. Many variations are available to create longitudinal passes and facing passes and for copy turning (a repeating finish pass). Regardless of the particular turning cycle or control, the basic programming process is as follows:

- Write the G-code to create a continuous, finished profile without regard to roughing.
- Place this code somewhere in the part program, and format it as a profile definition.
- Call the auto-turning cycle in the program, just as with any other canned cycle. The call to the canned cycle must indicate where the profile definition can be found.

There are many different methods for defining and calling an **auto-turning cycle** on a CNC turning center. Consult the operator's manual for specific information before attempting to use these cycles.

auto-turning cycle: Canned cycles that are used to automate stock removal for turning operations.

Example 12-9 shows the syntax and a programming example for auto-turning on the Okuma control.

Example 12-9

Canned Cycle Syntax: Okuma	Code Example
Rough	Example
`G85 NATnn U W D F`	`N100 G00 X1.00 Z.200 (pre-position)`
`Copy Turning`	`N110 G85 NAT01 U0.03 W0.015 D.125`

(continued)

Example 12-9 *(continued)*

G86 NATnn U W D F	F.010
	N120 G87 NAT01
Finish	...
G87 NATnn	...
	...
Profile Definitions	NAT01 G81 (START OF PROFILE DEFINITION)
G80, End of Profile	
G81, Longitudinal Turning	N200 G01 X.50 Z.200 F.007
G82, Transverse Facing	N210 G01 X.50 Z-1.0
	N220 G01 X1.0 Z-1.0
NAT = Definition Number	N230 G80 (END OF PROFILE)
U = Amount to leave on X-axis	...
W = Amount to leave on Z-axis	...
D = Depth of Cut (diameter value)	
F = Feed	

Auto-turning cycles all behave in roughly the same manner. On an Okuma turning center, there are several options for the cycle, but it is perhaps easiest to look at only rough and finish turning for now. The basic profile definition for longitudinal turning is defined by properly formatting the blocks of code that form a continuous toolpath. More than one profile definition can be created in a single program, so we must label the particular profile by replacing the standard line number with NATnn, when nn is the profile number. Next, we specify the direction of the passes with the G81 code for longitudinal passes or G82 for transverse passes. Finally, we end the definition by enclosing the profile with the G80 code, as in the following example:

```
NAT01 G81
...
...(Insert G-code to create a continuous profile)
...
G80
```

This section of code can be placed at any convenient place within the part program. As long as it is contained with the G81/G82 and G80 codes, it will be ignored during regular program execution.

To execute the auto-turning cycle, we must first position the tool to a starting point and then call the cycle with G85 for roughing or G87 for finishing. The G85 roughing cycle contains a number of variables that tell the control how to machine the workpiece, as in the following example:

```
G85 NAT01 U.03 W.015 D.10 F.012
```

NATnn tells the control which profile definition it should use. For example, if we use the code G85 NAT01, then the control will search for NAT01 and then execute the cycle.

U and W are used to specify the amount of material to be left on the workpiece for the finishing pass. The U-word indicates the finish allowance in the X-axis, and the W-word sets the finish allowance for the Z-axis. The finish allowance is typically between 0.005 and 0.015 depth of cut, but this value is largely dependent on the particular machining circumstances.

The D-word sets the amount of material that will be removed from the diameter with each pass. Again, this value is determined by the machining conditions and the rigidity of the setup. The cutting tool manufacturers usually recommend specific suggestions for the maximum radial depth of cut. A typical depth of cut is often around 0.08", which would translate to a D-value of 0.160". Exceeding this value may lead to premature tool failure.

Once we complete the roughing, then we are ready to run the finishing cycle by calling the G87 code. The finishing cycle simply executes the profile definition in the manner that it is written without having to reproduce the code in the program, as in the following example:

```
G87 NAT01
```

As you can see, no codes are needed other than the profile definition number. However, we should note that the feed for the finishing pass must be specified in the profile definition.

We can now look at a complete programming example in Listing 12-9, which uses an auto-turning cycle for roughing and finishing. In this example we use auto-turning cycles to produce the workpiece that is shown in Figure 12.21. Notice that cutter compensation is also used with this cycle. It is integrated into the profile definition.

Listing 12-9	Program Code	Explanation
	%	
	O1209 (Chapter 12, Example 9)	
	N10 G50 S1100	
	N20 G90 G95 G96 M41 S100	
	N30 G00 X10. Z10.	
	N40 T080808 M03 (Turning Tool)	
	N50 G00 X1.25 Z.25	Move to start position.
	N60 G85 NAT01 U0.03 W0.015 D.125 F.010	Call the roughing cycle to cut passes at a 0.125" depth. Leave a 0.015" finish allowance on each surface.
	N70 G87 NAT01	Call the finish cycle.
	NAT01 G81	Longitudinal profile definition number NAT01
	N50 G42 G01 X.50 Z.25 F.007	P1
	N60 G01 X.50 Z-.25	P2
	N70 G01 X1.0 Z-1.0	P3

(continued)

Listing 12-9 *continued*

```
N80 G40 G01 X1.25 Z-1.0        P4
N130 G80 (END OF
PROFILE)
N140 G00 X10. Z10.
N150 M30
%
```

These cycles are also found on Fanuc controls, but there are a few slight differences in the profile definition. On most Fanuc systems, the profile definition is designated with the beginning and ending line numbers of the profile. We have to be careful not to use the automatic renumbering feature on a code editor, or else the numbers will not match.

Example 12-10 specifies the syntax and gives a code example for one particular style of Fanuc auto-turning cycle. These codes may not represent every control, and the operator's manual should be consulted before programming. We will also notice that the direction of the cut (longitudinal or transverse) is specified in the call to the canned cycle with G71 or G72 codes. This is distinctly different from the Okuma, which specifies the direction in the profile definition.

Example 12-10

Canned Cycle Syntax: Fanuc	**Code Example**
Rough Longitudinal Turning	Example
G71 P Q U W D F	N100 G00 X1.00 Z.200 (pre-position)
Rough Facing	N110 G71 P200 Q220 U0.03 W0.015
G72 P Q U W D F	D630 F.010
	N120 G70 P200 Q220 F.007
Finish	...
G70 P Q F	...
	...
Profile Definitions	N200 G01 X.50 Z.200 F.007
Pnnnn, Starting Line Number	N210 G01 X.50 Z-1.0
Qnnnn, Ending Line Number	N220 G01 X1.0 Z-1.0
	...
U = Amount to leave on X-axis (signed value)	...
W = Amount to leave on Z-axis (signed value)	
D = Depth of Cut (radial value, 1/10,000")	
F = Feed	

12.5 SUMMARY OF TURNING CODES

Function	Okuma	Fanuc
Tool change	Tnnnn(nn)	Tnnnn
Spindle speed mode	G96, Constant surface speed G97, Constant RPM	G96 Constant surface speed G97 Constant RPM
Feed rate mode	G94 IPM, G95 IPR	G94 IPM, G95 IPR G98/G99 on some controls
Inch/metric values	Set in machine parameters	G20/G21
Return to home position	G20 (optional)	G28
Multi-pass threading cycle	G71	G76
Drilling canned cycle	G74	G74
Grooving canned cycle	G73	G75
Auto-turning cycle	G85, Rough turning G86, Copy turning G87, Finish pass	G71, Rough turning G72, Rough facing G70, Finish pass
Auto-turning profile definition	NATnn, Profile number designation G81, Start of longitudinal definition G82, Start of transverse definition G80, End of profile definition	P and Q are the starting and ending line numbers of the profile

Your Turn

Create an NC program to be used on the lathe for a new prototype pen or pencil.

CHAPTER SUMMARY

- Lathe and mill programming share many of the same concepts as G & M codes. However, there is less standardization with turning centers, so we must be careful to understand the differences before writing a part program for the turning.
- The X- and Z-axes are used for turning. The X-axis is often programmed as a diameter value, which means that it will effectively move half of the programmed distance. The Z-axis is always aligned with the spindle.
- Tool and workpiece setups on a CNC lathe are similar in concept to that of CNC milling centers. The setup person must communicate to the control the location of the workpiece origin and each of the cutting tools.
- Turning tools are often set up to be programmed at an imaginary tool tip rather than at the center of the tool radius. Tool tip programming can lead to errors, if we do not understand its limitations. Often the easiest way around these limitations is to use automatic cutter compensation.
- Turning centers support a number of canned cycles for threading, drilling, grooving, and automatic stock removal. The cycles can greatly improve programming efficiency, but the operator's manual should always be consulted before programming—many of these codes are not standardized.

BRING IT HOME

1. What are the two dominant control types used on CNC turning centers?
2. Where is the work zero typically located on the workpiece for turning operation?
3. How does the machine control know that a tool is cutting at a specified diameter?
4. Describe how one might establish the front face of the workpiece as the Z-axis origin.
5. What does it mean to"program on the diameter"? How is this different from programming with radial values?
6. How is tool tip programming different from tool center programming? Explain.
7. What is one source of error when turning a taper with tool tip programming? How can we compensate for this error?
8. What is one source of error when using circular interpolation with tool tip programming? Is there any way we can compensate for this error? Can we simply adjust the coordinates?
9. Why do we need to set the proper tool orientation in the register before attempting to use cutter compensation?
10. What is the difference between the constant surface speed and the constant RPM in relation to the speed of the spindle?
11. Are there any situations in which you would not want to use constant surface speed mode?
12. Why do we have to limit the maximum speed of the spindle? Give a code to limit this speed to 2,300 RPM.

EXTRA MILE

1. Write a complete part program to create the finished profile of the workpiece shown in *Figure 12.22*. Use the following parameters:

Figure 12.22 **Create a part program to machine this finished profile.**

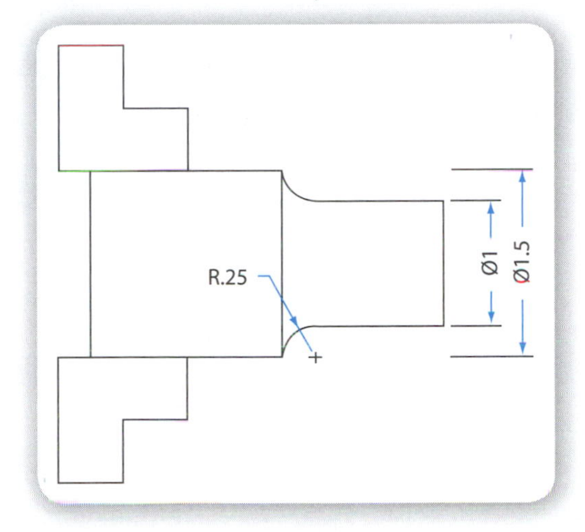

Material: aluminum
Tool: T0808 80° OD turning
Spindle speed: 300 constant SFPM
Feed: 0.010 IPR
Challenge: Revisit this programming problem, but this time use an auto-turning cycle to rough and finish the workpiece. Use the same parameters that were used the first time.

2. Write a complete part program using canned cycles to produce the drilled hole, groove, and thread on the workpiece that is shown in *Figure 12.23*, *Figure 12.24*. You may assume that the OD and chamfer are already finished. Use the following parameters:

Figure 12.23 **Use canned cycles for drilling, threading, and grooving to produce the desired features on this workpiece.**

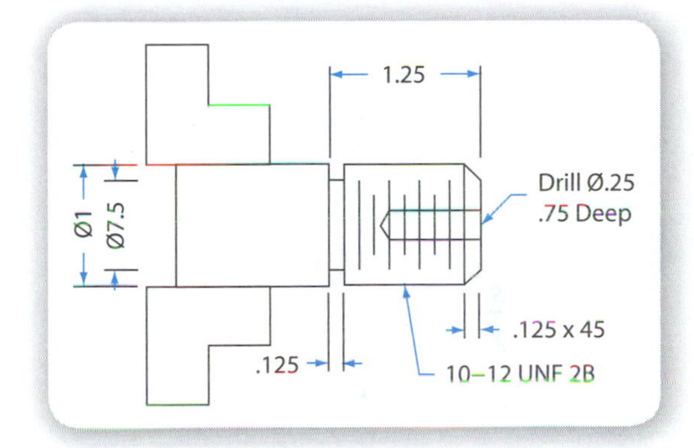

Material: 1018 steel
Tools:
Ø0.25 HSS drill, T0101
60 carbide threading tool, T0202
0.125 carbide grooving tool, T0303
Cutting speed: 300 SFPM for carbide and 100 SFPM for HSS tools. Use constant surface speed where appropriate.
Feed: 0.005 IPR except for thread

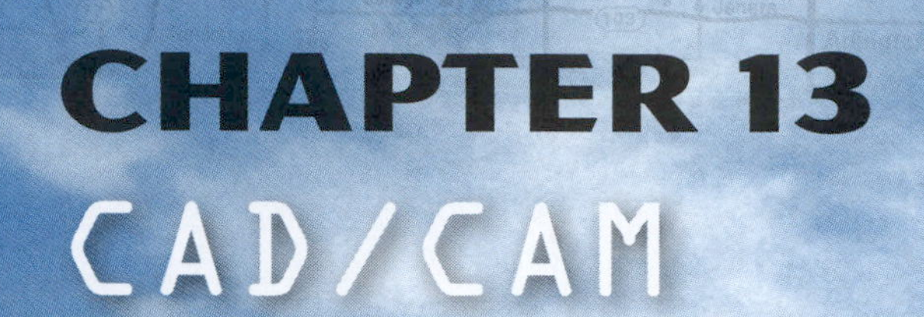

CHAPTER 13
CAD/CAM

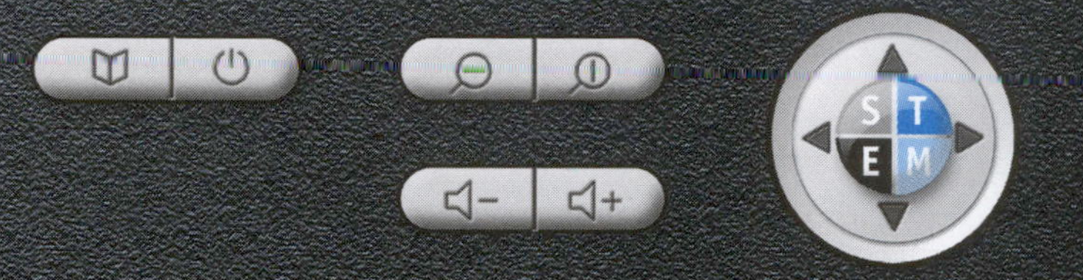

Menu

START LOCATION | DISTANCE | END LOCATION

Before You Begin

Think about these questions as you study the concepts in this chapter:

 What are some of the more common acronyms and abbreviations used today in manufacturing?

 How do computers generate complicated NC code that is too difficult to complete by hand?

 What are some of the general trends within the CAD/CAM industry?

 What are some of the common technologies that are common to Computer Integrated Manufacturing?

 What are some of the more common rapid prototyping processes seen today in manufacturing?

Key Terms

CAD/CAM
Computer-Aided Drafting (CAD)
Computer-Aided Engineering (CAE)
Computer-Aided Manufacturing (CAM)
Computer-Aided Part Programming (CAPP)
Computer Integrated Manufacturing (CIM)
ERP/MRP
Flexible Manufacturing System (FMS)
NURBS
Post-processing
Rapid Prototyping (RP)
STEP Models

13.1 COMPUTER-AIDED ANYTHING (CAA)

In today's age of rapidly evolving computer technology, there seems to be an abundance of industrial processes and techniques that want to attach *computer-aided* to their name. Half of the challenge for a beginner in this industry is to simply understand the language. A sea of acronyms, initials, and abbreviations has flooded the landscape in an attempt to obscure the essence of the ideas. Most of these terms were invented by marketing people and academics, and the fancy names do not necessarily reflect the intrinsic values of the idea that they represent. In this chapter, we will flesh out the essences of the computer-aided processes that are so prevalent in industry.

13.2 WHAT IS CAD/CAM SOFTWARE?

Many toolpaths are simply too difficult and expensive to program manually. For these situations, we need the help of a computer to write an NC part program.

The fundamental concept of CAD/CAM is that we can use a Computer-Aided Drafting (CAD) system to draw the geometry of a workpiece on a computer. Once the geometry is completed, then we can use a Computer-Aided Manufacturing (CAM) system to generate an NC toolpath based on the CAD geometry.

The progression from a CAD drawing all the way to the working NC code is illustrated as follows:

Computer-Aided Drafting/Computer-Aided Manufacturing (CAD/CAM):
A class of computer software that has the ability to create electronic representations, geometric entities, and model machining processes, and create NC code based upon the model.

STEP 1

The geometry is defined in a CAD drawing. This workpiece contains a pocket to be machined. It might take several hours to manually write the code for this pocket. However, we can use a CAM program to create the NC code in a matter of minutes. (***Figure 13.1***)

STEP 2

The model is next imported into the CAM module. We can then select the proper geometry and define the style of toolpath to create, which in this case is a pocket. We must also tell the CAM system which tools to use, the type of material, feed, and depth of cut information. (***Figure 13.2***)

Computer-Aided Drafting (CAD):
Computer software that is used to model geometric entities and visually represent the geometry.

STEP 3

The CAM model is then verified to ensure that the toolpaths are correct. If any mistakes are found, it is simple to make changes at this point. (***Figure 13.3***)

STEP 4

The final product of the CAD/CAM process is the NC code. The NC code is produced by *post-processing* the model to create NC code that is customized to accommodate the particular variety of CNC control. (***Figure 13.4***)

Computer-Aided Manufacturing (CAM):
A group of technologies that use computers to design, automate, control, and improve manufacturing operations.

Another acronym that we may run into is CAPP, which stands for Computer-Aided Part Programming. CAPP is the process of using computers to aid in the programming of NC toolpaths. However, the acronym CAPP never really gained widespread acceptance, and today we seldom hear this term. Instead, the more marketable CAD/CAM is used to express the idea of using computers to help generate NC part programs. This is unfortunate because CAM is an entire group of technologies related to manufacturing design and automation—not just the software that is used to program CNC machine tools.

Computer-Aided Part Programming (CAPP):
A class of computer software that is used to generate NC code.

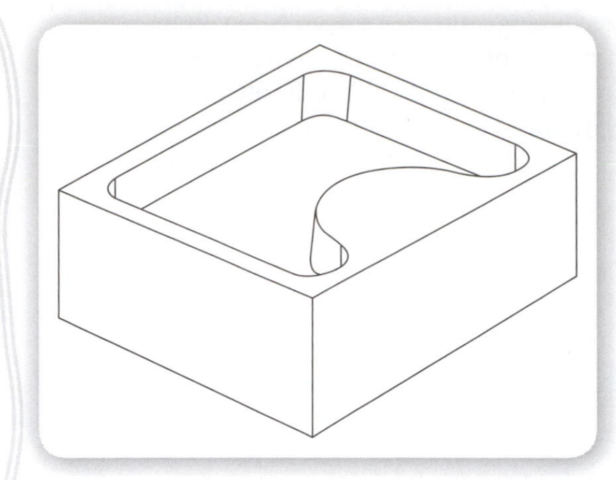

Figure 13.1 The geometry is defined in the CAD drawing.

Figure 13.2 The CAM system creates toolpaths that are bounded by the geometry.

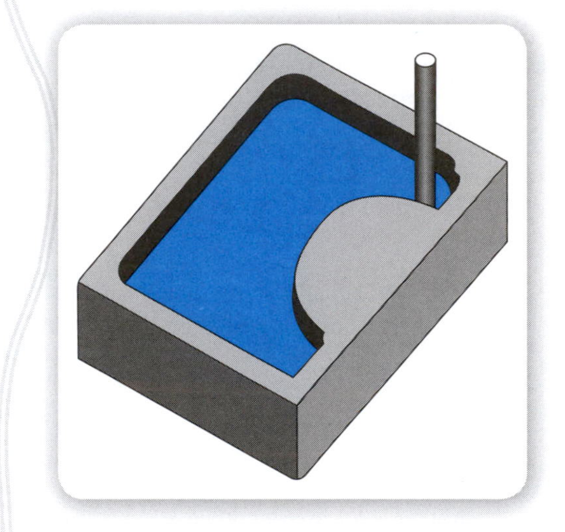

Figure 13.3 The model is verified.

Figure 13.4 Post processing produces the finished G & M-code.

```
N110 T1 M6
N112 G0 G90 X2.5949 Y.7796 A0. S6112 M3
N114 G43 H1 Z.1
N116 M8
N118 G1 Z–.5 F12.2
N120 Y.5446 F24.4
N122 G2 X2.45 Y.405 I–.1449 J.0054
N124 G1 X2.4124
N126 Y.8954
N128 G2 X2.2299 Y.9687 I.3376 J1.1046
N130 G1 Y.405
N132 X2.0475
N134 Y1.0832
N136 G2 X1.865 Y1.2579 I.7026 J.9168
```

13.3 DESCRIPTION OF CAD/CAM COMPONENTS AND FUNCTIONS

CAD/CAM systems contain both CAD and CAM capabilities—each of which has a number of functional elements. It will help to take a short look at some of these elements in order to understand the entire process.

CAD Module

The CAD portion of the system is used to create the geometry as a CAD model. The CAD model is an electronic description of the workpiece geometry that is mathematically precise. The CAD system, whether standalone or as part of a CAD/CAM package, tends to be available in several different levels of sophistication. The CAD software generally represents the geometry, as illustrated in *Figure 13.5*, by one of the following methods:

Case Study

Motorcycle Design with Mastercam

The *American Chopper* TV show lets viewers look over the shoulders of the designers and machinists at Orange County Choppers (OCC) as they conceive and build one-of-a-kind chopper motorcycles and accessories.

OCC is very high tech. They rely on the latest waterjet system capable of cutting through ceramic, glass, even a 12-inch block of steel with minimal heat and no distortion of the material. They use a mandrel tube bender to create unique snaking exhaust systems without leaving any ripples in the metal. Advanced CNC lathes and mills turn and cut these exotic and beautiful parts.

OCC imports its designs seamlessly from CAD files into Mastercam and translates them into manufacturing programs. Mastercam itself has substantial design capabilities and OCC also creates unique parts within the Mastercam graphic environment.

In 2006, OCC began expanding its manufacturing capacity. The first step was moving its bike manufacturing operations into a new 100,000 sq. ft. facility where it can produce selected bikes on a semiproduction basis. Lead Engineer Jim Quinn said that OCC will have to streamline its manufacturing procedures to achieve output targets.

Jim and his staff will be using Mastercam to refine or redesign tooling and fixtures for increased productivity. For example, an adjustable jig designed in Mastercam and used to fabricate custom bikes will be redesigned for production of limited-edition vehicles. The custom jig makes it possible to adjust the bike's "rise" and "stretch" to make each bike's dimensions proportional to the measurements of the person who will be riding it most often. It's a great timesaver in building custom bikes.

"Once we are ready to make hundreds of bikes with the same 'rake and ride' specifications, we already have that jig designed in Mastercam," said Quinn. "It will be quite easy to go back in and duplicate the dimensions we need, take out a lot of the adjustability, and generate the CNC programs to make our production jig." The same concept also holds for other tooling and fixtures.

The old plant will become the CNC production shop. OCC intends to "cherry-pick" its custom part, accessory, and wheel designs to identify ones that would be most appealing to fans and manufacture these on a production basis. "Nasty Wheel," the engraved Orange County Chopper logo air cleaner cover and the OCC Dagger Shield coil cover will be among the first.

Mastercam will have a central role in optimizing part manufacturing productivity. The OCC Dagger Shield coil cover, the company's signature accessory, is a prime example. It was designed in SolidWorks and imported into Mastercam. Quinn said, "Mastercam allows me to turn, manipulate, and move that part into whatever situation I need to hold it securely and machine it efficiently. Because the work piece is already cut to near net shape with only light cuts required around the perimeter of the shield, the part can be cut at higher speed with less force for a faster production cycle and better surface finish." Mastercam's 3D surfacing toolpath "is wonderful." The polishing they do to get parts ready for chroming is very minimal.

Quinn said that the first couple times this part was made on a prototype basis; the complete process took about six hours. Now after tweaking the various steps, they have the time down to one and a half hours. The part is production-ready.

In just six years, Discovery Channel viewers have watched OCC grow from three employees to 60, and boost its production to 80 commissioned custom choppers in 2005. The new plant will allow OCC to boost overall production to 120 bikes in 2006 and as many as 240 in 2007.

2-D LINE DRAWINGS Geometry is represented in two axes, much like drawing on a sheet of paper. Z-level depths will have to be added on the CAM end.

3-D WIREFRAME MODELS Geometry is represented in three-dimensional space by connecting elements that represent edges and boundaries. Wireframes can be difficult to visualize, but all Z-axis information is available for the CAM operations.

3-D SURFACE MODELS These are similar to wireframes except that a thin skin has been stretched over the wireframe model to aid in visualization. Inside, the model is empty. Complex contoured surfaces are possible with surface modelers.

Figure 13.5 Four possible variations of CAD modeling: (clockwise from the top) 2-D, 3-D Wireframe, Solid Model, and Surface Model. Notice that the corner has been sliced off the solid and surface model to expose the interior. The surface model reveals a hollow interior while the solid model contains a filled interior.

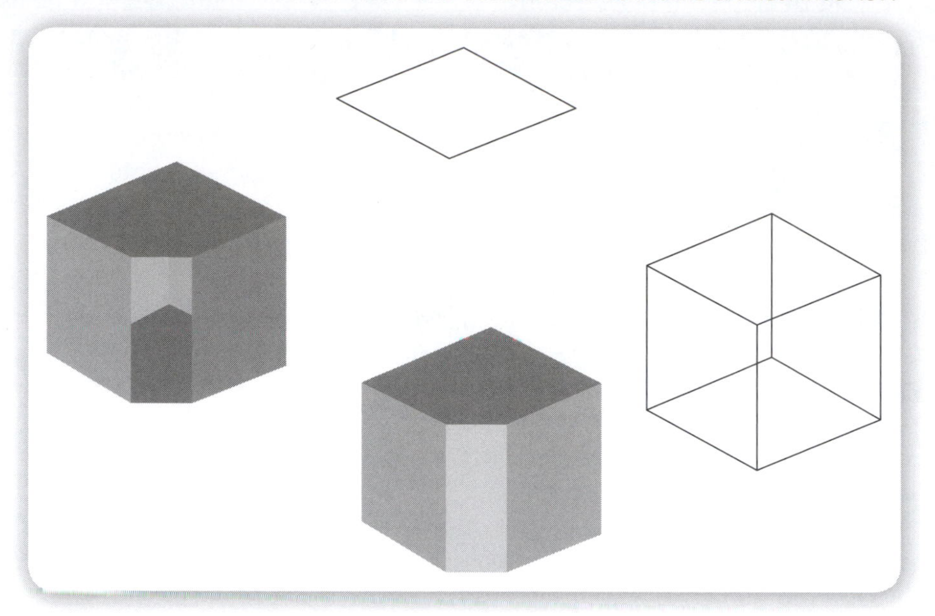

3-D SOLID MODELING This is the current state of the market technology that is used by all high-end software. The geometry is represented as a solid feature that contains mass. Solid models can be sliced open to reveal internal features and not just a thin skin.

CAM Module

The CAM module is used to create the machining process model based upon the geometry supplied in the CAD model. For example, the CAD model may contain a feature that we recognize as a pocket (as we saw in *Figure 13.1*). We could apply a pocketing routine to the geometry, and then all of the toolpaths would be automatically created to produce the pocket. Likewise, the CAD model may contain geometry that should be produced with drilling operations. We can simply select the geometry and instruct the CAM system to drill holes at the selected locations.

The CAM system will generate a generic intermediate code that describes the machining operations, which can later be used to produce G & M code or conversational programs. Some systems create intermediate code in their own proprietary language, while others use open standards such as APT for their intermediate files.

The CAM modules also come in several classes and levels of sophistication. First, there is usually a different module available for milling, turning, wire EDM, and fabrication. Each of the processes is unique enough that the modules are typically sold as add-ins. Each module may also be available with different levels of capability. For example, CAM modules for milling are often broken into stages as follows, starting with very simple capabilities and ending with complex, multi-axis toolpaths:

- 2½-axis machining
- Three-axis machining with fourth-axis positioning
- Surface machining
- Simultaneous five-axis machining

Each of these represents a higher level of capability that may not be needed in all manufacturing environments. A job shop might only require three-axis capability, while a mold shop might need full surfacing capability, as illustrated in ***Figure 13.6*** and ***Figure 13.7***. An aerospace contractor might need a sophisticated five-axis CAM package that is capable of complex machining. This class of software might start at $5,000 per installation, but the most sophisticated modules can cost $15,000 or more. Therefore, there is no need to buy software at such a high level that we will not be able to use it to its full potential.

Figure 13.6 **A typical surface that might be machined with a CAM system.**

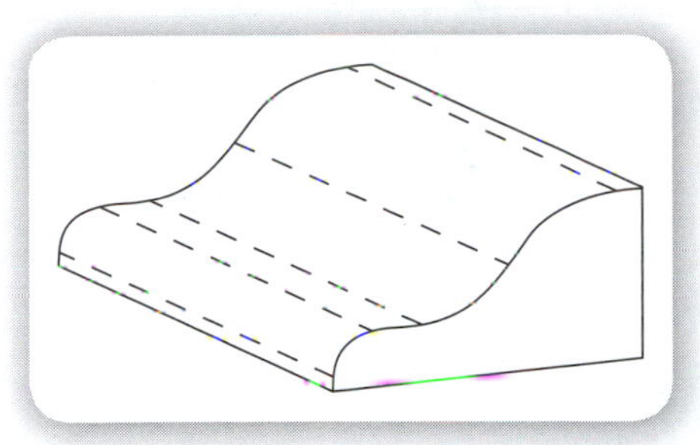

Figure 13.7 **The resulting toolpath that a ball end mill might follow to produce the surface.**

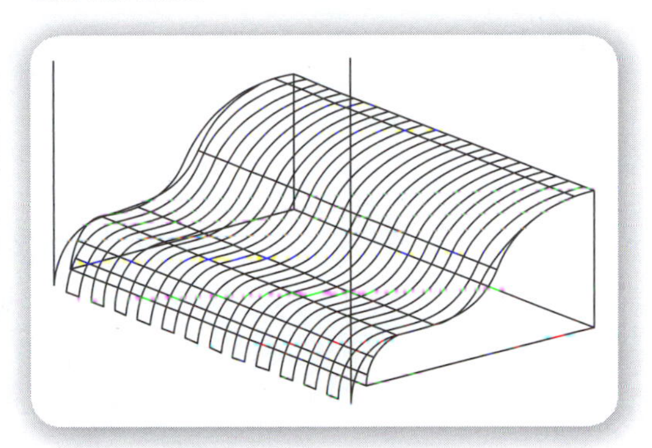

Geometry Versus Toolpath

One important concept we must understand is that the geometry represented by the CAD drawing may not be exactly the same geometry that is produced on the CNC machine tool. CNC machine tools are equipped to produce very accurate toolpaths as long as the toolpaths are either straight lines or circular arcs. CAD systems are also capable of producing highly accurate geometry of straight line and circular arcs, but they can also produce a number of other classes of curves. Most often these curves are represented as Non-Uniform Rational Bezier Splines (NURBS) as shown in ***Figure 13.8***. NURBS curves can represent virtually any geometry, ranging from a straight line or circular arc to complex lofted surfaces.

NURBS:
Non-Uniform Rational B-Splines are a class of curve that allows for complex geometry to be specified through simple "control points". NURBS are very useful in computer-aided design.

Take, for example, the geometric entity that we call an ellipse. An ellipse is a class of curve that is mathematically different from a circular arc. An ellipse is easily produced on a CAD system with the click of the mouse. However, a standard CNC machine tool cannot be used to directly produce an ellipse. It can only create lines and circular arcs. The CAM system will reconcile this problem by estimating the curve with line segments as in ***Figure 13.9***.

The CAM system will generate a bounding geometry on either side of the true curve to form a *tolerance zone*. It will then produce a toolpath from the line segment that stays contained within the tolerance zone. The resulting toolpath will not be mathematically correct—the CAM system will only be able to estimate the surface. This basic method is used to produce estimated toolpaths for both 2-D curves and 3-D surface curves.

Some CAM programs also have the ability to convert the line segments into arc segments. This can reduce the number of blocks in the program and lead to smoother surfaces.

The programmer can control the size of the tolerance zone to create a toolpath that is as accurate as is needed. Smaller tolerance zones will produce finer toolpaths and more numerous line segments, while larger tolerance zones will produce

Figure 13.8 *Complex surfaces, such as this dome, are often represented in the CAD model with a class of curves knows as NURBS. In order to machine the surface, the CAD/CAM system will create toolpaths that follow along the NURBS curve boundary. The CAD/CAM system will generate G & M-code for arcs and straight lines in order to estimate the surface.*

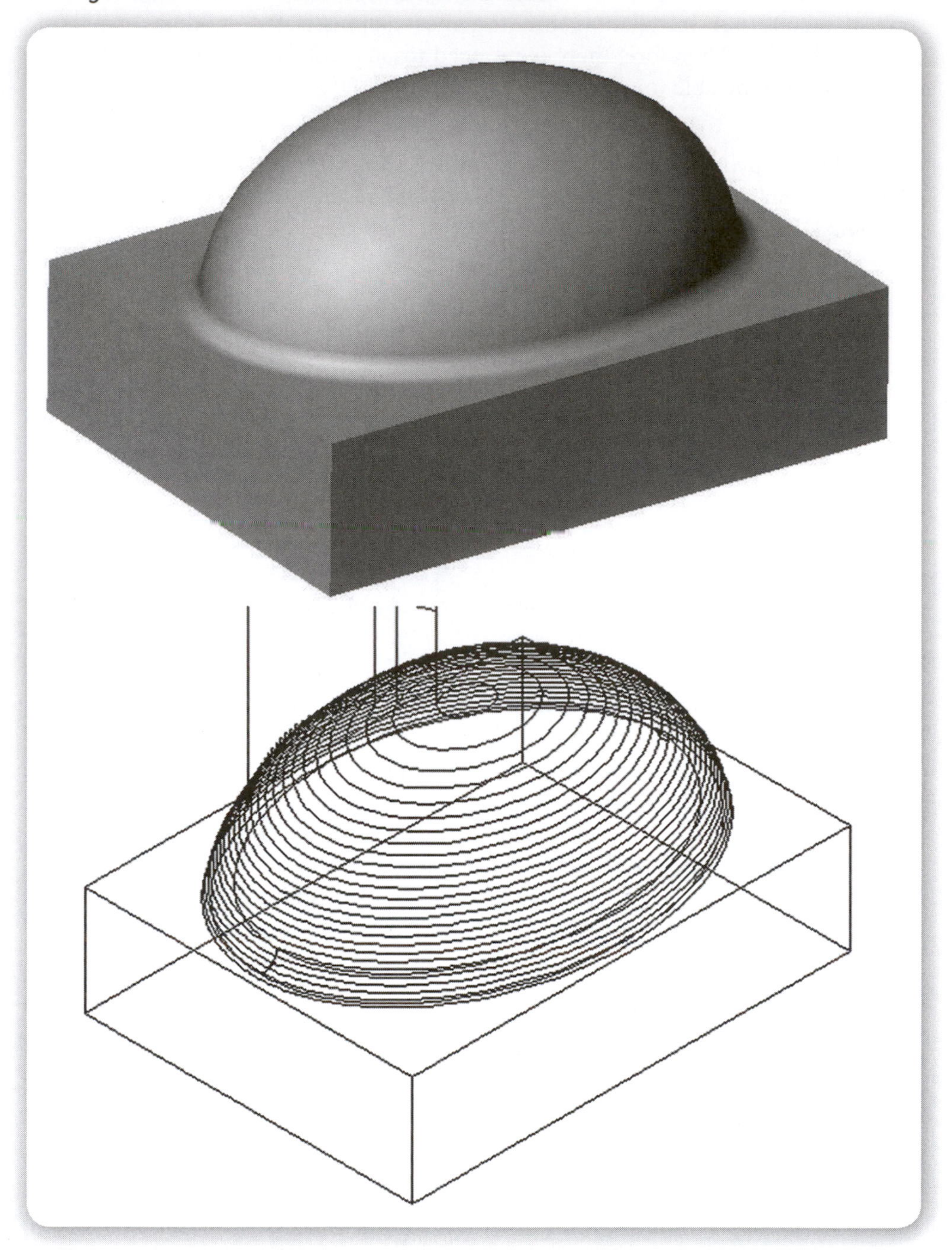

fewer line segments and coarser toolpaths. Each line segment will require a block of code in the NC program, so the NC part program can grow very large when using this technique.

We must use caution when machining surfaces. It is easy to rely on the computer to generate the correct toolpath, but finished surfaces are further estimated during machining with ball end mills. If we do not pay attention to the limitations of these techniques, then the accuracy of the finished workpiece may be compromised.

Figure 13.9 CNC machine tools usually only understand circular arcs or straight lines. Therefore, the CAM system must estimate curved surfaces with line or arc segments. The curve is this illustration is that of an ellipse and the toolpath generated consist of tangent line segments that are contained within a tolerance zone.

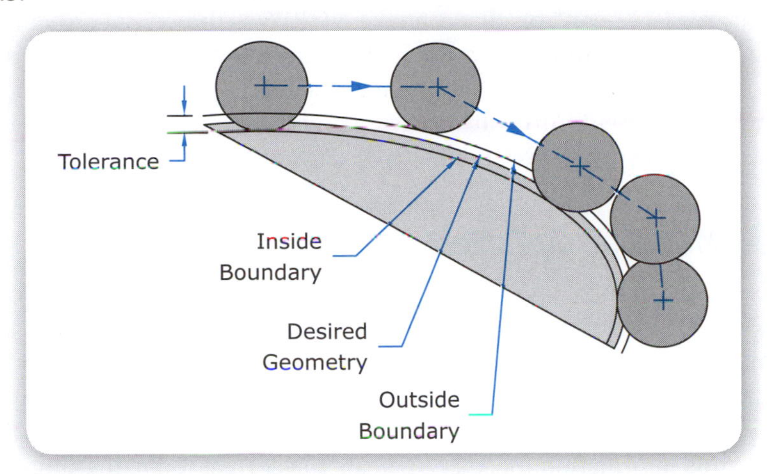

NURBS Curves

Previously we discussed that line segments and arcs are used to estimate curves. However, much of the new CAD/CAM software is choosing to use a mathematical class of curves called NURBS. NURBS is short for Non-Uniform Rational B-Splines.

NURBS use a series of control points to define a curve (***Figure 13.10***). While the curve itself passes through an infinite number of points in space, the curve can be defined with very little data. Because the curve is mathematically defined, the control parameters can simply be passed to another software (or CNC control) and it will be reproduced precisely. Also interesting is that NURBS can represent both straight line and circular arcs.

Figure 13.10 NURBS curves can accurately represent complex geometry in 2-D or 3-D space. The geometry can be interpreted precisely by CAD/CAM systems which understand the NURBS parameters and equations.

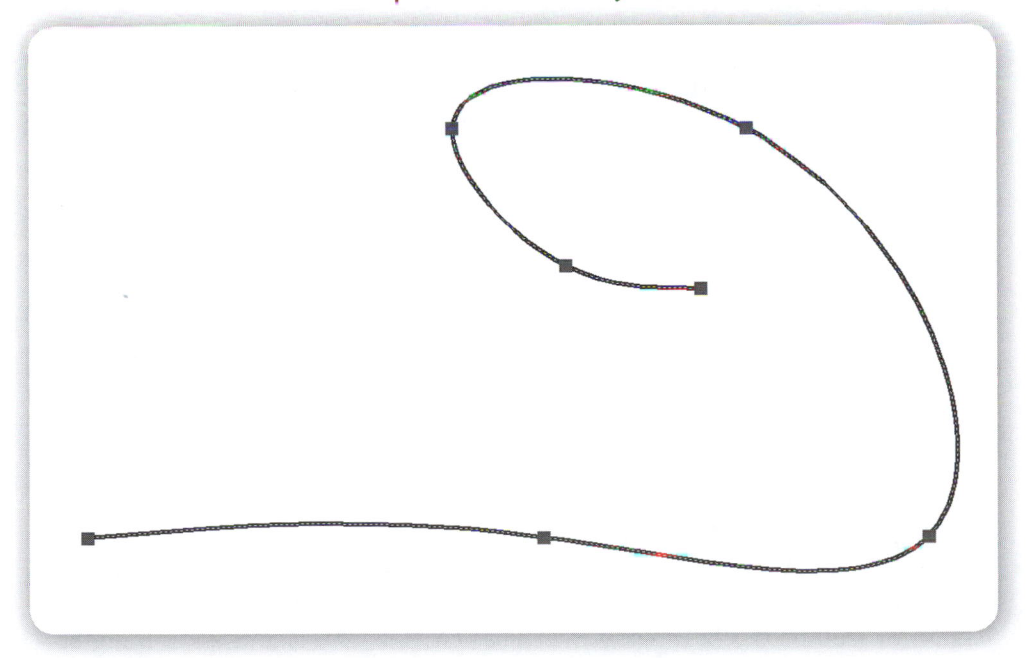

Much of the standard software today will use NURBS to represent all of the 2-D and 3-D curves in a design. For that matter arcs, spheres, lines, and planes are also represented internally as NURBS curves. However, most CNC machine tools only recognize lines and arcs. So at the code output stage, the CAD/CAM software will create lines and arc segments that estimate the curves. The technological trajectory is moving towards CNC controls that can take NURBS as direct input, that is NURBS interpolation. A few manufacturers are now producing CNC controls with NURBS interpolation; however, standardization and consumer acceptance has been slow. Furthermore, the CAD/CAM software companies have been slow to create a connection between their software and the CNC control.

Tool and Material Libraries

To create the machining operations, the CAM system will need to know which cutting tools are available and what material we are machining. CAM systems take care of this by providing customizable libraries of cutting tools and materials. Tool libraries contain information about the shape and style of the tool. Material libraries contain information that is used to optimize the cutting speeds and feeds. The CAM system uses this information together to create the correct toolpaths and machining parameters.

The format of these tool and material libraries is often proprietary and can present some portability issues. Proprietary tool and material files cannot be easily modified or used on another system. More progressive CAM developers tend to produce their tool and material libraries as database files that can be easily modified and customized for other applications.

Verification

CAM systems usually provide the ability to verify that the proposed toolpaths are correct. This can be via a simple backplot of the tool centerline or via a sophisticated solid model of the machining operations. The solids verifications is often a third-party software that the CAD/CAM software company has licensed. However, it may be available as a standalone package.

Post-Processor

post-processing:
The operation in which CAD/CAM machining operations will be converted to machine specific G & M-code.

The post-processor is a software program that takes a generic intermediate code and formats the NC code for each particular machine tool control. The post-processor can often be customized through templates and variables to provide the required customization.

Portability

Portability of electronic data is the Achilles' heel of CAD/CAM systems and continues to be a time-consuming concern. CAD files are created in a number of formats and have to be shared between many organizations. It is very expensive to create a complex model on a CAD system; therefore, we want to maximize the portability of our models and minimize the need for recreating the geometry on another system. Table 13.1 shows a few of the common formats for CAD data exchange.

CAM process models are not nearly as portable as CAD models. We cannot usually take a CAM model developed in one system and transfer it to another platform. The only widely accepted standard for CAM model interchange is a version of Automatically Programmed Tool (APT). APT is a programming language used to describe machining operations. APT is an open standard that is well

Table 13.1 *Common CAD File Formats*

Format	Description and Comments
DXF	■ Widely accepted and available exchange format for 2-D and 3-D wireframe geometry ■ Text, dimensions, and 3-D information are often lost due to version differences and incompatibility in the CAM system
DWG	■ A very popular format native to 2-D and 3-D AutoCAD models ■ CAM translators are usually available at additional cost and often only allow one-way exchanges
IGES	■ 3-D wireframe and surface model data ■ Widely accepted ■ Translation errors are common and CAD/CAM system might not be able to modify data
SAT	■ Solid models native to the ACIS modeling kernel ■ Works well particularly if both the CAD and CAM system use ACIS
STL	■ Solid data used for rapid prototyping ■ Widely accepted and available
Parasolids (X_T and X_B)	■ Native parasolids solid modeling format. Currently the dominant solids modeling kernel that is used by many CAD and CAD/CAM systems ■ Perfect data translation between CAD and CAD/CAM systems that incorporate the parasolids modeling kernel

Rapid Prototyping (RP):

A group of technologies that can create semi-functional products directly from CAD models without the use of traditional machining, casting or molding. Generally used for mock-ups or to verify designs.

documented and can be accessed by third-party software developers. A number of CAD/CAM systems can export to this standard, and the CAM file can later be used by post-processors and verification software.

There are some circumstances when the proprietary intermediate files created by certain CAD/CAM systems can be fed directly into a machine tool without any additional post-processing. This is an ideal solution, but there is not currently any standard governing this exchange.

One other option for CAD/CAM model exchange is to use a reverse post-processor. A reverse post-processor can create a CAD/CAM model from a G & M-code NC part program. These programs do work; however, the programmer must spend a considerable amount of time determining the design intent of the model and to separate the toolpaths from the geometry. Overall, reverse post-processing has very limited applications.

13.4 SOFTWARE ISSUES AND TRENDS

Throughout industry, numerous software packages are used for CAD and CAD/CAM. Pure CAD systems are used in all areas of design, and virtually any product today is designed with CAD software—gone are the days of pencil and paper drawings. There are hundreds of thousands of CAD seats installed in the United States alone.

CAD/CAM software, on the other hand, is more specialized. CAD/CAM is a small but important niche confined to machining and fabrication organizations, and it is found in much smaller numbers than its CAD big brother. The number of seats installed measures only in the tens of thousands.

CAD/CAM systems contain both the software for CAD design and the CAM software for creating toolpaths and NC code. However, the CAD portion is often weak and unrefined when compared to much of the leading pure CAD software. This mismatch sets up the classic argument between the CAD designers and the CAD/CAM programmer on what is the *best* way to approach CAD/CAM.

A great argument can be made for creating all geometry on an industry-leading CAD system and then importing the geometry into a CAD/CAM system. A business is much better off if its engineers only have to create a CAD model one time and in one format. The geometry can then be imported into the CAD/CAM package for process modeling. Furthermore, industry-leading CAD software tends to set an unofficial standard. The greater the acceptance of the standard, the greater the return on investment for the businesses that own the software.

The counter argument comes from small organizations that do not have the need or resources to own both an expensive, industry-standard CAD package and an expensive CAD/CAM package. They tend to have to redraw the geometry from the paper engineering drawing or import models with imperfect translators. Any original models will end up being stored as highly non-standardized CAD/CAM files. These models will have dubious prospects of ever being translated to a more standardized version.

Regardless of the path that is chosen, organizations and individuals tend to become entrenched in a particular technology. If they have invested tremendous effort and time into learning and assimilating a technology, then it becomes very difficult to change to a new technology, even when presented with overwhelming evidence of a better method. It can be quite painful to change. Of course, if we had a crystal ball and could see into the future, this would never happen; but the fact is that we cannot always predict what the dominant technology will be even a few years down the road.

STEP models:

A neutral CAD model format that allows the easy exchange of product information between engineers and manufacturers related to geometry, tolerances, processes, etc.

The result is technology entrenchment that can be very difficult and expensive to get out from under. About the only protection we can find is to select the technology that appears to be the most standardized (even if it is imperfect) and stay with it—then, if major changes appear down the road, we will be in a better position to adapt.

CIM

Computer Integrated Manufacturing (CIM):

A system of manufacturing that uses computer information systems to plan, monitor and control manufacturing processes.

Computer Integrated Manufacturing (CIM) is a term that is used to describe the computer control of manufacturing systems. Ideally, a CIM system is capable of initiating the manufacture of a product and monitoring the progress throughout the production through feedback loops. CIM is not a just single entity, but a system of smaller technologies working together. CIM encompasses all of the technology ranging from quality control systems to CAD/CAM. CIM is largely used to describe the technological artifacts and managerial philosophies of the modern manufacturing environment.

Flexible Manufacturing System (FMS):

Groups of CNC machine tools that are highly integrated with automated material handling and computerized control systems.

CIM is often associated with automation. Robotics and Flexible Manufacturing Systems (FMS) are commonly cited belonging to CIM. This can be true. An FMS is essentially a group of CNC machines that produce a mix of parts. They generally have some form of automated material handling (such as robotics) and an overall control scheme which might interface with a more broad-based system that manages an enterprise (see ERP/MRP in the following section).

STEP Solid Models—A New Paradigm for CAD/CAM

One exciting new idea in CAD/CAM is the concept of a data file format called STEP that can contain information about not only the geometry of the part but also the manufacturing, inspection and assembly of the workpiece. STEP is an acronym for Standard for the Exchange of Product data. The concept is elegant: create a single computer model that can be used by any CAD or CAD/CAM system (so called "format neutral") that also contains embedded information related to manufacturing.

CAD solid model files generally include only the theoretical shape and volume of a design which is constructed of features such as extrusions, revolution, holes, etc. Even simple information—like a size tolerance--is communicated at the detailing level. In other words, an engineer creates a solid model of the workpiece, but also has to create another document called an engineering drawing that gives the manufacturer all of the rules necessary to produce the part.

In the STEP model concept, all of the necessary information is embedded in the features of the model. For example, a tolerance can be attached to a feature such as a hole. Because this information about the tolerance is built-in to the model, a CAD/CAM system can now pick up the information and decide on how it should be manufactured. It's a hole. How many ways can we make a hole? Any machinist knows that it depends on the size, surface finish, and tolerance requirements of the feature. Some holes can be drilled and reamed, some milled, yet others are bored or even ground. It usually takes the intervention of an expert programmer or manufacturing engineer to decide how the hole gets made. With STEP, a CAD/CAM program could theoretically use the embedded information to decide how to make decision about the manufacturing operations to produce the workpiece. This automated process, called Autonomous STEP-compliant CNC (ASNC), will certainly lead to great productivity gains in the near future. To illustrate, imagine that the NC programmer simply receives the electronic file from the customer via email. He then imports it into a CAD/CAM system which intelligently selects the machining operation and tools to create toolpaths for the part. The programmer only needs to make modifications to the process based on special circumstances.

ERP/MRP

In this industry you will eventually be compelled to interface with the company's ERP/MRP system. This is another one of those acronyms a that is used to obfuscate a simple concept. Enterprise Resource Planning (ERP) and Manufacturing Requirements Planning (MRP) are a couple of fancy names for the software that an organization uses to run their business. To be fair, this incredible technology really is what we would expect today—one computer system that ties together all of the activities related to selling, producing and delivering a product. ERP/MRP systems tie together financial, purchasing engineering, manufacturing and delivery functions in an organization.

ERP/MRP:
Enterprise Resource Planning and Manufacturing Resource Planning are computer systems capable of integrating sales, engineering and manufacturing data to plan for and procure materials and resources to support production.

The typical ERP/MRP system is modular. There are financial and accounting modules that allow the management to predict and monitor cost and profits. Other modules will schedule machinery and personnel. The manufacturing modules usually allow for process planning which will in turn drive scheduling and cost; requisition cutting tools, and schedule preventative maintenance.

The ERP/MRP system connects all aspects of the manufacturing business. Imagine that the sales department manages to sell a number of new winch assemblies to a tractor manufacturer. It all starts with the order. What is the quantity promised quantity and delivery date? The order triggers material purchase orders. Stock items like bearing, paint and fasteners are inventoried to determine the quantity of each component that will need to be purchased. Other specialty parts will need to be produced in-house from raw stock and custom castings. The ERP/MRP system checks the production schedule to see when machines and personnel are available and then issues work orders to get the production started. There may be a bottleneck at a particular process. The system will alert a manager who will then make a human judgment about the priority of the competing work orders.

The ERP/MRP system is central to running a large organization that is fundamentally in the business of making things. For example, a manufacturing engineer might receive a production order for a particular part. From there, he will be able to link to the related engineering drawings and data. The manufacturing engineer will need this information for he will be responsible for planning the manufacturing operations for the order. These operations will then drive cost estimates and delivery schedules. Any change along the way will cascade down the line. The purchasing department will see that they will need to place orders for castings from the factory in India. The shipping and receiving department will see that they need to schedule trucks for to deliver the assemblies at the appointed time. Human Resources will see that they need to hire machinists and mechanics. So on and so forth. Everyone has access to the data they need which is all driven from on point.

Other Classes of Software

Much of what we know as CAD/CAM software can be classified as shop floor-level software. This is software that comes from a machine shop perspective and has been traditionally used by machinists, CNC programmers, and manufacturing engineers. The design tools are generally rudimentary and would not likely be used as a primary CAD system by designers and engineers. These systems were developed specifically to perform CAD/CAM.

Another class of software began life from a design perspective and then eventually entered the CAM arena. This software starts as the primary design software that is used throughout an organization and may be described as *enterprise-level* software. The trend in the industry is toward enterprise-level software solutions. Rather than having disparate software from many different vendors, larger organizations have chosen to make the move toward one software solution that every section of the organization can use.

Computer-Aided Engineering (CAE):
Computer software that is used to aid in the engineering of complex components and systems. This class of software generally contains a core CAD modeler, components for engineering analysis, and CAM.

Software of this class often starts with a core CAD system that is used to design individual mechanical components and assemblies. From the CAD core, modules are added not only for CAM, but also for Computer-Aided Engineering (CAE). This might include software modules for Finite Element Analysis (FEA), dynamic load and vibration analysis, heat transfer, and injection molding design.

The ideal behind this solution is to provide seamless portability between the CAD model and all the possible software that might be needed in the design, engineering, and manufacturing processes. However, the approach can take on a few different forms. One approach is for the software vendor to provide all of the pieces. This provides consistency, one-stop-shopping, and technical support. A second approach is for the software vendor to provide the modeling core and then provide an interface for third-party developer to create the different modules that operate as plug-ins within the main program. The latter method tends to lead to more innovation but comes at the expense of dealing with multiple vendors.

It can often be difficult to sort out the different software that is available. Table 13.2 lists some of the more popular software that is commercially available. The categories are general and in no way represent the great variety of available software.

Table 13.2 *Comparison of Commercial CAM Software*

Shop-floor CAD/CAM	Core CAD with Third-Party Plug-ins for CAM/CAE	Enterprise-level CAD/CAM/CAE
Mastercam	SolidWorks	Pro/Engineer
SurfCAM	Solid Edge	Ideas
EdgeCAM	Autodesk Inventor	CATIA
Esprit		Unigraphics

Your Turn

Create a mold for a candy bar using cad/cam software. The mold should incorporate many of the featured NC characteristics from previous chapters.

CHAPTER SUMMARY

- CAD/CAM is the term used to describe a class of software that is used to create NC toolpaths from CAD geometry.
- Post-processors are software programs that generate the G & M code for a specific machine tool based upon the CAM process model. The post-processor can often be customized to provide the desired output.
- CAD files can be translated into standard formats for data exchange between different systems. CAM model, tool, and material information are much less standardized and cannot be as easily moved between dissimilar systems.
- CAD systems use a variety of mathematical models to represent geometry. CAM systems must take these mathematical representations and translate them into lines and arcs that can be produced with G & M code.

BRING IT HOME

1. Why would we need to use CAD/CAM? Why not just write G & M code by hand?
2. What kind of problems can be encountered when transferring CAD models to a CAM system? Are there any solutions to these problems?
3. Explain the current state of standardization in CAD/CAM. Can you suggest any strategies to overcome some of these limitations?
4. Do CAM models always accurately represent toolpaths? Explain.
5. What is the difference between a shop-floor CAD/CAM system and an enterprise-level system?

Kemp
ALLEN
Westminster
Harrod
Maysville
Alger
McGuffey
HARDIN
Grant
Kenton
Forekar
Hepburn
Meeker
Marseilles
Harpster
Little Sandusky
Wyandot
New Winchester
Galion
Kirkpatrick
Morral
Caledonia
Martel
Iberia
MORROW
Denmark
Mount Gilead
AUGLAIZE
Uniopolis
Waynesfield
Cridersville
Hume
Blanchard
Big Island

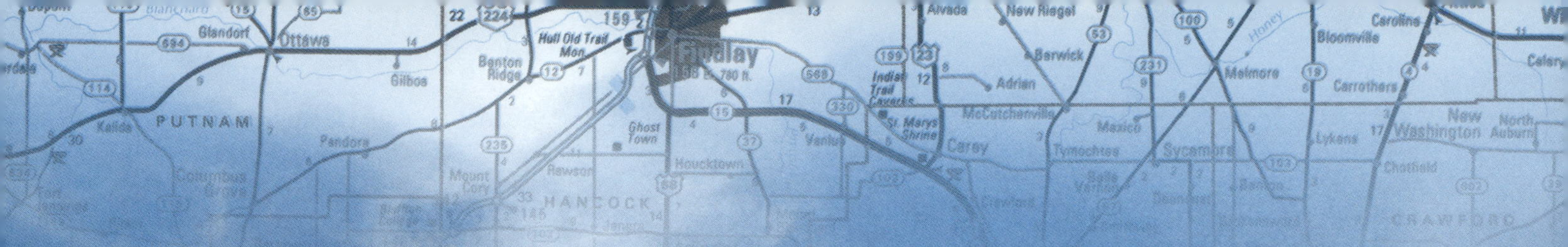

CHAPTER 14
Mathematics for NC Programming

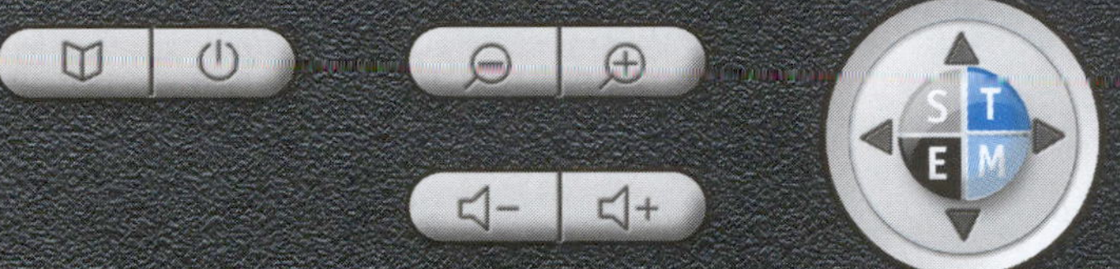

Menu

START LOCATION | DISTANCE | END LOCATION

Before You Begin

Think about these questions as you study the concepts in this chapter:

1. Why do we need to be able to convert English to metric and metric to English?
2. How is algebra used in calculating manufacturing information?
3. How is trigonometry used to solve right angle and oblique triangle problems?

Key Terms

Acute
Complementary
Constant
Equation
Oblique
Obtuse
Proportion
Ratio
Right Angle
Supplementary
Tangent
Variable

Machinists have always needed the ability to apply the principles of mathematics to solve manufacturing problems. This is also true of the CNC machinist and programmer, except that the calculations have become more frequent and complex. The conventional machinist often relies on his senses to hear, see, or feel that the cutting conditions are correct—the CNC machinist often does not have this luxury. The *feel* of metal cutting is gone, and by the time the CNC machinist *hears* that something is wrong, often the only thing left to *see* is a broken cutting tool and a scrap workpiece.

Geometry that was once difficult or impossible to produce with conventional machining is now common in CNC machining. Consequently, NC programmers are required to determine the tool positions in many unlikely situations.

Machinists and programmers tend to be people who have good visual/spatial capabilities; they can visualize relationships between physical elements. Fortunately, there seems to be a high correlation between visual thinking and being good with geometry. The trick is to be able to translate visual information into mathematical ideas and vice versa. In the following section, we will concentrate on this idea as well as review some of the more common ideas in mathematics that you will encounter on the job.

complementary: Describes angles that when added together equal 90°. The two acute angles of a right triangle are complementary.

14.1 BASIC CONCEPTS: SPEAKING THE LANGUAGE

The mathematics found in the machine trade is filled with various symbols and notations that can seem difficult to understand at first. Different sources may use dissimilar notations, which can add to the confusion. The key to understanding the technical information you will be confronted with is first to understand the language.

Two confusing notations are found with the arithmetic operators. First, the symbol for multiplication can be expressed in any of the following forms:

$$3 \times 3 = 9$$
$$3 \cdot 3 = 9$$
$$3(3) = 9$$

The symbol to indicate division can also have several forms:

$$9 \div 3 = 3$$
$$9/3 = 3$$
$$\frac{9}{3} = 3$$

Equalities and Equations

Equality is a short way of saying that two things are in the state of being equal to each other. For example, we might say that two is equal to one plus one. Equalities are used frequently to describe how expressions or terms are related, and they can serve as a definition for later operations. Some examples follow:

$$2 = 8/4$$
$$3 + 3 = 10 + 4 - 8$$
$$4^2 = 4 \times 4$$
$$2b = 8$$
$$a = b = c$$

There are other symbols to describe these relationships. Table 14.1 contains many frequently used symbols, so you might want to become familiar with them.

Table 14.1 Frequently Used Symbols

Symbol	Meaning	Example
$=$	Equality	$5 = 4 + 1$
$\neq$	Not equal	$5 \neq 4 + 2$
$\approx$	Approximately equal	$\sqrt{2} \approx 1.414$
$<$	Less than	$5 < 6$
$>$	Greater than	$6 > 5$
$\leq$	Less than or equal	$5 \leq 6, 6 \leq 6$
$\geq$	Greater than or equal	$6 \geq 5, 5 \geq 5$

equation:

A mathematical equality that contains known and unknown quantities. Equations often show a relationship between a number of variables, such as the relationship between tool diameter and RPMs for a constant cutting speed.

Equation is a term used to describe when two algebraic expressions are in a state of equality to each other. The machinist is confronted with many equations to find values for everything from cutting speeds to the angles in a triangle. Many of the equations are *plug-n-chug*; you simply fill in the blanks and out comes the correct answer. An equation that has been arranged to arrive at a useful answer without any further work is often called a formula. Take, for example, the formula for calculating the spindle speed for machining:

$$\text{RPM} = \frac{12 \times \text{Velocity}}{\pi \times \text{Diameter}}$$

With most formulas you are expected to already know the values to be filled in. You then perform the indicated operations and you get the correct answer. Of course, you have to know how to actually perform the operations in the right order. We can also do some rearranging of the equation to form an entirely new and useful formula to find some other value—these subjects are covered in later sections.

constant:

A value that does not change.

Equations use two different types of values: *constants* and *variables*. **Constants** are values within the equation that do not change. The following equation has two variables (a and b) and one constant (3).

$$3 + b = a$$

We can then *evaluate* the equation for a known value of b. For example, if $b = 7$, then we plug 7 in for b and perform the operation.

$$3 + b = a$$
$$3 + 7 = a$$
$$a = 10$$

You may have noticed that a and 10 ended up on different sides of the equation—this is the normal manner in which the results are displayed. However, $a = 10$ and $10 = a$ are mathematically identical.

variable:

The symbolic representation of an element in an equation without regard to an actual numeric value of the element.

Let's try evaluating our spindle speed formula the same way. We already know our two constants, 2 and π ($\pi \approx 3.14159$). The unknown **variable** is RPM, and the two variables we are expected to know are the velocity and diameter. Assume that you were given instructions to use a 1.0" diameter cutter and a velocity of 100 SFPM:

1. Write out the original formula:

$$\text{RPM} = \frac{12 \times \text{Velocity}}{\pi \times \text{Diameter}}$$

2. Plug in the values that you know:

$$\text{RPM} = \frac{12 \times 100}{3.14159 \times 1}$$

3. Perform the operations:

$$\text{RPM} = \frac{1200}{3.14159}$$

$$\text{RPM} = 381.97$$

Order of Operations

It is very important to understand that when working with any mathematical expression, you must complete the operations in a prescribed order. The result of your calculation will change if this order is not followed. Below is a list that describes the order in which mathematical operations must be performed.

1. Groups inside parentheses
2. Exponents and radicals
3. Multiplication and division (performed from left to right, whichever occurs first)
4. Addition and subtraction (performed from left to right, whichever occurs first)

Let's start with an example that follows rule 4. All of the operations are of equal precedence, so we proceed from left to right:

$$3 + 2 - 1 = y$$
$$5 - 1 = y$$
$$y = 4$$

If the expression contains multiplication or division operations, then we must perform them before addition or subtraction:

$$3 + 2 \times 2 - 1 = y$$
$$3 + 4 - 1 = y$$
$$7 - 1 = y$$

Multiplication and division can be performed in any order and the results will be the same:

$$3 \times 8/2 = y$$
$$3 \times 4 = y$$
$$y = 12$$

is the same as

$$3 \times 8/2 = y$$
$$24/2 = y$$
$$y = 12$$

Terms that are raised to a power have a higher precedence than multiplication/division or addition/subtraction:

$$1 + 2 \times 3^2 = y$$
$$1 + 2 \times 9 = y$$
$$1 + 18 = y$$
$$y = 19$$

Similarly, roots are taken before multiplication/division or addition/subtraction:

$$3 \times \sqrt{4} = y$$
$$3 \times 2 = y$$
$$y = 6$$

Finally, any operations within parentheses are performed before all other operations. Be sure to perform the operations within the parentheses according to the standard order of operations.

$$(1 + 2 \times 3) \times 2^2 = y$$
$$(1 + 6) \times 2^2 = y$$
$$7 \times 2^2 = y$$
$$7 \times 4 = y$$
$$y = 28$$

Examples A, B, and C illustrate how the answer will change when the parentheses are in different locations:

A. $1 + 2 \times 3^2 = 19$
B. $1 + (2 \times 3)^2 = 37$
C. $(1 + 2 \times 3)^2 = 49$

Parentheses can also be nested within other parentheses. In that case, operations within the innermost set or sets of parentheses are performed first:

$$((2 + 3) \times 4) + 1 = y$$
$$(5 \times 4) + 1 = y$$
$$20 + 1 = y$$
$$y = 21$$

Exponents and Radicals

Power functions are common in many mathematical formulas—even those found in metalworking. Any number can be raised to a power, as in 3^2, which we would read as "three to the second power." The number three is called the base, and the small superscript two is the exponent. This means that the number three is multiplied by itself two times:

$$3^2 = 3 \times 3$$

Raising a number to the second power is also called squaring, and we might say "three squared." Raising a number to the third power is called cubing and would be read as "three cubed." It does not end there; three can be raised to any number of powers. Fortunately, those expressions do not have a special name to remember.

$$3^3 = 3 \times 3 \times 3 = 27$$
$$3^4 = 3 \times 3 \times 3 \times 3 = 81$$
$$3^5 = 3 \times 3 \times 3 \times 3 \times 3 = 243$$

There are two special cases with exponents that would be useful for you to understand. First, any number raised to the power of zero is equal to one. Second, any number raised to the power of one is equal to itself. These two instances are not all too common, but you will encounter them from time to time.

$$3^0 = 1$$
$$3^1 = 3$$

With larger exponents, you will probably want to use a calculator to speed your calculations and improve accuracy. The typical pocket calculator has a special key for squaring and another key for all other exponents. The x^2 key is used to find the square of a number. Simply enter the number to be squared and then press x^2 the key. The calculator will display the result immediately; you do not have to press the = key. For example, pressing

$$4\ x^2$$

displays the result 16.

Squaring is a very common operation—that is why it has its own key. However, if any other exponent is needed, you can use the y^x key. This works a little differently. You usually enter the base, press the y^x key, then enter the exponent, and then finally press the = key. For example,

$$3\ y^x\ =$$

displays the result 243.

Another concept related to powers is roots. A root is an inverse function that works backward through the operation to determine the power to which a number was raised. Our first example raised three to the power of two, which we know equals nine. If we started out with the number nine and wanted to find its square root, then we would use the root function. This operation uses the following notation, called a radical:

$$\sqrt{9}$$

When you see the radical symbol, it means that you are looking for the base number that when raised to the second power will equal the term under the radical (in this case, 9). Incidentally, the number under the radical is called the *radicand*. Three times three is equal to nine, so the square root of nine is three.

$$3 \times 3 = 9 \text{ therefore, } \sqrt{9} = 3$$

$$4 \times 4 = 16 \text{ therefore, } \sqrt{16} = 4$$

$$5 \times 5 = 25 \text{ therefore, } \sqrt{25} = 5$$

There is also a square root key on your calculator that gives an immediate result. If we wanted to find the square root of 25, the following keystrokes would be used:

$$25\sqrt{x}$$

which displays the result 5.

Square roots are probably the most common root to find, but roots of other orders may be needed. For instance, we may need to find the third root of a number. The third root would be a number that when raised to the third power would equal the term under the radical. To be clear about which order we are looking for, we place the number on the left-hand side of the radical. For example, if we were looking for the third root of 27, the expression would look like the following:

$$\sqrt[3]{27}$$

Again, when you see this symbol, it means that you are looking for the base number that when raised to the indicated power will equal the term under the radical (in this case, 27):

$$3 \times 3 \times 3 = 27 \text{ therefore, } \sqrt[3]{27} = 3$$

Other roots are expressed in a similar manner:

$$4 \times 4 \times 4 = 64 \text{ therefore, } \sqrt[3]{64} = 4$$

$$2 \times 2 \times 2 \times 2 \times 2 = 32 \text{ therefore, } \sqrt[5]{32} = 2$$

Very few roots are perfect whole numbers, so finding a root often requires trial-and-error, and the result is usually an estimation. For all practical matters, a pocket calculator is the easiest way to find a root, and the results are more than accurate enough for our purposes. There is a special key used to find square roots, which is usually accessed by first pressing a key to access the second set of functions for the keypad (2nd or sometimes *Inv*). For example, to find $\sqrt[5]{32}$, we would use the following keystrokes:

$$2^{nd}\ 32\ 5\ =$$

This displays the result 2.

Numbers can also be expressed using negative exponents. Numbers with a negative exponent will always be smaller than one. The negative exponent means that the number is really part of a fraction and that our base and exponent are on the bottom of that fraction—except with a positive exponent. Take, for example, the number 3^{-2}.

$$3^{-2} = \frac{1}{3^2}$$

This is equal to 1/9 = 0.11111111.

On a calculator,

$$3\, y^X\, 2 \,{}^{+}\!/\!_{-} =$$

displays the result 0.11111111.

The last important variation of exponents and radicals is a number with a fractional exponent. Many computer programs must symbolize any radical, such as a square root, as a fractional exponent, so understanding their meaning is important. Working with fractional exponents is not difficult; we simply have to put the exponent in a form that is familiar. In the following example, a radical is put over the 27 and we find the third root of 27.

$$27^{1/3} = \sqrt[3]{27}$$

The fractional exponent may also have a number other than the one in the numerator. For example, $27^{2/3}$ has the fractional exponent 2/3. In such a case, the denominator still becomes the order, but then the whole expression is raised to the second power.

$$27^{2/3} = \sqrt[3]{27^2}$$

It is important to keep notation and order of operations straight. First, we find the third root of 27, then we raise the result to the second power. The result is that $27^{2/3}$ is equal to nine:

$$\sqrt[3]{27} = 3 \text{ and then } 3^2 = 9$$

The main purpose of this discussion, in the context of the machine trades, is to be able to recognize instances of fractional exponents and to be able to crunch the results in a calculator or spreadsheet. Imagine being confronted with the following expression:

$$\sqrt[4]{120^5}$$

This would be tedious to solve by trial-and-error, so you would probably want to use a calculator or spreadsheet. First, we put it in a form that is easier to work with, and then we express the fraction in its decimal form:

$$\sqrt[4]{120^5} = 120^{5/4} = 120^{1.25}$$

Next, we plug it into our calculator:

$$120 \; y^x \; 1.25 =$$

This displays the result 397.17.

Operations with Exponents

Occasionally we may run into a situation where operations must be performed on numbers that are raised to some power. In ordinary situations we must follow the order of operations and take the powers or roots first. However, there are several useful exceptions when terms of the same base are multiplied or divided, or when a power is raised to a power. The three basic rules are as follows:

1. Terms with the same base can be multiplied by adding their exponents together:

$$2^2 \times 2^3 = 2^5$$

2. Terms with the same base can be divided by subtracting their exponents:

$$2^{10} \div 2^8 = 2^2$$

3. When a power is raised to another power, the exponents are multiplied:

$$(2^2)^3 = 2^6$$

Scientific Notation

Scientific notation is often used to describe very large or very small numbers without having to include all of the leading and trailing zeros. The basic idea behind scientific notation is that any number is some factor of ten. For example,

$$120 = 1.2 \times 10 \times 10$$

Since 10×10 is also 10^2, we could write 120 as follows:

$$1.2 \times 10^2$$

Granted, writing out 120 is not going to use too much extra ink, but scientific notation is useful for larger numbers. For example, 1,650,000 can be factored as $1.65 \times 1{,}000{,}000$, and then written in scientific notation as

$$1.65 \times 10^6$$

Scientific notation can also be used for very small numbers that would have a confusing number of zeros after the decimal point. For example,

$$0.0015 = 1.5 \times 0.001 = 1.5 \times 10^{-3}$$

We know from our previous discussion that negative exponents will give results that are smaller than one. You might correctly guess that 0.001 is a power of 10 with a negative exponent.

$$0.001 = \frac{1}{1000}$$

$$0.001 = \frac{1}{10^3}$$

$$0.001 = 10^{-3}$$

Other examples of small numbers that can be expressed in scientific notation are as follows:

$$0.125 = 1.25 \times 10^{-1}$$
$$0.0000451 = 4.51 \times 10^{-5}$$
$$0.000000008 = 8.0 \times 10^{-9}$$

A common example of very small numbers found in the machine trades is the values used with surface finish. Surface roughness is expressed in microinches—millionths of an inch. Rather than writing out that the surface finish is 0.000008, we could use scientific notation to write 8×10^{-6}.

If you are having trouble grasping this idea of scientific notation, especially with negative exponents, you might try another approach to your thinking. One easy way is simply to move the decimal point. If you have a small number and you want to convert it to scientific notation, just write out the number and move the decimal point to the right until you are one place past the first nonzero digit. Next, count the number of places you moved and use that as your exponent. For example, follow these steps:

1. Write down the number—0.00312.
2. Move the decimal point three places to the right—3.12.
3. Write the number in scientific notation—3.12×10^{-3}.

Numbers larger than one can be dealt with in a similar manner:

1. Write down the number—15,600.0.
2. Move the decimal point four places to the left—1.56.
3. Write the number in scientific notation—1.56×10^4.

You can use the same technique if you have started with scientific notation and you wish to convert to decimal form. Let's take the previous example (1.56×10^4) and work backwards:

1. Write down the number in scientific notation—1.56×10^4.
2. Take the coefficient portion of the number by itself—1.56.
3. Move the decimal point four places to the right—15,600.

The technically correct way to express a number in scientific notation is with a coefficient between one and ten. Although each of the three examples following is mathematically correct expressions of 12,000, only 1.2×10^4 is in the proper scientific notation.

$$12 \times 10^3$$
$$1.2 \times 10^4$$
$$0.12 \times 10^5$$

However, it is common practice in engineering, publications, and tables to use an exponent that is a multiple of three (see the bold items in Table 14.2). This may not be the absolutely proper way to write it, but it is done nonetheless. In addition, scientific notation is not usually used if the number is between 999 and 0.01.

A pocket calculator can also handle scientific notation. Anytime numbers of a very small or large magnitude are entered into a calculator, there is a high probability of making a mistake. A misplaced decimal point can lead to a scrap workpiece and leave you with some embarrassing explaining to do. For example,

Table 14.2 Scientific Notation

10^9	**1,000,000,000**
10^8	100,000,000
10^7	10,000,000
10^6	**1,000,000**
10^5	100,000
10^4	10,000
10^3	**1,000**
10^2	100
10^1	10
10^0	1
10^{-1}	0.1
10^{-2}	0.01
10^{-3}	**0.001**
10^{-4}	0.0001
10^{-5}	0.00001
10^{-6}	**0.000001**
10^{-7}	0.0000001
10^{-8}	0.00000001
10^{-9}	**0.000000001**

you might come across a formula to calculate the linear expansion of a material as the temperature changes. The formula involves a very small coefficient that is multiplied by a much larger number (the original length of the material). The first step is to write out the formula, and then convert the small number to scientific notation:

$$5.125'' \times 0.0000093 = \text{thermal expansion per degree}$$
$$5.125'' \times (9.3 \times 10^{-6}) = \text{thermal expansion per degree}$$

Next, enter the formula into your calculator with the following keystrokes:

$$5.125'' \times 9.3\ \mathit{EE}\ 6\ {}^{+}\!/\!_{-} = \text{thermal expansion per degree}$$

This gives the result 0.00004766 or 47.66×10^{-6} inches.

14.2 ESSENTIAL ALGEBRA

We are now at the point where we can begin discussing algebra. You may be surprised and frightened by the fact that you will actually need to use algebra in many situations as a machinist or programmer. Do not worry if your past experiences with algebra have left a bad taste in your mouth. We are only going to concentrate on a few basic rules that can be applied to everyday situations.

It may take you a couple of hours to learn some simple techniques to manipulate and solve algebraic equations, but doing so can bring you a lifetime of mathematical independence.

We already touched lightly on algebra when we saw equations and formulas with variables. Any equation that uses one or more variables is said to be *algebraic*:

$$a + 4 = 7 \text{ is algebraic}$$
$$3 + 4 = 7 \text{ is not algebraic}$$

Variables can be represented by any symbol, including letters, special characters, and even words or abbreviations. Variables are used to represent any quantity that can change. Take the formula for finding the area of a rectangle:

$$a = b \times h$$

We have to define each of the variables before this formula will have any meaning to us at all. We can say that a represents the area, b represents the length of the base, and h represents the height. We could have given the variable any symbol we wanted to, but as long as we define which value the variable represents, the formula will remain correct. For example, we can use the following alternative symbols as long as their meaning is defined:

$$d = e \times f$$

Sometimes a formula is easier to understand if we skip the symbols altogether and just use words. You will find that this is a useful technique if the symbols start to cloud your understanding of the equation.

$$\text{Area} = \text{Base} \times \text{Height}$$

It may be useful at this point to remind you that in algebra the multiplication sign is not normally used to show multiplication between variables. Any time two variables are placed next to each other, multiplication is implied. The following examples are all correct, but the first one will not be used from this point forward unless needed for clarity:

$$a = b \times h$$
$$a = bh$$
$$a = b(h)$$

Our formula for the area of a rectangle is already in the proper form so that we can solve it in terms of the base and height. Nevertheless, what if we already know the area and height, but we really need to find the length of the base? We have to get b on one side of the equation all by itself in order for the answer to make any sense:

$$b = a/h$$

This is called transposing (rearranging) the equation, and it is one of the most useful skills in algebra. We can transpose known formulas to solve for other values within them. The basic idea behind this is that we will add and subtract, multiply and divide until the equation meets our needs. There are just a few simple rules to learn first, and we must remember that any manipulation we perform on one side of the equation must be performed on the other side to maintain equality.

1. To remove a term that is added, subtract it from both sides.
2. To remove a term that is subtracted, add it to both sides.
3. To remove a term that is multiplied, divide both sides by that term.
4. To remove a term that divides another term, multiply both sides by that term.
5. To remove a power, take the root of both sides.
6. To remove a root, raise both sides by its power.

Table 14.3 Algebraic Manipulations

Rule One	**Example**
Write the equation.	$a + 2 = 6$
Subtract 2 from each side.	$a + 2 - 2 = 6 - 2$
Perform the operations.	$a = 4$
Rule Two	
Write the equation.	$a - 2 = 6$
Add 2 to each side.	$a - 2 + 2 = 6 + 2$
Perform the operations.	$a = 8$
Rule Three	
Write the equation.	$3a = 12$
Divide both sides by 3.	$\frac{3a}{3} = \frac{12}{3}$
Simplify and perform any arithmetic.	$a = 4$
Rule Four	
Write the equation.	$\frac{a}{5} = 3$
Multiply both sides by 5.	$\frac{5 \times a}{5} = 3 \times 5$
Simplify.	$a = 15$
Rule Five	
Write the equation.	$a^2 = 16$
Take the root of both sides.	$\sqrt{a^2} = \sqrt{16}$
Simplify.	$a = 4$
Rule Six	
Write the equation.	$\sqrt{a} = 3$
Square both sides.	$(\sqrt{a})^2 = 3^2$
Simplify.	$a = 9$

To illustrate these rules, let's look at a few short examples in Table 14.3 and, in each case, try to solve the equation for a.

The concepts above can be applied to situations that, at first glance, look a little different. For example, we may be confronted with the following formula and need to find x:

$$\frac{2}{x+6} = 10$$

In this case we will treat $x + 6$ as one unit and then apply rule 4:

$$\frac{2}{(x+6)} = 10$$

$$2 = 10(x + 6)$$

Now we have come to a problem that we have not yet encountered. Somehow we have to get x out of $10(x + 6)$. We can do this by multiplying each term inside the parentheses by 10:

$$2 = 10x + 60$$

Next, we can apply rules 1 and 3 to isolate x:

- Subtract 60 from both sides:

$$2 - 60 = 10x$$

- Divide both sides by 10:

$$\frac{2-60}{10} = x$$

- Perform the arithmetic and put into standard form:

$$x = -5.8$$

We can then check our answer by plugging it into the original equation to see if it checks out. If the answer is true, then we have not made a mistake.

$$\frac{2}{-5.8+6} = 10 \Rightarrow \frac{2}{0.2} = 10 \Rightarrow 10 = 10$$

We could go on at length about algebra, but it is getting beyond the scope of this section. At this point, you should be starting to understand that there is a lot more to learn and that a full course in basic algebra would be an appropriate and practical plan for anyone involved in manufacturing.

14.3 UNIT CONVERSIONS

Unit conversions are another common problem faced by programmers and machinists. For example, we may have to convert inches to millimeters, gallons to liters, or horsepower to kilowatts. We can often find a formula in a handbook that will allow us to plug in the appropriate values to calculate the answer. Formulas have a place, but they lack the flexibility to be applied to novel situations. Furthermore, formulas are often not available, leaving us with only a definition.

One easy solution to this problem is to put the definition into a **proportion.** Proportions are fractions that act like mathematical analogies that describe the relationship between values—A is to B as C is to D. The power and beauty of proportions is that it does not matter what type of mathematical analogy we set up, as long as it is the same relationship on both sides.

For example, we might have to convert 75 mm to inches. We look in a handbook and find that 1 in. equals 25.4 mm. How then do we perform the conversion? All we have to do is put one relationship on one side of the equation and the other relationship on the other (x is to 1 as 75 is to 25.4). Then, we solve the equation for the unknown:

$$\frac{x \text{ in.}}{1 \text{ in.}} = \frac{75 \text{ mm}}{25.4 \text{ mm}} \Rightarrow x \text{ in.} = \frac{75 \text{ in.} \times 1}{25.4 \text{ mm}} \Rightarrow x = 2.9527$$

Proportions can be applied to virtually any conversion problem. Additionally, they free us from having to rely on formulas. As long as we understand one of the relationships, we can find the other. For example, imagine that we notice on a package that the weight is given as 18 ounces, or 510 grams. How many grams are there in 16 ounces (1 pound)? We could solve this problem with a proportion:

$$\frac{x\text{ g}}{16\text{ oz.}}=\frac{510\text{ g}}{18\text{ oz.}}\Rightarrow x\text{ g}=\frac{510\text{ g}\times 16\text{ oz.}}{18\text{ oz.}}\Rightarrow x=453$$

14.4 GEOMETRY

Geometry is obviously very important in machining, but often we may not be aware of some of the basic relationships and rules. CNC programming often requires the machinist to make many calculations that are based on the relationship between lines, angles, and circles. Therefore, we might find it useful to review some of the more useful principles. Some other common relationships and principles are available in Appendix B.

Properties of Intersecting Lines

Figure 14.1 illustrates the following properties:

- Two lines that intersect each other will always form two sets of identical angles.
- The two angles on the same side of a line are called **supplementary** and they always add up to 180°. This property can be used to find an unknown angle when the other is known.
- Angles opposite each other will be equal. This property is useful when extending lines to imaginary intersections and can help to solve a number of programming problems.
- A line that is parallel to A will form corresponding angles that are identical to the first intersection.

supplementary: Describes two angles whose sum equals 180°. The adjacent angles of two intersecting lines are supplementary.

Figure 14.1 Intersecting lines (A and B) always form two sets of identical angles. Angles opposite each other will be equal. Likewise, a parallel line (C) will form identical angles.

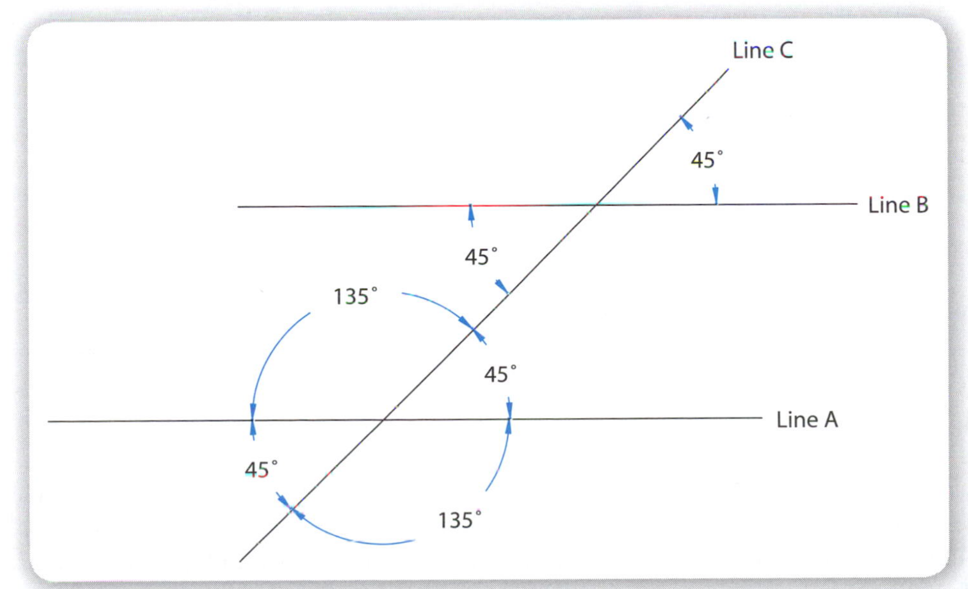

Tangency

tangent:

The condition created when an arc touches another arc, curve, or line at only one point.

- A point of tangency is created when a line touches a circular arc—the line is said to be **tangent** to the arc, as shown in ***Figure 14.2***. Tangent relationships are important for planning toolpaths, because the tool is always tangent to the finished surface as it is cutting. Tangent relationships are also used for inspecting angular surfaces when we cannot reliably measure from a corner. See the *Illustrated Applications* section for an example.
- A line drawn from the tangent point to the center of the arc will be perpendicular to the tangent line. This property allows us to calculate the location of a tangent point when a circle of a known diameter is used.
- When a circular arc is set tangent to two intersecting lines, a line drawn from the center of the circle to the intersection point will perfectly bisect (divide in two) the angle between the lines, as illustrated in ***Figure 14.3***. Again, this is useful for inspection and for planning toolpaths. We can use the known radius and the calculated angles to create a triangle that is coincident to the center, tangent point, and the intersections. We can then use this information and trigonometry to calculate the coordinates of toolpaths and inspection points.

Figure 14.2 A point of tangency is created when a circular arc touches a line. A line drawn from the tangent point to the center of the arc will be perpendicular.

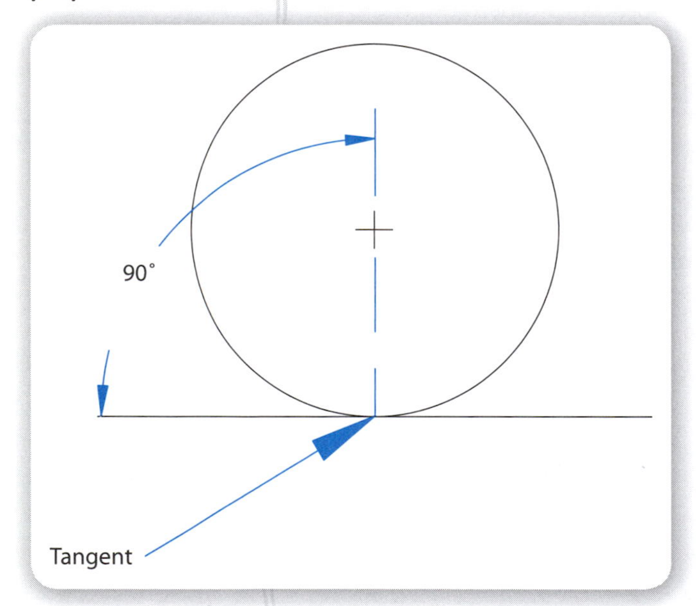

Figure 14.3 When a circular arc is set tangent to two intersecting lines, a line drawn from the center of the circle to the intersection point will bisect the angle and form a perpendicular line from the center of the arc to the tangent point. This concept can be used for layout, inspection and for calculating the location of toolpaths.

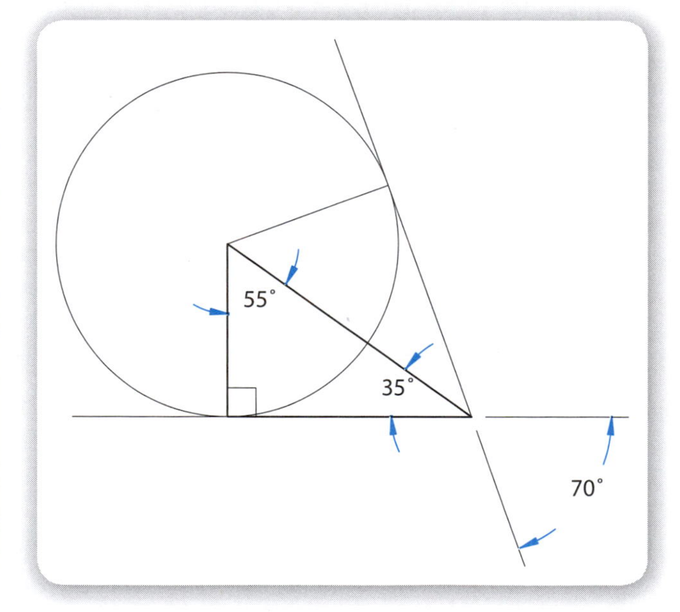

14.5 TRIGONOMETRY

Trigonometry is a branch of mathematics devoted to the measurement of triangles. It is probably the most important area of mathematics for the machinist and CNC programmer and is used in everyday calculation for setups, programming, and inspection. If you have not studied trigonometry before, you may think that it is shrouded in mystery and lives only in the realms of textbooks and nerds. Do not be concerned. The kind of trigonometry we will work with is highly visual and highly practical. Furthermore, by learning a few simple rules, you will find it second nature.

We will work with two categories of trigonometry in this section. The first is called right triangle trigonometry, which is only concerned with triangles that contain one 90° angle (a **right angle**). The next is **oblique** triangle trigonometry, which is only concerned with triangles that *do not* contain a right angle. Oblique triangle trigonometry is more difficult; however, a majority of manufacturing problems can be reduced to right angle trigonometry.

right angle:

A 90° angle.

oblique:

A description of triangles that do not contain a 90° angle.

The Pythagorean Theorem—the Precursor to Trigonometry

Before we start a serious discussion of trigonometry, it might be useful to review the Pythagorean Theorem. This theorem states an algebraic relationship between the lengths of the sides of a right triangle but will not directly help us to find the angles. Nonetheless, there are situations when the Pythagorean Theorem can give us the bits of information needed for further calculations.

Below is a simple equation for the Pythagorean Theorem, where the variables refer to the sides of the triangle in ***Figure 14.4***. We should note that *c* is always the hypotenuse and *a* and *b* can be either of the legs.

$$a^2 + b^2 = c^2$$

To use this equation we must know two of the sides. For example, if side a is 1.5" and side b is 2", and then we can find side c as shown in Table 14.4.

We can solve any side of the triangle as long as two sides are known. However, we must perform some algebra on the equation to isolate the unknown side. You can find a transposed version of the equation in Appendix B that will allow you to simply plug the known values in and quickly find the answer. We will work through one more example in Table 14.5 to find side a when side c is equal to 2.25" and side b is equal to 2".

Notice that the value for a is the value that was given in the previous example. This is a good way to check to make sure that no errors have been made.

Figure 14.4 Use the Pythagorean Theorem to find the hypotenuse.

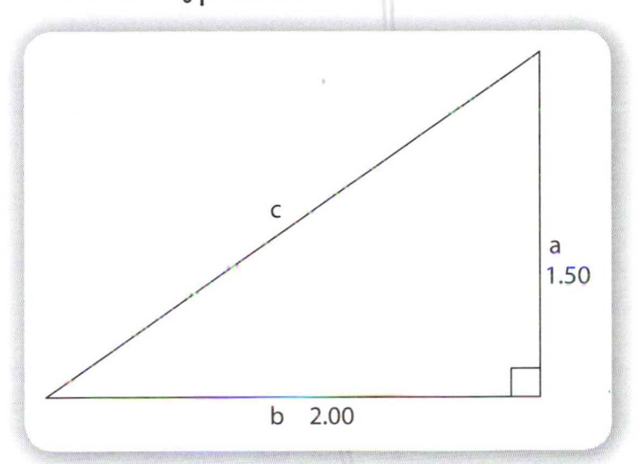

Table 14.4 Use the Pythagorean Theorem to Find the Hypotenuse (Side *c*)

Write the equation.	$a^2 + b^2 = c^2$
Fill in the known quantities.	$1.5^2 + 2^2 = c^2$
Perform the squaring operation.	$2.25 + 4 = c^2$
Perform the addition.	$6.25 = c^2$
Take the square root of each side to remove the exponent.	$\sqrt{6.25} = \sqrt{c^2}$
Simplify and put into standard form.	c = 2.5"

Table 14.5 Use the Pythagorean Theorem to Find Side a

Write the equation.	$a^2 + b^2 = c^2$
Transpose the equation for the unknown variable.	$a^2 = c^2 - b^2$
Fill in the known quantities.	$a^2 = 2.5^2 - 2^2$
Perform the squaring operation.	$a^2 = 6.25 - 4$
Perform the subtraction.	$a^2 = 2.25$
Take the square root of each side to remove the exponent.	$\sqrt{a^2} = \sqrt{2.25}$
Simplify and put into standard form.	a = 1.5"

Right Triangle Trigonometry

The short story of right triangle trigonometry is that it is a tool to help us to determine any angle or side of a triangle as long as we know any two other sides or angles. Trigonometry uses something called a trigonometric function to describe the relationship between these sides and the angle between them. We can use these well-understood relationships to determine an angle or length of a side by applying the trigonometric function in the proper formula. For example, the triangle described in the previous example had the following dimensions: a = 1.5", b = 2", and c = 2.5".

By pressing the proper keys on our calculator, we can quickly determine that the interior angles are 53.13° and 36.87°.

But how does this work? Trigonometry points out that similar triangles (triangles with identical interior angles but different side-lengths) will have exactly the same proportions. Let's scale the previous triangle bigger by a factor of two and then look at the relationship between the lengths of its legs, as illustrated in *Figure 14.5*.

Figure 14.5 Similar triangles have the exact same proportions. The relationships between any two legs of the triangle are known as trigonometric functions. Knowing the measurement of a particular angle allows us to predict the lengths of the other sides and the measurements of the other angles.

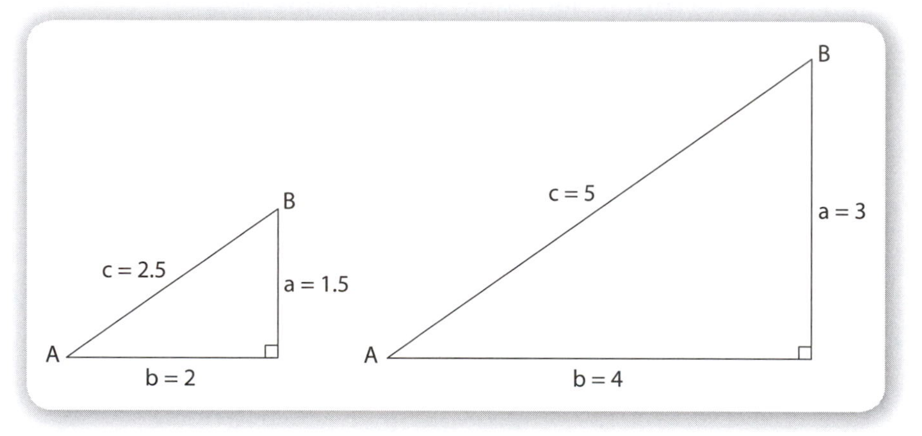

If we compare the lengths of side a and side b of the first triangle by dividing a by b we find that

$$1.5 \div 2 = 0.75$$

If we then look at the second triangle and make the same comparison, we find that the same relationship exists:

$$3 \div 4 = 0.75$$

This relationship between the sides of the triangle is called a trigonometric function. As we might expect, triangles with interior angles that are different from our example will have different proportions. This is not a problem. Mathematicians have developed methods of finding the value of these functions for any interior angle, and we can find the values in a handbook or in our pocket calculator. We can then use these functions to find the values of any other side or angle as long as we know two other values.

When we see a statement like sin 30°, it is read "the sine of 30 degrees." *Do not make the mistake of thinking that this means sine times 30 degrees.* The sine of 30 degrees is actually a constant, and each angle has a unique value for any given trigonometric functions. For example, sin 30°, sin 35°, and sin 40° are all different values. We will use several other trigonometric functions such as cosine (cos) and tangent (tan), which we will learn about shortly.

Before we go on, it would be appropriate to learn how to use our pocket calculator to find the values of trigonometric functions. There are keys on your calculator that read *Sin*, *Cos* and *Tan*. For example, we could find the sine of 30° by simply keying 30 on our calculator and then pressing the *Sin* key. (We do not need to press the = key.) The other functions can be found in a similar manner:

- To find the sine of 30° on a calculator, key in the following:

  30 *Sin*, which gives the result 0.5000.

- To find the cosine of 30° on a calculator, key in the following:

  30 *Cos*, which gives the result 0.8660.

- To find the tangent of 30° on a calculator, key in the following:

  30 *Tan*, which gives the result 0.5774.

You might want to go ahead and try a few more numbers to make sure you understand:

- Try to find the values for sin 20°, cos 45°, and tan 22.5°.

  The answers are 0.3420, 0.7071, and 0.4142, respectively.

We only need to understand the definitions of three basic trigonometric functions to solve right triangles; these are given in the definitions box that follows. We will look at these definitions and then see how they are used.

Definitions

Function	Definition
sine (sin)	$\sin \alpha$ = Opposite / Hypotenuse
cosine (cos)	$\cos \alpha$ = Adjacent / Hypotenuse
tangent (tan)	$\tan \alpha$ = Opposite / Adjacent

These definitions refer to the triangle in ***Figure 14.6***. It is important to note that the sides of a triangle are named in relation to the angle in question, called alpha (α). (The hypotenuse always stays in the same place—opposite the 90° angle.) The side *opposite* must be opposite the angle, and the side *adjacent* must be next to the angle. In Figure 14.6 the angle is displayed in two different positions; therefore, the sides have been labeled to reflect the proper orientation.

Figure 14.6 Trigonometric functions are defined by the sides in relation to the angle in question (α). The hypotenuse is always the same, but the sides adjacent and opposite will change depending on the angle used.

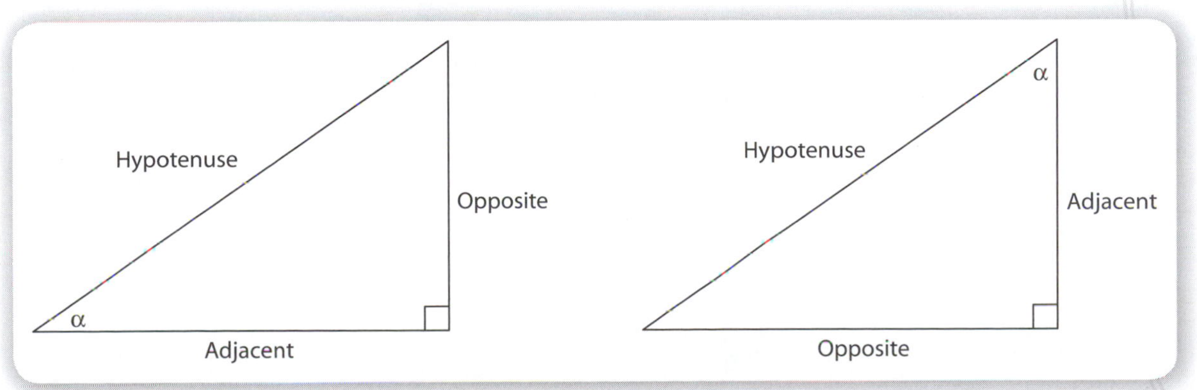

We can use trigonometric functions to find unknown components of a triangle by simply selecting the proper definition. We can then plug the known quantities into the definition and then solving for the unknown value. Take for example the triangle in ***Figure 14.7***. The angle is given as 35° and the side opposite is given as 1.00". How then can we find the hypotenuse and the side adjacent?

Figure 14.7 Find the hypotenuse and side adjacent.

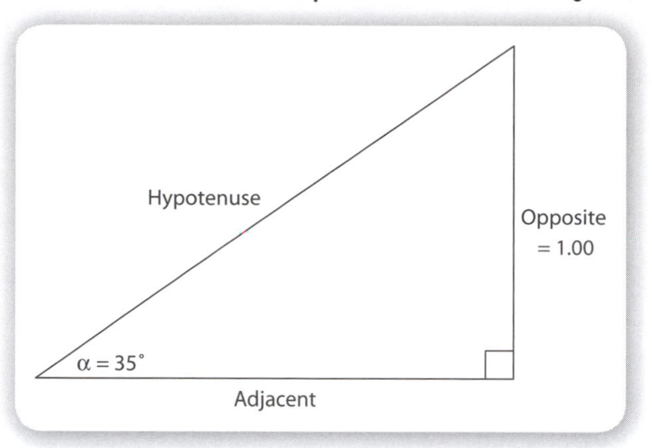

Memorizing Trigonometric Function Definitions

Trigonometry is used so commonly in CNC programming that it is critical to be able to quickly recall the three function definitions. One method of making this task easier is to come up with a mnemonic or other memory device to help you remember them. For example, SOH CAH TOA is an easy way to remember the definitions. (We just have to make sure to remember where to put the "equals" and "division" signs.)

sinα = **O**pposite / **H**ypotenuse
cosα = **A**djacent / **H**ypotenuse
tanα = **O**pposite / **A**djacent
= SOH CAH TOA

If you prefer, you can make up a phrase to help you remember. See if you can come up with a better phrase than the one to follow:

Some **O**range-**H**eaded **C**ats **A**lways **H**ave **T**houghts **O**f **A**nxiety

The solution is to first select the proper equation that contains only the factors we are concerned with and then solve the equation for the unknown. Let's start by finding the hypotenuse by working through this problem in Table 14.6.

Table 14.6 Find the Hypotenuse in Figure 14.7

Select the proper equation. We selected this equation because we already know the side opposite, and we want to find the hypotenuse. This is the only equation that contains both sides.	sin α = Opposite / Hypotenuse
Use algebra to transpose the equation and isolate the unknown.	Hypotenuse = Opposite / sin α
Fill in the known values.	Hypotenuse = 1.00 / sin 35°
Find the value for sin 35° with a calculator.	Hypotenuse = 1.00 / 0.5736
Perform the arithmetic.	Hypotenuse = 1.7434"

It is that simple. We only have to understand how to select the proper equation and perform the proper operation to find the unknown values of the triangle. If you are having any trouble with the transposition of the equations, there are two ways to remedy this. First, review the section on algebra and learn how to perform these operations. Second, consult the table in Appendix B. It contains the transposed equations that will allow you to select the proper equation based upon the known and unknown quantities.

Let's now find the side adjacent in the same triangle (see Table 14.7). This time we know the side opposite and wish to find the side adjacent. Therefore, we select the only equation that contains both of these values:

$$\tan \alpha = \text{Opposite} / \text{Adjacent}$$

Let's try one more example. The triangle in ***Figure 14.8*** has a 30° angle, and the hypotenuse is given as 2.00"; we will need to find the side adjacent and side opposite. Let's start by finding the side opposite in Table 14.8.

Table 14.7 Find the Adjacent Side in Figure 14.7

Select the proper equation.	$\tan \alpha = \text{Opposite} / \text{Adjacent}$
Use algebra to transpose the equation and isolate the unknown.	$\text{Adjacent} = \text{Opposite} / \tan \alpha$
Fill in the known values.	Adjacent = 1.00 / tan 35°
Find the value for tan 35° with a calculator.	Adjacent = 1.00 / 0.7002
Perform the arithmetic.	Adjacent = 1.4281"

Figure 14.8 Find the opposite and adjacent sides.

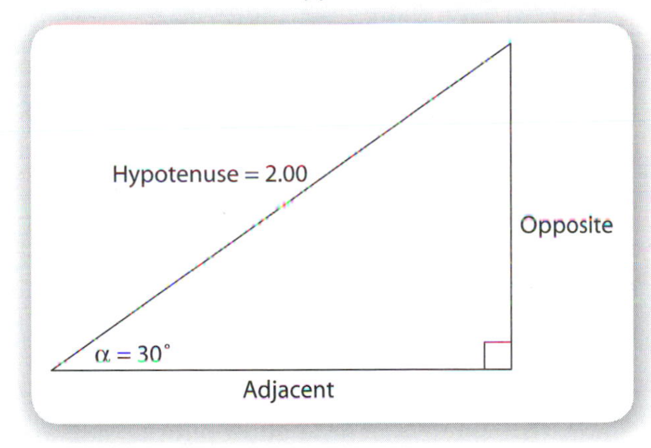

Now, we will find the adjacent side in Table 14.9.

Table 14.8 Find the Opposite Side in Figure 14.8

Select the proper equation.	$\sin \alpha = \text{Opposite} / \text{Hypotenuse}$
Use algebra to transpose the equation and isolate the unknown.	$\text{Opposite} = \text{Hypotenuse} \times \sin \alpha$
Fill in the known values.	Opposite = 2.00 × sin 30°
Find the value for sin 30° with a calculator.	Opposite = 2.00 × 0.500
Perform the arithmetic.	Opposite = 1.00"

Table 14.9 Find the Adjacent Side in Figure 14.8

Select the proper equation.	cos α = Adjacent / Hypotenuse
Use algebra to transpose the equation and isolate the unknown.	Adjacent = Hypotenuse × cos α
Fill in the known values.	Adjacent = 2.00 × cos 30°
Find the value for cos 30° with a calculator.	Adjacent = 2.00 × 0.8660
Perform the arithmetic.	Adjacent = 1.7321"

Solving an Unknown Angle

We can also use trigonometry to find an unknown angle if we already know two of the sides. However, we will need to use an inverse trigonometric function to accomplish this. An inverse function simply works in the reverse direction as a standard trigonometric function.

You may remember from the definitions of the trigonometric functions that the definition of the sine function was

sin α = Opposite / Hypotenuse

ratio:
an comparison between the dimension of unit and the dimensions of another.

By its very definition, the **ratio** between the two legs of the triangle gives us the sine of the angle, but not the angle directly. To find the angle we need to use the inverse function on our calculator in conjunction with the correct function key:

- We have already seen how to find the sine of 45°:

  45 *Sin* gives the result 0.7071.

- When we start with 0.07071—the sine of 45°—we can then apply an inverse function to find the original angle. Use the following keystrokes:

  0.7071 *2nd Sin*, which verifies our result of 45°.

The inverse functions, which are also called arcsine, arccosine, and arctangent, may be displayed on your calculator as $\sin^{-1}$, $\cos^{-1}$, and $\tan^{-1}$. These inverse functions are usually accessed by first pressing the *2nd* or sometimes the *Inv* key on your calculator.

Take, for example, the triangle in ***Figure 14.9***. The hypotenuse is given as 2.00", the side opposite is 1.00", and the side adjacent is 1.73205". Let's find the angle (α) by using only the hypotenuse and the side opposite.

First, we select the proper equation. From the definitions we know that the hypotenuse and opposite side are related to the sine function, as the sine equation is the only one of our definitions that contains both of these sides. We then find the angle as follows in Table 14.10:

Figure 14.9 Find angle α when the hypotenuse and side opposite are known.

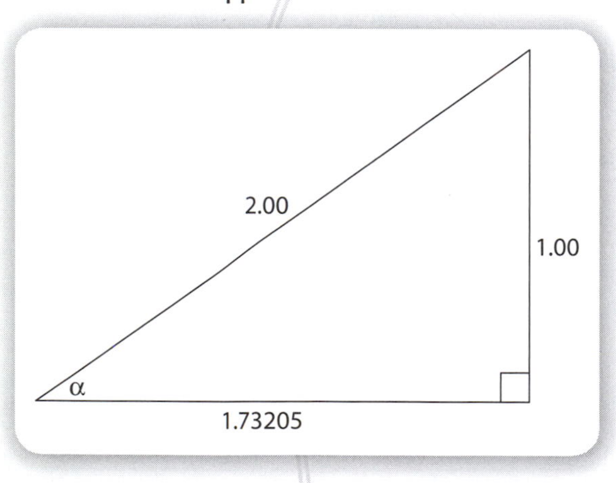

Table 14.10 Find Angle α in Figure 14.9. When the Hypotenuse and Side Opposite Are Known

Select the proper equation.	sin α = Opposite / Hypotenuse
Fill in the known values.	sin α = 1.00 / 2.00
Perform the arithmetic.	sin α = 0.500
Take the inverse sine to find the angle (key .500 *2nd Sin* into your calculator).	α = 30°

We could also the find the angle by using any of the two sides and applying the proper inverse function. For example, we will find the angle by using the side opposite and the side adjacent as detailed in Table 14.11.

Table 14.11 **Find Angle α in Figure 14.9. When the Sides Adjacent and Opposite Are Known**

Select the proper equation.	tan α = Opposite / Adjacent
Fill in the known values.	tan α = 1.00 / 1.73205
Perform the arithmetic.	tan α = 0.57735
Take the inverse tangent to find the angle (key .57735 *2nd Tan* into your calculator).	α = 30°

We could have also used the hypotenuse and side adjacent to find the angle and we would have had the same result. You may want to try it for yourself—use the following equation:

$$\cos \alpha = \text{Adjacent} / \text{Hypotenuse}$$

Oblique Triangle Trigonometry

We mentioned earlier that oblique angle trigonometry is concerned with triangles that do not contain any right angles. Two examples of oblique triangles can be seen in ***Figure 14.10***. You might note that we do not use the terminology of hypotenuse, side adjacent, or side opposite when working with oblique triangles. The only nomenclature to understand is that angles are always paired with a side that is directly *opposite* the angle. The labeling of any angle/side combination is strictly arbitrary.

Figure 14.10 ***Oblique triangles do not contain a right angle.***

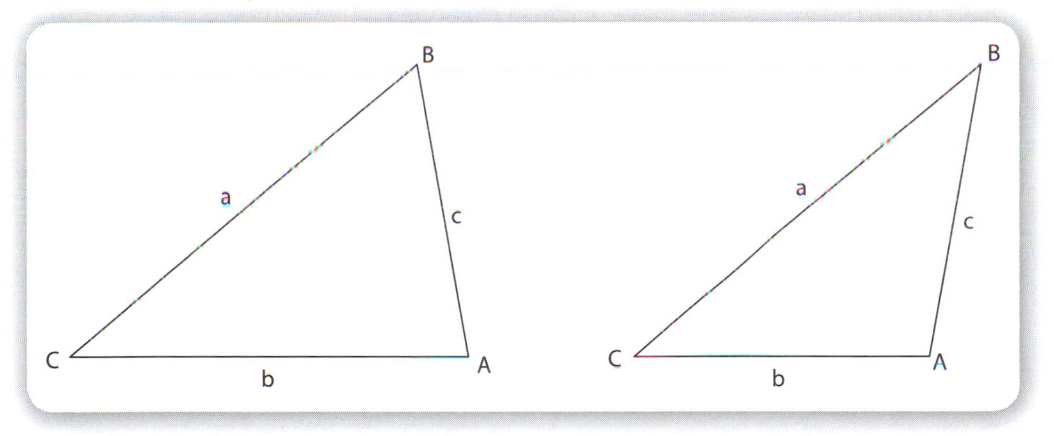

There are two laws we must learn in order to apply trigonometry to oblique triangle problems in manufacturing: the Law of Sines and the Law of Cosines. Each of these laws can be distilled into an equation containing four variables. We will need to know three of these variables to find the unknown angle or side.

Definitions

Law of Sines $$\frac{a}{\sin A} = \frac{b}{\sin B} = \frac{c}{\sin C}$$

Figure 14.11 Use the Law of Sines to find side A.

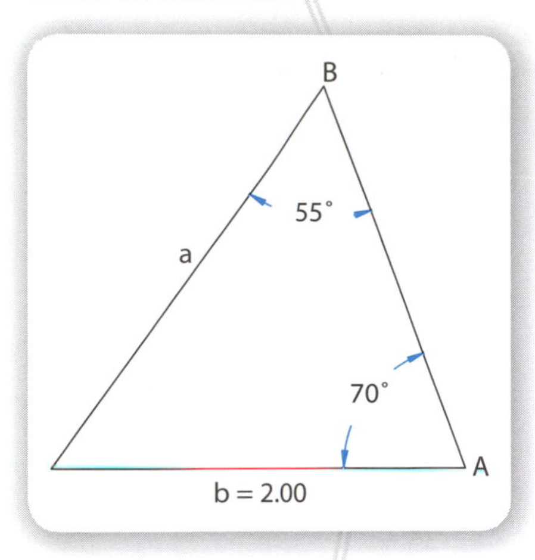

Figure 14.12 Use the Law of Sines to find angle B.

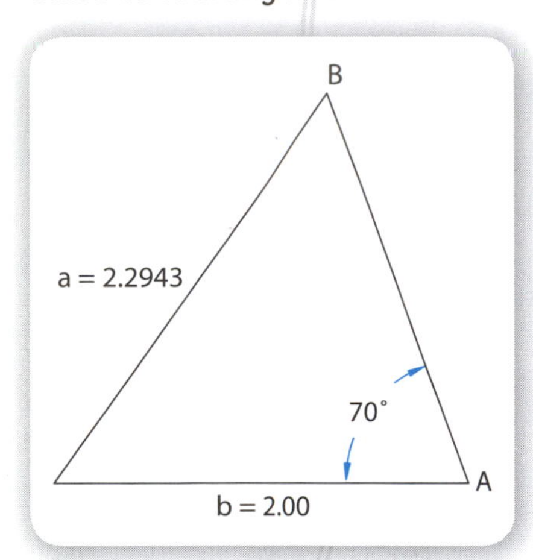

The Law of Sines

The Law of Sines is used anytime we are concerned with the values of two angle/side pairs. It is not difficult to use; we only need to plug in the values that we know and then calculate the unknown. For example, take the triangle in *Figure 14.11*. Upon inspection we see that this triangle meets the criteria for the Law of Sines: it is concerned with the pairs a/A and b/B. The unknown value in the triangle, side a, can be found via the process shown in Table 14.12.

Table 14.12 Use the Law of Sines to Find Side a in Figure 14.11

Select the proper equation.	$\frac{a}{\sin A} = \frac{b}{\sin B}$
Use algebra to transpose the equation and isolate the unknown.	$a = \frac{b \sin A}{\sin B}$
Fill in the known values.	$a = \frac{2.00 \sin 70^\circ}{\sin 55^\circ}$
Find the values for the sine of 70° and 55°.	$a = \frac{2.00 \times 0.9397}{0.8192}$
Perform the arithmetic.	a = 2.2943"

We can also use the Law of Sines to find an unknown angle. In the example in Table 14.13, we will use the side a that we calculated previously and use the Law of Sines to find angle B of the triangle in *Figure 14.12*.

Table 14.13 Use the Law of Sines to Find Angle B in Figure 14.12

Select the proper equation.	$\frac{a}{\sin A} = \frac{b}{\sin B}$
Use algebra to transpose the equation and isolate the unknown.	$\sin B = \frac{b \sin A}{a}$
Fill in the known values.	$\sin B = \frac{2.00 \sin 70^\circ}{2.2943}$
Find the values for the sine of 70°.	$\sin B = \frac{2.00 \times 0.9397}{2.2943}$
Perform the arithmetic.	sin B = 0.8192
At this point, we only know the sine of a. To find the angle, we need to take the inverse (key in 0.8192 *2nd Sin*).	B = 55°

There are situations when a triangle can have two possible solutions for a given set of known values. For example, both triangles in *Figure 14.13* have the same values for sides a and c and angle C. However, we can see that one triangle has an 80° angle and the other triangle has a 100° angle.

If we were to solve these triangles using the Law of Sines, we would find that the sine of angle A is the same in each problem, as verified in Table 14.14. This can obviously lead to some confusion. It turns out that for any given value of sine,

Figure 14.13 Triangles with the same known values can have different solutions. Care must be taken when an obtuse angle is present in the triangle.

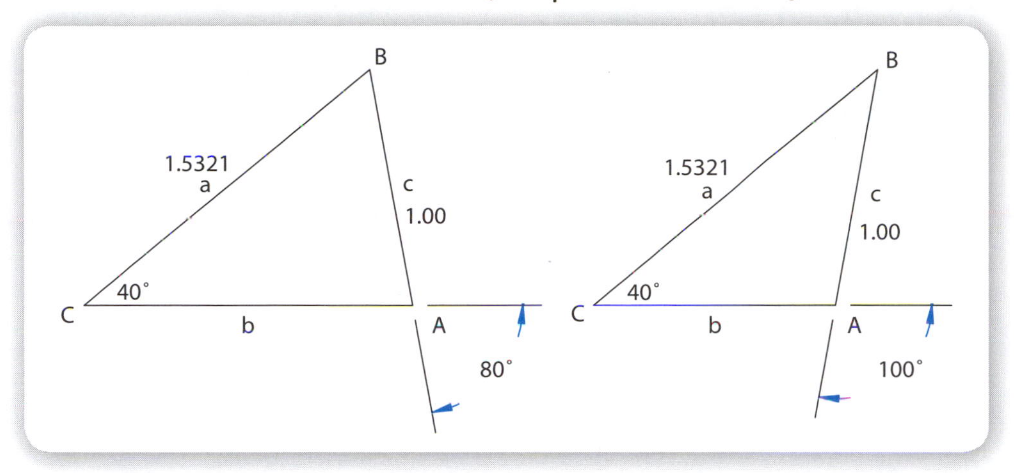

there are at least two corresponding angles. Angles that are mirror images of each other will have the same values for their sine function. Take the following for example:

- The sines of both 60° and 120° equal 0.8660.
- The sines of both 45° and 135° equal 0.7071.
- The sines of both 30° and 150° equal 0.5000.

Table 14.14 Find Angle A for Both Triangles in Figure 14.13

$\sin A = \frac{a \sin C}{c}$
$\sin A = \frac{1.5321 \sin 40^\circ}{1.00}$
$\sin A = \frac{1.5321 \times 0.6428}{1.00}$
$\sin A = 0.9848$

When we made the calculation, we simply ended up with the value for the sine of angle A. To then find the angle we have to take the inverse sine. However, when we take the inverse sine on 0.9848, we get the answer 80°. But what about the second triangle? Most calculators are not sophisticated enough to provide an alternate solution, so it is up to us to be aware of the situation and decide whether or not the angle should be **acute** or **obtuse.** If the angle is obtuse, we can find the corresponding angle by subtracting the calculator answer from 180°. For example, when we took the inverse sine of 0.9848, we got the answer 80°. The other possible solution is found by subtracting that answer from 180°:

- $180^\circ - 80^\circ = 100^\circ$
- $\sin 100^\circ = 0.9848$

Therefore, the other solution to angle A would be an obtuse angle of 100°, which has exactly the same sine value as 80°.

acute:
An angle smaller than 90°.

obtuse:
Describes an angle that is greater than 90°.

Definitions

Law of Cosines	$a^2 = b^2 + c^2 - (2bc \cos A)$

The Law of Cosines

The Law of Cosines is used to solve triangles when we are concerned with three sides and an angle. Notice in the equation that it contains sides a, b, and c, and the angle A, which is paired with side a. If we know any three of the values, we can find the fourth. The labels are again arbitrary, and it makes no difference to the calculations which side is which. Just make sure that the angle and its side are paired.

Let's look at a simple example of the Law of Cosines by solving for side a in the triangle in ***Figure 14.14***. This problem is solved in Table 14.15 by filling in the known values and then performing the arithmetic.

Figure 14.14 The Law of Cosines used to find side a.

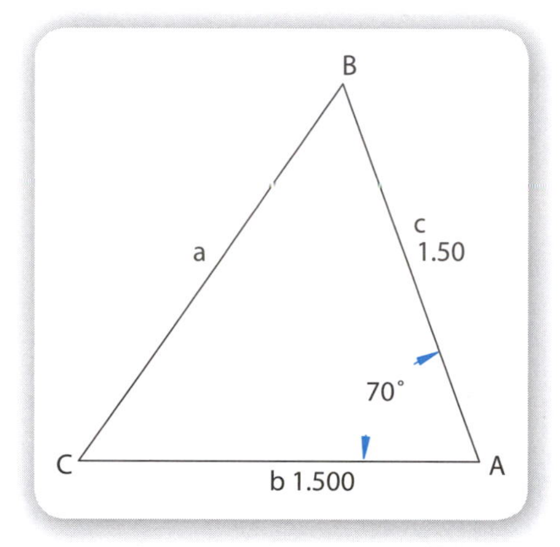

Table 14.15 Use the Law of Cosines to Find Side a in Figure 14.14

Write the equation.	$a^2 = b^2 + c^2 - (2bc \cos A)$
Fill in the known values.	$a^2 = 1.5^2 + 1.5^2 - (2 \times 1.5 \times 1.5 \times \cos 70°)$
Square the two sides and find the cosine of 70°.	$a^2 = 2.25 + 2.25 - (2 \times 1.5 \times 1.5 \times 0.3420)$
Perform the arithmetic.	$a^2 = 2.0609$
Take the square root.	$a = 1.7207$

We can also use the Law of Cosines to find an angle when three sides are known. Let's use the same triangle for the next example, but this time we will start by knowing only the sides and then find angle A, as in ***Figure 14.15***. The transposition of this equation can be a little tricky, so be careful with your algebra. The solution is given in Table 14.16. You can also find a table in Appendix B that contains formulas for finding each of the unknowns.

Solving an Oblique Triangle

We could also use the Law of Cosines to find sides b or c, but it can be difficult to rearrange the equation to isolate any term other than side a or angle A. In such a situation, the Law of Sines is much easier and faster to use. In fact, in most situations, we will use a combination of methods to find every element in an oblique triangle. We will next look at the triangle in ***Figure 14.16*** and find the three unknowns in Table 14.17 by applying what we have learned about oblique triangles.

Figure 14.15 The Law of Cosines used to find angle A.

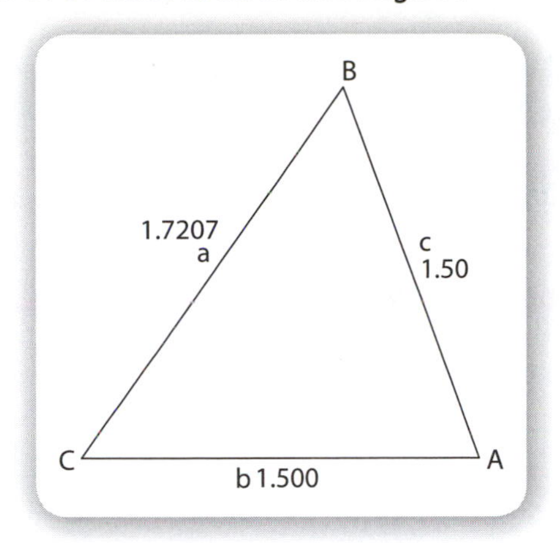

Table 14.16 Use the Law of Cosines to Find Side a in Figure 14.15

Write the equation.	$a^2 = b^2 + c^2 - (2bc \cos A)$
Transpose the equation to isolate the cosine of angle A	$2bc \cos A = b^2 + c^2 - a^2$
Transposition continued	$\cos A = \frac{b^2 + c^2 - a^2}{2bc}$
Fill in the known values.	$\cos A = \frac{1.5^2 + 1.5^2 - 1.7207^2}{2 \times 1.5 \times 1.5}$
Square the terms and perform the arithmetic.	$\cos A = 0.3420$
Take the inverse cosine to find the angle.	$A = 70°$

Table 14.17 Find the Three Unknowns in Figure 14.16

The three known values make this triangle a candidate for the Law of Cosines. We will use it to find side a.	$a^2 = b^2 + c^2 - (2bc \cos A)$ $a^2 = 1.25^2 + 1.75^2 - (2 \times 1.25 \times 1.75 \times 0.5)$ $a = 1.5612$
Now that side a is known, we can find angle C with the Law of Sines.	$\frac{a}{\sin A} = \frac{c}{\sin C}$ $\sin C = \frac{c \sin A}{a}$ $\sin C = \frac{1.75 \times 0.866}{1.5612}$
The inverse sine gives us angle C.	$\sin C = 0.9707$
	$C = 76.10°$
The interior angles must sum to 180°. We can use this principle to then find angle B.	$B = 180° - A - C$ $B = 180° - 60° - 76.1°$ $B = 43.9°$

Figure 14.16 Find the sides of this oblique triangle.

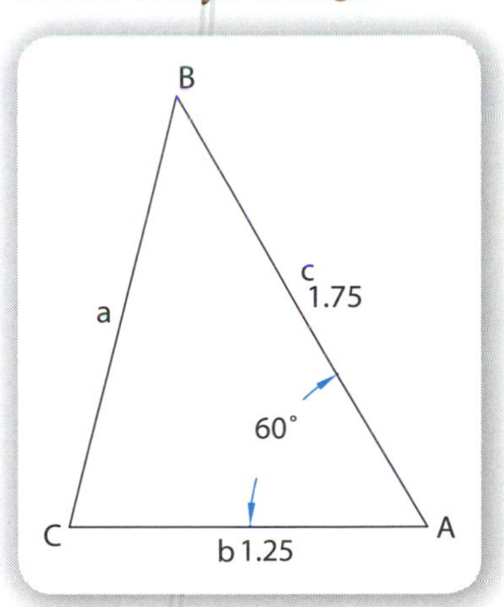

14.6 ILLUSTRATED APPLICATIONS

In this section, we will look at some common shop applications that will use some of the mathematical concepts that we have discussed in this chapter.

Speeds and Feeds

PROBLEM We notice while milling an unfamiliar material that the tool tends to perform the best when turning at 1200 RPM with a ø 0.750" end mill. We will have this material to machine often, so we decide that it would be a good idea to document the cutting velocity for future reference. What is the cutting velocity?

SOLUTION

- Transpose the formula for spindle speed to reflect the cutting velocity.

$$\text{RPM} = \frac{12 \times \text{Velocity}}{\pi \times \text{Diameter}}$$

$$\text{Velocity} = \frac{\text{RPM} \times \pi \times \text{Diameter}}{12}$$

- Next, plug the current values into the formula to find the current cutting velocity. This will be the cutting velocity that we can use for the next job and for cutting tools of a different diameter.

$$\text{Velocity} = \frac{1200 \times 3.14159 \times 0.750}{12}$$

$$\text{Velocity} = 236 \text{ feet per minute}$$

Coolant to Add to a Tank

PROBLEM The coolant in a CNC machining center looks a bit thin. Upon inspection, it is found to be a ratio of 35 parts water to 1 part cutting fluid—a little short of the manufacturer's recommended ratio of 19:1. If there are currently 15 gallons of mixed coolant in the tank, how much concentrated cutting fluid should we add to achieve a 19:1 mixture?

SOLUTION Perhaps the easiest way to approach this problem is to look at the final concentration: 19 parts water to 1 part concentrate would indicate that the concentrate makes up 1/20 or 5% of the final mixed coolant. We can also see that of the mixed coolant in the tank, the concentrate makes up 1/36 of this mixture or about 0.4166 gallons of the 15 total gallons.

If we let x equal the amount of coolant to add in gallons, we can establish an equation that describes the problem:

$$\frac{0.4166 + x}{15 + x} = 0.05$$

In English, when the current volume of concentrate plus the added volume is divided by the total volume plus the added volume, the resulting sum will equal 5%.

Solving this problem requires us to isolate x on one side by itself, as done in Table 14.18.

We can check our answer by plugging the values into the original equation to see if the final mix contains 5% concentrated cutting fluid:

$$\frac{0.4166 + 0.3509}{15 + 0.3509} = 0.05$$

Table 14.18 Solution to Coolant Problem

$\frac{0.4166 + x}{15 + x} = 0.05$
$0.4166 + x = 0.05\ (15 + x)$
$0.4166 + x = 0.75 + 0.05x$
$x - 0.05x = 0.75 - 0.4166$
$0.95x = 0.3334$
$x = \frac{0.3334}{0.95}$
$x = 0.3509$ gal.

Surface Finish Conversions

PROBLEM The blueprint for the part we are working on specifies that the surface roughness of our workpiece should be 1.6 microns. (A micron is 10^{-6} meters) However, our roughness gage measures in micro-inches (10^{-6} inches). What is the equivalent measurement in micro-inches?

SOLUTION From a handbook, we determine that one inch is equal to 2.54×10^{-2} meters or (0.0254 meters). We can then set up a proportion to convert 1.6 microns to micro-inches as follows (letting x be the micro-inch equivalent). If you are not comfortable with scientific notation, the decimal form is displayed in the right column of Table 14.19.

Table 14.19 Solution to Surface Finish Problem

$\frac{x}{1.6 \times 10^{-6}} = \frac{1}{2.54 \times 10^{-2}}$	$\frac{x}{0.0000016} = \frac{1}{0.0254}$
$x = \frac{1.6 \times 10^{-6} \times 1}{2.54 \times 10^{-2}}$	$x = \frac{0.0000016}{0.0254}$
$x = 63 \times 10^{-6}$ or 63 μ in.	$x = 0.000063$ in.

Inspecting an Angle

PROBLEM The v-groove in ***Figure 14.17*** must have a depth of 0.750 inches. We cannot accurately measure to the point, so how might we inspect the depth?

SOLUTION We could place a gage pin in the groove and then measure from the top of the pin to the top of the workpiece. To find this measurement, we will use a Ø0.750" pin and calculate the theoretical distance from the vertex of the angle to the top of the pin. The size of the gage pin is arbitrary as long as it makes tangential contact with the surfaces of the angle.

Figure 14.17 **Inspecting the depth of an angle with a gage pin.**

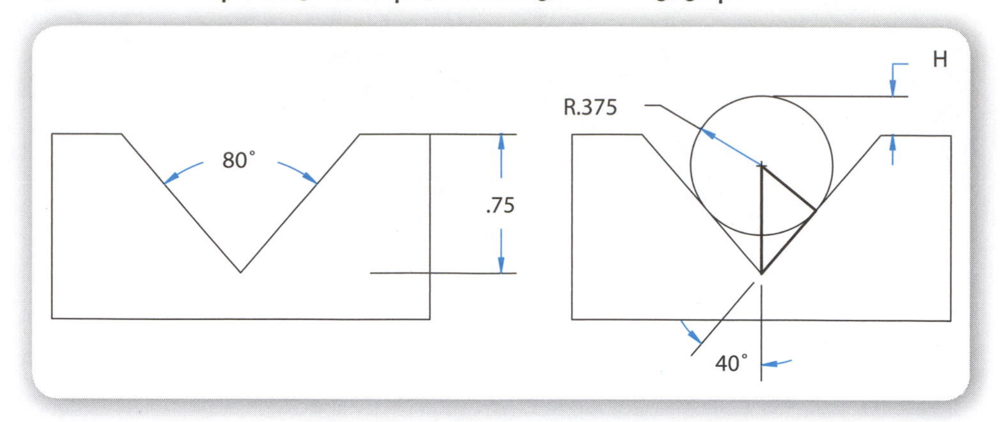

- We start by finding the hypotenuse of the triangle between the vertex and the center of the pin. We know the angle is 40° and that the side opposite is the radius (0.375"):

$$\sin \alpha = \text{Opposite} / \text{Hypotenuse}$$
$$\sin 40^\circ = 0.375" / \text{Hypotenuse}$$
$$\text{Hypotenuse} = 0.375" / \sin 40^\circ$$
$$\text{Hypotenuse} = 0.5834"$$

- Add the hypotenuse to the radius to find the distance from the vertex to the top of the pin. Then subtract the specified depth (0.750") to find the distance from the top of the pin to the top of the workpiece (H). We could then inspect this v-groove by measuring the actual H.

$$\text{Total height} = 0.5834" + 0.375" = 0.9584"$$
$$\text{Height to top} = 0.9584" - 0.750" = 0.2084"$$

Center Distance for Holes on an Angle

PROBLEM The workpiece we are working on in ***Figure 14.18*** must have a bolt hole pattern drilled into it. Find the distance between the two holes indicated by dimension a.

Figure 14.18 **Find the distance (a) between the two upper holes.**

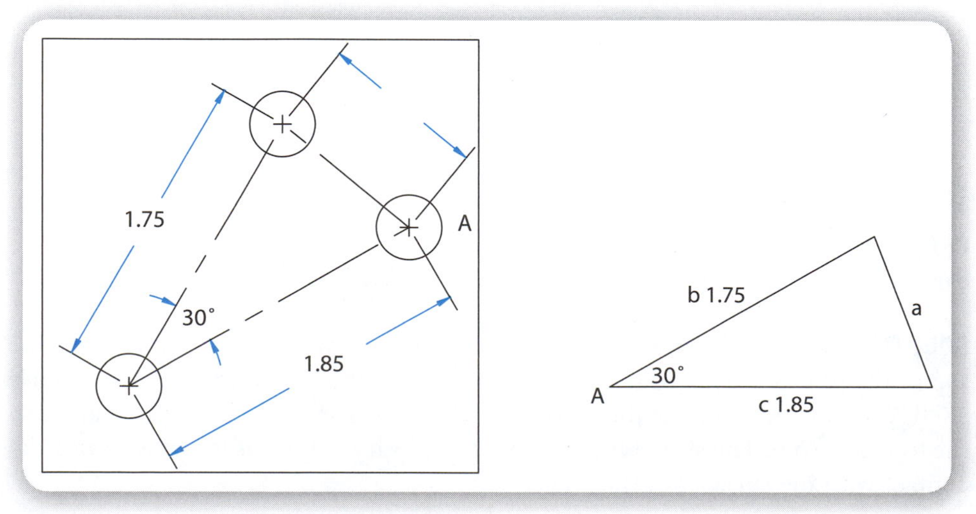

SOLUTION The distance between the two holes is most easily found with the Law of Cosines. We can use the triangle that is coincident to the three holes and then solve for the unknown side as is detailed in Table 14.20.

Table 14.20 Solution to Inspection Problem

$a^2 = b^2 + c^2 - (2bc\ \cos A)$
$a^2 = 1.75^2 + 1.85^2 - (2 \times 1.75 \times 1.85 \times 0.866)$
$a^2 = 0.87749$
$\sqrt{a} = \sqrt{0.87749}$
$a = 0.9367$

Your Turn

Students can work from pre-created parts given to them by the instructor to measure and then calculate distances and angles using algebraic formulas.

CHAPTER SUMMARY

- The order of operation is important to understand when performing arithmetic or algebraic operations. Failure to follow the proper order will lead to incorrect results.
- Many problems in manufacturing are expressed algebraically. Therefore, it is important to be able to use basic algebra. Besides algebra, it is also helpful to have a command of the various notations and operations that are found in common formulas.
- Algebra is similar to arithmetic except that it uses variables. Variables are terms within the equation whose value can change. We solve an equation by isolating the variable on one side of the equation with the rules of algebra.
- A proportion is an easy way to solve unit conversion problems. It can be set up as an equation showing two fractional relationships and then solved algebraically.
- Right triangle trigonometry is used to solve triangles that contain one 90° angle. Triangles that do not contain a right angle are called oblique, and we must use a separate set of laws to solve these triangles.

BRING IT HOME

Show all your work on a separate sheet of paper.

1. Solve the following equations for x by using the proper order of operations.

$x = 2(3)$	$x = 4 \div (1 + 3)$	$x = \sqrt{36}$
$x = 5(5 + 3)$	$x = (3 + 4)(3 + 4)$	$x = \sqrt{36 - 11}$
$x = 1 + (3 + 3 \times 5)$	$x = (3 + 4)^2$	$x = \dfrac{10+6}{\sqrt{16}}$
$x = 3^2 + 4$	$x = (6 - 1 \times 3)^3$	$x = \sqrt[3]{27}$

2. Put the following numbers into proper scientific notation or convert from scientific notation to decimal form.

1600	0.0056	20,000
2.54×10^{-2}	1.45×10^6	3.1×10^3

3. Solve the following equations for x.

$1 = 2x$	$x + 2 = 3 \times 4$	$6 = 12 \div x$
$4 = 4x + 2$	$x^2 = 16$	$6 = 12 \div (x + 2)$

4. Convert the following values to the indicated units. You may need to consult a handbook or Appendix B for the requisite information.

1.50 m to in.	0.0125 in. to mm	68° F to °C
0.065 mm to in.	0.5 gal. to liters	35°C to °F
50.8 mm to in.	25 liters to gal.	15 hp to kW
2 in. to mm	1000 kg to lbs.	12 kW to hp
0.375 in. to mm	35 lbs. to kg	

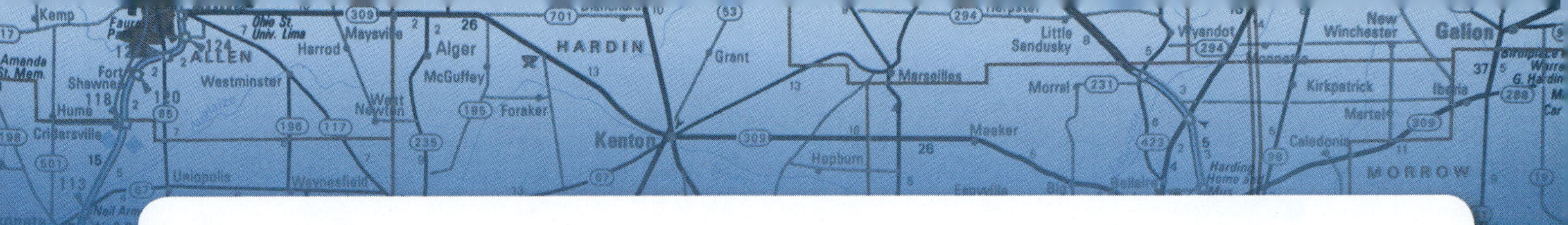

5. Find every side and angle in the triangle in *Figure 14.19*. Be sure to sketch the triangles and label each element on a separate sheet of paper.

Figure 14.19 Find each unknown side and the angle of triangles A through F.

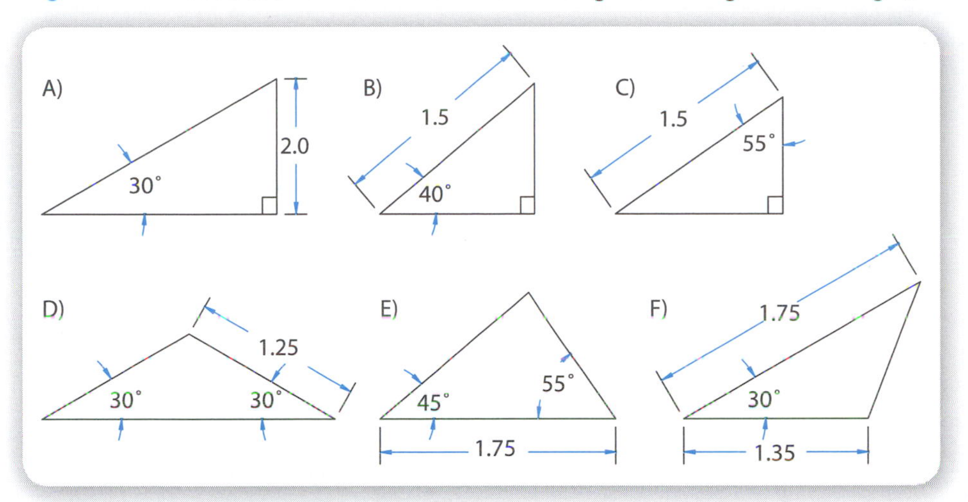

APPENDIX A
G & M Code Reference

Code	Definition and Syntax
G-codes	
G00	Rapid traverse G00 X Y Z
G01	Linear interpolation G01 X Y Z F
G02	Circular interpolation—clockwise G02 X Y Z I J (or R) F
G03	Circular interpolation—counterclockwise G03 X Y Z I J (or R) F
G04	Programmed dwell
G10	Automatically load tool and work offsets
G17	X-Y arc plane selection
G18	X-Z arc plane selection
G19	Z-Y arc plane selection
G20	Inch programming units
G21	Millimeter programming units
G28	Return to machine zero through intermediate point (G91)G28 X Y Z
G29	Return from machine zero to an intermediate point
G40	Cancel cutter compensation
G41	Cutter compensation left
G42	Cutter compensation right
G43	Select tool height offset
G49	Cancel tool height offset
G54–59	Workplane selection
G73	Drill cycle with chip breaker

G74	Tapping cycle (LH)
G76	Boring cycle with orient and rapid retract
G80	Canned cycle cancel
G81	Drilling cycle G81 X Y Z F R L
G82	Drilling cycle with dwell G82 X Y Z F R L P
G83	Peck drilling cycle G83 X Y Z F Q R L
G84	Tapping cycle G84 X Y Z F R L
G85	Boring cycle G85 X Y Z F R L
G86	Boring cycle—rapid retract
G87	Boring cycle—manual retract
G88	Boring cycle—manual retract with dwell
G89	Standard boring cycle with dwell
G90	Absolute coordinate system
G91	Incremental coordinate system
G92	Shift work coordinate system
G98	Return tool to initial plane after canned cycle
G99	Return tool to retract plane after canned cycle
M-codes	
M00	Unconditional program stop
M01	Optional program stop
M02	Program end
M03	Spindle on clockwise
M04	Spindle on counterclockwise
M05	Spindle off
M06	Automatic tool change
M07	Coolant on—mist
M08	Coolant on—flood (default)
M09	Coolant off
M19	Oriented spindle stop
M30	Program end and reset
M97	Execute a local subroutine
M98	Execute an external subprogram
M99	Return control to main program after subprogram or subroutine

Letters	
A	Angular coordinate around X-axis
B	Angular coordinate around Y-axis
C	Angular coordinate around Z-axis
D	Tool diameter offset designation
F	Linear feed rate
G	Preparatory code
H	Tool height offset designation
I	Incremental distance to arc center along X-axis from start
J	Incremental distance to arc center along Y-axis from start
K	Incremental distance to arc center along Z-axis from start
L	Number of times to loop a canned cycle
M	Miscellaneous code
N	Line number
O	Program name
P	Dwell time
Q	Peck depth
R	Z-coordinate of retract plane
S	Spindle speed (RPM)
T	Tool number designation
X	Coordinate on X-axis
Y	Coordinate on Y-axis
Z	Coordinate on Z-axis

APPENDIX B
Reference Information

B-1 SPEEDS AND FEEDS

Speed and Feed Formulas

- N = the number of teeth on the cutting tool
- Chipload = the feed per tooth per revolution
- Velocity = the cutting speed in surface feet per minute.
- IPM = linear feed in inches per minute
- IPR = linear feed in inches per revolution

$$\text{RPM} = \frac{12 \times \text{Velocity}}{\pi \times \text{Diameter}}$$

$$\text{RPM} = \frac{4 \times \text{Velocity}}{\text{Diameter}}$$

$$\text{Velocity} = \frac{\text{RPM} \times \pi \times \text{Diameter}}{12}$$

$$\text{Feed}_{\text{IPR}} = \text{Chipload} \times \text{N}$$

$$\text{Feed}_{\text{IPM}} = \text{Chipload} \times \text{N} \times \text{RPM}$$

or

$$\text{Feed}_{\text{IPM}} = \text{Feed}_{\text{IPR}} \times \text{RPM}$$

Machining Data

Table B-1 Tap Drill Sizes

Unified Thread			ISO Thread		
Thread	Tap Drill	Tap Drill Size (in.)	Thread	Tap Drill Size (mm)	Tap Drill Size (in.)
0-80	3/64	0.0469	M1 × 0.25	0.75	0.0295
1-64	53	0.0595	M1.1 × 0.25	0.85	0.0335
1-72	53	0.0595	M1.2 × 0.25	0.95	0.0374
2-56	51	0.0700	M1.4 × 0.3	1.10	0.0433
2-64	49	0.0730	M1.6 × 0.35	1.25	0.0492
3-48	5/64	0.0785	M1.8 × 0.35	1.45	0.0571
3-56	46	0.0820	M2 × 0.4	1.60	0.0630
4-40	43	0.0890	M2.2 × 0.45	1.75	0.0689
4-48	42	0.0935	M2.5 × 0.45	2.05	0.0807
5-40	38	0.1015	M3 × 0.5	2.50	0.0984
5-44	37	0.1040	M3.5 × 0.6	2.90	0.1142
6-32	36	0.1065	M4 × 0.7	3.30	0.1299
6-40	33	0.1130	M4.5 × 0.75	3.70	0.1457
8-32	29	0.1360	M5 × 0.8	4.20	0.1654
8-36	29	0.1360	M6 × 1	5.00	0.1969
10-24	25	0.1495	M7 × 1	6.00	0.2362
10-32	21	0.1590	M8 × 1.25	6.80	0.2677
12-24	16	0.1770	M9 × 1.25	7.80	0.3071
12-28	14	0.1820	M10 × 1.5	8.50	0.3346
1/4-20	7	0.2010	M11 × 1.5	9.50	0.3740
1/4-28	3	0.2130	M12 × 1.75	10.20	0.4016
5/16-18	F	0.2570	M14 × 2	12.00	0.4724
5/16-24	I	0.2720	M16 × 2	14.00	0.5512
3/8-16	5/16	0.3125	M18 × 2.5	15.50	0.6102
3/8-24	Q	0.3320	M20 × 2.5	17.50	0.6890
7/16-14	U	0.3680	M22 × 2.5	19.50	0.7677
7/16-20	25/64	0.3906	M24 × 3	21.00	0.8268
1/2-13	27/64	0.4219	M27 × 3	24.00	0.9449
1/2-20	29/64	0.4531	M30 × 3.5	26.50	1.0433
9/16-12	31/64	0.4844	M33 × 3.5	29.50	1.1614
9/16-18	33/64	0.5156	M36 × 4	32.00	1.2598
5/8-11	17/32	0.5312	M39 × 4	35.00	1.3780
5/8-18	37/64	0.5781	M42 × 4.5	37.50	1.4764
3/4-10	21/32	0.6562	M45 × 4.5	40.50	1.5945
3/4-16	11/16	0.6875	M48 × 5	43.00	1.6929
7/8-9	49/64	0.7656	M52 × 5	47.00	1.8504
7/8-14	13/16	0.8125	M56 × 5.5	50.50	1.9882
1-8	7/8	0.8750	M60 × 5.5	54.50	2.1457
1-12	59/64	0.9219	M64 × 6	58.00	2.2835

Table B-2 Drill Tip Heights

Drill Diameter	Decimal	118°	135°
1/16	0.0625	0.019	0.013
3/32	0.0938	0.028	0.019
1/8	0.1250	0.038	0.026
5/32	0.1563	0.047	0.032
3/16	0.1875	0.056	0.039
7/32	0.2188	0.066	0.045
1/4	0.2500	0.075	0.052
9/32	0.2813	0.084	0.058
5/16	0.3125	0.094	0.065
11/32	0.3438	0.103	0.071
3/8	0.3750	0.113	0.078
13/32	0.4063	0.122	0.084
7/16	0.4375	0.131	0.091
15/32	0.4688	0.141	0.097
1/2	0.5000	0.150	0.104
17/32	0.5313	0.160	0.110
9/16	0.5625	0.169	0.116
19/32	0.5938	0.178	0.123
5/8	0.6250	0.188	0.129
21/32	0.6563	0.197	0.136
11/16	0.6875	0.207	0.142
23/32	0.7188	0.216	0.149
3/4	0.7500	0.225	0.155
25/32	0.7813	0.235	0.162
13/16	0.8125	0.244	0.168
27/32	0.8438	0.253	0.175
7/8	0.8750	0.263	0.181
29/32	0.9063	0.272	0.188
15/16	0.9375	0.282	0.194
31/32	0.9688	0.291	0.201
1.00	1.0000	0.300	0.207

Table B-3 Speeds and Feeds for Common Materials

Material Type	Cutting Speed[a] SFPM: HSS Tool	Cutting Speed[a] SFPM: Carbide Tool	Feeds[b] Milling Feeds Expressed in Thousandths of an Inch Per Tooth, Per Revolution — Milling, Depth of Cut = 1/2 Tool Diameter, Tool Diameter: 1/4	1/2	3/4	1	Turning[c] 0.001" Per Rev.	Drilling 0.001" Per Revolution, Tool Diameter: 1/8	1/4	1/2	3/4
Free-machining Steel	130	500	0.6	1.4	2.3	3.2	10.0	1.0	2.1	4.4	6.8
Low-carbon Steel	110	400	0.6	1.3	2.1	2.9	9.1	0.9	1.9	4.0	6.2
Medium-carbon Steel	100	350	0.5	1.1	1.8	2.5	7.7	0.8	1.6	3.4	5.2
High-carbon Steel	90	300	0.4	0.9	1.5	2.2	6.7	0.7	1.4	2.9	4.5
Alloy Steel	80	300	0.3	0.8	1.3	1.8	5.6	0.6	1.2	2.4	3.8
Stainless Steels	50	180	0.4	1.0	1.6	2.3	7.1	0.7	1.5	3.1	4.9
Tool Steels	50	200	0.3	0.7	1.2	1.7	5.3	0.5	1.1	2.3	3.6
Cast Iron	70	250	1.0	2.3	3.8	5.4	16.7	1.7	3.5	7.3	11.3
Aluminum	300	750	1.2	2.8	4.6	6.5	20.0	2.0	4.2	8.8	13.6
Brass	250	500	1.0	2.3	3.8	5.4	16.7	1.7	3.5	7.3	11.3
Copper	100	200	0.6	1.4	2.3	3.2	10.0	1.0	2.1	4.4	6.8
Magnesium	400	900	1.2	2.8	4.6	6.5	20.0	2.0	4.2	8.8	13.6
Titanium	50	200	0.6	1.3	2.1	2.9	9.1	0.9	1.9	4.0	6.2

Notes

[a]Approximate speeds for milling annealed material. Speed may be increased 10% for turning and reduced 10% for drilling.

[b]Cutting conditions and surface finish requirements may warrant lower or higher feeds.

[c]Values are for rough turning. Finish feed is based largely on surface finish requirements—see Table 4.

Table B-4 Approximate Surface Finish for Turning

Specified Roughness R_a μ (in.)[b]	Feed Required to Produce Roughness Value[a] (0.001" per revolution) — Tool Nose Radius (in.): 0.008	1/64	1/32	3/64	1/16
4	1.0	1.4	2.0	2.4	2.8
8	1.4	2.0	2.8	3.5	4.0
16	2.0	2.8	4.0	4.9	5.7
32	2.8	4.0	5.7	6.9	8.0
64	4.0	5.7	8.0	9.8	11.3
125	5.6	7.9	11.2	13.7	15.8
250	7.9	11.2	15.8	19.4	22.4

Notes

[a]$R_a \approx \text{Feed}^2/32$ Radius

[b]To convert to μ in micrometers, multiply by 0.0254.

B-2 UNITS AND CONVERSION FACTORS

Table B-5 *Conversion Factors*

Definition (* denotes exact values)	Conversion Example
Length	
1 in. = 25.4 mm*	4.125 in. × 25.4 = 104.775 mm
1 in. = 2.54 cm*	4.125 in. × 2.54 = 10.4775 cm
1 in. = 0.0254 m*	4.125 in. × 0.0254 = 0.104775 m
1 mm = 0.03937 in.	50 mm × 0.03937 = 1.9685 in.
1 cm = 0.3937 in.	50 cm × 0.3937 = 19.685 in.
1 m = 39.37 in.	50 m × 39.37 = 1968.5 in.
1 m = 3.28 ft.	50 m × 3.28 = 164.04 ft.
Area	
1 in.^{2} = 6.452 cm^2	6 in^2 × 6.452 = 38.71 mm^2
1 cm^2 = 0.155 in.^{2}	120 cm^2 × 0.155 = 18.6 in.^{2}
Volume	
1 in.^{3} = 16.387cm^3	350 in.^{3} × 16.387 = 5735 mm^3
1 cm^3 = 0.061 in.^{3}	1900 cm^3 × .061 = 115.9 in.^{3}
1 gal. = 3.785 liter	5 gal. × 3.785 = 18.925 liter
1 liter = 0.264 gal.	4 liter × 0.264 = 1.056 gal.
Force or Weight[a]	
1 lb. = 0.454 kg	180 lbs. × 0.454 = 81.72 kg
1 lb. = 454 g	0.75 lbs. × 454 = 340.5 g
1 kg = 2.2 lbs.*	25 kg × 2.2 = 55 lbs.
1 g = 0.0022 lbs.*	500 g × 0.0022 = 1.1 lbs.
Power	
1hp = 0.746 kW	15 hp × 746 = 11.19 kW
1 kW = 1.34 hp	20 kW × 1.34 = 26.8 hp
Torque	
1 lb.-ft. = 1.356 N-m	75 lb.-ft. × 1.356 = 101.7 Newton-meters
1 N-m = 0.738 lb.-ft.	100 N-m × 0.738 = 73.8 lb.-ft.
Temperature	
°C = 5/9(°F – 32)	5/9 × (212 °F – 32) = 100 °C
°F = 9/5°C + 32	9/5 × 100°C + 32 = 212 °F

Notes

[a]Newton is technically the measurement of force or weight in SI units. A Newton is defined as a 1 kg mass accelerated at a rate of 1 m/s^2. On earth, a 1kg mass would weigh 9.81 Newtons. However, the common usage is to mistakenly specify the mass when we really mean weight (e.g.,"I would like five kilos of potatoes, please").

B-3 MATHEMATICS

Table B-6 Properties of Polygons

Entity	Property
Circle	$A = \pi \times r^2$ $C = 2 \times \pi \times r$
Rectangle	$A = b \times h$
Triangle	$A = \frac{1}{2} \times b \times h$
Sphere	$V = \frac{4}{3} \times \pi \times r^3$
Cylinder	$V = \pi \times r^2 \times h$
Rectangular prism	$V = b \times h \times d$
Regular polygon, interior angles	Interior angles as illustrated in ***Figure B-1***: measure of angle A = 360/N° measure of angle B = 180° - (measure of angle A)

A = Area
B = Length of Base
C = Circumference
D = Depth (thickness)
H = Height
V = Volume
N = Number

Figure B-1 Angles of a polygon.

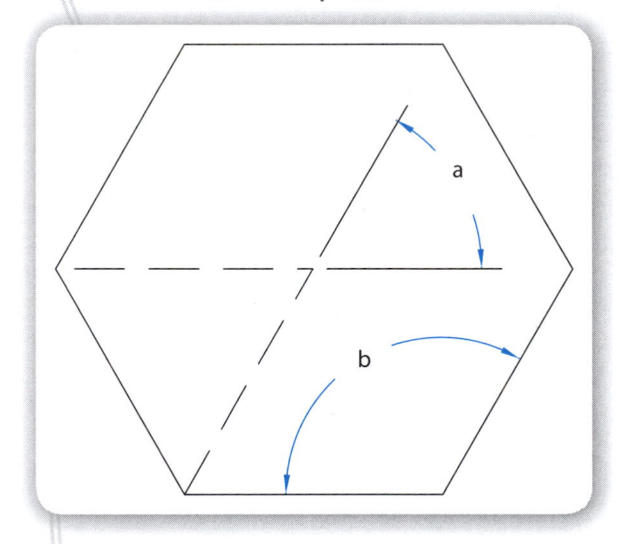

Figure B-2 Sides labeled for use with the Pythagorean Theorem.

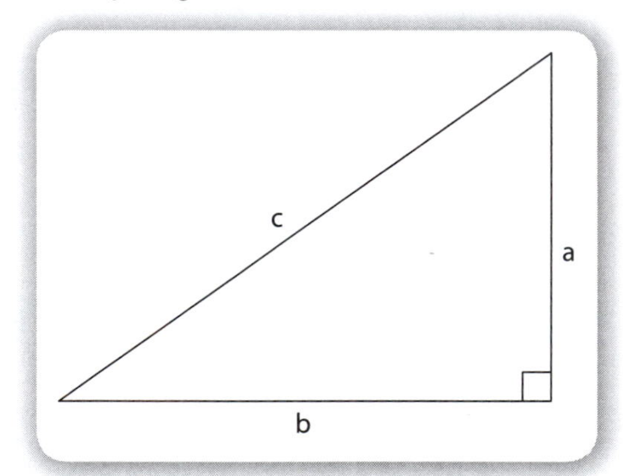

Pythagorean Theorem

$$a^2 + b^2 = c^2$$

The equations and labels in this section refer to the triangle illustrated in *Figure B-2*.

To solve for side a:

$$a = \sqrt{b^2 - c^2}$$

To solve for side b:

$$b = \sqrt{c^2 - a^2}$$

To solve for side c:

$$c = \sqrt{a^2 + b^2}$$

Trigonometric Functions

Table B-7 Trigonometric Functions (Refers to the Sides and Angles Illustrated in Figure B-3)

sine α = Opposite/Hypotenuse
cosine α = Adjacent/Hypotenuse
tangent α = Opposite/Adjacent
cotangent α = Adjacent/Opposite
secant α = Hypotenuse/Adjacent
cosecant α = Hypotenuse/Opposite

Law of Sines

$$\frac{a}{\sin A} = \frac{b}{\sin B} = \frac{c}{\sin C}$$

We use a different notation to describe oblique triangles. The equations for the Law of Sines and the Law of Cosines refer to the angles and sides shown in *Figure B-4*.

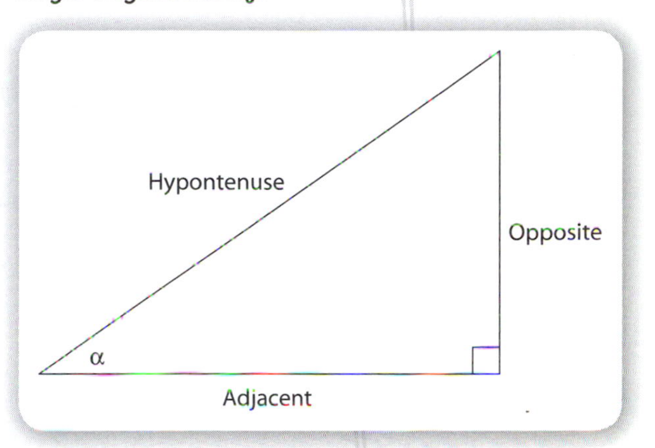

Figure B-3 The standard triangle used for right angle trigonometry.

Law of Cosines

$$a^2 = b^2 + c^2 - (2bc \cos A)$$

To fine side a where A is the angle and a is the side opposite of that angle, and b and c are known, use the following:

$$a = \sqrt{b^2 + c^2 - (2bc \cos A)}$$

To find the cosine of A where A is the angle and a is the side opposite of that angle, and b and c are known, use the following. (Take the arccosine to find the angle.)

$$\cos A = \frac{b^2 + c^2 - a^2}{2bc}$$

Figure B-4 The Law of Sines and Law of Cosines are used to solve oblique triangles.

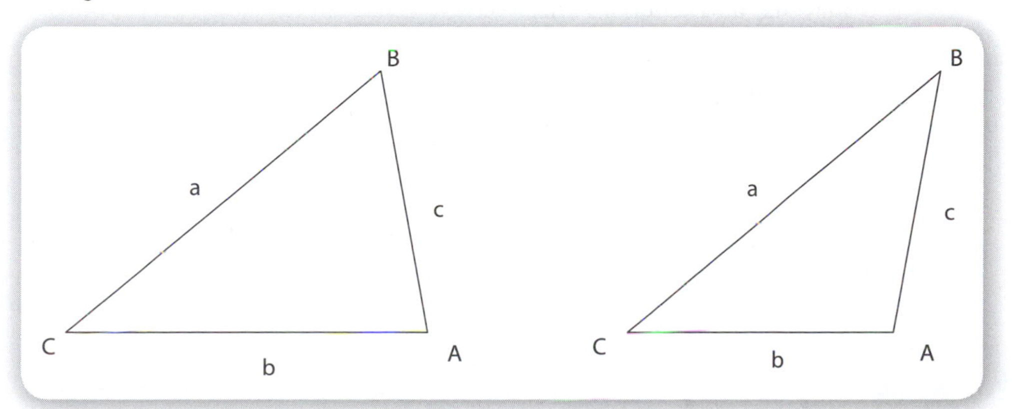

Statistics

Calculation of the mean and standard deviation of a sample:

Starting with a small set of data (X1, X2. . .): 5.6, 5.7, 5.6, 6.0, 6.1, 5.9

Calculate the sample mean ($\overline{X}$)

1. Add each value in the data set to get the Sum (Σ)

$$5.6 + 5.7 + 5.6 + 6.0 + 6.1 + 5.9 = 34.9$$

2. Divide the sum by the number of items in the data set (N)

$$34.9 \div 6 = 5.817 \text{ or } \Sigma \div N = 5.817$$

3. The mean of this sample is: $\overline{X} = 5.817$

Calculate the sample standard deviation (STD)

1. Review the formula:

$$STD = \sqrt{\frac{n * \Sigma x^2 - (\Sigma x)^2}{n * (n-1)}}$$

2. First we will find the Σx^2 quantity which means "the sum of the squared values of each individual data value."
3. Take all of the X values in the table and then square each value.
4. Find the sum of the squared values.

X	X^2
5.6	31.36
5.7	32.49
5.6	31.36
6.0	36.00
6.1	37.21
5.9	34.81
Sum of the "X Squareds" (Σx^2)	**203.23**

5. Next find the find the sum of all the X values and then square this quantity

X	
5.6	
5.7	
5.6	
6.0	
6.1	
5.9	
34.9	Sum of X's
1218.01	**Sum of the X's Squared**

6. Finally solve the equation one step at a time to find the sample standard deviation (STD) by substituting the calculated values.

$$STD = \sqrt{\frac{n * \Sigma x^2 - (\Sigma x)^2}{n * (n-1)}}$$

$$STD = \sqrt{\frac{6*203.23-1218.01}{6*(6-1)}}$$

$$STD = \sqrt{\frac{1.370}{30}}$$

$$STD = \sqrt{.045667}$$

$$STD = .2137$$

Scientific Calculator for Statistics Calculations

Most scientific calculators will have built-in statistical functions. Look on your calculator to be sure you have the button as shown in ***Figure B-5***

The first step is to enter the data (x_1, x_2, x_3. . .) into the calculator. This process is similar on many makes and models of the calculator.

On this particular calculator the [Σ+] button is used to enter individual values. The keystrokes for our data set are as follows:

5.6 [Σ+] 5.7 [Σ+] 5.6 [Σ+] 6.0 [Σ+] 6.1 [Σ+] 5.9 [Σ+]

Now that the data is entered, the built in statistics functions can be used to report the values. Here are the keystrokes for some common statistics.

Number of Items in Data Set: [2nd] [n] gives the result 6.0000

Mean (Average) of the Data Set: [2nd] [X̄] gives the result 5.8167

Sample Standard Deviation: [2nd] [σxn-1] gives the result .2137

Figure B-5 A scientific calculator will often have statistics functions.

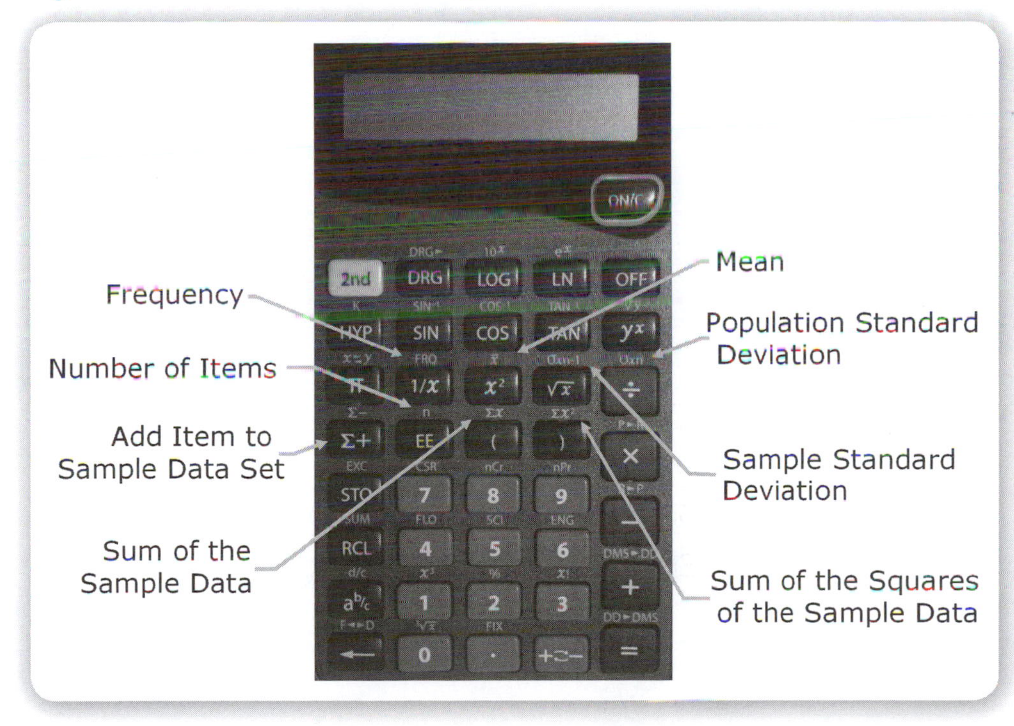

B-4 ENGINEERING MATERIALS

Table B-8 Physical Properties of Common Materials

Material Type	Density[a] (lbs./in.^{3})	Coefficient of Linear Expansion[b] (μ in./in./°F)	Melting Point[c] (°F)
Aluminum Alloys	0.10	13.0	1220
Brass	0.31	11.0	1700
Cast Iron	0.26	6.5	2100
Copper	0.32	9.8	1980
Steel	0.28	8.3	2500
Stainless (Austenitic)	0.29	9.5	2800
Stainless (Martensitic)	0.28	5.6	2800
Titanium	0.16	5.3	3025

Notes

[a]To convert to g/cm^3, multiply by 27.74.

[b]To convert to μm/m/°C, multiply by 1.8.

[c]To convert to Celsius, subtract 32 and then multiply by 0.5555.

Table B-9 Symbols for Important Elements in Metalworking

Al—Aluminum	Ni—Nickel
B—Boron	P—Phosphorous
C—Carbon	Pb—Lead
Cr—Chromium	Sn—Tin
Co—Cobalt	Si—Silicon
Cu—Copper	Ta—Tantalum
Fe—Iron	Ti—Titanium
Mg—Magnesium	V—Vanadium
Mn—Manganese	W—Tungsten
Mo—Molybdenum	Z—Zinc

Table B-10 Manufacturing Operations Form

Name:		Date:	Project Name:				Material:	
Op.#	**Description**	**Cutting Tools**	**Velocity**	**D.O.C**	**Feed**	**Surface Finish**	**MRR or Cutting Time**	**Graphic Description**

Table B-11 *Programming Manuscript Form*

Name:			Date:		Program Name:			Page ___ of ___
Line #	Code	X	Y	Z	I (R)	J (K)	D, F, H, R, S, Other	Comments

B-5 DOCUMENTATION TOOLS

Table B-12 *Setup Sheet*

Job #:		
Setup #:		
Date:		
File Name:		
Material:		
Notes		

Tool	Description	Offset Number
1		
2		
3		
4		
5		
6		
7		
8		
9		
10		
11		
12		
13		
14		

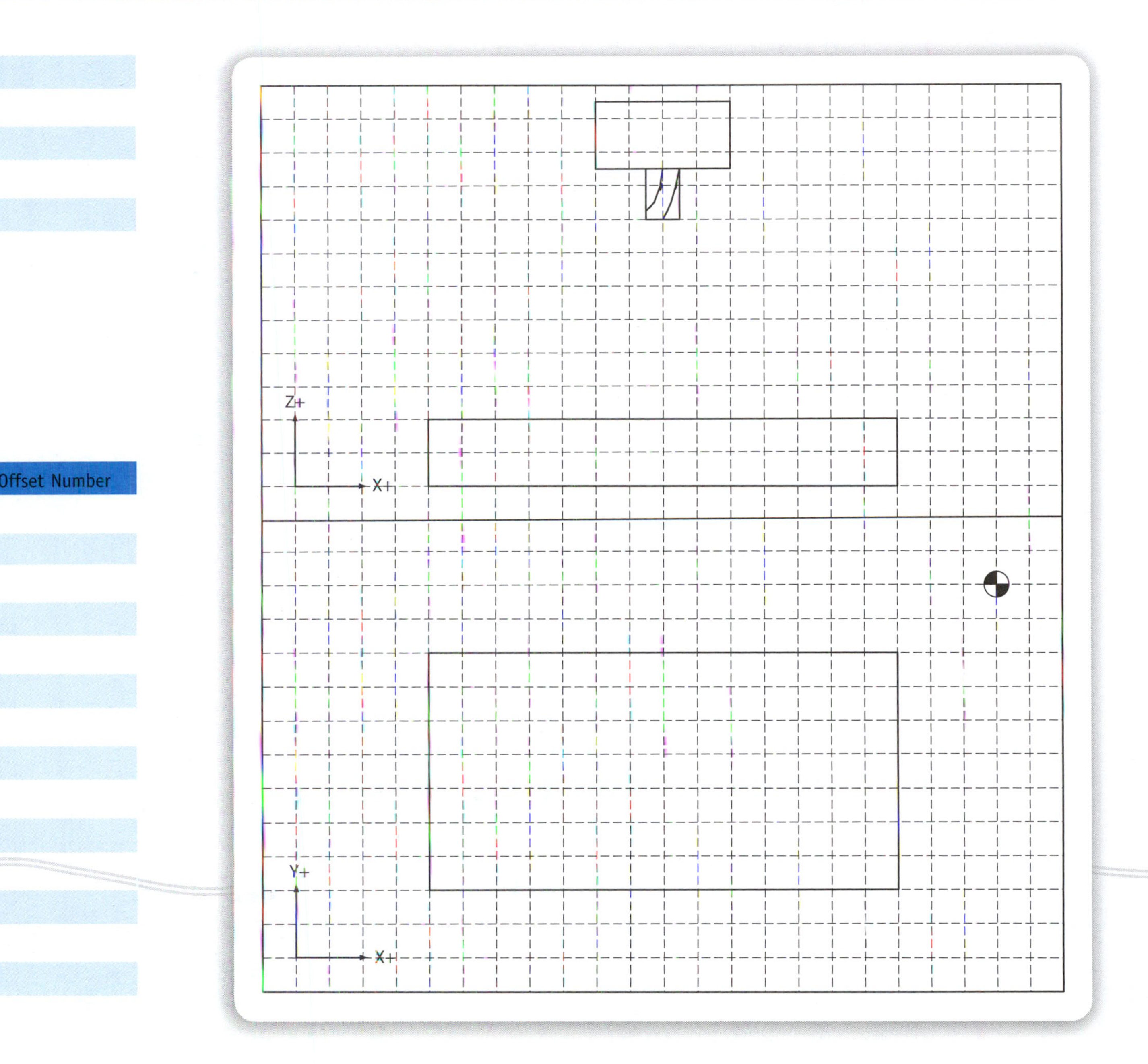

GLOSSARY

A

accuracy: The difference between a specified dimension to the actual dimension that is produced.

acute: An angle smaller than 90°.

address: A letter used in G & M code programming to designate a class of functions. Examples include G, M, X, Y, and Z. Letters are never used alone, but instead are combined with numbers to form words.

ASCII encoding: American Standard Code for Information Interchange. A standard encoding scheme for simple text files with nearly universal portability.

Automatic Tool Changer (ATC) A device used to automatically insert the required cutting tools into the spindle or cross slide position as the part program executes on a CNC machine tool.

auto turning cycle: Canned cycles that are used to automate stock removal for turning operations.

axis: A single linear direction used in the Cartesian coordinate system to locate points in space (e.g. left-right, in-out up-down).

B

backplot: Simple graphical representations of a toolpath that is made up of lines and arcs and used to verify the accuracy of an NC part program.

ball screw: A specialized lead screw that uses close-fitting ball bearings to reduce friction and backlash between the screw and nut. Ball screws are used to transmit motion from the servomotors to the machining table.

block delete: A function used to cause the control to ignore a block of code when the block-delete switch is active. The block is proceeded with the "/" character.

block: A single line of code in an NC part program.

C

canned cycle: Preprogrammed subroutine that helps to automate machining tasks. Examples include drilling and roughing cycles.

carbide: Very hard materials made from metallic carbides such as tungsten titanium and tantalum carbides. Used extensively for cutting tools.

cartesian coordinate system: A system used to define points in space by establishing directions (axes) and a reference position (origin). Coordinate systems can be rectangular or polar.

cemented carbide: Metallic carbides that have been *cemented* together with a binding metal through a sintering process. The amount and composition of the binder can dramatically affect the characteristics of the cutting tool.

chip breaker: A geometrical feature of a cutting tool that encourages the chip to break rather than remain intact.

communication software: Application programs that are designed to control the flow of information though, primarily, a serial port.

compensation: The act of moving the tool centerline away from the part edge in order to maintain the correct dimensions.

complementary: Describes angles that when added together equal 90°. The two acute angles of a right triangle are complementary.

Computer Integrated Manufacturing (CIM): A system of manufacturing that uses computer information systems to plan, monitor and control manufacturing processes.

Computer Numerical Control (CNC): A form of electromechanical motion control used on machine tools, whereby a computer and computer program are used to perform machining operations.

Computer-Aided Drafting (CAD): Computer software that is used to model geometric entities and visually represent the geometry.

Computer-Aided Drafting/Computer-Aided Manufacturing (CAD/CAM): A class of computer software that has the ability to create electronic representations, geometric entities, and model machining processes, and create NC code based upon the model.

Computer-Aided Engineering (CAE): Computer software that is used to aid in the engineering of complex components and systems. This class of software generally contains a core CAD modeler, components for engineering analysis, and CAM.

Computer-Aided Manufacturing (CAM): A group of technologies that use computers to design, automate, control, and improve manufacturing operations.

Computer-Aided Part Programming (CAPP): A class of computer software that is used to generate NC code.

constant surface speed: A mode of operation that allows a lathe spindle to continuously adjust the spindle RPM as the tool moves through different diameters. This mode is used to provide an improved surface finish and tool life.

constant: A value that does not change.

continuous improvement: A belief and method by which systems can be improved incrementally through rational analysis and observation.

control: The computer that operates a CNC machine tool; the machine control unit (MCU).

conventional cutting: Tool rotation that opposes the material feed.

conventional machine tool: A manually operated machine tool.

coordinate: A distance along an axis or axes from the origin of a Cartesian coordinate system. Any point in space can be located relative to a known position with the use of coordinates.

cutter compensation: An automatic method used to offset the cutting tool from the finished edge. Cutter compensation allows the tool diameter to change without having to rework the part program. Also called tool radius compensation or tool nose compensation.

cutting tool: Any hard tool that is used to remove material. Examples include end mills, drills, saw blades, and turning tools.

D

datum: A reference plane or point that is defined on the engineering drawing from which other dimensions are established.

dedicated tooling: Workholding tools that are designed for one particular job and are not intended to be used elsewhere.

diameter offset: A value stored in the offsets register in the control that will be used for automatic cutter compensation. Each tool will have its own diameter offset value.

dimension: The measure of geometrical attributes. In manufacturing, dimensions are used to specify the linear size of features and angular relationships.

Distributed Numerical Control (DNC): A distribution architecture that allows numerous machine tools to communicate with a centralized computer. Not to be confused with Direct Numerical Control, which attempts to run many computers from a central controller.

distribution: A concept in statistics that describes how a group of measurements will have variation from the average. A "normal distribution" is a common pattern in nature as a result of chance.

dwell: A short pause in machine movement usually to allow a cut to be completed.

E

EIA/ISO encoding: An encoding scheme that is similar to ASCII and is most commonly found on CNC controls.

encoder: Any device that is used to give positioning feedback to the control. Digital linear and rotary encoders are common on CNC machine tools.

End of Block (EOB): The termination of one complete line of code in a G & M code NC part program.

engineering drawing: A technical drawing that formally describes the shape, size and specifications of a product. Sometimes called a blueprint.

engineering material: Materials that are selected for their favorable mechanical properties such as strength, hardness and formability.

equation: A mathematical equality that contains known and unknown quantities. Equations often show a relationship between a number of variables, such as the relationship between tool diameter and RPMs for a constant cutting speed.

ERP/MRP: Enterprise Resource Planning and Manufacturing Resource Planning are computer systems capable of integrating sales, engineering and manufacturing data to plan for and procure materials and resources to support production.

exact stop: A method of CNC tool control that forces the machine to decelerate and stop without over-traveling before proceeding to the next tool move.

F

feedback loop: Electronic signals that are sent back to the control to indicate actual position, velocity, or state of the machine tool. The control will then compare the actual condition to the desired position and make adjustments.

ferrous: Materials that contain a substantial amounts of iron.

file: The smallest unit of storage for related information on a computer system. Each individual NC part program is usually stored as an individual file.

fixture offset: Work offset.

fixture: Workholding tools that are designed to clamp and position a workpiece for machining.

Flexible Manufacturing System (FMS): Groups of CNC machine tools that are highly integrated with automated material handling and computerized control systems.

H

height offset: The difference between the length of a cutting tool and a reference value. Tools that are different lengths will have different height offset values.

histogram: A graphical representation of statistical distribution. Individual measurements are grouped into "bins" of a specified size range. The quantity in each "bin" is then displayed on the vertical axis.

I

imaginary tool tip: The point of intersection formed by two lines that are tangent to a tool nose radius and aligned with the major axes of a lathe. This is a common programming mode for CNC turning.

indexable insert: Disposable, multi-tipped cutting tools that are designed to be held in a durable tool holder. The insert can be *indexed* to a fresh corner whenever the current corner becomes dull.

interpolation: Tool movements that travel through all theoretical points of a programmed path. Linear and circular interpolation are common functions of CNC machine tools.

isometric: A common form of pictorial representation in technical drawings which represents an object as viewed by rotating about each axis equally (effectively a corner view).

L

lead in/lead out: The act of entering or exiting a finished toolpath at a shallow angle or arc to ensure a consistent finish and to prevent gouging that can be caused by tool deflection.

lean manufacturing: The practice, tools and philosophy of manufacturing efficiency first developed in Japan by Toyota Motors.

linear: Along a straight line.

Local Area Network (LAN): A method of connecting local computers so that they may share information with each other. The network is closed and private.

M

machinabilty index: A scale that attempts to compare the energy (and resulting force, heat and tool wear) that are required to perform machining operations on various materials.

Machine Control Unit (MCU): The main control computer of a CNC system; control or controller.

machine home: A physical location established by the machine tool builder from which all other measurements are referenced. Also called machine zero.

machine tool: A machine that can be used to produce another machine. The basic equipment used in precision metalworking.

machine zero: A reference position that is established at machine startup. The machine zero is a *hard* position that is set by tripping a sensor or limit switch.

macro: A computer language tool in NC programming that provides the user with some flexibility and automation through the use of variables, loops and decision structures.

manual machining: Machining that is performed on non-automated machinery. The motion of the tool is guided by the hands of a skilled machinist.

mean: The average of a set of measurements.

miscellaneous code (M-code): Codes that control machine functions other than tool movements. For example, M-codes are used to control coolant and spindle rotation.

modal: Describes codes or values that stay active until changed by another code or value. Most G & M codes are modal.

modular tooling: Workholding tooling that is manufactured in standard mounting units and designed to be quickly reconfigured for other setups.

N

Numerical Control (NC): The early, programmable electronic control systems that were developed before the availability of inexpensive and compact computers. The grandfather of CNC.

NURBS: Non-Uniform Rational B-Splines are a class of curve that allows for complex geometry to be specified through simple "control points". NURBS are very useful in computer-aided design.

O

oblique: A description of triangles that do not contain a 90° angle.

obtuse: Describes an angle that is greater than 90°.

offline programming: NC programming that is completed on a computer system, which is separate from the CNC control.

offset register: The storage location in the control that contains the values for the tools and fixture offsets. These values are established during setup and are used by the control during automatic operation.

online programming: The act of entering code on the console of the CNC control.

orient: The act of moving the spindle to a known angular position. This is required for functions such as threading and boring, where the angular position must be accounted for.

origin: The point where two or more axes of a coordinate system intersect. The origin will have the coordinates of zero.

orthographic: A method of representation for three-dimensional objects by which the object is displayed at a right angle to the viewer. The viewer must construct a mental picture of the object by looking straight at the object from several different orientations.

P

part program: The instructions written by the programmer to produce a workpiece.

part zero: The location on the workpiece that serves as the origin for the programmed coordinates in the part program.

peck drilling: Drilling operations that reciprocate in and out of the drilled hole to clear and break chips while machining.

population: An entire group of objects to be measured.

post-processing: The operation in which CAD/CAM machining operations will be converted to machine specific G & M-code.

precision: The degree of certainty in a measurement.

preparatory code: Codes that carry out machining operations or establish machine settings; G-codes.

process capability: The measure of a process'ability to produce products within the specified tolerances.

program stop: An M-code that causes the program execution to pause until started again by the operator. This stop can be conditional (M01) or unconditional (M00).

R

Rapid Prototyping (RP): A group of technologies that can create semi-functional products directly from CAD models without the use of traditional machining, casting or molding. Generally used for mock-ups or to verify designs.

ratio: an comparison between the dimension of unit and the dimensions of another.

reference position: A known position or datum from which other measurements can be taken.

reference tool: See Zero Setting tool.

repeatability: The ability of a machine to return accurately to the same position time after time.

resolution: The degree to which measurements can be differentiated from one size to another. The scale of the measuring instrument.

right angle: A 90° angle.

rotary table: A device that is mounted on the milling table to provide programmable rotation of a workpiece.

rule of ten: The concept that a measuring instrument should have ten times the resolution of the tolerance of the measured feature.

S

sample: In statistics, a small group of individuals that are selected for measurement from a much larger population in order to make generalizations about the dimensions of that population.

section view: A drawing view that simulates the cutting and removal of a portion of the object in order to visualize details within the interior.

serial port: The standard RS-232 communications port that is found on most desktop computers and CNC machine tools. The RS-232 can send and receive data over a distance of 100′, which is much farther than a parallel port. The port may be accessed through a 9-pin or 25-pin sub-D connector.

servomotors: A specialized motor used in motion control systems that can deliver continuous motion at various speeds.

single block: A mode of operation that causes the CNC control to stop at the end of each block of code.

solids verification: Computer generated three-dimensional representations of CNC toolpaths and the resulting workpiece. Used to aid in visualization of the machining process.

spindle tooling: Tooling that adapts cutting tools to the rotating spindle of a machine tool. Examples include end mill adapters, collet chucks, and boring heads.

standard deviation: A specific statistical measurement that shows the variability of a sample (or population) from the average.

Statistical Process Control (SPC): A system of dimensional data gathering and statistical analysis used to ensure the quality of a manufacturing process.

statistics: Measurements that describe the attributes of a group.

STEP models: A neutral CAD model format that allows the easy exchange of product information between engineers and manufacturers related to geometry, tolerances, processes, etc.

stepper motor: A specialized motor used in low-end motion control systems that rotates a predefined angle with every electrical pulse.

subprogram: A external program referenced by the main program in order to perform a specific programming task.

subroutine: Blocks of code within the main program used to repeat a specific programming task.

supplementary: Describes two angles whose sum equals 180°. The adjacent angles of two intersecting lines are supplementary.

syntax: The rules of structure that must be followed when writing in a specific language; grammar.

T

tangent: The condition created when an arc touches another arc, curve, or line at only one point.
text editor: Application programs that are capable of saving simple text files.
text file: A computer file that contains only raw text and line spacing that is usually encoded in the ASCII format. NC part programs are often created as text files prior to being moved to a CNC control.
tolerance: The deviation in size that is allowed from a specified dimension as measured from maximum to minimum.
tool radius compensation: Cutter compensation.

V

variability: The degree to which a group of measurements might be spread out from the average. The "spread of the curve".
variable: The symbolic representation of an element in an equation without regard to an actual numeric value of the element.
vertex: A fixed point of rotation of an angle.

W

word: The programming expression formed when a letter (address) is combined with a number. Examples: G01, X3.500, M30.
work offset: The distance from the machine zero to the work zero along each axis.
work zero: The datum on a workpiece from which all coordinates are programmed and measured; the origin.
workplane: A virtual coordinate system that allows the definition of more than one series of work offsets. Workplanes are called from within a part program with G54, G55, G56, etc.

Z

zero setting tool: A cutting tool used to establish the axial work-zeros of a workpiece from which all other cutting tools are referenced through the use of tool offsets.

INDEX

Numerics

A

B

C

D

E

F

G

M

N

O

P

Q

R

S

U

V

W

X

Y

Z